Contribution of Colloidal Materials to Air, Water and Soil Environmental Sustainability

From Basic Concepts to Advanced Applications

Contribution of Colloidal Materials to Air, Water and Soil Environmental Sustainability
From Basic Concepts to Advanced Applications

Edited by

Izabel C. Riegel-Vidotti
Federal University of Paraná (UFPR), Brazil
Email: izabel.riegel@ufpr.br

Lizandra Maria Zimmermann
Fundação Universidade Regional de Blumenau, Brazil
Email: lmz@furb.br

and

Heloise R. de Barros
Department of Applied Science and Technology (DISAT), Politecnico di Torino, Italy
Email: heloise.barros@polito.it

ROYAL SOCIETY OF CHEMISTRY

Hardback ISBN: 978-1-83767-344-5
PDF ISBN: 978-1-83767-614-9
EPUB ISBN: 978-1-83767-613-2

A catalogue record for this book is available from the British Library

The Royal Society of Chemistry is a charity, registered in England and Wales, Number 207890, and a company incorporated in England by Royal Charter (Registered No. RC000524), registered office: Burlington House, Piccadilly, London W1J 0BA, UK, Telephone: +44 (0)20 7437 8656.

For further information see our website at www.rsc.org

For general enquiries, please contact books@rsc.org

For EU product safety enquiries, please email books@rsc.org or contact Royal Society of Chemistry Worldwide (Germany) GmbH, Römischer Hof, Unter den Linden 10, 10117 Berlin.

Printed in the United Kingdom by CPI Group (UK) Ltd, Croydon, CR0 4YY, UK

Preface

Love, urgency, and inspiration. This book was conceived on these three pillars; love for science, urgency to make smart decisions towards sustainable technologies, and inspiration to connect within one publication, air, water, and soil sustainably issues through colloid science. The topics of colloids and environmental sustainability are both extremely relevant, especially in the context of current technological challenges, taking aspects of rapid expansion of industrial activities, water scarcity, contamination of natural resources, and what is being witnessed through climate change. All these global concerns are no longer isolated but interconnected issues that demand immediate attention from the scientific community.

This book comprehensively explores how colloidal technologies, alongside advances in nanotechnology, can offer innovative solutions for these pressing environmental problems. In this sense, contributions are made to achieve the sustainable development goals (SDGs), particularly SDG 3 (good health and well-being), SDG 11 (sustainable cities and communities), and SDG 13 (climate action). Other modern concepts presented in this book will shed light on environmental, social, and governance (ESG) practices, offering insights to guide industry regulations, urban planning, environmental policies, and academic research, in pursuit of sustainable development and corporate responsibility.

Although recurrently treated as separate topics, applying colloidal materials to mitigate environmental issues has gained importance over the years. In this context, we highlight the need for a multidisciplinary approach to engage professionals in many areas to address the subject efficiently.

This book project started some time ago and was validated on many different terms. After many years of researching and teaching colloids and surface chemistry to undergraduate and graduate students, the editors have gathered information on various aspects of the subject, with a special focus on the gap between colloids and environmental science. Herein, the reader will find valuable contributions connecting many different areas towards

Contribution of Colloidal Materials to Air, Water and Soil Environmental Sustainability: From Basic Concepts to Advanced Applications
Edited by Izabel C. Riegel-Vidotti, Lizandra Maria Zimmermann and Heloise R. de Barros

Published by the Royal Society of Chemistry, www.rsc.org

one goal: using colloidal materials in sustainable technologies. This book provides fundamental and applied knowledge of environmental problems in air, water, and soil through colloidal science technologies. By covering several innovative and advanced technologies, the material serves as a useful reference for professionals in a variety of industries and fields.

We are confident that this book delivers valuable insights and critical information, appealing to both academic and industrial researchers, as well as professionals across various sectors of industry.

Furthermore, considering the lack of similar literature that includes all the proposed topics in one publication, this book will be very appealing to the academic and non-academic communities. It covers the areas of chemistry, materials science, physics, and engineering.

We have organized this book into three main parts: air, water, and soil. Each part begins by discussing environmental issues and the impact of major pollutants associated with the topic. The other chapters cover key areas such as the intricate synthesis of colloidal systems and their immobilization in devices designed for air, water, or soil applications. Special emphasis is placed on the novel uses of these materials in technological solutions, ranging from sensors that detect airborne contaminants to filters that purify water and enhance soil health.

We hope this book contributes to a broad understanding of how colloidal materials can be used both now, and in the future, primarily to prevent and solve environmental problems.

We would like to thank our mentors, students, and research partners who have, over the years, contributed to our personal and professional growth and have inspired us in so many ways to accomplish more.

We extend our deepest gratitude to all the contributors of this book who wholeheartedly believed in this project and chose to be a part of it.

The Editors

Contents

Contribution of Colloidal Materials to Air, Water and Soil Environmental Sustainability: From Basic Concepts to Advanced Applications
Edited by Izabel C. Riegel-Vidotti, Lizandra Maria Zimmermann and Heloise R. de Barros

Published by the Royal Society of Chemistry, www.rsc.org

Section II: Water

Section I: Air

CHAPTER 1

Introduction

I. C. RIEGEL-VIDOTTI,*[a] L. M. ZIMMERMANN,[b] H. B. BARROS[c] AND L. F. BALCEWICZ[a]

[a] Department of Chemistry, Federal University of Paraná, Brazil; [b] Department of Chemistry, Regional University of Blumenau (FURB), Brazil; [c] Department of Applied Science and Technology (DISAT), Politecnico di Torino, Italy
*Email: izabel.riegel@ufpr.br

According to IUPAC, the term colloidal 'refers to a state of subdivision, implying that the molecules or polymolecular particles dispersed in a medium have at least in one direction a dimension roughly between 1 nm and 1 μm, or that in a system discontinuities are found at distances of that order.'[1]

Colloids are unique systems with important specific and distinguished properties governed by surface phenomena, that were established even before the recognition of colloid existence as it is understood nowadays. The principles that govern colloidal behavior are related to the investigations of capillary rise (Robert Hook – 1661), intermolecular forces being of shorter range than gravity (John Freind – 1675–1728 and Young in 1805 who estimated the range of intermolecular forces as 0.2 nm), and the equation of meniscus curvature (Young and Laplace – 1805). Figure 1.1 depicts a timeline of important contributions to colloid science from the starting point of the recognition of colloids.[2–5]

The high surface-to-volume ratio and high energy at the interface are the key properties that make colloidal systems highly size-dependent, qualifying them as excellent candidates for many technological applications, as will be demonstrated throughout this book.

Contribution of Colloidal Materials to Air, Water and Soil Environmental Sustainability: From Basic Concepts to Advanced Applications
Edited by Izabel C. Riegel-Vidotti, Lizandra Maria Zimmermann and Heloise R. de Barros

Published by the Royal Society of Chemistry, www.rsc.org

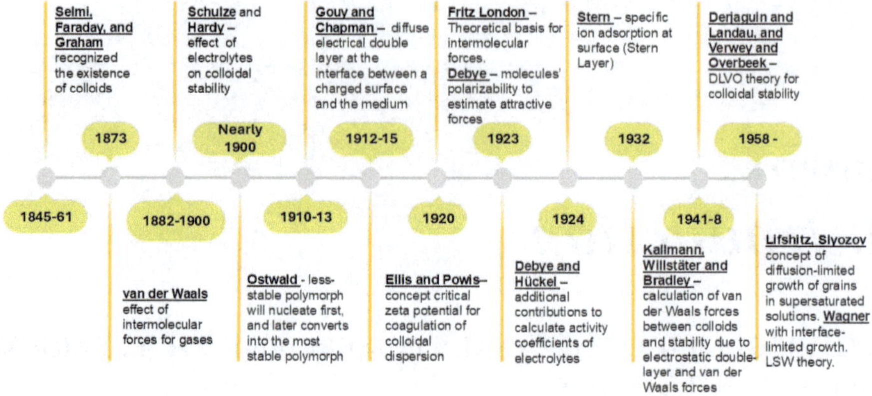

Figure 1.1 Timeline of colloidal foundation concepts.

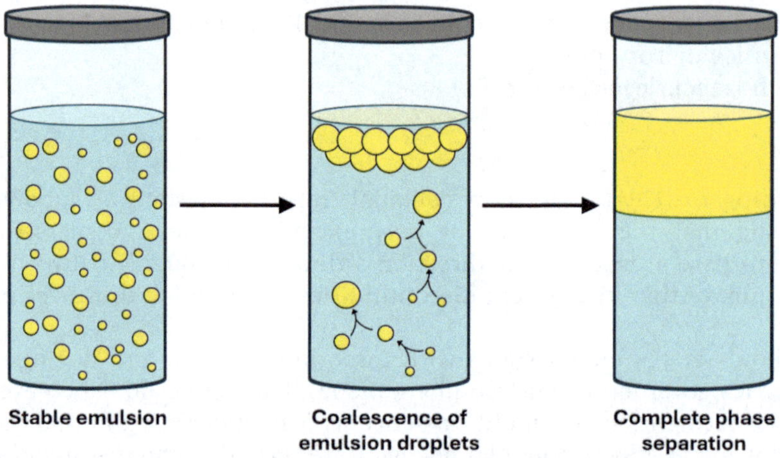

Figure 1.2 Destabilization of oil-in-water emulsion through coalescence.

When suspended in a medium, the stability of colloidal systems is primarily influenced by electrostatic repulsion, Brownian motion, and the nature of the interactions between particles. According to the nature of the colloidal particles and the interactions established with the dispersing medium, colloids can be classified as lyophilic or lyophobic. Due to their chemical affinity with the solvent, lyophilic colloids are thermodynamically stable and well dispersed. Lyophobic colloids are usually said to be unstable, even under the condition of minimizing interfacial free energy, they require external energy for dispersion and are prone to sedimentation and aggregation. An oil-in-water emulsion is a good example of a lyophobic colloid, as shown in Figure 1.2. Due to the incompatibility between the liquids, the oil droplets tend to aggregate to reduce the contact area with the aqueous dispersing phase. This destabilizes the dispersion, leading to complete phase separation and thus dissolving the colloid. Lyophobic colloids are more sensitive to the presence of electrolytes, and this is the reason why small quantities of electrolytes can flocculate them.

As shown, the knowledge that particles are prevented from aggregating by electrical repulsing forces was uncovered between 1882 and 1890 and is a concept that is extremely useful today. A general approach to colloid stability, considering the electric double layer and van der Waals interactions, was proposed as the Derjaguin–Landau–Verwey–Overbeek (DLVO) theory. This approach is particularly important in predicting colloid stability and equilibrium morphologies.

Apart from thermodynamic factors, aggregation kinetics is a matter of concern, especially considering real-life applications. Kinetic models, such as the Smoluchowski aggregation equation, describe the aggregation processes in colloids at sufficiently low volume fractions. However, denser systems, which are of particular industrial interest, at higher solid volume fractions, add further complexity to the aggregation process. Sandkühler *et al.* proposed a dimensionless model based on the population balance equation (PBE) to study colloidal aggregation under non-sheared conditions. Experimental data showed a strong match with the model, revealing an aggregation master curve for average radii as a function of dimensionless time, applicable to different stability ratios (W) and volume fractions. The model was also extended to describe gelation kinetics. Additional experiments demonstrated that surface chemistry significantly impacts both the stability and aggregation kinetics of colloidal dispersions.[6]

Finally, there are many aspects of colloidal systems to be considered. It is important to consider that these systems are generally under metastable conditions with respect to bulk states. This means the system is in an equilibrium state at a thermodynamic potential with a local minimum given certain constraints (*e.g.* T, p). The activation barrier between the colloidal and bulk states can be overcome under the synthesis conditions, especially at high temperatures.[7,8]

The ability to control thermodynamic and kinetic factors in colloid formation allows the use of colloidal systems in environmental applications, such as in decontamination. The particles can serve as containers for further separation and detection or can also be used as microreactors for pollutant degradation. Figure 1.3 demonstrates how high-porosity materials can be used for contaminant immobilization through adsorption. The colloidal-scale cavities in materials like silica micro- and nanoparticles are particularly advantageous due to their high surface area, which significantly enhances contaminant removal. Understanding the physicochemical properties of both contaminants and colloidal formulations is crucial to promoting favorable interactions between them. Self-assembled structures, such as the micelles shown in Figure 1.4, are highly relevant for water remediation. The inward-facing carbon chains of surfactants within the micelle form a hydrophobic core, which is well-suited for trapping nonpolar contaminants.

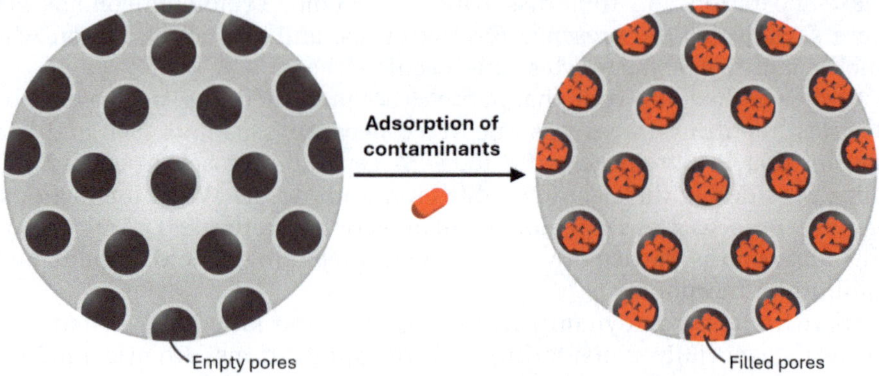

Figure 1.3 Adsorption of contaminants in high porosity materials.

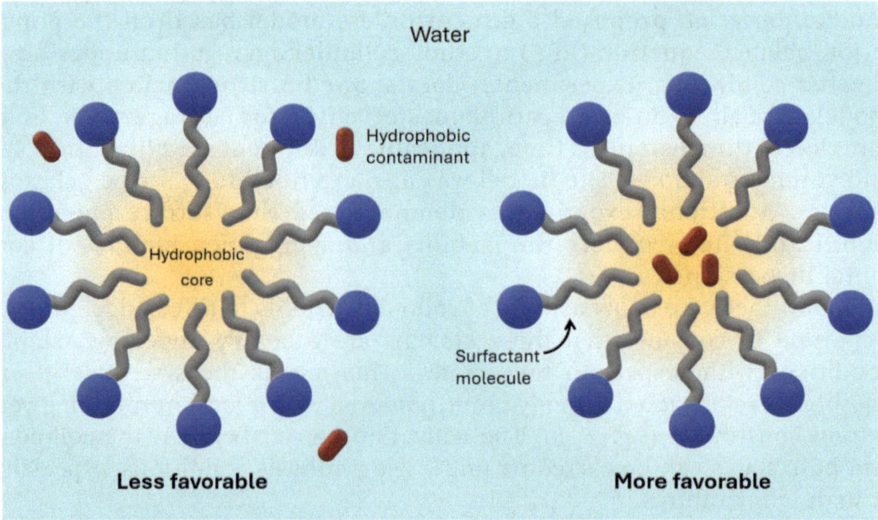

Figure 1.4 Removal of hydrophobic contaminants in water using micelles. Surfactant molecules are composed of a hydrophilic head, in blue, and a hydrophobic tail, in gray.

Transitioning colloidal formulations from research facilities to industrial production involves numerous challenges. Scaling up procedures should take into account the fundamental aspects of colloidal science towards morphology controlling long term stability and performance. Once these challenges are addressed, colloidal materials can be effectively applied to promote environmental sustainability and support a circular economy (Figure 1.5).

First acknowledging the main contaminating issues in air, water and soil, this book explores multiple features of colloidal systems as tools for environmental monitoring, protection and remediation (Figure 1.6).

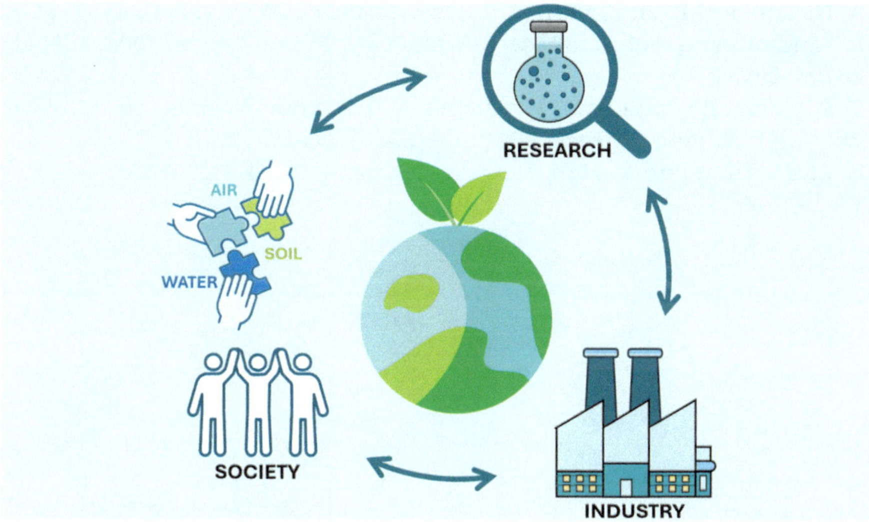

Figure 1.5 Schematic representation of circular economy approaches.

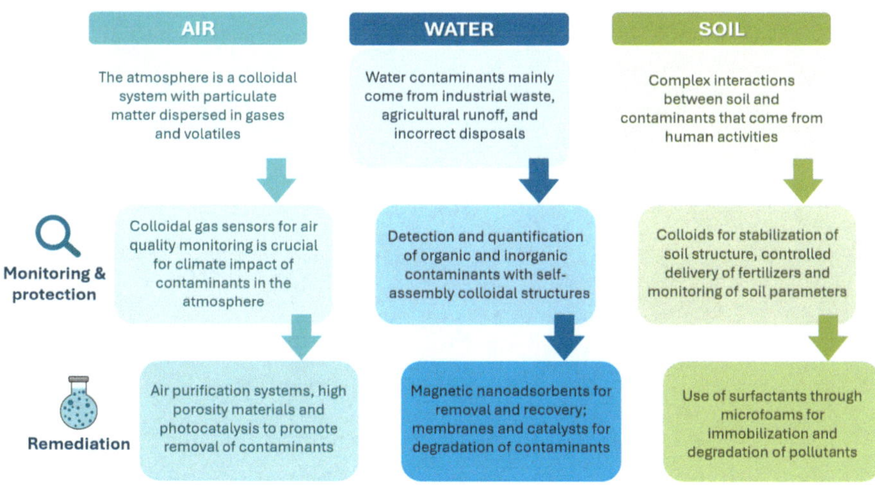

Figure 1.6 Summary of the book's contents.

References

1. Colloidal, 3.0.1 edn, International Union of Pure and Applied Chemistry (IUPAC), 2019.
2. R. M. Pashley and M. E. Karaman, *Applied colloid and surface chemistry*, John Wiley & Sons, 2021.
3. P. Sennett and J. Olivier, *Ind. Eng. Chem.*, 1965, **57**(8), 32–50.
4. V. A. Shneidman, *Phys. Rev. E: Stat., Nonlinear, Soft Matter Phys.*, 2013, **88**(1), 010401.

5. A. Nelson and L. H. Friedman, *J. Phys. Chem. C*, 2022, **126**(4), 2161–2178.
6. P. Sandkühler, J. Sefcik and M. Morbidelli, *J. Phys. Chem. B*, 2004, **108**(52), 20105–20121.
7. F. P. García de Arquer, D. V. Talapin, V. I. Klimov, Y. Arakawa, M. Bayer and E. H. Sargent, *Science*, 2021, **373**(6555), eaaz8541.
8. B. Li, D. Zhou and Y. Han, *Nat. Rev. Mater.*, 2016, **1**(2), 1–13.

CHAPTER 2

The Behaviour of Colloidal Particles at Environmental Compartments

R. H. M. GODOI,*[a] F. BAGLIOLI,[a] H. EVANGELISTA,[b] S. J. GONÇALVES JUNIOR,[b] A. F. L. GODOI,[a] Y. J. PARRA,[a] T. PAULIQUEVIS,[c] G. POLEZER,[d] B. G. SEBBEN[a] AND C. I. YAMAMOTO[e]

[a] Environmental Engineering Department, Federal University of Paraná, Curitiba, Paraná, Brazil; [b] Radioecology and Global Change Laboratory (LARAMG), Rio de Janeiro State University, Rio de Janeiro, Brazil; [c] Department of Environmental Sciences, Federal University of São Paulo, Diadema, São Paulo, Brazil; [d] Department of Technology, State University of Maringa, Umuarama, Paraná, Brazil; [e] Chemical Engineering Department, Federal University of Paraná, Curitiba, Paraná, Brazil

*Email: rhmgodoi@ufpr.br

2.1 General Aspects

Colloids are mixtures where one substance is dispersed evenly throughout another. In an atmospheric system, an aerosol is generally defined as a suspension of liquid or solid particles in a gas, with particle diameters in the range from 10^{-9} to 10^{-3} meters, suspended in the atmosphere. This allows them to remain airborne for extended periods. Due to their size, colloidal particles, called aerosols when in the atmosphere, have a large surface area per unit volume or per unit mass of material. The properties of this surface, or more specifically, the interfacial region between the solid colloid and the bulk liquid

Contribution of Colloidal Materials to Air, Water and Soil Environmental Sustainability: From Basic Concepts to Advanced Applications
Edited by Izabel C. Riegel-Vidotti, Lizandra Maria Zimmermann and Heloise R. de Barros

Published by the Royal Society of Chemistry, www.rsc.org

(water) phases, govern the behaviour of a colloidal suspension in the air. The primary parameters that determine the environmental and health effects of aerosol particles are their formation process, concentration, size, structure, chemical composition and interaction with solar radiation. Atmospheric aerosol particles originate from a wide variety of natural and anthropogenic sources. Primary particles are directly emitted as liquids or solids from sources such as biomass burning, incomplete combustion of fossil fuels, traffic-related suspension of road dust, soil, and mineral dust, sea salt, and bioaerosols (particles of biological origin in the atmosphere, including bacteria, viruses, fungal spores, pollen, and plant debris). Secondary particles, on the other hand, are formed by gas-to-particle conversion in the atmosphere. Specifically, secondary organic aerosols (SOAs) are formed through the chemical reactions of volatile organic compounds (VOCs) with atmospheric oxidisers like ozone and hydroxyl radicals. These processes contribute to the formation of photochemical smog and fine particulate matter. Once formed, particles can collide and stick together in a process known as coagulation, forming larger particles, which can settle more quickly or interact with other atmospheric components. Additionally, large particles can also act as nuclei for condensation, which is particularly important in urban atmospheres where it contributes to cloud formation and the growth of aerosols. The natural removal of aerosols from the atmosphere can occur through wet or dry deposition. Wet deposition involves particles being removed from the atmosphere through precipitation, while dry deposition involves particles settling onto surfaces due to gravitational settling and turbulence.

Overall, the predominant chemical components of aerosols are sea salt, crustal elements, sulphate, nitrate, ammonium, organic compounds, black carbon, trace metals, and biogenic particles (such as carbohydrates, proteins, and lipids). The relative abundance of these chemical components can vary significantly depending on location, time, meteorological conditions, and particle size fractions. Anthropogenic emissions leading to atmospheric aerosols have increased dramatically over the past century and have been implicated in human health effects, in visibility reduction in urban areas, and in perturbing the Earth's radiation balance. Aerosols are of central importance for atmospheric chemistry and physics, the biosphere, climate, and public health. They influence the Earth's energy balance, the hydrological cycle, atmospheric circulation, and the abundance of greenhouse and reactive trace gases. This chapter outlines the research perspectives on the properties and interactions of atmospheric aerosols and their effects on climate and human health in three distinct environments: the Amazon rainforest, urban cities, and the Antarctic environment.

2.2 Urban Planning and Colloidal Pollution: Health Implications in Brazilian Environments

Atmospheric pollution can be defined as contamination of the environment by any means of chemical, physical, and biological forcing, in a way that alters the natural characteristics of the atmosphere.[1] Based on this definition, since the dawn of mankind, it has been possible to identify contaminating activities due to a myriad of misplaced influences on the atmospheric environment even from ancient times. In this context, the beginning of the Industrial Revolution left an important mark, with an unprecedented increase in the magnitude of air pollution due to the huge employment of coal in industry. The textile industry and fabrication of metal ores, in particular, stand out in this scenario. Over the last centuries, industrial and technological development has directly improved the quality of life. However, this statement must be approached with caution, as sometimes it can be indirectly contradictory. That is, the impact of emissions may cause health effects on the population. A good example might be the improvement of transportation technologies, which has quickened travel between remote parts of the globe. However, these technologies are among the main sources of atmospheric pollutants worldwide, which have been proven to be dangerous to human health. As a matter of fact, according to *The Lancet*,[2] exposure to atmospheric pollution is considered the main avoidable cause of death globally. To determine the extent of air pollution's impact on the planet, the Global Burden of Disease (GBD) report estimated that approximately 7.1 million worldwide deaths in 2016 were related to atmospheric pollution.[2,3]

Nowadays, the fast pace of global urbanization is associated with the expansion of modern industries and non-agricultural sectors, especially in low- and middle-income countries.[4] The urgent need to take better care of the environment in cities has recently been manifested in the Sustainable Development Goal (SDG) 11, which addresses urban planning and development. One specific SDG indicator, 11.6.2, is the annual mean level of $PM_{2.5}$ (one fraction of PM size) weighted by population for which a global database was developed by the World Health Organization.[5] Sustainable and responsible urban development is co-dependent with air quality. At least two goals of the 2030 Agenda for Sustainable Development recognise this importance: SDG3—Ensure healthy lives and promote well-being for all at all ages, and SDG11—Make cities and human settlements inclusive, safe, resilient, and sustainable.[6] Clearly, air pollutant inventories, specifically of $PM_{2.5}$ (including its composition and levels), in small- and medium-sized cities of middle- and low-income countries are greatly beneficial as tools to formulate policies to improve the well-being and quality of life for those most at risk.[2] Within cities, anthropic activities contribute to human exposure and to the budget of the colloids in the atmosphere known as particulate matter or PM. Consequently, since these particles may contain a variety of toxic organic and inorganic compounds, these activities also contribute to the exposure of the population to such species. In this way, in any city, the air quality is determined by the balance between the emissions of these colloids and the capacity to disperse and remove those air pollutants. Because of that, in the last few decades, researchers have been dedicated to assessing the composition of PM and identifying the compounds responsible for the adverse effects.

To better address these questions, particulate matter is classified according to its aerodynamic size. Three main size fractions are considered: PM_{10}, $PM_{2.5}$, and $PM_{0.1}$. These correspond to particles with sizes smaller than or equal to 10 μm, 2.5 μm, and 0.1 μm, respectively. It is important to note that these fractions are nested within one another. In other words, $PM_{0.1}$ is considered along with $PM_{2.5}$, and the two smaller fractions are considered along with PM_{10}. The relevance of this size distribution lies both in the assessment of atmospheric emission sources and the evaluation of their potential health impacts. Usually, PM mass is reduced alongside particle size, meaning that the larger fractions contain higher masses than the finer fractions, and the nano fractions contain even less, under the same sampling conditions. In contrast, surface area and particle count in a sample are inversely proportional to particle size. Thus, the fine and nano fractions correspond to most of the atmospheric suspended particles.[7] Also, given their smaller masses, the finest particles can be transported over longer distances and for longer periods. Because of that, their toxicity is a relevant concern in relation to public health issues. Additionally, nanometric particles (<0.1 μm) pose a significant health risk due to their ability to penetrate the alveolar region of the lungs and enter the bloodstream. Unlike larger particles, which are typically filtered out in the upper respiratory tract, these ultrafine particles can reach parts of the body that larger particles cannot, potentially leading to adverse health effects in organs beyond the respiratory system.

Identification of PM sources can be used to detect and manage these emissions.[8] Different PM-emitting activities release trace compounds that, separately or combined, can generate distribution profiles of the emitted aerosols. In general, investigation techniques work with the concentration of these compounds and multivariate statistical analyses to determine the possible sources, in a kind of "reversal approach". Conversely, the region of interest can be analysed to construct its aerosol composition profile with respect to the sources present in the area. A diverse range and combination of pollutants can be found, given physical, meteorological, and socio-economic characteristics. There is a relationship between the size of emitted particles and their source. Finer particles, with diameters between 1.0 and 2.5 μm, are usually generated during combustion processes. Processes of a more mechanical nature, such as resuspension, are responsible for heavier and larger particles, with sizes above 2.5 μm. On the other hand, particles with radii smaller than 1.0 μm are strongly related to secondary aerosols and processes like coagulation, condensation and new particle formation (nucleation). Global assessments of air pollution have revealed that the highest $PM_{2.5}$ exposure levels are found in Asia, whereas lower exposure occurs in the US.[5] However, substantial health risks are present even at pollutant concentrations below the new updated WHO AQGs, highlighting the general unspoken hypothesis that no level of $PM_{2.5}$ is safe.[9] Additionally, Burnett *et al.*[10] showed that a reduction of 1 $\mu g\,m^{-3}$ in $PM_{2.5}$ concentrations provides a more significant health benefit in a city with relatively low levels (*i.e.*, in the range of 10–30 $\mu g\,m^{-3}$), as compared to cities with much higher mean concentrations, close to or above 100 $\mu g\,m^{-3}$.

2.2.1 Impact on Human Health

When considering toxic substances, the three main pathways for human contamination are dermal contact, ingestion, and inhalation. Consequently, there is a need to assess the bioaccessibility of these substances using biological fluids, such as saliva, blood, and lung fluid, to determine the extent to which these toxicants can be absorbed and pose health risks. In the respiratory system, PM can be deposited in different locations based on its aerodynamic size. Particles with diameters larger than 10 μm can be inhaled, but they are promptly filtered in the upper respiratory airways. In fact, only PM with a diameter of less than 10 μm can reach the deeper structures of the respiratory system.[11] Given their sizes, those particles can lodge in the trachea and bronchi. Even so, the larger ones (2.5 to 10 μm) are observed to have short residence times in the respiratory system. Instead, they are removed by mucociliary clearance, being expelled by coughing or swallowed into the gastrointestinal system.[12] In this way, only PM with a diameter of less than 2.5 μm is capable of penetrating the pulmonary regions where gaseous exchange occurs (alveoli region).

When these PM compounds reach the alveoli region, some of them are capable of solubilizing in the pulmonary fluid, due to their chemical characteristics. Because of this, they may pass through the respiratory system, and by traveling through the cardiovascular system, reach other organs. On the other hand, the portion of these $PM_{2.5}$ species that remains deposited in the respiratory system can induce pulmonary outcomes, such as oxidative stress.[7,13] However, the main outcomes of $PM_{2.5}$ are heart-related ones, due to the overload of the respiratory system on the circulatory system. As these particles dissolve in the alveoli[14] and enter the bloodstream, they pose risks to various parts of the body. Elevated atmospheric concentrations of these particles are associated with systemic effects, including an increased relative risk of cerebrovascular accidents, or strokes.[15] Particulate matter that dissolves in physiological fluids is referred to as bioaccessible.

Besides particle size, characteristics like the chemical composition of PM will determine the dissolution rates and extent of the particles. Evidently, these characteristics are influenced by emission sources, geographic and meteorological conditions, and mitigation strategies. Recent studies have also indicated differences in the composition and solubility of PM depending on its residence time in the atmosphere. Polezer *et al.*[16] suggest that the elements contained in the particles can undergo chemical transformation during and/or after their emission. Hence, there is ongoing research focused on the mass and elemental concentration of these particles. In recent years, studies have shown that the soluble fraction of these elements poses higher direct risks to human health than the total exposure concentration.[7] Considering these concepts, different pathways pose different risks to human health. Namely, ingestion and inhalation can result in more severe and quick effects when compared to dermal contact. Also, when considering atmospheric pollution specifically, the inhalation pathway is the most critical. The gases and aerosols present in the air can be more quickly transported throughout the body *via* the processes mentioned above. In this context, we can evaluate how the urban development of two different Brazilian cities has played a crucial role in their respective air qualities. Even considering that Manaus (AM) is located in the allegedly pristine Amazon, the effects of the Free Economic Zone of the municipality have profoundly impacted sources and concentrations of inhalable PM. The industrial activities and lack of stringent environmental regulations in Manaus contribute to higher emissions of particulate matter, including metals like Mn and Pb, which have significant health implications due to their bioaccessibility and potential to cause oxidative stress and other health issues.

In contrast, the city of Curitiba (PR) has adopted a different approach to urban development by integrating sustainable practices aimed at improving the well-being of its residents. Curitiba's focus on green urban planning, efficient public transportation, and pollution control measures has led to a reduction in emissions from vehicular and industrial sources. These sustainable practices contribute to the unique $PM_{2.5}$ profile observed in Curitiba, which differs not just in mass concentration but also in elemental composition and bioaccessibility when compared to Manaus. As a result, while it might seem surprising that the pollutant levels in both cities are similar despite their dissimilar geographical constraints, it is crucial to understand that these similar mass concentrations do not equate to similar health risks. The differences in urban development strategies have resulted in distinct chemical compositions of the particulate matter in each city, influencing their bioavailability and the associated health risks. For instance, the higher solubility of certain metals in Curitiba's $PM_{2.5}$ suggests that while pollutant levels might be comparable, the potential health impact in Manaus could be greater due to the differences in the chemical makeup of the particles. Thus, these findings underscore the importance of considering not just pollutant levels but also their sources, composition, and bioaccessibility when assessing the impact of urban pollution on public health.

2.3 Atmospheric Colloids in Antarctica: Sources, Composition, and Climatic Implications

The Antarctic continent, covering an expanse of approximately 14 million km^2, plays a crucial role in the global environmental system by hosting the majority of the world's ice and freshwater reserves. The Antarctic environment is the coldest, driest, and most sensitive to both local and global perturbations.[17] The Antarctic ice sheet, a key component of the continent, is vital to the Earth's climate system, serving as gateways to the world's oceans.[18] The Antarctic continent is also home to a unique and diverse range of terrestrial life, challenging the perception that it is merely a frozen desert. Many factors influence this extreme environment, including atmospheric circulation patterns, air temperature, humidity, aerosol concentrations, solar radiation, and moisture transport processes. Unique atmospheric phenomena, such as katabatic winds and tropopause folding events, contribute to the distinct climatic conditions in the Antarctic region.[19]

The Antarctic atmosphere has demonstrated a crucial role in global processes, with its characteristics impacting not only local conditions but also climate dynamics in a broader perspective. Despite the potential for pollutants to reach the region through long-range atmospheric transport, the Antarctic environment has maintained its low levels of anthropogenic contamination, highlighting its pristine nature. The extreme cold and dry conditions in Antarctica make it an ideal location for studying a myriad of atmospheric processes under extreme though pristine conditions, including aerosol processes and life cycle, cloud microphysics, and radiative transfer.[20] Adding to this scenario its vast areas covered by the ice sheets and glaciers, as well as their substantial volume, the Antarctic continent is one of the planet's major heat sinks, owing to the high reflection of solar radiation.

As a remote region, Antarctica has been the focus and object of a deeper understanding of natural processes and the impacts of human influence on this environment. The lack of a significant human population in Antarctica has provided valuable insights into the complex interplay between the continent's physical geography,[21] marine ecosystem,[22] atmospheric chemistry,[23,24] and global climate patterns.

One key aspect of the Antarctic environment that has received significant attention from the scientific community is the composition and behaviour of airborne particles in the continent's atmosphere. The Antarctic atmosphere is characterised by very low aerosol concentrations compared to other regions of the world, owing to the continent's isolation from major sources of particulate matter and the relatively clean air masses that dominate the region. However, the aerosols present in the Antarctic atmosphere can have a profound impact on the local and global climate. Understanding the complex processes governing the formation, transport, and transformation of atmospheric colloidal systems in Antarctic air is crucial for accurately predicting the continent's response to ongoing climate change. Factors such as long-range atmospheric transport, *in situ* chemical reactions, and the influence of local biological activity all contribute to the unique characteristics of the Antarctic aerosol composition. By studying the chemical composition, physical properties, and reactivity of these colloidal systems, researchers can gain valuable insights into the role they play in the region's climate, including their effects on cloud formation, surface albedo properties, radiative balance, and the potential impacts of climate change on the Antarctic cryosphere.[25]

Research on colloidal systems in the Antarctic atmosphere can provide important information to support the development of more accurate climate models and inform policy decisions related to the long-term management and conservation of this critical global environmental system.

2.3.1 Sources and Composition of Atmospheric Aerosols in Antarctica

Atmospheric aerosols in Antarctica originate from a variety of natural and anthropogenic sources. Natural sources include sea salt, mineral dust, and biogenic emissions from the ocean and terrestrial ecosystems. Anthropogenic sources encompass the long-range transport of pollutants from distant continents, as well as local activities such as research station operations and tourism.

Studies on the detailed composition of aerosols in Antarctica are limited, primarily focusing on coastal regions where most research stations are concentrated,[25–27] while fewer studies have investigated the central areas of the continent.[24,28,29] Aerosols also are important for the ice nuclei and cloud condensation nuclei formation in the Antarctic atmosphere, playing crucial roles in the formation and properties of clouds, as well as influencing the climate. A growing body of research has shed light on the chemical processes and reactivity of these colloidal systems in the Antarctic atmosphere.

The chemical composition of Antarctic aerosols is complex, with both inorganic and organic components present, and it is heavily influenced by the continent's geographic isolation. In addition to the vast ice reserves, the Antarctic continent also contains various forms of liquid freshwater, including permanent non-freezing surface lakes, subglacial shelf water, and subsurface waters recharged by melting ice.[30] These water bodies can serve as sources of precursor materials for the formation of aerosols, which can then interact with the region's atmospheric chemistry and climate dynamics.

Aerosol loadings in Antarctica exhibit seasonal variations, with increased concentrations in January and decreased levels in February and March.[31] Correia *et al.* (1998)[32] analysed the aerosol concentrations at King George Island (Antarctic Peninsula, 62°05′ S; 58°23.5′ W) over a nine-year period, and demonstrated these annual seasonal variations. The collected particles were categorised into two size fractions: fine particles (aerodynamic diameter <2 μm) and coarse particles (2–10 μm) (Figure 2.1).

During the Antarctic summer and spring, the particle number size distribution is dominated by recently formed particles from the gas phase, despite the absence of external sources of condensable gases. The seasonal cycle is closely linked to the biological activity in the surrounding oceans, as well as the dynamics of the atmospheric transport patterns.[25] The increased phytoplankton productivity in the Southern Ocean leads to the release of dimethyl sulphide (DMS), a precursor for the formation of sulphate aerosols.[25] These biogenic sulphate particles can then contribute to the formation of cloud condensation nuclei, influencing cloud properties and the regional radiative balance. In contrast, the winter months are characterised by a predominance of aged and long-range transported aerosols from surrounding continents.[33] The breakdown of the Antarctic inversion layer during these periods can also lead to the intrusion of sea salt aerosols from

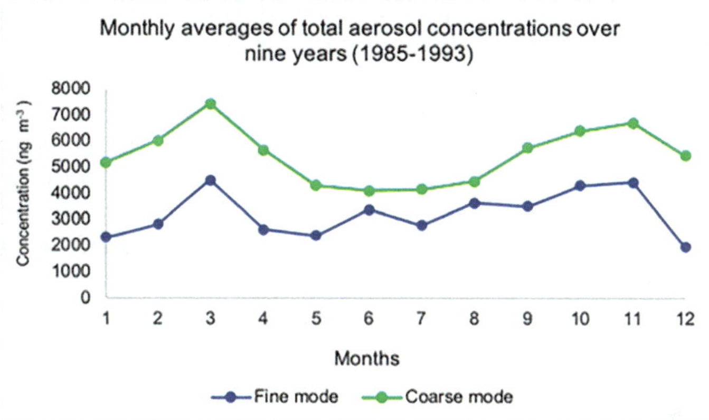

Figure 2.1 Seasonal variation in monthly mean total aerosol concentration on King George Island (Antarctic Peninsula) over nine years according to the data of Correia *et al.* (1998).[32]

the coastal regions, further modifying the aerosol composition and properties.[33] Studies have shown that the Antarctic near-coastal sub-micrometer aerosol particles in summer are characterised by their hygroscopicity, size distributions, volatility, and chemical composition.[34]

2.3.1.1 Natural Sources

Sea salt aerosols are a prominent constituent of the Antarctic atmosphere, originating from the surrounding Southern Ocean.[29] These particles play a vital role in the region's atmospheric chemistry and climate dynamics, influencing cloud formation, precipitation patterns, and the overall radiative balance of the Antarctic system. At the global level, marine aerosols constitute one of the most important sets of natural aerosols, playing a significant role in tropospheric chemistry.

The ocean produces a large amount of sea salt particles annually. The interaction of wind stress with the ocean surface results in the mechanical production of these aerosols, where there is a combination of inorganic sea salt and organic matter. As a component of marine spray, these aerosols are produced through the process of wave breaking, where the bursting of bubbles gives rise to sub-micron particles.[35] Sea salt is highly hygroscopic, meaning it has a high capacity to absorb water vapor, and it constitutes an important source of cloud condensation nuclei in remote marine areas.[29] Thus, once present in the atmosphere, the droplets formed by the disruption of bubbles can be transported over long distances due to the action of air masses.

As the droplets begin to evaporate, they become a concentrated saline solution, containing more or less water content depending on the ambient humidity. Sea salt aerosols cover sizes ranging from 100 nm to tens of micrometers, and the main component is sodium chloride (NaCl). Sea salt is commonly in a complex mixture containing other species, which may include inorganic cations and anions.[36,37] In the troposphere, mixtures with particles containing ammonia and crustal particles make sea salts the most important alkaline aerosols, as they also contain and react with acidic species, such as sulphuric acid, nitric acid, and methanesulphonic acid (CH_3SO_3H).[13,19,20]

Another prominent group of particles present in Antarctic aerosols is the sulphate-containing particles, which can either be of marine or terrestrial origin. A significant portion of the sulphate aerosols in Antarctica originates from the oxidation of DMS, a biogenic compound emitted by phytoplankton in the surrounding Southern Ocean. Once released into the atmosphere, DMS can undergo a series of oxidation reactions to form sulphate particles, often internally mixed with sea salt. The sulphate-containing particles originating from terrestrial sources, on the other hand, can be transported from distant continental regions, such as South America, by cyclonic systems passing between the Antarctic Peninsula and southernmost South America.[38] Furthermore, volcanic activity can contribute to the atmospheric sulphate load in Antarctica.

Mineral dust particles are another important component of the Antarctic aerosol population, originating from the continent's huge ice-free areas and glacial outwash plains. These aerosols, which are soil particles suspended in the atmosphere, play a fundamental role in the radiative balance of the atmosphere and the hydrological cycle, mainly due to their influence on cloud condensation nuclei.[39] The friction exerted by the wind on continental surfaces can cause the suspension of soil particles into the atmosphere. Also known as 'dust', these particles mainly originate from arid or semi-arid regions, and their emissions depend greatly on environmental and meteorological conditions, such as reduced soil moisture, low cohesive force between the particles and the soil, and regions with little vegetation and low precipitation. Like marine aerosols, mineral dust particles cover a range of sizes. Recent studies have identified the Patagonian desert and South America as two of the largest contributors to the flow of mineral dust particles to Antarctica,[40] highlighting this continent as an archive of environmental variability in South America, especially Patagonia.[38]

Volcanic eruptions emit both solid particles and gases, primarily SO_2, CO_2, and water vapor, among others. Volcanic activity is estimated to be responsible for the release into the atmosphere of 15 to 90 Tg per year of particles.[41] Volcanic ash, composed of particles with a diameter of less than 2 μm, has the potential to remain suspended in the atmosphere for extended periods, ranging from days to weeks, depending on its size.[42] This characteristic allows volcanic ash to be transported over vast distances, with fine ash particles capable of traveling hundreds to thousands of kilometers. During large-scale eruptions, volcanic ash can be carried several tens of kilometers high into the atmosphere, reaching even the most remote and iron-starved oceanic regions.[43] At high altitudes, volcanic eruptions emit large quantities of sulphur compounds that oxidise to form SO_2, which is later coated with H_2SO_4 droplets, a process that unfolds in the stratosphere in the weeks and months following the eruption. The H_2SO_4 condenses under all atmospheric conditions, due to its low vapor pressure, forming sulphate particles. Additionally, volcanic aerosols can act as ice nucleation particles in the upper troposphere, thereby interfering with the Earth's radiative balance in another way.

In Antarctica, the effects of volcanic eruptions on atmospheric aerosol loading and composition can have significant and long-lasting impacts. Volcanic aerosols, including sulphuric acid particles originating from volcanic ejecta, can reach Antarctica through the stratosphere or troposphere. The introduction of volcanic aerosols to the Antarctic atmosphere can lead to changes in air and precipitation chemistry.[44] Studies based on Antarctic ice core records have revealed the historical impact of volcanic eruptions on the region. Ice core data from Antarctica have shown evidence of major volcanic events, with volcanic sulphate aerosols and fine ash deposits being preserved in polar ice sheets and glaciers.[45] These records indicate that violent volcanic eruptions can inject sulphate aerosols into the stratosphere, resulting in acid precipitation that is recorded in Antarctic snow layers.[46] Furthermore, ice core records from Antarctica and Greenland have provided valuable insights into the role of volcanic eruptions in driving past variations in Earth's climate.[47]

2.3.1.2 Anthropogenic Sources

The anthropogenic components of Antarctic aerosols typically originate from long-range transport from other continents, as there are no major industrial or urban sources within the continent itself. One important constituent to consider is black carbon (BC) aerosols. Carbonaceous aerosols represent one of the main fractions of tropospheric aerosols, and they can constitute a complex mixture of different chemical compounds. Within this class, BC particles can be described as a solid material with a dark appearance, produced during the incomplete combustion of fossil fuels and biomass. BC has a graphitic microstructure and is the main light-absorbing aerosol in the atmosphere. The climate effects of BC aerosols strongly depend on their physical and chemical properties, thus play an important role in the Earth's atmosphere, mainly in terms of radiative forcing.[48] In general, bright, coloured, or translucent particles tend to reflect radiation in all directions and back into space, while darker aerosols can absorb significant amounts of light. Some authors consider BC emissions as the second most important anthropogenic emission related to climate forcings, with enormous potential to warm the atmosphere. Recent studies have suggested that anthropogenic climate forcings related to carbonaceous aerosols are a significant factor, as BC forcing from fossil fuel combustion ranges from +0.1 to 0.3 $W\,m^{-2}$.[49]

Due to BC particles being emitted predominantly in the fine mode of aerosols, their residence time in the atmosphere is long (from 1 to 4 weeks), which allows these particles to be deposited in locations very distant from their emission sources. Some studies have highlighted the importance of the transport of these particles and the need for long-term measurements, prioritising the analysis of the physical and optical properties of BC aerosols. These studies also evidence the presence of Antarctic aerosols on spatial and temporal scales, originating from different regions of the globe,[50] *e.g.* South America and Africa.[51] In fact, the long-term dispersion of aerosols is an important mechanism and should be considered in investigative discussions and arguments, but the lack of knowledge about the regional effects of BC particles, local sources, and, consequently, their environmental implications in remote locations has not been discussed in depth.

Regarding the global distribution and importance of aerosols, some studies have specifically addressed the influence of BC in Antarctica with respect to deposition in ice, air mass transport, concentration levels, and implications for climate change.[38,50,52] BC particles identified in remote polar regions imply a widespread climatic effect. These radiation-absorbing aerosols have a significant effect due to their effect on the increase of snow albedo (*i.e.*, a measure of surface reflectivity), as the environment has a vast cover of snow, sea ice, and glacial ice. The bright surface of snow and ice, which reflects an average of 90% of the incident shortwave radiation, makes the polar environment more susceptible to optical effects. The deposition of these carbonaceous particles on the snow also can substantially contribute to ice loss, caused by faster melting due to radiation absorption induced heating.[53] The albedo effects will be discussed below.

The intensification of human activities in Antarctica, especially on the continental coast, such as scientific expeditions and tourism, has increased the emissions of anthropogenic species, particularly during the austral summer period.[38] The local impact of pollution has grown, accentuating the concentrations of BC due to various activities. These include the incineration of scientific station waste, fossil fuel emissions for power generation, and emissions from motor vehicles, such as cars used in the region, ships, and aircraft surrounding Antarctica.[38,54] Since the 1980s, Antarctic tourism has experienced a period of rapid growth, and the number of visitors to the land portion has increased sharply each year. More than 40 000 people visit Antarctica every year, and the numbers tend to grow, with the main means of transportation to Antarctica being by air or sea.[54]

2.3.2 Snowpack Chemistry and Photochemical Reactions

Due to the process of dry or wet deposition, aerosols are deposited onto the Earth's surface, oceans, and ice. Particularly on ice, layers are formed and/or accumulated over time, agglomerating particles with different compositions. The relationship between sunlight and snow plays an important role in atmospheric aerosols, as particles deposited in the snow can absorb solar radiation, reducing the albedo of the ice layers. The presence of snow disrupts the composition of polar air since it forms a porous medium, offering good conditions for interaction with atmospheric gases. Photochemical reactions on the snow and ice sheets can lead to the production of volatile compounds which are then released into the atmosphere, modifying the chemical composition of aerosols.[55]

The photochemical reactions of various chemical components occur in ice and snow, generating species that significantly impact the atmosphere and the local environment. A variety of low-molecular-weight organic compounds are emitted from the snow cover, primarily from the upper ice layers, due to the direct or indirect photo-oxidation of natural organic materials present in the snow. By controlling the oxidative capacity of the atmosphere through the production of O_3, NO_x has received special attention, frequently reported in studies focused on this topic in Antarctica and the Arctic.[56]

The upper snow pack layers are a source of chemical products for the atmosphere, with elevated emissions of NO and OH observed during summer, potentially explaining why the boundary layer composition above the snow layers is considerably different from that predicted by atmospheric models[56] showing that snow is a natural reservoir of nitrate and the primary source of NO_x. Albedo is a critical factor in Antarctica's surface energy budget and deposited aerosols in the surface strongly influence albedo in polar regions. This is an important and significant mechanism that will certainly influence the local and regional climate by altering the radiative balance.

The photochemical reactivity of the Antarctic atmosphere is particularly important due to the increased UV radiation caused by the depletion of the ozone layer.[24] This creates optimal conditions for heterogeneous photochemical reactions on the surface of aerosol particles, which can significantly influence the composition and properties of these particles. Modelling, laboratory, and field studies suggest that snow pack processes release gases[56] into the atmosphere that can undergo chemical transformations on sea salt and other types of aerosols in the remote Antarctic regions.

2.4 Atmospheric Colloids in the Amazon: Sources, Variability, and Climate Influence

The Amazon rainforest is considered a natural laboratory for studying surface–atmosphere interactions. As one of the most intact tropical forests, it constitutes nearly 40% of the global tropical forest cover and covers close to 60% of the total area of Brazil.[57] In recent decades, extensive research has focused on the Amazon region due to the interactions between the Earth's surface and the atmosphere and their impact on atmospheric composition.[58] Understanding these exchange processes is fundamental for regulating the global climate and maintaining biodiversity. However, despite significant research efforts, essential gaps in scientific knowledge remain, particularly regarding the detailed mechanisms of surface–atmosphere interactions, the effects of deforestation, the role of the region in global carbon and water cycles, the magnitude of trace gas fluxes, chemical speciation, and the role of haze. Addressing these gaps is essential for advancing climate science and ensuring the sustainable management of the Amazon.[59]

The South American Amazon rainforest plays a critical role in regulating the global climate, mainly through its influence on regional and global water cycles. One of the most vital processes in the region is evapotranspiration, during which water is released from soil and plants into the atmosphere. This process is crucial for maintaining the rainforest's microclimate and driving cloud formation.[60] Clouds formed in the Amazon rainforest are essential for transporting moisture across South America, providing vital rainfall to other regions. This water transport, known as 'flying rivers', underscores the Amazon's importance in supporting ecosystems and agriculture across the continent.[60] Cloud formation is directly influenced by atmospheric composition, which typically includes aerosols originating from local forest emissions and long-distance transport from other regions. These natural emissions contribute to the complex interactions between the forest and the atmosphere, affecting cloud formation, rainfall patterns, and overall climate.[58,60]

Throughout the year, the Amazon is characterised as a source of bioaerosols, including biogenic particulate matter (BPM) and volatile organic compounds (VOCs).[57] However, the Amazon's aerosol composition is dynamic, varying with seasonality and external influences such as biomass burning, urban pollution, and dust transported from distant regions.[60] This complex mixture of locally produced and transported aerosols highlights the need to understand the natural and anthropogenic factors that shape atmospheric processes in the region.

During the rainy season, Amazonian regions, especially pristine areas such as the Alta Torre Amazon Observatory (ATTO), experience a near-pristine atmosphere.[61] This is due to the effective removal of aerosols through precipitation and the minimal presence of anthropogenic aerosol sources. These conditions make the Amazonian atmosphere an ideal benchmark for studying the natural atmosphere, allowing scientists to compare polluted regions to these reference concentrations. At other times of the year, in addition to biogenic emissions from forests, the Amazon receives significant inputs of dust and aerosols transported across the Atlantic from Africa. This atmospheric transport, originating mainly from the Sahara desert and biomass burning on the African continent, brings micronutrients such as calcium, potassium, and iron. These nutrients, in the form of microparticles, play an essential role in the Amazon's nutrient cycles, contributing to forest fertilisation and supporting the development and health of the region.[62,63] This transport highlights the interconnectedness of global ecosystems and highlights the importance of understanding transcontinental atmospheric processes in managing forest ecosystems and global climate patterns.

Much of the Amazon rainforest is considered pristine, but it is increasingly affected by human activities as urban expansion and industrialisation grow in the region.[64] Scientists have reported the presence of anthropogenic compounds, such as pollutants from urban centres and deforestation activities, in the Amazonian atmosphere.[65,66] These impacts are most pronounced during the dry season when the atmosphere becomes more polluted due to factors such as increased fire activity and reduced air dispersion. During months of scarce rainfall, the combined effects of reduced precipitation and increased emissions from biomass burning (linked to deforestation and agricultural practices) significantly increase air pollution and worsen air quality across the region.[65,66] The resulting increase in particulate matter prolongs its presence in the atmosphere, affecting cloud formation and rainfall patterns.[67] These changes underscore the escalating impact of human activities on the Amazon and highlight the urgent need for comprehensive monitoring and targeted mitigation strategies to protect this vital region.

In addition to seasonal influences, the atmospheric composition of the forest varies throughout the day,[68] with the biogenic fraction often dominating over other types of particles.[69] During the day, high solar incidence intensifies evapotranspiration, loading the atmosphere with water vapor. Additionally, the biogenic composition is dominated by giant PBAs, such as canopy debris, pollen, fern spores, and large fungal spores.[68] The emission of volatile organic compounds is also predominantly diurnal, facilitating the formation of secondary organic aerosols (SOAs) through their oxidation.

At night, fungal spores exhibit enhanced emission, which is well-documented and associated with active release under conditions of high relative humidity.[70] In the Amazon, spores are considered the most abundant type of bioaerosol, showing significant predominance in regions near the forest canopy, surpassing bacteria in number.[71] Biogenic particles are known to act as giant cloud condensation nuclei and can even participate in the formation of ice clouds.[72] In the forest, these bioaerosols are transported horizontally and vertically, potentially impacting regions far from their source or actively participating in cloud formation as Giant Cloud Condensation Nuclei (GCCN).[73] The vertical distribution of PBAs observed in the Amazon shows high concentrations at the forest–atmosphere interface, with significant biogenic material reaching up to 300 m in height, reflecting the critical role of vertical transport in dispersing these materials.[73]

2.4.1 Impacts and Outcomes in a Changing Climate Scenario

The effects of climate change are evident in the emission and transport of bioaerosols. For instance, the transmission of fungal spores depends heavily on high humidity. A water imbalance in forest ecosystems, such as the Amazon, can significantly affect the reproduction of these fungi and, consequently, their ecological functions, thereby influencing the biotic system.[74] In contrast, drier and warmer conditions lead to increased pollen emissions. Rising global temperatures extend the growing season of plants, leading to a longer emission period and a more intense release of pollen grains, which may increase by up to 40%. Additionally, drier periods promote the persistence of pollen grains in the atmosphere, allowing them to remain suspended for longer durations.[75]

Projections for the coming years indicate a marked extension of pollen seasons. These changes have already been observed in USA[75] and in Brazil.[76] In Curitiba, a city in southern Brazil, the effects of climate change have been evident in the grass pollen emission seasons. Researchers have observed an intensification in pollen concentration and an earlier peak. Between 2018 and 2019, the most intense pollen season lasted from August to April, peaking in September. In comparison, earlier studies showed peaks in November, with the emission period spanning from October to April.[76] Consequently, there is an increase in allergy cases among the population, with symptoms being more prolonged and occurring earlier.[75] There is also an aggravation of conditions such as asthma and allergy attacks. The interaction of pollen grains with air pollutants can also enhance their potential to trigger allergic reactions. Despite its importance to public health, monitoring air pollen concentrations is uncommon. Pollutants such as ozone and particulate matter can bind to pollen grains, potentially increasing their ability to trigger allergic reactions in sensitive individuals.[77]

References

1. WHO, Air Pollution, https://www.who.int/health-topics/air-pollution#tab=tab_1 (accessed 1 September 2024).
2. P. J. Landrigan, R. Fuller, N. J. R. Acosta, O. Adeyi, R. Arnold, N. Nil Basu, A. B. Baldé, R. Bertollini, S. Bose-O'Reilly, J. I. Boufford, P. N. Breysse, T. Chiles, C. Mahidol, A. M. Coll-Seck, M. L. Cropper, J. Fobil, V. Fuster, M. Greenstone, A. Haines, D. Hanrahan, D. Hunter, M. Khare, A. Krupnick, B. Lanphear, B. Lohani, K. Martin, K. V. Mathiasen, M. A. McTeer, C. J. L. Murray, J. D. Ndahimananjara, F. Perera, J. Potočnik, A. S. Preker, J. Ramesh, J. Rockström, C. Salinas, L. D. Samson, K. Sandilya, P. D. Sly, K. R. Smith, A. Steiner, R. B. Stewart, W. A. Suk, O. C. P. van Schayck, G. N. Yadama, K. Yumkella and M. Zhong, *Lancet*, 2018, **391**, 462.
3. WHO, *WHO global air quality guidelines: particulate matter (PM2.5 and PM10), ozone, nitrogen dioxide, sulfur dioxide and carbon monoxide*, WHO European Centre for Environment and Health, 2021.
4. G. Chen, Y. Guo, M. J. Abramson, G. Williams and S. Li, *Sci. Total Environ.*, 2018, **622–623**, 721.
5. WHO, *Ambient Air Pollution: A Global Assessment of Exposure and Burden of Disease*, World Health Organization Press, Geneva, 2016.
6. United Nations Department of Economic and Social Affairs, https://sdgs.un.org/ (accessed 1 September 2024).

7. J. Niu, P. E. Rasmussen, N. M. Hassan and R. Vincent, *Water, Air, Soil Pollut.*, 2010, **213**, 211.
8. WHO, *WHO Guidelines for Indoor Air Quality: Selected Pollutants*, WHO, Regional Office for Europe, Copenhagen, 2010.
9. G. Polezer, S. Potgieter-Vermaak, A. Oliveira, L. D. Martins, J. C. Santos-Silva, C. A. B. Moreira, T. Pauliquevis, A. F. L. Godoi, Y. Tadano, C. I. Yamamoto and R. H. M. Godoi, *Environ. Geochem. Health*, 2023, **45**, 1841.
10. R. T. Burnett, C. Arden Pope, M. Ezzati, C. Olives, S. S. Lim, S. Mehta, H. H. Shin, G. Singh, B. Hubbell, M. Brauer, H. Ross Anderson, K. R. Smith, J. R. Balmes, N. G. Bruce, H. Kan, F. Laden, A. Prüss-Ustün, M. C. Turner, S. M. Gapstur, W. R. Diver and A. Cohen, *Environ. Health Perspect.*, 2014, **122**, 397.
11. P. A. Tanner, H. L. Ma and P. K. N. Yu, *Environ. Sci. Technol.*, 2019, **42**, 71111.
12. C. Puls, A. Limbeck and S. Hann, *Atmos. Environ.*, 2012, **55**, 213.
13. C. Angelé-Martínez, C. Goodman and J. Brumaghim, *Metallomics*, 2014, **6**, 1358–1381.
14. M. Guney, C. M. J. Bourges, R. P. Chapuis and G. J. Zagury, *Sci. Total Environ.*, 2017, **579**, 1358.
15. L. F. C. Nascimento, J. B. Francisco, M. B. R. Patto and A. M. Antunes, *Cad. Saude Publica*, 2012, **28**, 1319.
16. G. Polezer, Y. S. Tadano, H. V. Siqueira, A. F. L. Godoi, C. I. Yamamoto, P. A. de André, T. Pauliquevis, M. de, F. Andrade, A. Oliveira, P. H. N. Saldiva, P. E. Taylor and R. H. M. Godoi, *Environ. Pollut.*, 2018, **235**, 394.
17. G. E. Shaw, *Rev. Geophys.*, 1988, **26**, 89.
18. U. C. Herzfeld, R. Stosius and M. Schneider, *Ann. Glaciol.*, 2000, **30**, 76.
19. E. Vignon, F. Hourdin, C. Genthon, B. J. H. Van de Wiel, H. Gallée, J.-B. Madeleine and J. Beaumet, *J. Adv. Model. Earth Syst.*, 2018, **10**, 98.
20. C. Tomasi, A. A. Kokhanovsky, A. Lupi, C. Ritter, A. Smirnov, N. T. O'Neill, R. S. Stone, B. N. Holben, S. Nyeki, C. Wehrli, A. Stohl, M. Mazzola, C. Lanconelli, V. Vitale, K. Stebel, V. Aaltonen, G. de Leeuw, E. Rodriguez, A. B. Herber, V. F. Radionov, T. Zielinski, T. Petelski, S. M. Sakerin, D. M. Kabanov, Y. Xue, L. Mei, L. Istomina, R. Wagener, B. McArthur, P. S. Sobolewski, R. Kivi, Y. Courcoux, P. Larouche, S. Broccardo and S. J. Piketh, *Earth-Sci. Rev.*, 2015, **140**, 108.
21. S. C. Cox, B. Smith Lyttle, S. Elkind, C. Smith Siddoway, P. Morin, G. Capponi, T. Abu-Alam, M. Ballinger, L. Bamber, B. Kitchener, L. Lelli, J. Mawson, A. Millikin, N. Dal Seno, L. Whitburn, T. White, A. Burton-Johnson, L. Crispini, D. Elliot, S. Elvevold, J. Goodge, J. Halpin, J. Jacobs, A. P. Martin, E. Mikhalsky, F. Morgan, P. Scadden, J. Smellie and G. Wilson, *Sci. Data*, 2023, **10**, 250.
22. L. R. Pertierra, F. Santos-Martin, K. A. Hughes, C. Avila, J. O. Caceres, D. De Filippo, S. Gonzalez, S. M. Grant, H. Lynch, C. Marina-Montes, A. Quesada, P. Tejedo, T. Tin and J. Benayas, *Ecosyst. Serv.*, 2021, **49**, 101299.
23. M. M. Frey, H. K. Roscoe, A. Kukui, J. Savarino, J. L. France, M. D. King, M. Legrand and S. Preunkert, *Atmos. Chem. Phys.*, 2015, **15**, 7859.

24. S. J. G. Junior, J. Weis, S. China, H. Evangelista, T. H. Harder, S. Müller, M. Sampaio, A. Laskin, M. K. Gilles and R. H. M. Godoi, *Sci. Total Environ.*, 2021, **758**, 143586.
25. M. R. Giordano, L. E. Kalnajs, A. Avery, J. D. Goetz, S. M. Davis and P. F. DeCarlo, *Atmos. Chem. Phys.*, 2017, **17**, 1.
26. J. Anzano, E. Abás, C. Marina-Montes, J. del Valle, D. Galán-Madruga, M. Laguna, S. Cabredo, L.-V. Pérez-Arribas, J. Cáceres and J. Anwar, *Atmosphere*, 2022, **13**, 1621.
27. A. Correia, P. Artaxo and W. Maenhaut, *Ann. Glaciol.*, 1998, **27**, 560.
28. S. J. Gonçalves Junior, H. Evangelista, J. Weis, T. H. Harder, S. China, S. Müller, M. M. Marques, N. de Magalhães Neto, H. R. Passos, M. Sampaio, J. C. Simões, B. V. Ximenes de Oliveira, C. I. Yamamoto, A. Laskin, M. K. Gilles and R. H. M. Godoi, *Commun. Earth Environ.*, 2023, **4**, 1.
29. R. Udisti, U. Dayan, S. Becagli, M. Busetto, D. Frosini, M. Legrand, F. Lucarelli, S. Preunkert, M. Severi, R. Traversi and V. Vitale, *Atmos. Environ.*, 2012, **52**, 109.
30. B. Neumann, A. Mikoleit, J. S. Bowman, H. W. Ducklow and F. Müller, *Front. Environ. Sci.*, 2019, 7, 1.
31. C. Truzzi, L. Lambertucci, S. Illuminati, A. Annibaldi and G. Scarponi, *Ann. Chim.*, 2005, **95**, 867.
32. A. L. Correia, Aerossóis atmosféricos na Antártica: sazonalidade, composição elementar e relação com El Niño. 1998. Dissertation – Universidade de São Paulo, São Paulo, 1998. Available in: https://repositorio.usp.br/item/000971161.
33. F. Parungo, B. Bodhaine and J. Bortniak, *J. Aerosol Sci.*, 1981, **12**, 491.
34. E. Asmi, A. Frey, A. Virkkula, M. Ehn, H. E. Manninen, H. Timonen, O. Tolonen-Kivimäki, M. Aurela, R. Hillamo and M. Kulmala, *Atmos. Chem. Phys.*, 2010, **10**, 4253.
35. R. J. Hopkins, Y. Desyaterik, A. V. Tivanski, R. A. Zaveri, C. M. Berkowitz, T. Tyliszczak, M. K. Gilles and A. Laskin, *J. Geophys. Res.: Atmos.*, 2008, **113**, D04209.
36. A. Laskin, R. C. Moffet, M. K. Gilles, J. D. Fast, R. A. Zaveri, B. Wang, P. Nigge and J. Shutthanandan, *J. Geophys. Res.: Atmos.*, 2012, **117**, D15302.
37. A. P. Ault, T. L. Guasco, O. S. Ryder, J. Baltrusaitis, L. A. Cuadra-Rodriguez, D. B. Collins, M. J. Ruppel, T. H. Bertram, K. A. Prather and V. H. Grassian, *J. Am. Chem. Soc.*, 2013, **135**, 14528.
38. E. B. Pereira, H. Evangelista, K. C. D. Pereira, I. F. A. Cavalcanti and A. W. Setzer, *J. Geophys. Res.*, 2006, **111**, D03303.
39. V. A. Karydis, P. Kumar, D. Barahona, I. N. Sokolik and A. Nenes, *J. Geophys. Res.: Atmos.*, 2011, **116**, D23204.
40. S. Gassó, A. Stein, F. Marino, E. Castellano, R. Udisti and J. Ceratto, *Atmos. Chem. Phys.*, 2010, **10**, 8287.
41. T. A. Mather, D. M. Pyle and C. Oppenheimer, in *Volcanism and the Earth's Atmosphere*, American Geophysical Union (AGU), 2004, p. 189.

42. F. Pardini, S. Corradini, A. Costa, T. Esposti Ongaro, L. Merucci, A. Neri, D. Stelitano and M. de' Michieli Vitturi, *Atmosphere*, 2020, **11**, 359.
43. S. Duggen, N. Olgun, P. Croot, L. Hoffmann, H. Dietze, P. Delmelle and C. Teschner, *Biogeosciences*, 2010, **7**, 827.
44. R. Delmas and C. Boutron, *J. Geophys. Res.: Oceans*, 1980, **85**, 5645.
45. J. Cole-Dai, D. G. Ferris, J. A. Kennedy, M. Sigl, J. R. McConnell, T. J. Fudge, L. Geng, O. J. Maselli, K. C. Taylor and J. M. Souney, *J. Geophys. Res.: Atmos.*, 2021, **126**, 1.
46. J. Ren, C. Li, S. Hou, C. Xiao, D. Qin, Y. Li and M. Ding, *J. Geophys. Res.: Atmos.*, 2010, **115**, D11301.
47. M. Sigl, J. R. McConnell, M. Toohey, M. Curran, S. B. Das, R. Edwards, E. Isaksson, K. Kawamura, S. Kipfstuhl and K. Krüger, *Nat. Clim. Change*, 2014, **4**, 693.
48. M. Z. Jacobson, *J. Geophys. Res.*, 2001, **106**, 1551.
49. M. Schulz, C. Textor, S. Kinne, Y. Balkanski, S. Bauer, T. Berntsen, T. Berglen, O. Boucher, F. Dentener, S. Guibert, I. S. A. Isaksen, T. Iversen, D. Koch, A. Kirkeva, J. E. Penner, G. Pitari, S. Reddy, Ø. Seland, P. Stier and T. Takemura, *Atmos. Chem. Phys.*, 2006, **22**, 5225–5246.
50. S. J. G. Junior, N. Magalhães, R. C. Charello, H. Evangelista and R. H. M. Godoi, *An. Acad. Bras. Cienc.*, 2022, **94**, e20210805.
51. E. A. dos Santos, H. Evangelista, C. de Morisson Valeriano, C. C. A. Neto, A. Castagna and M. Heilbron, *Aeolian Res.*, 2020, **45**, 100596.
52. H. Evangelista, J. Maldonado, R. H. M. Godoi, E. B. Pereira, D. Koch, K. Tanizaki-Fonseca, R. Van Grieken, M. Sampaio, A. Setzer, A. Alencar and S. C. Gonçalves, *J. Atmos. Chem.*, 2007, **56**, 225.
53. O. Ogunro, S. Elliott, N. Jeffery, F. Hoffman, H. Wang and O. Wingenter, *arXiv*, 2017, preprint, arXiv:1702.03325.
54. H. J. Lynch, K. Crosbie, W. F. Fagan and R. Naveen, *Antarct. Sci.*, 2010, **22**, 123.
55. S. G. Warren, *Philos. Trans. R. Soc., A*, 2019, **377**, 20180161.
56. E. W. Wolff, A. E. Jones, T. J. Martin and T. C. Grenfell, *Geophys. Res. Lett.*, 2002, **29**, 51.
57. A. Baccini, S. J. Goetz, W. S. Walker, N. T. Laporte, M. Sun, D. Sulla-Menashe, J. Hackler, P. S. A. Beck, R. Dubayah, M. A. Friedl, S. Samanta and R. A. Houghton, *Nat. Clim. Change*, 2012, **2**, 182.
58. S. T. Martin, M. O. Andreae, P. Artaxo, D. Baumgardner, Q. Chen, A. H. Goldstein, A. Guenther, C. L. Heald, O. L. Mayol-Bracero, P. H. McMurry, T. Pauliquevis, U. Pschl, K. A. Prather, G. C. Roberts, S. R. Saleska, M. A. Silva Dias, D. V. Spracklen, E. Swietlicki and I. Trebs, *Rev. Geophys.*, 2010, **48**, RG2002.
59. P. Artaxo, F. Gerab and M. L. C. Rabello, *Nucl. Instrum. Methods Phys. Res.*, 1993, **75**, 277.
60. D. Zemp, C. F. Schleussner, H. Barbosa and R. van der Ent, *Resilience of Tropical Ecosystems – Future Challenges and Opportunities*, Zurich, 2015.
61. M. O. Andreae, O. C. Acevedo, A. Araùjo, P. Artaxo, C. G. G. Barbosa, H. M. J. Barbosa, J. Brito, S. Carbone, X. Chi, B. B. L. Cintra, N. F. Da Silva,

N. L. Dias, C. Q. Dias-Júnior, F. Ditas, R. Ditz, A. F. L. Godoi, R. H. M. Godoi, M. Heimann, T. Hoffmann, J. Kesselmeier, T. Könemann, M. L. Krüger, J. V. Lavric, A. O. Manzi, A. P. Lopes, D. L. Martins, E. F. Mikhailov, D. Moran-Zuloaga, B. W. Nelson, A. C. Nölscher, D. Santos Nogueira, M. T. F. Piedade, C. Pöhlker, U. Pöschl, C. A. Quesada, L. V. Rizzo, C. U. Ro, N. Ruckteschler, L. D. A. Sá, M. De Oliveira Sá, C. B. Sales, R. M. N. Dos Santos, J. Saturno, J. Schöngart, M. Sörgel, C. M. De Souza, R. A. F. De Souza, H. Su, N. Targhetta, J. Tóta, I. Trebs, S. Trumbore, A. Van Eijck, D. Walter, Z. Wang, B. Weber, J. Williams, J. Winderlich, F. Wittmann, S. Wolff and A. M. Yáñez-Serrano, *Atmos. Chem. Phys.*, 2015, **15**, 10723.

62. J. A. Rizzolo, C. G. G. Barbosa, G. C. Borillo, A. F. L. Godoi, R. A. F. Souza, R. V. Andreoli, A. O. Manzi, M. O. Sá, E. G. Alves, C. Pöhlker, I. H. Angelis, F. Ditas, J. Saturno, D. Moran-Zuloaga, L. V. Rizzo, N. E. Rosário, T. Pauliquevis, R. M. N. Santos, C. I. Yamamoto, M. O. Andreae, P. Artaxo, P. E. Taylor and R. H. M. Godoi, *Atmos. Chem. Phys.*, 2017, **17**, 2673.
63. D. Moran-Zuloaga, F. Ditas, D. Walter, J. Saturno, J. Brito, S. Carbone, X. Chi, I. Hrabě De Angelis, H. Baars, R. H. M. Godoi, B. Heese, B. A. Holanda, J. V. Lavrič, S. Martin, J. Ming, M. L. Pöhlker, N. Ruckteschler, H. Su, Y. Wang, Q. Wang, Z. Wang, B. Weber, S. Wolff, P. Artaxo, U. Pöschl, M. Andreae and C. Pöhlker, *Atmos. Chem. Phys.*, 2018, **18**, 10055.
64. I. Kourtchev, B. G. Sebben, S. Brill, C. G. G. Barbosa, B. Weber, R. R. Ferreira, F. A. F. D'Oliveira, C. Q. Dias-Junior, O. A. M. Popoola, J. Williams, C. Pöhlker and R. H. M. Godoi, *Sci. Total Environ.*, 2024, **944**, 173918.
65. Y. J. Parra, G. M. Pereira, D. Custódio, S. B. de Figueiredo, C. Alves and P. de Castro Vasconcellos, *Air Qual., Atmos. Health*, 2024, **17**, 599.
66. C. Gonçalves, B. R. Figueiredo, C. A. Alves, A. A. Cardoso, R. da Silva, S. H. Kanzawa and A. M. Vicente, *Atmos. Res.*, 2016, **182**, 102.
67. L. Nagy, P. Artaxo and B. R. Forsberg, *Interactions Between Biosphere, Atmosphere and Human Land Use in the Amazon Basined*, Springer, Berlin Heidelberg, 2016.
68. S. China, S. M. Burrows, B. Wang, T. H. Harder, J. Weis, M. Tanarhte, L. V. Rizzo, J. Brito, G. G. Cirino, P. L. Ma, J. Cliff, P. Artaxo, M. K. Gilles and A. Laskin, *Nat. Commun.*, 2018, **9**, 4793.
69. L. Wu, X. Li, H. Kim, H. Geng, R. H. M. Godoi, C. G. G. Barbosa, A. F. L. Godoi, C. I. Yamamoto, R. A. F. De Souza, C. Pöhlker, M. O. Andreae and C. U. Ro, *Atmos. Chem. Phys.*, 2019, **19**, 1221.
70. N. Löbs, C. G. G. Barbosa, S. Brill, D. Walter, F. DItas, M. De Oliveira Sá, A. C. De Araújo, L. R. De Oliveira, R. H. M. Godoi, S. Wolff, M. Piepenbring, J. Kesselmeier, P. Artaxo, M. O. Andreae, U. Pöschl, C. Pöhlker and B. Weber, *Atmos. Meas. Tech.*, 2020, **13**, 153.
71. F. F. C. Souza, D. V. Rissi, F. O. Pedrosa, E. M. Souza, V. A. Baura, R. A. Monteiro, E. Balsanelli, L. M. Cruz, R. A. F. Souza, M. O. Andreae, R. A. Reis, R. H. M. Godoi and L. F. Huergo, *Sci. Total Environ.*, 2019, **688**, 83.

72. B. H. Matthews, A. N. Alsante and S. D. Brooks, *ACS Earth Space Chem.*, 2023, **7**, 1207.
73. C. G. G. Barbosa, P. E. Taylor, M. O. Sá, P. R. Teixeira, R. A. F. Souza, R. I. Albrecht, H. M. J. Barbosa, B. Sebben, A. O. Manzi, A. C. Araújo, M. Prass, C. Pöhlker, B. Weber, M. O. Andreae and R. H. M. Godoi, *npj Clim. Atmos. Sci.*, 2022, **5**, 1.
74. S. Mota de Oliveira, E. Duijm, M. Stech, J. Ruijgrok, M. Polling, C. G. G. Barbosa, G. R. Cerqueira, A. H. M. Nascimento, R. H. M. Godoi, P. E. Taylor, S. Wolff, B. Weber and J. Kesselmeier, *Front. Ecol. Evol.*, 2022, **10**, 1.
75. N. G. Brasil, Cientistas alertam: a temporada de alergias está prestes a piorar, https://www.nationalgeographicbrasil.com/ciencia/2024/07/cientistas-alertam-a-temporada-de-alergias-esta-prestes-a-piorar (accessed 24 July 2024).
76. J. Camargo, S. Brill, C. Souza, B. Sebben, T. Pauliquevis, P. Taylor and R. Godoi, *J. Allergy Clin. Immunol.*, 2021, **147**, AB85.
77. H. Amini, M. Amini and R. Wright, *Immunol. Allergy Clin. North Am.*, 2024, **44**, 1.

CHAPTER 3

Development of Air Quality Sensors

DEUBER LINCON DA SILVA AGOSTINI,*[a]
LIZANDRA MARIA ZIMMERMANN,*[b]
MARÍA FERNANDA HERNÁNDEZ RODRIGUEZ,[b]
JÉSSICA MANTELATO BOMFIM CORRÊA,[a]
BRUNO HENRIQUE DE SANTANA GOIS,[a]
VITOR HUGO UZELOTO FERNANDES MINGRONI,[a]
PEDRO LEONARDO SILVA[a] AND GABRIEL DA CRUZ DIAS[a]

[a] Universidade Estadual Paulista (UNESP) – Department of Physics, Rua Roberto Símonsen, 305, Presidente Prudente, SP, 19060-900, Brazil;
[b] Universidade Regional de Blumenau (FURB) – Department of Chemistry, Rua Antônio da Veiga, 140, Blumenau, SC, 89030-903, Brazil
*Emails: deuber.agostini@unesp.br; lmz@furb.br

3.1 Introduction

The growing world population has led to unprecedented industrial intensification in food, pharmaceutical, and textile production sectors.[1,2] This expansion has led to environmental degradation and a scarcity of natural resources, including arable land, increasing per capita resource consumption, greenhouse gas (GHG) emissions, and a high level of harmful urban air pollutants. GHGs, due to human activity, such as agriculture and industry, include carbon dioxide (CO_2), methane (CH_4), and nitrous oxide (N_2O), the three major gases that exacerbate climate change.[1–3] GHGs act by cooling the stratosphere due to the energy re-radiated back to the surface, which is absorbed by these gases due to their absorption intrinsic capacity,

Contribution of Colloidal Materials to Air, Water and Soil Environmental Sustainability: From Basic Concepts to Advanced Applications
Edited by Izabel C. Riegel-Vidotti, Lizandra Maria Zimmermann and Heloise R. de Barros

Published by the Royal Society of Chemistry, www.rsc.org

consequently increasing the atmosphere's temperature.[4] Industrial intensification has also increased air pollution from compounds such as particulate matter ($PM_{2.5}$ and PM_{10}, ≤2.5 and 10 micrometers in diameter, respectively),[5] volatile organic compounds (VOCs), nitrogen dioxide (NO_2), sulfur dioxide (SO_2), and ammonia (NH_3).[6] Troposphere ozone (O_3) also belongs to the harmful class of gases, triggering various health problems.[7] It primarily originates from anthropogenic and natural sources, with the main contribution from the former source.[8,9] Carbon monoxide (CO) emissions come from various sources, especially related to incomplete combustion. CO is a serious indoor air pollutant due to its potential to cause poisoning, which can be fatal.[10] The World Health Organization (WHO) states that the above pollutants harm human health.[3,11–13] Therefore, air quality monitoring is crucial for developing action plans to control and reduce these gases' environmental impact.

Atmospheric measurements are crucial for monitoring air quality, weather forecasting, energy production, agriculture, domestic and industrial safety, transportation, and public health.[11] Cases of asphyxiation, explosions, and deaths worldwide constantly remind us that gas sensors for measuring air quality are essential for these applications, as they provide information on the concentration of the various gases in the air.[14,15]

From another perspective, gas sensing is closely related to advances in emerging energy production. The detection and quantification of hydrogen (H_2) take part in green hydrogen generation technologies. Additionally, H_2 gas sensors play a vital role in safety, as they help mitigate the risk of explosions when H_2 is exposed to air.[16]

Controlling vapors from substances like ethanol, methanol, and acetone significantly reduces accidents and health risks. Detecting these vapors drives the development of innovative smart devices that contribute to various fields, including biomedical, forensic, biological, and environmental monitoring.[17–19] Acetone, for example, has been known as a diabetes biomarker when detected in very low concentrations in breath.[20]

Over the last decade, a wide range of technologies have been developed to improve the quality of gas sensors. Aspects such as sensitivity, response/recovery time, selectivity, lower limit of detection (LOD) and limit of quantification (LOQ), lifetime, and accessibility have been considered. These technologies include electrochemical-based gas sensors,[21–23] metal oxides,[24–27] catalytic,[28–30] optical,[31–33] among others.[34–37] However, many of these gas sensors have significant limitations. For instance, some respond slowly or recover poorly at room temperature.[38] Optical sensors are prone to spectral interference and calibration problems, while catalytic sensors can be susceptible to poisoning.[39,40] In addition, most gas sensors exhibit cross-sensitivity to other gases and are affected by temperature variations.[41] Even commercially available gas analyzers have limitations in operating temperature, offline sensitivity, and the need for expensive equipment and skilled personnel.[41] In this context, the recent use of colloid-based materials for gas detection has attracted particular interest due to their unique and promising properties, including the possibility of *in situ* measurements.[42–45]

A colloid is a system consisting of two phases: a dispersed phase and a continuous phase. The dispersed phase generally consists of nano-sized particles evenly distributed in the dispersion medium or the continuous phase.[46,47] Due to their size, colloidal particles possess unique properties, including a high surface-to-volume ratio, stability under various environmental conditions, and the ability to undergo surface functionalization.[48–50] These properties make colloids ideal for a wide range of technological applications.[51–56]

The versatility of colloids in developing next-generation sensing technologies can overcome the limitations of conventional gas sensing technologies, offering more robustness, miniaturization, and adaptability under different operating conditions.[57] Advanced nano detection technologies, especially nanostructured metal oxide gas sensors, are promising techniques for monitoring low gas concentrations in complex gas mixtures on the order of parts per million (ppm) and parts per billion (ppb).[58] A wide range of innovative classes and syntheses of nanomaterials, such as dispersed and immobilized nanoparticles (NPs) in polymeric materials and colloidal quantum dots (CQDs), are essential in developing gas sensors.[59–61]

In colloidal technologies for gas sensing, various methods have been used to synthesize NPs and CQDs in porous materials such as graphene, single-walled carbon nanotubes (SWCNTs), or polymers such as arylene polyether (PAE). In these systems, electrostatic, electrochemical, and chemisorptive fields are typically used for gas detection.[62] For example, a chemiresistive sensor designed to detect H_2 under ambient conditions was developed using palladium colloidal NPs synthesized within PAE. The PAE restricts NP growth and stabilizes their dispersion within the SWCNTs of the sensor, where the detection mechanism relies on redox processes, enabling a reversible response to H_2.[63] Another example involves a gas sensor based on cellulose and copper oxide (CuO) nanocomposites for detecting hydrogen sulfide (H_2S) at low temperatures (40 to 80 °C). The detection mechanism is predicated on the absorption and desorption of oxygen molecules on the sensor's surface. Exposure to H_2S alters the concentration of charge carriers on the CuO interface, leading to a reversible change in the resistance of the sensor.[64] Additionally, a notable response to acetone was achieved using Cu-doped p-type zinc oxide (ZnO) nanostructures at room temperature, where the sensor's resistance increases in response to acetone concentrations starting from 1 ppm.[64]

This chapter provides an overview of colloidal technologies for gas sensing. The reader should gain a comprehensive understanding of the basic principles, classes of colloid-based sensors, sensing strategies, recent advances, and current challenges in this area of research.

3.2 Basic Principles of Gas Sensing Operation Using Colloidal Technologies

Gases can interact with the surface of sensor materials through two main mechanisms: physisorption and chemisorption. Physisorption involves the physical adsorption of gas molecules, typically induced by van der Waals forces, which are relatively weak and reversible. This interaction occurs at low binding energies, causing the gas molecule to be retained on the surface without forming a strong chemical bond.[65] In contrast, chemisorption entails the formation of chemical bonds between gas molecules and the sensor surface, resulting in a higher binding energy and, consequently, a generally irreversible process.[66]

These interactions are essential for the sensitivity and selectivity of gas sensors. Physisorption, for instance, allows for a quick and reversible response, which is desirable for repeated and continuous detections.[65] On the other hand, chemisorption, while providing greater stability in detection, can lead to a slower and less reversible response, making it more suitable for detecting specific gases where stability and detection permanence are critical.[67,68]

Understanding these interaction mechanisms is essential to control the key parameters that affect sensor performance, starting from the design phase. Sensor material selection is determining, and factors such as composition, particle size, physicochemical properties, morphology, and stability must be considered. Additionally, the type of energy involved—whether light, current, resistance, or acoustic signals—plays a significant role in the way the sensor interacts with gases and the nature of the subsequent reactions that occur on the surface.

In the following section, we explore various gas sensing strategies that leverage colloidal systems, focusing on using nanomaterials to enhance gas detection capabilities.

3.3 Sensing Strategies with Colloid-based Sensors for Air Monitoring

Colloid-based sensors can be broadly categorized according to their operating principles, as described in the subsequent section, whether they are sensors made up of metal oxides with chemiresistive approaches or semiconductors in confinement effect with optical responses, electrochemical sensors, and others. These materials may emerge as promising technologies for detecting various types of gases, meeting an ever-growing demand for air quality monitoring and industrial applications. Figure 3.1 depicts a general view of gas sensing design using colloids and different types of outputs that can be correlated with quantitative analyses previously established in analytical curves.

The performance of colloid-based gas sensors is enhanced by the unique properties of colloidal materials, which offer high surface area, tunable optical and electrical characteristics, and improved reactivity. These features make colloids particularly advantageous for developing highly sensitive and selective sensors for detecting gases in complex environments.

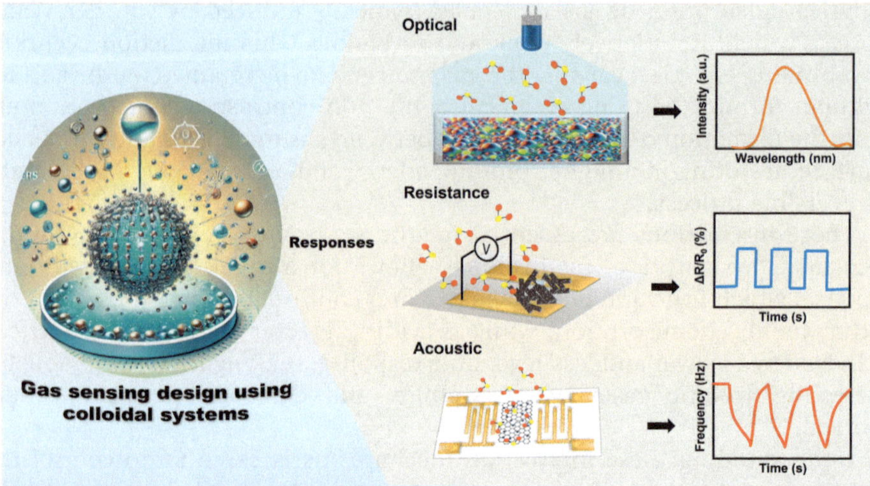

Figure 3.1 Design and sensing strategies for gas sensors based on colloidal systems. Sensors built with colloids can give current, absorption/photoluminescence, and frequency responses when interacting with gas molecules under an external stimulus.

3.3.1 Optical Sensors for Gas Detection

Optical gas sensors work by detecting changes in the optical properties of materials when gas molecules interact with their surfaces. This interaction, combined with the absorption or emission of light, produces a measurable optical signal (Figure 3.1). Furthermore, optical sensing signals can be measured by reflectivity, colorimetry, fluorescence, surface plasmon resonance (SPR), Raman scattering, absorbance, refractive index, and non-dispersive infrared (NDIR).[69,70] These sensors offer significant advantages, such as high sensitivity, fast response, and the ability to operate under various environmental conditions. Recent advances in nanotechnology, particularly integrating nanomaterials, have substantially enhanced their performance.[71–73] One of the most influential advances is using quantum dots (QDs), where quantum confinement effects at the nanoscale change the behavior of charge carriers (electrons and holes). As the size of the material decreases, the movement of these charge carriers is restricted, resulting in discrete rather than continuous energy levels, which changes the optical properties of the material.[74,75]

CQDs are widely explored, considering their spectral response, which can be tuned simply by adjusting the size of CQDs. Also, they have a large optical absorption spectrum and narrow emission. Upon absorption of photons with energy $(E) \geq$ band gap energy, electron–hole pairs are generated. Electrons are promoted to the conduction band, leaving behind carriers created at the valence band (holes). Electron–hole interband recombination can generate photons, giving rise to an excitonic emission (Figure 3.2). Another radiative recombination can compete with excitonic emission due to

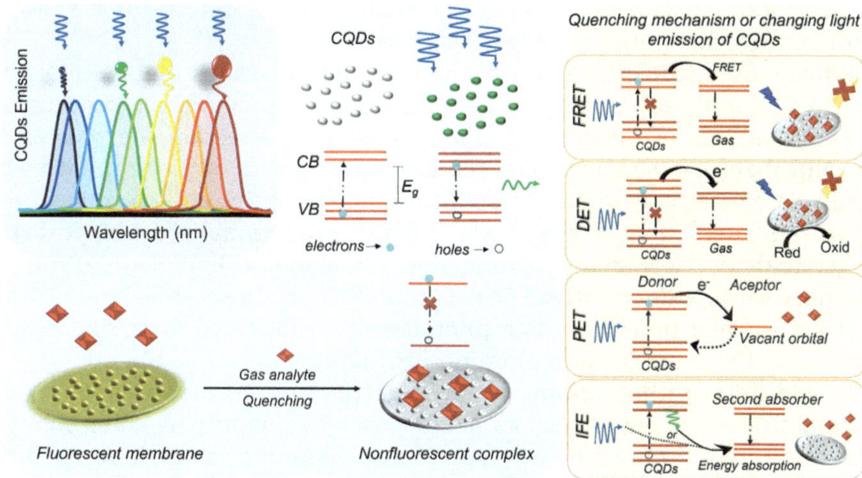

Figure 3.2 When light with sufficient energy irradiates CQDs, electrons are excited from the valence band (VB) to the conduction band (CB). The subsequent electron–hole recombination generates photons (excitonic emission), with the emission color directly dependent on the size of the CQDs. For gas sensing, the main mechanisms of photoluminescence (PL) quenching of CQDs were depicted for Förster resonance energy transfer (FRET), photoinduced electron transfer (PET), Dexter energy transfer (DET), and internal filter effect (IFE).

structural defects, serving as radiative-recombination centers. Excitonic emission is of higher energy than emission due to structural defects. Both are related to radiative recombination of charge carriers, relaxing electrons from excited states to a lower energy state. Finally, non-radiative relaxation modes are due to trapping states within the bandgap. Traps are usually related to dangling bonds for an unpassivated surface.[76] The kinetics of each relaxation mode determines which event will prevail.[77,78]

When CQDs are applied in devices for gas sensing purposes, usually solid-state light-emitting is used as a quantitatively correlated response. As a general example, the operating mode can be a gradual luminescence quenching. The mechanisms for this kind of photophysical behavior can be related to the population of the surface states and how processes like charge separation, redox reaction at the interface, or dissociation of part of the excitons occur under exposure of gas molecules.[79,80] The main quenching mechanisms related to CQDs are presented in Figure 3.2.

These optical properties, which are intrinsically linked to the high surface-to-volume ratio of CQDs, play a crucial role in the efficacy of gas sensors. This characteristic enables them to exhibit high reactivity with various chemical species, providing more adsorption sites for interaction with gases than other materials. Consequently, the functionalization of their surfaces results in enhanced selectivity and sensitivity.[81] CQDs are notable for their solution processability and ease of use across different substrates.

Furthermore, sensors based on these materials can detect gases at room temperature, facilitating a range of applications.[81–83]

When combined with a photodetector on the same substrate, a solid-state quantum dot photoluminescence response can be practical for gas sensing, as is the case for a compact optical sensor for trace detection of explosives in air. Nitrobenzene was used as a representative gas to evaluate the device's ability to generate an optical response, quenching the photoluminescence upon the interaction of the ethylenediamine-functionalized PbS with the target analyte. The system was excited with a blue LED light source, and the response was obtained with a LOD of around 2 ppb. The architecture utilized CQDs with dual functions: as a photodetector fabricated on a silicon substrate and the sensitive photoluminescent layer.[84]

In addition to photoluminescence, colorimetric sensors are another detection strategy.[85] These sensors detect gases by quantifying color changes that occur upon interacting with specific analytes or as a result of pH shifts in the gaseous environment.[86,87] An innovative example is the development of smart labels based on cellulose nanofibrils and carbon dots that combine colorimetric and fluorescent detection for real-time monitoring of food freshness by detecting acidic and alkaline gases.[86]

Another highly sensitive detection strategy involves SPR, which arises from the collective oscillation of electrons near the surface of the metal (*e.g.*, gold and silver) when coupled with incident electromagnetic radiation typically for thin films or nanoparticles.[88] Two primary plasmonic modes include localized surface plasmon resonance (LSPR), where the morphology of the nanomaterial and the dispersion medium largely determines its effectiveness,[89] and surface plasmon polaritons (SPPs), which allow energy to propagate along the material interface.[90] One example of this technology is the development of sensors based on plasmonic silver (Ag) NPs deposited on porous, thin layers of indium oxide (In_2O_3). The Ag NPs absorb blue light and transfer excited electrons to the In_2O_3 surface. This enables high external quantum efficiency (EQE) and a remarkable response to low NO_2 concentrations (1 ppm) with a minimum power consumption of 63 nW.[91]

NDIR gas detection operates on the principle of gas molecular absorption spectroscopy. Infrared light passing through a gas is selectively absorbed within specific spectral bands, corresponding to the vibrational frequencies of the gas molecules.[92] In NDIR gas detection, colloids can play a role by enhancing the performance of sensors. Liu *et al.* proposed a new nanoantenna-integrated narrowband photodetector by applying PbS CQDs. Au nanodisks were deposited on the PbS CQD layer, creating a nanoantenna effect that enhanced the spectral selectivity of the photodetector and gave a programmable resonance wavelength. The choice of PbS CQDs was due to their tunable absorption toward near-infrared light. NDIR gas-sensing based on the nanoantenna-integrated narrowband photodetector showed selectivity toward CH_4 and NH_3 with a LOD of 170 and 481 ppm, respectively.[93]

Moving forward from the principles that guide strategies for optical sensors, attention is now directed to another key category in gas detection: resistive sensors, which utilize changes in electrical resistance to detect gases.

3.3.2 Chemiresistive Gas Sensors

Semiconductor gas sensors operate based on the reversible change in electrical resistance when exposed to different gaseous analytes. This change is due to the interactions between gas molecules and the surface of the sensitive layer of the semiconductor, leading to an alteration in the concentration of the primary charge carriers, depending on the content of a specific component in the gas phase.[94]

The gas sensitivity can be defined from eqn (3.1), where the sensor signal S can be determined in various ways, for example, the ratio between the sensor resistance in the presence of the analyte gas (R_{gas}) and the resistance in the background atmosphere (R_{air})[95,96] or as the ratio between the increase in sensor resistance during the detection of the analyte gas and the sensor resistance in the background atmosphere (eqn (3.2)).[97]

$$S = \frac{R_{gas}}{R_{air}} \text{ or } S = \frac{R_{air}}{R_{gas}} \tag{3.1}$$

$$S = \frac{(R_{gas} - R_{air})}{R_{air}} \tag{3.2}$$

The most common type of graph used to represent gas detection is presented in Figure 3.3, showing the gas sensitivity relative to exposure time.[95]

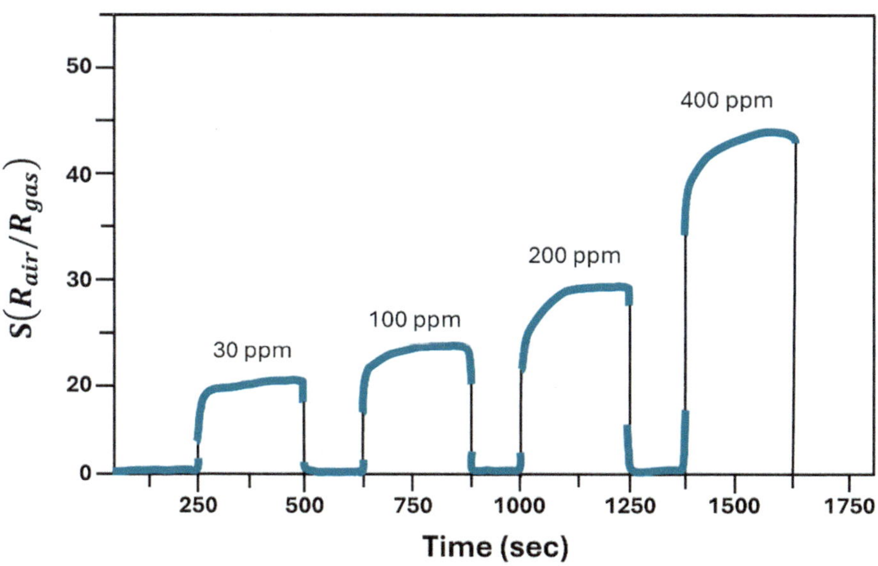

Figure 3.3 Typical representation of a gas sensor using variation in resistance.[95]

There are two mechanisms of the receptor function: (i) the mechanism of ionosorption and (ii) the mechanism of oxygen vacancies.[98] Environmental factors such as temperature and humidity affect the sensor operation in both cases. In general, the main description is valid for each case:

(i) Ionosorption mechanism: the chemisorption of various oxygen species on the metal oxide semiconductor's (MOS) surface.[98] Ionosorption is influenced by various factors, with temperature being a key factor. Temperature plays a crucial role in the efficiency of ionosorption, as both the quantity and type of adsorbed oxygen species change with the sensor's temperature. High and low temperatures affect adsorption differently: O_2^- predominates below 420 K; O^- becomes dominant between 420 and 670 K; and above 670 K, O^{2-} begins to form in parallel, with immediate incorporation into the lattice beyond 870 K.[98,99]

The sensor's response to a reducing gas (red) occurs due to the oxidation of the gas molecules by chemisorbed oxygen species (O_n^{m-}) (eqn (3.3)).[100]

$$\mathrm{Red_{(g)}} + \mathrm{O}_n{}^{m-}{}_{\mathrm{(ads)}} \rightarrow \mathrm{RO}_{n\mathrm{(g)}} + m\mathrm{e}^- \tag{3.3}$$

where RO_n is the oxidation product. The release of chemisorbed oxygen species allows the trapped electrons to return to the semiconductor, decreasing the surface potential barrier and thereby enhancing conductivity. For example, reducing gases like (eqn (3.4)–(3.11)):

a. Ammonia[101]

$$2\mathrm{NH_{3(g)}} + 3\mathrm{O^-_{(ads)}} \rightarrow \mathrm{N_{2(g)}} + 3\mathrm{H_2O} + 3\mathrm{e}^- \tag{3.4}$$

$$2\mathrm{NH_{3(g)}} + 5\mathrm{O^-_{(ads)}} \rightarrow \mathrm{N_{2(g)}} + 3\mathrm{H_2O} + 5\mathrm{e}^- \tag{3.5}$$

$$2\mathrm{NH_{3(g)}} + 4\mathrm{O^-_{(ads)}} \rightarrow \mathrm{N_{2(g)}} + 3\mathrm{H_2O} + 4\mathrm{e}^- \tag{3.6}$$

b. Carbon monoxide[102]

$$2\mathrm{CO_{(g)}} + \mathrm{O^-_{(ads)}} \rightarrow 2\mathrm{CO_{2(g)}} + 4\mathrm{e}^- \tag{3.7}$$

c. Hydrogen[103]

$$\mathrm{O_2{}^-_{(ads)}} \rightarrow 2\mathrm{H_2O} + 4\mathrm{e}^- \tag{3.8}$$

d. Hydrogen sulfide[104]

$$2\mathrm{H_2S} + 3\mathrm{O_2{}^-_{(ads)}} \rightarrow 2\mathrm{H_2O} + 2\mathrm{SO_2} + 3\mathrm{e}^- \tag{3.9}$$

e. VOCs[105,106]

$$\mathrm{CH_3CH_2OH} + \mathrm{O^{2-}_{(ads)}} \rightarrow \mathrm{CH_3CHO} + 2\mathrm{H_2O} + 2\mathrm{e}^- \tag{3.10}$$

$$\mathrm{CH_3COCH_3} + \mathrm{O^-_{(ads)}} \rightarrow \mathrm{CH_3COC^+} + \mathrm{OH^-} + \mathrm{e}^- \tag{3.11}$$

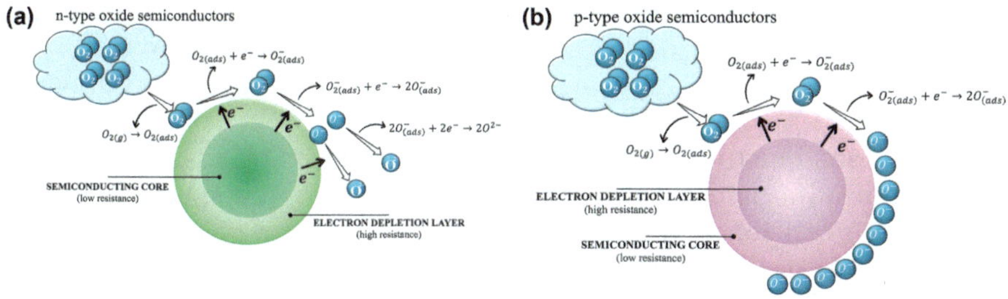

Figure 3.4 Formation of electronic core–shell structures in (a) n-type and (b) p-type oxides.[110]

(ii) Oxygen vacancy mechanism: changes in the electrical properties of the MOS are induced by an oxidation/reduction reaction on its surface. These reactions lead to the creation or elimination of oxygen vacancies. These defects (vacancies) in the crystalline structure of the metal oxide affect the sensor's electrical conductivity.

The most common type of gas sensors based on MOS utilize a resistive transducer to detect gases by the change in resistance. The chemical interaction on the sensor surface alters the charge concentration, which is influenced by the presence of electrons or holes, categorizing MOS into n-type or p-type.

The most traditional materials for resistive-type sensors are n-type semiconductor metal oxides with a wide bandgap, such as (M_xO_y), *i.e.* ZnO, SnO_2, TiO_2, and WO_3.[107,108] They play important roles in practical applications due to their excellent characteristics such as high sensibility, low cost, rapid response, and fast recovery. In n-type materials, free electrons are the main charge carriers, whereas in p-type materials, holes play this role. For n-type sensors, the detection process involves the adsorption and desorption of oxygen molecules, for example, on the surface of the sensing material, as shown in Figure 3.4.

In an atmospheric environment, oxygen molecules adsorbed on the surface of n-type semiconductor materials capture electrons from the conduction band, forming ionized oxygen species such as O_2^-, O^-, and O^{2-}.[109] These oxygen species induce a space-charge region on the surface of the sensing material, creating a potential barrier among the adjacent grains and hindering the transport of charge carriers. This process results in the formation of an electron depletion layer, which increases electrical resistance (Figure 3.4(a)).[109,110] Sensors that operate below 100 °C use oxygen ions (O_2^-), and temperatures between 100 and 300 °C convert O_2^- into O^- as it captures some electrons, while temperatures above 300 °C convert O^- ions into O^{2-}. The underlying chemical reactions for this process can be represented by the following eqn (3.12)–(3.15):[109]

$$O_{2(g)} \rightarrow O_{2(ads)} \tag{3.12}$$

$$O_{2(ads)} + e^- \rightarrow O_2{}^-{}_{(ads)} \quad T < 100\ °C \tag{3.13}$$

$$O_2{}^-{}_{(ads)} + e^- \rightarrow 2O^-{}_{(ads)}\ 100\ °C < T < 300\ °C \tag{3.14}$$

$$O^-{}_{(ads)} + e^- \rightarrow O^{2-}{}_{(ads)}\ T > 300\ °C \tag{3.15}$$

These reactions illustrate how the capture of electrons by adsorbed oxygen leads to ionization and the formation of different oxygen species, altering the material's resistance. A notable example is the use of bismuth vanadate ($BiVO_4$) colloids to detect H_2S *via* a type-n detection mechanism. When the sensor is exposed to air, oxygen molecules are adsorbed onto the surface of the sensing material, capturing electrons and forming oxygen species such as $O_{(ads)}$, O^-, $O_2{}^-$, and O^{2-}, depending on the working temperature, as presented in the above equations. When H_2S gas, a reducing agent at the working temperature of 75 °C, comes into contact with the analytical layer, the H_2S molecules react with the adsorbed oxygen species, releasing the trapped electrons back to the $BiVO_4$.[111] The detection reaction is described in eqn (3.16).

$$2H_2S_{(ads)} + O_2{}^-{}_{(ads)} \rightarrow 2H_2O_{(g)} + 2SO_{2(g)} + 6e^- \tag{3.16}$$

Detection of CH_4 and CO gases using Pd–SnO_2 involves the adsorption of O^- on the surface of SnO_2, as described in eqn (3.14). This adsorbed oxygen reacts with the reducing molecules CH_4 and CO, thereby increasing the resistance of the sensing layer,[112] described as follows (eqn (3.17) and (3.18)):

$$CH_4 + 4O^- \rightleftharpoons CO_2 + 2H_2O + 4e^- \tag{3.17}$$

$$CO + O^- \rightleftharpoons CO_2 + e^- \tag{3.18}$$

Nanowires of tin oxide (SnO_2) were able to detect H_2, ranging from 13 to 40 ppm at 250 °C. However, after doping with palladium (Pd), the sensor's operating temperature decreased to 150 °C, with response and recovery times reduced from 13 s/15 s to 3 s/6 s. The Pd doping effectively enhances the adsorption energy of H_2 on SnO_2, with H_2 molecules adsorbed on SnO_2 as donor surface state, resulting in an n-type doping effect that decreases the material's resistance and enhances gas response.[113]

As n-type semiconductors, the variation in SnO_2 resistance arises from the capture of electrons through the reaction between adsorbed oxygen species and H_2 molecules. This interaction can be described as follows (eqn (3.19) and (3.20)):

$$H_{2(gas)} \rightarrow H_{2(ads)} \tag{3.19}$$

$$H_{2(ads)} + O^-{}_{(ads)} \rightarrow H_2O_{(gas)} + e^- \tag{3.20}$$

Chemisorbed oxygen species increase resistance by withdrawing electrons from the conduction band of SnO_2, which are then trapped by the oxygen species. This process leads to a thicker electron depletion layer on the surface of the semiconductor.[113] Pd doping effectively enhances the adsorption energy of H_2 on SnO_2, resulting in an n-type doping effect that decreases the material's resistance and enhances the gas response. When the material is exposed to an H_2 atmosphere, H_2 reacts with surface-adsorbed oxygen and the system releases electrons, narrowing the depletion layer and reducing material resistance, as shown in Figure 3.5. Figure 3.5(a) illustrates the interaction mechanism between pure SnO_2 and H_2, while Figure 3.5(b) shows the interaction mechanism between SnO_2 and H_2 after doping with Pd.[113]

When a p-type semiconductor comes into contact with ambient air, oxygen molecules are adsorbed onto its surface, extracting electrons from the conduction band. This adsorption process leads to the formation of an accumulation layer of holes on the material's surface, as illustrated in Figure 3.4(b).

A notable distinction between p-type and n-type structures lies in the variation of resistance between the external and internal layers of the materials. In n-type materials, the external layer exhibits higher resistance than the core. Upon exposure to a gas, these materials experience an increase in resistance, causing the charges to behave similarly to a series circuit. Conversely, in p-type materials, the higher resistance is located in the core, leading to the accumulation of charges primarily on the surface (external layer). This distribution results in behavior akin to a parallel circuit, thereby reducing the system's equivalent resistance.

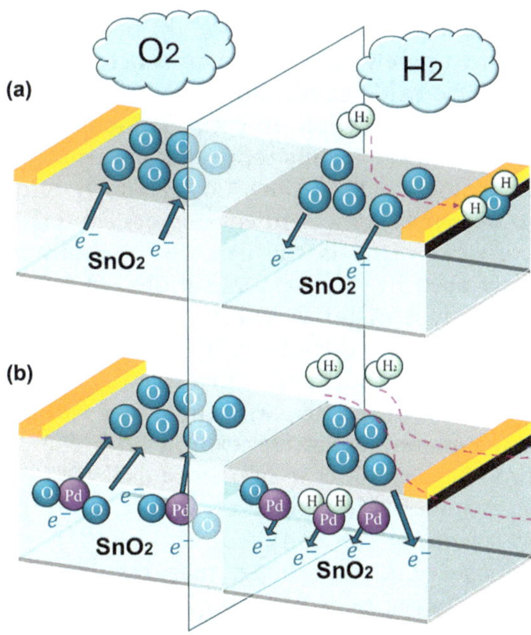

Figure 3.5 (a) Representation of the interaction between SnO_2 and H_2, and (b) representation of the interaction between Pd/SnO_2 and H_2.[113]

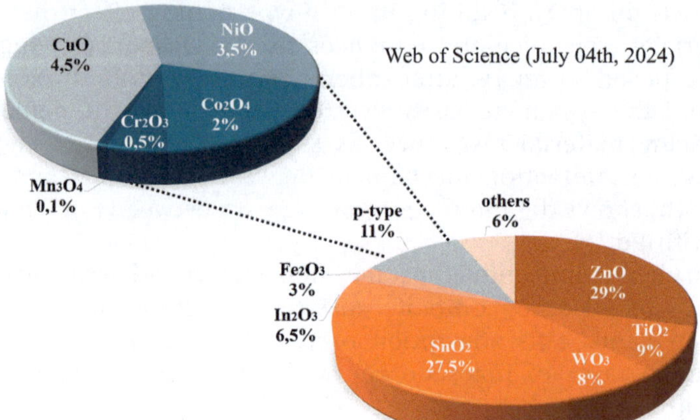

Figure 3.6 Graphical distribution of gas sensors by semiconductor types, showing the predominance of n-type oxides (on the right) compared to p-type oxides (on the left).

According to a bibliometric analysis conducted in the Web of Science database on July 4, 2024, using the keywords "oxide semiconductors gas sensors" in titles, abstracts, or indexing, 13 479 results were found. These results were categorized into publications in the form of articles, review articles, and books based on the search filters. Figure 3.6 shows the results of a subsequent search that included the chemical formula of the oxide material along with the keywords "gas sensor" in titles, abstracts, or indexing, revealed that approximately 89% of the works belong to the group of gas sensors that utilize n-type oxide semiconductors (such as SnO_2, ZnO, TiO_2, WO_3, In_2O_3, and Fe_2O_3) for detection. In contrast, as the literature analysis revealed, only about 11% of the results represent gas sensors that use p-type oxide semiconductors (such as NiO, CuO, Co_3O_4, Cr_2O_3, and Mn_3O_4).

It is important to note that some semiconductors can exhibit both p-type and n-type characteristics. This is the case for titanium oxide (TiO_2), a semiconductor widely used in various applications, including gas sensors, and is intrinsically an n-type material. Through doping processes, TiO_2 can begin to conduct holes, a characteristic of p-type semiconductors. The doping processes for TiO_2 are still under investigation, and other dopants may emerge, making them competitive with the n-type version.[114] According to Yoon *et al.*, another material that can exhibit both characteristics is indium arsenide (InAs) semiconductor nanocrystals (NCs). Generally, InAs NCs display n-type characteristics; however, with the use of the reducing agent diethylzinc, p-type InAs NCs can be synthesized.[115]

Despite the wide use of metal oxides, they require high energy consumption and high operating temperatures (200–500 °C) for detection applications, such as in the case of SnO_2. The limitation of high temperatures can be overcome by carbon-based materials, which are highly effective in gas detection at low temperatures. Graphene, in particular, has emerged as an extremely attractive and intensively investigated material for gas sensor applications due to its unique thermal and electrical properties. Its high electronic mobility of approximately 200 000 $cm^2 V^{-1} s^{-1}$ and large specific surface area of about 2630 $m^2 g^{-1}$ endow graphene and its derivatives with promising properties for the development of gas sensors, such as high sensitivity, low-temperature operation, and relative ease of fabrication. In this context, chemiresistive sensors based on graphene are widely mentioned in current literature, highlighting their main qualities, including fast detection and good sensitivity. This is due to the ability of graphene to modify its electrical conductivity and resistivity during interaction with adsorbates, influenced by its large specific surface area and susceptibility to the adsorption of external molecules, making it a potential candidate for monitoring toxic pollutant gases. However, the challenge remains in developing graphene-based devices that exhibit excellent detection responses and good selectivity.[116] In addition to this, its capacity to interact with gas molecules can enhance sensor performance. By integrating with polymeric materials or colloidal systems, its durability and sensitivity can be optimized to fulfill specific detection requirements.[117]

Studies on NO_x detection have explored various sensitive nanomaterials like graphene. Compared to electrochemical cells, field-effect transistors, and other types of gas sensors, graphene-based chemiresistive gas sensors exhibit low noise and high mechanical strength. The NO_x adsorbed on the graphene surface continuously extracts electrons and extends the hole accumulation zone (main carriers) on the graphene surface to decrease the resistance. The sensor's sensitivity is thus represented as the variation in the contact resistance of graphene with NO_x, where the total resistance is given by total $R_{total} = R_{contact} + R_{material} + R_{electrodes}$.[118]

However, graphene-based gas sensors without appropriate surface modification show low sensitivity and selectivity due to the limited number of active sites and non-selective molecular adsorption. Reduced graphene oxide (rGO) has prevailed over different graphene derivatives due to its remarkable characteristics such as excellent response characteristics, chemical stability, large specific surface area, high carrier mobility, and good mechanical strength. All these features make rGO an ideal candidate for gas detection.

Various detection mechanisms have been explored to explain the response of polymer–graphene composite films in gas detection. In polymerized vinyl acetate (PVAc) and rGO PVAc–rGO composite films made of plasma, the detection can result from a combination of redox reactions, charge transfer, and polymer expansion. The rGO has unreduced oxygen molecules at the edges of its planes, attracting electrons from ammonia and amines due to its high electronegativity as seen in Figure 3.7. This increases the PVAc–rGO composite film's ability to capture ammonia/amine gas molecules. Ammonia/amine molecules are also retained in the fragmented chains of PVAc through van der Waals forces between PVAc and amine groups. During the gas detection process, PVAc–rGO films capture oxygen, interact with ammonia/amine gas

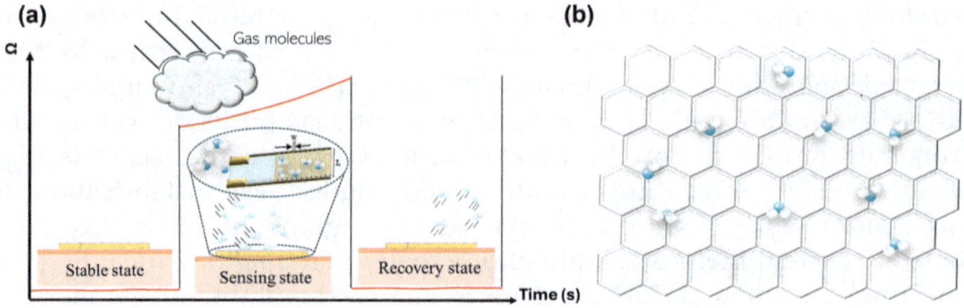

Figure 3.7 (a) Gas detection process of ammonia/amine, and (b) gas molecule trapping by rGO.[119]

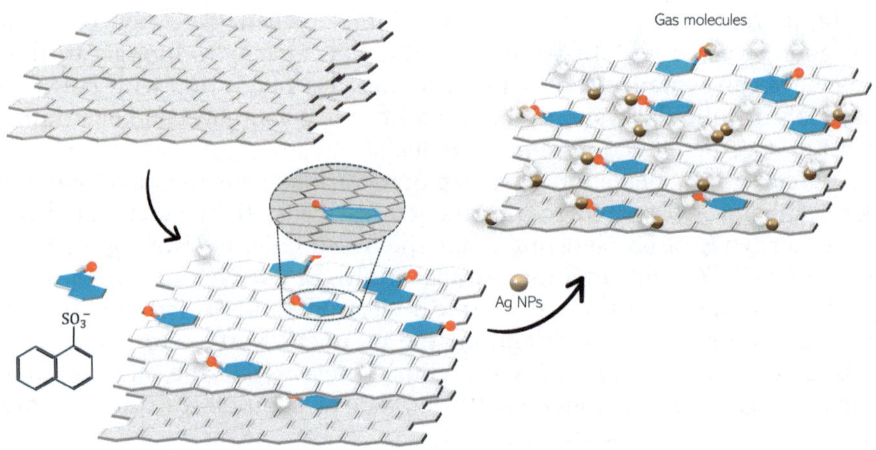

Figure 3.8 Schematic illustration employing Ag–NA-rGO for NO_2 detection schematic illustration.[96]

molecules, and release their electrons into the sensor film. The recombination of holes in the PVAc–rGO film with electrons from the analyte gas reduces the carrier concentration, increasing the film's resistance during detection. Electron-donating analyte gases increase the delocalization of π-conjugated electrons in PVAc–rGO sensor films, forming a neutral polymeric structure in the PVAc–rGO composite film, which reduces the conductivity during gas detection.[119]

Chen *et al.*[96] conducted a study in which reduced rGO was prepared with sulfophenyl ($Ph–SO_3^-$), a typical electron-withdrawing group for NO_2. This group was introduced *via* supramolecular assembly of naphthalene-1-sulfonic acid with rGO through π–π stacking, resulting in NA-rGO. This compound was used to study the gas-sensing characteristics of NO_2. To further enhance the adsorption capacity for NO_2 molecules, silver NPs were introduced into the system through electrostatic interaction as exceptionally hydrophilic "hosts," resulting in Ag–NA-rGO (Figure 3.8).

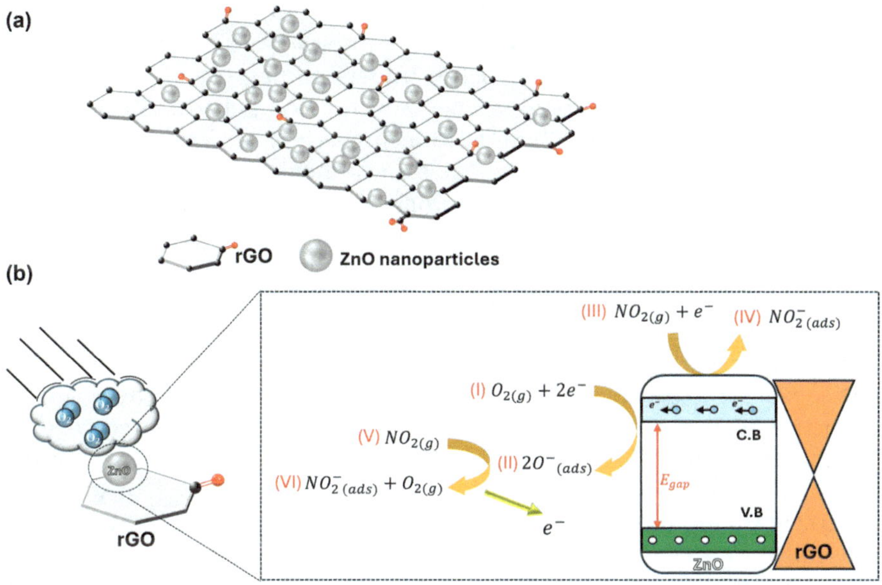

Figure 3.9 Diagram of the sensitivity mechanism of the material affected by NO_2 gas.

The change in material sensitivity is explained by the carrier transition mechanism and the surface chemical reaction. As shown in Figure 3.9, oxygen adsorbs on the material's surface and captures electrons from the valence band, as described in eqn (3.21). When NO_2 gas molecules are adsorbed on the material's surface, they are chemically adsorbed as NO_2^- (eqn (3.22)). Simultaneously, some NO_2 molecules react with O^- (eqn (3.23)), leading to a change in carriers in the material and consequently altering the material's resistance.[120]

$$O_{2(g)} + 2e^- \rightarrow 2O^-_{(ads)} \tag{3.21}$$

$$NO_{2(g)} + e^- \rightarrow NO_2^{-}{}_{(ads)} \tag{3.22}$$

$$NO_{2(g)} + 2O^-_{(ads)} \rightarrow NO_2^{-}{}_{(ads)} + O_{2(g)} + e^- \tag{3.23}$$

Besides sensors that operate by altering the resistance of the active layer, gas sensors can also modify the refractive index of this layer. In the detection of acetone vapors, the carbonyl functional group is crucial for identifying the substance. A colloidal solution of PVA with polydimethylsiloxane (PDMS) can be employed to detect acetone, as both the –OH group in PVA and the $-OSi(CH_3)_2^-$ group in PDMS interact with acetone, causing changes in the refractive indices of these materials and resulting in a shift in the output spectrum of the sensor structure. The double bond of acetone accepts electrons from the OH ions in PVA, forming $-C{=}O\cdots H{-}O$ intermolecular bonds, which reduces the refractive index of PVA.[121]

Advancing future research, exploring colloids can unveil crucial insights into adsorbent–adsorbate interactions in complex systems. Further investigation can direct efforts towards better understanding the selectivity in adsorption of different adsorbate species and the competitive effects among them in colloidal systems. These advancements are critical for developing more effective environmental remediation strategies, optimizing colloidal capacity to adsorb and retain contaminants. In this way, they significantly contribute to the protection and sustainability of natural resources and global ecosystems, while paving the way for advanced application of purification and separation technologies across various industries and environmental settings.

Mitri *et al.* reported the development of highly efficient chemiresistive PbS CQD sensors for the detection of NO_2 at room temperature. These sensors demonstrated the capability to selectively detect NO_2, compared to other atmospheric pollutants such as CH_4, CO, and CO_2, within a concentration range of 2.8 to 100 ppm. The developed sensor exhibited excellent sensitivity to NO_2, with rapid responses and complete recovery after the gas was released.[122]

Another study on NO_2 detection was conducted by Li *et al.*, who used SnS CQDs as an alternative to PbS. The use of PbS raises concerns due to its lead content, which poses environmental and health risks. This device exhibited high sensitivity to low concentrations (1–10 ppm) and good selectivity compared to NH_3, SO_2, CO, and H_2. Additionally, the sensor showed response times of 180 s and recovery times of 466 s while operating at room temperature. CQDs have been employed in the development of sensors that operate at room temperature. This enables everyday applications since many of the developed materials are sensitive to gases at high temperatures.[123]

Chemical sensors formed by colloidal hierarchical nanoflowers of WS_2 are used for NH_3 detection, exhibiting good response and selectivity. During gas detection, an electrical charge is transferred between the gas (target analyte) and the active material (WS_2 nanoflowers), causing changes in the electrical properties of the sensor material. This occurs because the lone pair of electrons from NH_3 is transferred to the sensor's conduction band after adsorption, resulting in a reduction in electrical conductance. This suggests that positive holes are the main charge carriers on the surface of WS_2 nanoflowers, indicating a p-type doping behavior. This phenomenon arises due to the depletion of positive charge carriers (holes) on the sensor surface by the negative charge carriers from NH_3.[124]

In addition to CQDs, various colloidal materials are utilized in gas sensing. Notably, aminated polystyrene (a-PS) colloids and intrinsically conducting polymers, such as polyaniline (PANI), have effectively detected toxic gases. These materials provide distinct properties that complement existing advancements in the field.

Jin and Park utilized aminated polystyrene colloids (a-PS) to develop a hydrogen chloride (HCl) gas sensor, a compound extremely toxic and harmful to health when inhaled in high concentrations. In their study, the a-PS colloids were adjusted with different amination densities, with the a-PS(1.0) colloids standing out by showing the highest binding capacity (102 $\mu g\,mg^{-1}$) and an efficient reuse capacity. At HCl concentrations below 10 ppm, the a-PS(1.0) colloids exhibited a sensitivity of 3.88 $Hz\,ppm^{-1}$ (indicating the sensor's response frequency changes by 3.88 Hertz per part per million of HCl) and a LOD of 5.002 ppm. Additionally, the adsorption coefficient (k^*) for HCl in the a-PS(1.0) colloids was higher than that obtained for formaldehyde (HCHO), suggesting their potential application as sorbents in gas column filters.[125]

The device developed by Syrový *et al.* utilizes PANI colloids as a chemoresistive sensor for ammonia detection, a toxic and harmful gas. In this study, PANI colloids are applied as a sensitive layer through the rotogravure technique on a flexible polyethylene terephthalate (PET) substrate, with an interdigitated electrode (IDE) structure based on carbon. The sensors fabricated with colloids demonstrated an effective response to ammonia concentrations, with the lowest detectable level being 500 ppb. The results of this study indicate that the rotogravure of PANI colloids is a cost-effective and efficient approach for developing flexible sensors. This flexible sensor, as suggested by the author, can be applied both for monitoring maximum human exposure limits in workplaces and for monitoring atmospheric pollution.[126]

3.3.3 Acoustic Sensors

Acoustic gas sensors are devices that use waves (mechanical or acoustic) to detect the presence of various gases. There are several types of acoustic sensors, such as mass-sensitive film bulk acoustic resonators (FBAR), quartz crystal microbalance (QCM), and surface acoustic wave (SAW) sensors. Among these, the latter two are the most prominent acoustic gas sensors.[127]

3.3.3.1 Quartz Crystal Microbalance

QCM sensors operate based on the principle of piezoelectricity. Initially, QCM sensors were used as mass sensors; when a mass was deposited, there was a change in the crystal frequency, and they have only recently been used as gas sensors. The gas detection mechanism is based on the adsorption of molecules onto the quartz crystal surface, which causes a change in the crystal mass, altering the frequency in the same way as the mass sensor mechanism.[128] QCM sensors have advantages such as high sensitivity and real-time monitoring, but they have the drawback of selectivity.[129] Recent research has explored using colloids, such as metal NPs or polymers, as sensitive layers in QCM sensors to enhance their selectivity and sensitivity. An example of QCM application is the study by Xie *et al.*, which developed sensors based on ZnO colloidal microspheres demonstrating high efficiency in detecting alcohol vapors such as methanol and ethanol. When applied to

Table 3.1 Colloid-based gas sensors. Data collected from the literature considering aspects of sensing materials, strategies, target analytes, operation conditions, sensitivity, response and recovery.[a]

Types of sensors	Sensing strategies	Colloids materials	Target gas	Operating temperature (°C)	Concentration of gas	Sensitivity	Response (s)	Recovery (s)	Ref.
Plastic Optical fiber	Optical	Ag@Fe3O$_4$	H_2S	RT	50 ppm	NR	15	40	133
Optical/SMO	Chemiresistive/ Optical	PS colloid crystal	Ethanol	RT	100 ppm	22.8 (R_{air}/R_{gas})	23	33	134
SMO	Chemiresistive	$Ti_3C_2T_x$ MXene	NH_3	RT	500 ppm	75% ($\Delta R/R_0$)	109	342	135
SMO	Chemiresistive	Au:CdO	H_2S and NO_2	<30	NR	67.81% (NO_2) 93.5% (H_2S)	18.9 (NO_2) 12.6 (H_2S)	143.1 (NO_2) 188.1 (H_2S)	136
SMO	Chemiresistive	SnO_2 Pd/SnO_2	H_2	250 (SnO_2) 150 (Pd/SnO_2)	150 ppm (SnO_2) 40 ppm (Pd/SnO_2)	54 (SnO_2) 8.5 (Pd/SnO_2)	19 (SnO_2) 6 (Pd/SnO_2)	45 (SnO_2) 3 (Pd/SnO_2)	113
SMO	Chemiresistive	Indium-tin oxide (ITO)	H_2	RT	70 ppm	4%	NR	NR	137
SMO	Chemiresistive	$BiVO_4$	H_2S	75	25 ppm	22 (S_A/S_B)	NR	NR	111
SMO	Chemiresistive	WO_3 CQDs	H_2S	80	10 ppm 50 ppm	7 (10 ppm) (R_a/R_g) 57 (50 ppm) (R_a/R_g)	68 (10 ppm) 47 (50 ppm)	165 (10 ppm) 126 (50 ppm)	138
SMO	Chemiresistive	ZnO CQDs	H_2S	RT	50 ppm	113.5 (R_{air}/R_{gas})	16	820	139
SMO	Chemiresistive	Ag–TiO_2	Ethanol and acetone	350	30 ppm acetone and 50 ppm ethanol	NR	<5	<50	140
Optical fiber	Acoustic (WGA)	PVA	Acetone	20–38	151.97 ppm	−6.34 pm ppm^{-1} ($\Delta\lambda/\Delta C$)	640 (8 min)	NR	121

QCM	Acoustic	PPy/PVP	Alcohol	20 ± 2	Ethanol 36–179 mg L^{-1} 1-Propanol 36–183 mg L^{-1} 1-Butanol 37–184 mg L^{-1}	Ethanol 0.497 Hz mg^{-1} L 1-Propanol − 0.519 Hz mg^{-1} L 0.571 Hz mg^{-1} L 1-Butanol 0.571 Hz mg^{-1} L	NR	NR	141
QCM	Acoustic	Graphene oxide (GO)	NO_2, SO_2, CO, and NH_3	RT	NO_2 – 15 ppm SO_2 – 15 ppm CO – 150 ppm NH_3 – 750 ppm	NO_2 – 0.84 Hz ppm^{-1} SO_2 – 1.0 Hz ppm^{-1} CO – 0.04 Hz ppm^{-1} NH_3 – 0.02 Hz ppm^{-1}	<90	NR	142
QCM	Acoustic	ZnO	Methanol, ethanol, and isopropanol	25	15–300 ppm	NR	NR	NR	143
Organic	Redox reaction, charge transfer, and polymer swelling	rGO	Ammonia/ amines	30 to 150	50–3000 ppm	NR	For 100 ppm: 470	40	119
Organic	Chemoresistor	PANI	NH_3	23	NR	NR	NR	NR	144
Thermoelectric	Combustion catalyst	Pd/Al_2O_3	CH_4	400	1 ppm	34 μV ppm^{-1} ($\Delta V/c$)	NR	NR	145

[a] NR – not reported. SMO – semiconductor metal oxides. RT – room temperature. WGM – whispering gallery mode.

quartz crystal microbalance sensors, these microspheres exhibit excellent chemical sensitivity influenced by the concentration of the analyte vapor.[130]

3.3.3.2 *Surface Acoustic Wave*

SAW sensors are based on the principle of variations in the propagation of surface acoustic waves within a piezoelectric material. The three fundamental components of a SAW sensor are a piezoelectric substrate, interdigital transducers, and a sensing surface (*e.g.*, metal oxides). These components can be modified according to the specific detection requirements.[131] For example, Li *et al.*'s research explored the incorporation of colloidal PbS quantum dots into surface acoustic wave sensors, highlighting that treatment with lead nitrate ($Pb(NO_3)_2$) provided a faster response and a positive frequency shift due to the interaction of NO_2 molecules with the porous structure of the film. The sensor monitored NO_2 in a concentration range of 0.5 to 30 ppm, operating at room temperature with good sensitivity and selectivity.[132]

There are different types of gas sensors, and recent developments have considered both materials and detection strategies. Several materials have been explored in this process, some of which have been discussed in this section. Table 3.1 presents a compilation of colloid-based gas sensors, highlighting the materials used, the gases studied, concentrations, detection responses, operating temperature, and sensitivity on which the sensors were developed.

3.4 Concluding Remarks

This chapter has highlighted the significant role of colloidal technologies in the crucial task of gas monitoring, both indoors and outdoors. At the forefront of the new generation of air quality monitoring devices, these technologies hold immense potential to enhance air quality. This potential serves as a driving force for all those involved in the field. Most devices are designed with architectures that utilize semiconductor oxides, either doped or undoped with noble metals and other materials, to improve their sensor performances. These capabilities, such as sensitivity to ppb or ppm levels, real-time sensing, repeatability, stability, fast response, and low cost, were thoroughly examined. The primary strategies explored are based on chemiresistive responses and some of the most common optical sensing signals (photoluminescence, non-dispersive infrared, and surface plasmon resonance). Furthermore, a diverse range of investigations also utilize various electrochemical techniques, piezoelectric effect, and surface acoustic waves, that were highlighted from literature studies.

The advancement of gas sensor technologies is a complex task, riddled with challenges and limitations. These include the requirement for high-quality and expensive techniques, the operation of some sensors at high temperatures (*e.g.*, 100 to 300 °C for metal oxide semiconductors), and the interference of humidity levels, which can lead to cross-response to other gases. Most colloid technologies for gas detection depend on a design that involves substrates and matrices for the distribution of the actives that give specific output signals under external stimuli. Therefore, uniform distribution, porosity, and optimal conditions for diffusion are decisive to ensure the next steps, which depend on adsorption/desorption or specific interaction processes at the interface with the active sites.

While the challenges in gas sensor technology are significant, there are promising advances in recognizing target gaseous analytes and transduction signals, driving real-time response technologies, such as wearable devices, which are important for increasing safety in industrial, domestic, and outdoor environments. Semiconductor-based sensors, such as those using ZnO and SnO_2, offer advantages in sensitivity and cost-effectiveness, but they face limitations, including low selectivity and susceptibility to environmental factors like humidity. However, innovations involving nanohybrid materials are pushing the boundaries, improving sensor performance and enabling real-time sensing. Optical sensors, especially those with CQDs, present a promising alternative with their fast response, high sensitivity, and minimal drift. However, their successful implementation requires meticulous design to avoid cross-responses and ensure reliability.

Moving forward, the focus should be on developing novel materials and device architectures to enhance sensing capabilities while addressing issues such as temperature dependence, material stability, and selectivity. These advancements will pave the way for more effective, real-time gas monitoring solutions, benefiting both researchers and industry practitioners by offering improved accuracy and reliability in diverse applications.

Abbreviations

NC	Nanocrystals
a-PS	Aminated polystyrene
PL	Photoluminescence
PDMS	Polydimethylsiloxane
CQDs	Colloidal quantum dots
GHG	Greenhouse gas
VOCs	Volatile organic compounds
PM	Particulate matter
WHO	World Health Organization
LOD	Limit of detection
LOQ	Limit of quantification
ppm	Parts per million
ppb	Parts per billion
NPs	Nanoparticles
SWCNTs	Single-walled carbon nanotubes
PAE	Arylene polyether
SPR	Surface plasmon resonance

NDIR	Non-dispersive infrared
QDs	Quantum dots
LSPR	Localized surface plasmon resonance
SPPs	Surface plasmon polaritons
EQE	External quantum efficiency
VB	Valence band
CB	Conduction band
FRET	Förster resonance energy transfer
PET	Photoinduced electron transfer
DET	Dexter energy transfer
IFE	Internal filter effect
R_{gas}	Resistance of gas
R_{air}	Resistance of air
MOSs	Metal oxide semiconductors
Red	Reduction
rGO	Reduced graphene oxide
FBAR	Film bulk acoustic resonators
SAW	Surface acoustic wave
QCM	Quartz crystal microbalance
IDE	Interdigitated electrode
PVAc	Poly(vinyl acetate)
PANI	Polyaniline
PET	Polyethylene terephthalate
NR	Not reported
RT	Room temperature
WGM	Whispering gallery mode

Acknowledgements

The authors would like to thank CNPq (407774/2021 and 308846/2022-2), INEO/INCT and FAPESP (2014/50869-6 and 2024/00932-5), and CAPES-PRINT-UNESP – Unesp Network for the development of organic and flexible electronics – ORGFLEX, LNNano (Proc. 20230903). This study was financed in part by the Coordenação de Aperfeiçoamento de Pessoal de Nível Superior – Brasil (CAPES) – Finance Code 001.

References

1. M. L. Brusseau, A. D. Matthias and A. C. Comrie, in *Environmental and Pollution Science*, ed. A. D. Matthias and A. C. Comrie, Elsevier, Amsterdam, 3rd edn, 2019, pp. 293–309.
2. R. Maier, L. Hörtnagl and N. Buchmann, *Sci. Total Environ.*, 2022, **849**, 157541.
3. P. Sicard, E. Agathokleous, S. C. Anenberg, A. De Marco, E. Paoletti and V. Calatayud, *Sci. Total Environ.*, 2023, 160064.

4. N. J. Bassous, A. C. Rodriguez, C. I. L. Leal, H. Y. Jung, C. K. Lee, S. Joo, S. Kim, C. Yun, M. G. Hahm, M. Ahn, S. Kim, Y. S. Oh and S. R. Shin, *Adv. Sens. Res.*, 2024, **3**, 202300094.
5. Q. Yang, F. L. Zhang, Z. M. Jiang, C. S. Li and J. G. Song, *Sci. Total Environ.*, 2024, **947**, 174467.
6. A. Chlebnikovas, D. Paliulis, J. Bradulienė and T. Januševičius, *Environ. Sci. Pollut. Res.*, 2023, **30**, 81950.
7. P. Kittipornkul, T. Piyatida, H. Bua, S. Irawan, L. Prasert, A. Darakul and M. Prakobwan, *Environ. Res.*, 2023, **234**, 116566.
8. D. Nuvolone, D. Petri and F. Voller, *Environ. Sci. Pollut. Res.*, 2018, **25**, 8074.
9. A. Kumar and P. K. Singh, *J. Phys. Chem. A*, 2023, **127**, 10016.
10. T. Salthammer, *Environ. Sci.: Atmos.*, 2024, **4**, 291.
11. R. S. Sokhi, N. Moussiopoulos, A. Baklanov, J. Bartzis, I. Coll, S. Finardi, R. Friedrich, C. Geels, T. Grönholm, T. Halenka, M. Ketzel, A. Maragkidou, V. Matthias, J. Moldanova, L. Ntziachristos, K. Schäfer, P. Suppan, G. Tsegas, G. Carmichael, V. Franco, S. Hanna, J. P. Jalkanen, G. J. M. Velders and J. Kukkonen, *Atmos. Chem. Phys.*, 2022, **22**, 4615.
12. E. R. Kulick, J. D. Kaufman and C. Sack, *Stroke*, 2023, **54**, 882.
13. B. Alahmad, H. Khraishah, K. Althalji, W. Borchert, F. Al-Mulla and P. Koutrakis, *Can. J. Cardiol.*, 2023, **39**, 1182.
14. A. Piccoli, V. Agresti, M. Bedogni, G. Lonati and G. Pirovano, *Urban Clim.*, 2024, **55**, 101893.
15. P. Kumar, A. B. Singh, T. Arora, S. Singh and R. Singh, *Sci. Total Environ.*, 2023, 162163.
16. I. Darmadi, I. Östergren, S. Lerch, A. Lund, K. Moth-Poulsen, C. Müller and C. Langhammer, *Acc. Chem. Res.*, 2023, **56**, 1850.
17. Y. Qin, X. Wang and J. Zang, *Sens. Actuators, B*, 2021, **340**, 129959.
18. E. S. Guillén-López, F. López-Urías, E. Muñoz-Sandoval, M. Courel-Piedrahita, M. Sanchez-Tizapa, H. Guillén-Bonilla, V. M. Rodríguez-Betancourtt, O. Blanco-Alonso, A. Guillén-Bonilla and J. P. Morán-Lázaro, *Mater. Today Commun.*, 2021, **26**, 102138.
19. J. P. Morán-Lázaro, F. López-Urías, E. Muñoz-Sandoval, M. Courel-Piedrahita, A. Carreon-Alvarez, V. M. Rodríguez-Betancourtt, I. Zamudio-Torres, E. S. Guillén-López and A. Palafox-Corona, *Electron. Mater. Lett.*, 2023, **19**, 66.
20. H. J. Cho, S. J. Choi, N. H. Kim and I. D. Kim, *Sens. Actuators, B*, 2020, **304**, 127350.
21. A. Paul, S. Muthukumar and S. Prasad, *J. Electrochem. Soc.*, 2020, **167**, 037511.
22. A. K. Farquhar, G. S. Henshaw and D. E. Williams, *Sens. Actuators, A*, 2023, **354**, 114254.
23. D. E. Williams, *Curr. Opin. Electrochem.*, 2020, **22**, 145–153.
24. Q. Li, W. Zeng and Y. Li, *Sens. Actuators, B*, 2022, **354**, 131579.
25. R. Shaik, R. K. Kampara, A. Kumar, C. S. Sharma and M. Kumar, *Coord. Chem. Rev.*, 2022, **452**, 214752.

26. M. V. Nikolic, V. Milovanovic, Z. Z. Vasiljevic and Z. Stamenkovic, *Sensors*, 2020, **20**, 6694.
27. A. Mirzaei, H. R. Ansari, M. Shahbaz, J. Y. Kim, H. W. Kim and S. S. Kim, *Chemosensors*, 2022, **10**, 289.
28. N. Lu, S. Fan, Y. Zhao, B. Yang, Z. Hua and Y. Wu, *Sens. Actuators, B*, 2021, **347**, 130603.
29. I. C. Weber and A. T. Güntner, *Sens. Actuators, B*, 2022, **356**, 131346.
30. D. Spirjakin, A. Baranov, S. Akbari, C. T. Phong and N. N. Tuan, *2021 IEEE Sensors Applications Symposium (SAS)*, 2021, DOI: 10.1109/SAS51076.2021.9530079.
31. C. Tong, C. Sima, M. Chen, X. Zhang, T. Li, Y. Ai and P. Lu, *Sensors*, 2023, **23**, 15130.
32. A. Hänsel and M. J. R. Heck, *JPhys Photonics*, 2020, **2**, 012002.
33. B. Tsizh, O. Aksimentyeva, H. Yulia and R. Holyaka, *Mol. Cryst. Liq. Cryst.*, 2023, **159**, 166.
34. D. J. Buckley, N. C. G. Black, E. G. Castanon, C. Melios, M. Hardman and O. Kazakova, *2D Mater.*, 2020, **7**, 032002.
35. J. Dai, O. Ogbeide, N. Macadam, Q. Sun, W. Yu, Y. Li, B. L. Su, T. Hasan, X. Huang and W. Huang, *Chem. Soc. Rev.*, 2020, **49**, 1756.
36. M. Wusiman and F. Taghipour, *Crit. Rev. Solid State Mater. Sci.*, 2022, **47**, 416.
37. S. Srivastava, *Mater. Today: Proc.*, 2020, **37**, 3709–3712.
38. Y. X. Lu, C. T. Lin, M. H. Tsai and K. C. Lin, *Micromachines*, 2022, **13**, 40509.
39. S. Cimino and L. Lisi, *Catalysts*, 2019, **9**, 668.
40. A. G. Ramu, A. Umar, S. Gopi, H. Algadi, H. Albargi, A. A. Ibrahim, M. A. Alsaiari, Y. Wang and D. Choi, *Environ. Res.*, 2021, **201**, 112012.
41. X. Liu, R. Jayaratne, P. Thai, T. Kuhn, I. Zing, B. Christensen, R. Lamont, M. Dunbabin, S. Zhu, J. Gao, D. Wainwright, D. Neale, R. Kan, J. Kirkwood and L. Morawska, *Environ. Res.*, 2020, **185**, 109438.
42. S. Kang, K. Zhao, D. G. Yu, X. Zheng and C. Huang, *Adv. Fiber Mater.*, 2022, **4**, 404.
43. S. H. Kim, H. Jeong, B. Sharma and J. H. Myung, *ACS Appl. Mater. Interfaces*, 2022, **14**, 18275.
44. L. Zhu, J. Wang, J. Liu, Z. Xu, M. S. Nasir, X. Chen, Z. Wang, S. Sun, Q. Ma, J. Liu, J. Feng, J. Liang and W. Yan, *Sens. Actuators, B*, 2022, **354**, 131206.
45. Z. You, M. Zhao, H. Lu, H. Chen and Y. Wang, *ACS Appl. Mater. Interfaces*, 2024, **16**, 19359.
46. J. Deng, M. Molaei, N. G. Chisholm, T. Yao, A. Read and K. J. Stebe, *Curr. Opin. Colloid Interface Sci.*, 2022, **62**, 101629.
47. A. Chiolerio and M. B. Quadrelli, *Energy Technol.*, 2019, **7**, 1800580.
48. X. Chen, L. Lin, Z. Li and H. B. Sun, *Adv. Funct. Mater.*, 2022, **32**, 2104649.

49. M. Aguirre, N. Ballard, E. Gonzalez, S. Hamzehlou, H. Sardon, M. Calderon, M. Paulis, R. Tomovska, D. Dupin, R. H. Bean, T. E. Long, J. R. Leiza and J. M. Asua, *Macromolecules*, 2023, **56**, 2579.
50. A. Al Harraq, B. D. Choudhury and B. Bharti, *Langmuir*, 2022, **38**, 3001.
51. F. U. Din, S. Saleem, F. Aleem, R. Ahmed, N. U. Huda, S. Ahmed, N. Khaleeq, K. U. Shah, I. Ullah, A. Zeb and W. Aman, *Cogent Med.*, 2018, **5**, 1480572.
52. M. F. Nazar, M. A. Saleem, H. Basharat, A. Nasrullah, H. Asif, M. Ashfaq and R. Jamil, *Colloids Interface Sci. Commun.*, 2021, **45**, 100536.
53. Q. Zhang, Y. Yan, J. Liu, Y. Wu and Q. He, *Curr. Opin. Colloid Interface Sci.*, 2022, **61**, 101642.
54. Y. Cui, J. Wang, J. Liang and H. Qiu, *Small*, 2023, **19**, 202207609.
55. X. Zhao, H. Ma, H. Cai, Z. Wei, X. Tang and T. Qin, *Materials*, 2023, **16**, 5790.
56. D. G. Gulevich, I. R. Nabiev and P. S. Samokhvalov, *Mater. Today Chem.*, 2024, **30**, 101837.
57. G. G. Yang, J. Ko, H. J. Choi, D. H. Kim, K. H. Han, J. H. Kim, M. H. Kim, C. Park, H. M. Jin, I. D. Kim and S. O. Kim, *ACS Nano*, 2022, **16**, 18767–18776.
58. X. Chen, M. Leishman, D. Bagnall and N. Nasiri, *Nanomaterials*, 2021, **11**, 1927.
59. F. Mitri, A. De Iacovo, M. De Luca, A. Pecora and L. Colace, *Sci. Rep.*, 2020, **10**, 69478.
60. N. A. Isaac, I. Pikaar and G. Biskos, *Microchim. Acta*, 2022, **189**, 254.
61. Y. Tang, Y. Zhao and H. Liu, *ACS Sens.*, 2022, **7**, 3582–3597.
62. N. R. Tanguy, K. Khorsand Kazemi, J. Hong, K. C. Cheung, S. Mohammadi, P. Gnanasekar, S. S. Nair, M. H. Zarifi and N. Yan, *Carbohydr. Polym.*, 2022, **278**, 118920.
63. S. X. L. Luo, W. Yuan, M. Xue, H. Feng, M. J. Bezdek, T. Palacios and T. M. Swager, *ACS Nano*, 2023, **17**, 2679.
64. S. Brahma, Y. W. Yeh, J. L. Huang and C. P. Liu, *Appl. Surf. Sci.*, 2021, **564**, 150351.
65. A. Calvi, A. Ferrari, L. Sbuelz, A. Goldoni and S. Modesti, *Sensors*, 2016, **16**, 731.
66. Q. Li, H. Ghadiani, V. Jalilvand, T. Alam, Z. Farhat and M. A. Islam, *Materials*, 2024, **17**, 40965.
67. S. Das, S. Mojumder, D. Saha and M. Pal, *Sens. Actuators, B*, 2022, **354**, 131066.
68. W. Bulowski, R. Knura, R. P. Socha, M. Basiura, K. Skibińska and M. Wojnicki, *Electronics*, 2024, **13**, 3409.
69. M. Baharfar, J. Lin, M. Kilani, L. Zhao, Q. Zhang and G. Mao, *Nanoscale Adv.*, 2023, **5**, 5997.
70. S. Hooshmand, P. Kassanos, M. Keshavarz, P. Duru, İ Kayalan and M. K. Bayazit, *Sensors*, 2023, **23**, 8648.
71. D. Lun and K. Xu, *Micromachines*, 2022, **13**, 60919.

72. L. Y. Cheng, F. Ming, F. Zhao, L. Ye and D. Wang, *Front. Phys.*, 2022, **17**, 1178.
73. P. Surekha, A. D. Varshney, E. Jerusha, B. Pant and A. S. Rajesh, *Mater. Today: Proc.*, 2022, **62**, 2034.
74. M. G. Giordano, G. Seganti, M. Bartoli and A. Tagliaferro, *Molecules*, 2023, **28**, 62772.
75. V. Galstyan, *Anal. Chim. Acta*, 2021, **1151**, 67.
76. A. Issac, S. Al-Maskari, R. G. S. Sofin, A. R. Ibrahim and O. K. Abou-Zied, *J. Lumin.*, 2021, **239**, 118402.
77. S. Bin Hafiz, M. Scimeca, A. Sahu and D. K. Ko, *Nano Convergence*, 2019, **6**, 7.
78. A. R. C. Osypiw, S. Lee, S. M. Jung, S. Leoni, P. M. Smowton, B. Hou, J. M. Kim and G. A. J. Amaratunga, *Mater. Adv.*, 2022, **3**, 6773.
79. O. Dimitriev, D. Kysil, A. Zaderko, O. Isaieva, A. Vasin, Y. Piryatinski, M. Fahlman and A. Nazarov, *Nanoscale Adv.*, 2024, **6**, 2185.
80. A. Sharma and J. Das, *J. Nanobiotechnol.*, 2019, **17**, 92.
81. H. Liu, W. Zhang, H. Yu, L. Gao, Z. Song, S. Xu, M. Li, Y. Wang, H. Song and J. Tang, *ACS Appl. Mater. Interfaces*, 2016, **8**, 840–846.
82. M. Li, H. K. Jin, J. Li, Z. Huang and H. Liu, *IEEE Sens. J.*, 2022, **22**, 3017–3023.
83. F. Mitri, A. De Iacovo, M. De Luca, A. Pecora and L. Colace, *Sci. Rep.*, 2020, **10**, 69478.
84. F. Mitri, A. De Iacovo, S. De Santis, D. Quarta, C. Giansante, M. Orsini and L. Colace, *Nanotechnology*, 2022, **33**, 8814.
85. M. S. Freire, H. J. B. Silva, G. M. Albuquerque, J. P. Monte, M. T. A. Lima, J. J. Silva, G. A. L. Pereira and G. Pereira, *Sci. Total Environ.*, 2024, **931**, 172848.
86. Y. Wu, J. Feng, G. Hu, E. Zhang and H. H. Yu, *Sensors*, 2023, **23**, 52749.
87. H. H. Cho, R. Dutta, J. K. Kwak, C. Moon, M. J. Kim, S. J. Suh, D. H. Kim, J. H. Lee and S. Kim, *EcoMat*, 2023, **5**, 12389.
88. M. Zarei, S. M. Hamidi and K. W.-A. Chee, *Crystals*, 2023, **13**, 363.
89. J. Jana, M. Ganguly and T. Pal, *RSC Adv.*, 2016, **6**, 86174.
90. S. Sarkar and T. A. F. König, *Adv. Sens. Res.*, 2024, **3**, 2300054.
91. I. Cho, Y. C. Shim, K. Lee, M. Choi, J. Park, M. Kim, K. Kim, S. Kim, C. Choi, B. Choi, J. Yoo-Hyun and C. In-Pyo, *Nano Micro Small*, 2023, **19**, 27.
92. J. G. Liang, Y. Jiang, J. K. Wu, C. Wang, S. von Gratowski, X. Gu and L. Pan, *Sens. Actuators, A*, 2023, **356**, 114318.
93. P. Chen, Y. Mei, H. Y. Li, J. Hou, H. Chen, R. Mao, Z. Jian, S. Xu and H. Liu, *Sens. Actuators, B*, 2024, **417**, 136065.
94. M. V. Nikolic, V. Milovanovic, Z. Z. Vasiljevic and Z. Stamenkovic, *Sensors*, 2020, **20**, 26694.
95. X. H. Zhang, J. Chen, Y. Wu, Z. Xie, J. Kang and L. Zheng, *Colloids Surf., A*, 2011, **384**, 580.
96. Z. Chen, A. Umar, S. Wang, Y. Wang, T. Tian, Y. Shang, Y. Fan, Q. Qi, D. Xu and L. Jiang, *Nanoscale*, 2015, **7**, 10259.

97. X. Chen, M. Leishman, D. Bagnall and N. Nasiri, *Nanomaterials*, 2021, **11**, 81927.
98. J. Jońca and I. Sówka, *Zesz. Nauk. SGSP*, 2023, **85**, 7.
99. Y. Lin and Z. Fan, *Mater. Sci. Semicond. Process.*, 2020, **104**, 820.
100. A. Marikutsa, M. Rumyantseva, E. A. Konstantinova and A. Gaskov, *Sensors*, 2021, **21**, 72554.
101. B. Yang, X. Li, W. Yuan, Z. Li, N. Lu, S. Wang, Y. Wu, S. Fan and Z. Hua, *ACS Sens.*, 2020, **5**, 1838.
102. A. Dey, *Mater. Sci. Eng. B*, 2018, **229**, 206.
103. H. Gu, Z. Wang and Y. Hu, *Sensors*, 2012, **12**, 5517.
104. C. Ge, C. Jin, M. Wang, L. Bai, S. Hussain, G. Qiao, E. J. Kim and G. Liu, *Sens. Actuators, B*, 2022, **366**, 131991.
105. A. K. Nayak, R. Ghosh, S. Santra, P. K. Guha and D. Pradhan, *Nanoscale*, 2015, **7**, 12460.
106. J. Sethi, E. Sarlin, S. S. Meysami, R. Suihkonen, A. R. S. Santha Kumar, M. Honkanen, P. Keinänen, N. Grobert and J. Vuorinen, *Composites, Part A*, 2017, **102**, 305–313.
107. H. R. Kim, K. Il Choi, K. M. Kim, I. D. Kim, G. Cao and J. H. Lee, *Chem. Commun.*, 2010, **46**, 5061.
108. C. N. Wang, Y. L. Li, F. L. Gong, Y. H. Zhang, S. M. Fang and H. L. Zhang, *Chem. Rec.*, 2020, **20**, 1553.
109. X. Kang, S. P. Yip, Y. Meng, W. Wang, D. Li, C. Liu and J. C. Ho, *Nanoscale Adv.*, 2021, **3**, 6254.
110. H. J. Kim and J. H. Lee, *Sens. Actuators, B*, 2014, **193**, 607.
111. X. Qiao, Y. Xu, K. Yang, C. Li, H. Wang and L. Jia, *J. Hazard. Mater.*, 2020, **392**, 122471.
112. B. Yang, Z. Zhang, C. Tian, W. Yuan, Z. Hua, S. Fan, Y. Wu and X. Tian, *Sens. Actuators, B*, 2020, **321**, 128567.
113. S. Lu, Y. Zhang, J. Liu, H. Y. Li, Z. Hu, X. Luo, N. Gao, B. Zhang, J. Jiang, A. Zhong, J. Luo and H. Liu, *Sens. Actuators, B*, 2021, **345**, 130334.
114. V. C. Anitha, A. N. Banerjee and S. W. Joo, *J. Mater. Sci.*, 2015, **50**, 7495–7536.
115. J. Il Yoon, H. Kim, M. Kim, H. Cho, Y. Albert Kwon, M. Choi, S. Park, T. Kim, S. Lee, H. Jo, B. Kim, J. Ho Cho, J. Park, S. Jeong and M. Sung Kang, *Sci. Adv.*, 2023, **9**, eadj8276.
116. R. Ghosh, M. Aslam and H. Kalita, *Mater. Today Commun.*, 2022, **31**, 103182.
117. S. M. Sara, A. M. Al-Dhahebi and M. S. Mohamed Saheed, *Polymers*, 2022, **14**, 5235125.
118. L. Yang, G. Zheng, Y. Cao, C. Meng, Y. Li, H. Ji, X. Chen, G. Niu, J. Yan, Y. Xue and H. Cheng, *Microsyst. Nanoeng.*, 2022, **8**, 41378.
119. B. Nadekar, Y. B. Khollam, S. F. Shaikh, D. Wavhal, P. Varshney, B. Pandit and P. S. More, *Surf. Interfaces*, 2023, **42**, 103453.
120. D. Li, J. Lu, X. Zhang, D. Jin and H. Jin, *Nanomaterials*, 2023, **13**, 917.
121. B. Han, F. Xia, Z. W. Zhang, M. Wang and Y. N. Zhang, *Opt. Fiber Technol.*, 2023, **80**, 103383.

122. F. Mitri, A. De Iacovo, M. De Luca, A. Pecora and L. Colace, *Sci. Rep.*, 2020, **10**, 69478.
123. H. Li, M. Li, H. Kan, C. Li, A. Quan, C. Fu, J. Luo, X. Liu, W. Wang, Z. Yang and Y. Fu, *Surf. Coat. Technol.*, 2019, **362**, 78.
124. S. S. Gqoba, R. Rodrigues, S. L. Mphahlele, Z. Ndala, M. Airo, P. O. Fadojutimi, I. A. Hümmelgen and N. J. Moloto, *Processes*, 2021, **9**, 1491.
125. Y. J. Jin and J. Park, *Polymers*, 2020, **12**, 1591.
126. T. Syrový, P. Kuberský, I. Sapurina, S. Pretl, P. Bober, L. Syrová, A. Hamáček and J. Stejskal, *Sens. Actuators, B*, 2016, **225**, 510.
127. A. Kumar and R. Prajesh, *Sens. Actuators, A*, 2022, 113498.
128. A. Alassi, M. Benammar and D. Brett, *Sensors*, 2017, **17**, 2799.
129. J. Xie, L. Zhang, H. Xing, P. Bai, B. Liu, C. Wang, K. Lei, H. Wang and S. Yang, *Sens. Actuators, B*, 2020, **305**, 127479.
130. J. Xie, H. Wang and M. Duan, *Sens. Actuators, B*, 2014, **203**, 239.
131. I. Constantinoiu and C. Viespe, *Sensors*, 2020, **20**, 5118.
132. M. Li, H. Kan, S. Chen, X. Feng, H. Li, C. Li, A. Quan, H. Sun, J. Luo, X. Liu, W. Wang, H. Liu and Y. Fu, *Sens. Actuators, B*, 2019, **287**, 241.
133. J. D. Lopez-Vargas, A. Dante, R. C. Allil, I. Del Villar, I. R. Matias and M. M. Werneck, *Sens. Actuators, B*, 2024, **401**, 135054.
134. H. Gong, G. He, J. Xue, A. Li, Y. Ji and F. Sun, *Surf. Interfaces*, 2024, **48**, 104313.
135. H. Yu, L. Dai, Y. Liu, Y. Zhou, P. Fan, J. Luo and A. Zhong, *J. Alloys Compd.*, 2023, **962**, 171170.
136. E. M. Sulaiman, U. M. Nayef and F. S. A. Mutlak, *Opt. Laser Technol.*, 2022, **154**, 108336.
137. G. Rahamim, M. Mirilashvili, P. Nanikashvili, E. Greenberg, H. Shpaisman, D. Grinstein, S. Welner and D. Zitoun, *Sens. Actuators, B*, 2020, **310**, 127845.
138. H. Yu, Z. Song, Q. Liu, X. Ji, J. Liu, S. Xu, H. Kan, B. Zhang, J. Liu, J. Jiang, L. Miao and H. Liu, *Sens. Actuators, B*, 2017, **248**, 1029.
139. B. Zhang, M. Li, Z. Song, H. Kan, H. Yu, Q. Liu, G. Zhang and H. Liu, *Sens. Actuators, B*, 2017, **249**, 558.
140. X. Cheng, Y. Xu, S. Gao, H. Zhao and L. Huo, *Sens. Actuators, B*, 2011, **155**, 716.
141. N. L. Torad, I. M. Minisy, H. M. Sharaf, J. Stejskal, Y. Yamauchi and M. M. Ayad, *Synth. Met.*, 2021, **282**, 116935.
142. V. C. Van, N. X. Dinh, V. Ngoc Phan, A. T. Le, M. H. Nam, V. D. Lam, T. Van Dang and N. V. Quy, *Mater. Today Commun.*, 2020, **25**, 101682.
143. J. Xie, H. Wang and M. Duan, *Sens. Actuators, B*, 2014, **203**, 239.
144. S. Y. Park, M. S. Bae, I. D. Jeon and J. J. Lee, *Microelectron. Eng.*, 2011, **88**, 3035.
145. D. Nagai, M. Nishibori, T. Itoh, T. Kawabe, K. Sato and W. Shin, *Sens. Actuators, B*, 2015, **206**, 488.

CHAPTER 4

Advancing Air Sustainability: Light-induced Removal of Airborne Pollutants

DAVID MARIA TOBALDI* AND MASSIMO CUSCUNÀ

Consiglio Nazionale delle Ricerche, Istituto di Nanotecnologia – CNR NANOTEC, Via Monteroni, 73100 Lecce, Italy
*Email: david.tobaldi@nanotec.cnr.it

4.1 Introduction

Air pollution (indoor and outdoor) poses a significant threat to the health of millions of people worldwide, necessitating urgent action to alleviate its pervasive impacts. Not only does addressing this issue promise an enhancement in both life expectancy and quality of life, it significantly contributes to the pursuit of a more equitable and sustainable global environment.[1] According to a definition of the World Health Organisation (WHO), air pollution is: "*Contamination of the indoor or outdoor environment by any chemical, physical or biological agent that modifies the natural characteristics of the atmosphere*".[2] In recognition of the critical significance of this issue, the WHO has established a dedicated unit focused on "Air Quality and Health".[3] Their data reveal that a staggering 99% of the global population breathes air surpassing WHO guideline limits, exposing them to elevated pollutant levels. It is not surprising that the highest exposures are observed in low- and middle-income countries. Moreover, the connection between air quality and the planet's climate and ecosystems is crystal clear. Enhancing air quality has the potential to improve natural carbon sequestration by plants.[4] Many of the factors driving air pollution, such as the combustion of

Contribution of Colloidal Materials to Air, Water and Soil Environmental Sustainability: From Basic Concepts to Advanced Applications
Edited by Izabel C. Riegel-Vidotti, Lizandra Maria Zimmermann and Heloise R. de Barros

Published by the Royal Society of Chemistry, www.rsc.org

fossil fuels, also represent primary sources of greenhouse gas emissions.[2] Consequently, implementing policies, including those exploited through the use of colloidal materials to reduce air pollution, presents a powerful win–win strategy, offering dual benefits for both climate and health.[5]

Indoor and outdoor air pollutants with the most compelling evidence for public health concern include, but are not limited to, particulate matter (PM), carbon monoxide (CO), ozone (O_3), nitrogen oxides ($NO_x = NO + NO_2$), sulfur[6] dioxide (SO_2), and volatile organic compounds (VOCs). Adverse health effects may arise from both short- and long-term exposure to these diverse pollutants, emphasising the critical importance of addressing their impact on air quality. For a comprehensive description of these pollutants, readers are referred to the WHO global air quality guidelines.[7]

It has to be stressed that indoor air pollution is regarded as an "*invisible enemy*", as it is estimated that around 3.8 million people worldwide prematurely die each year from illnesses attributable to poor household indoor air quality.[8] This challenge has been further intensified by the lockdowns implemented in response to the emergence of the respiratory virus SARS-CoV-2 (COVID-19).[9] This was identified as potentially disastrous for public health, with consequences involving concurrent infectious and (indoor) air pollution-mediated diseases.[10]

Given this context, combatting air pollution is an imperative challenge of the 21st century. Various measures are employed to address indoor air pollution. Adsorption, electrochemical oxidation processes, and the utilisation of heating, ventilation, and air conditioning (HVAC) systems, along with air filtration systems, contribute to creating a well-ventilated indoor environment.[11,12] Outdoor air quality management often involves regulatory measures, such as emission standards and vehicle emission controls.[13] While these strategies are necessary to combat air pollution, they only offer a partial answer to it.

On the other hand, mineralisation of these pollutants into non-hazardous, or at least less hazardous, compounds through photocatalysis has great promise as a more feasible solution for their total or substantial removal.[14] The photocatalytic process effectively degrades various air pollutants using either solar or artificial light and the reactions take place under mild operating conditions.[15] It shares intrinsic similarities with the self-cleaning mechanism in the Earth's atmosphere, relying on indirect (sensitised) photo-oxidation to generate *in situ* oxidants within the air.[16] Basically, a photocatalyst absorbs photons, generating pairs of electrons and holes that subsequently react with dioxygen, water, and surface hydroxyl groups. This interaction results in the production of reactive oxygen species (ROS), which serve as the primary oxidants responsible for decomposing air pollutants.[17] While many semiconductors have the potential to exhibit photocatalytic behaviour, their efficiency and suitability for specific applications can vary, depending on factors such as bandgap, surface area, and recombination rates. Amongst them, titanium dioxide (TiO_2) is definitely the most studied photocatalyst, capturing scientific attention since the seminal *Nature* paper by Fujishima and Honda.[18] Subsequently, the work, also published in *Nature*, by O'Regan and Grätzel led to a shift in research focus towards electrodes based on high surface area nanostructured materials.[19] As particle size plays a crucial role in the photonic efficiency of a photocatalyst, controlled synthesis is crucial for fabricating photocatalysts with optimised properties. At the beginning of the 2000s, non-aqueous synthesis of nanoparticles emerged as a highly promising strategy for the controlled fabrication of metal–oxide nanoparticles with precisely adjustable particle size.[20] With the use of various solvents, phase-pure metal–oxide photocatalytically active nanoparticles with narrow size dispersion could be obtained.[21,22] However, for many applications, it is more desirable for nanoparticles to exist in the form of thin films, fabricated from dispersions only if the nanoparticles are colloidally stable in a suitable solvent.[23]

This chapter will first address the description of physical removal methods for air pollutants. Light-induced mineralisation technologies to convert these pollutants will then be explored and discussed, with a particular view toward colloidal materials.

4.2 Adsorption and HVAC Technologies

4.2.1 Adsorption

According to an IUPAC technical report, *adsorption* is defined as "*the enrichment in the concentration of molecules, atoms, or ions present in a fluid phase in the vicinity of an interface*".[24] Adsorption manifests itself in two forms: physical (physisorption) or chemical (chemisorption), dictated by the strength of the interaction between the adsorbate and the adsorbent surface. Physisorption relies on intermolecular forces involving induced or permanent dipoles and/or quadrupoles. In contrast, chemisorption involves a transformation in the electronic structure of both adsorbent (the solid material), and adsorbate (the adsorbed molecules), leading to the formation of a chemical bond. *Adsorbents* can be described as materials or solid substances that can effectively adsorb and remove gases or liquids due to their large surface area, suitable pore structure, and strong adsorption capacity.[25]

4.2.2 Adsorbents

While an ideal adsorbent is unlikely to exist, Choi and co-authors emphasise the need for comparison based on the specific process to which they will be adapted.[26] They also highlight the importance of ranking adsorbent parameters in relation to the specific constraints of the system. If we were to generalise, physisorbents, such as activated carbons and zeolites, demonstrate competitive adsorption capacities.

The former are well-known adsorbent materials; as they benefit from a wide range of available sources, their industrial-scale production is relatively low-cost.[27] Typical sources include coals, industrial by-products, and various biomasses, including wood.[28] The preparation of activated carbons involves two main stages: carbonisation and activation. In the carbonisation stage, precursors undergo heating and pyrolysis in an inert atmosphere, producing a carbonaceous material, the so-called char, which undergoes a subsequent activation process (Figure 4.1(a)). Physical activation involves the partial gasification of chars using steam, CO_2, air, or a combination at high temperatures (around 1100 to 1250 K), enhancing porosity and active sites.[29] Chemical activation, performed at lower temperatures, employs chemical substances to open micropores and develop a well-structured carbon

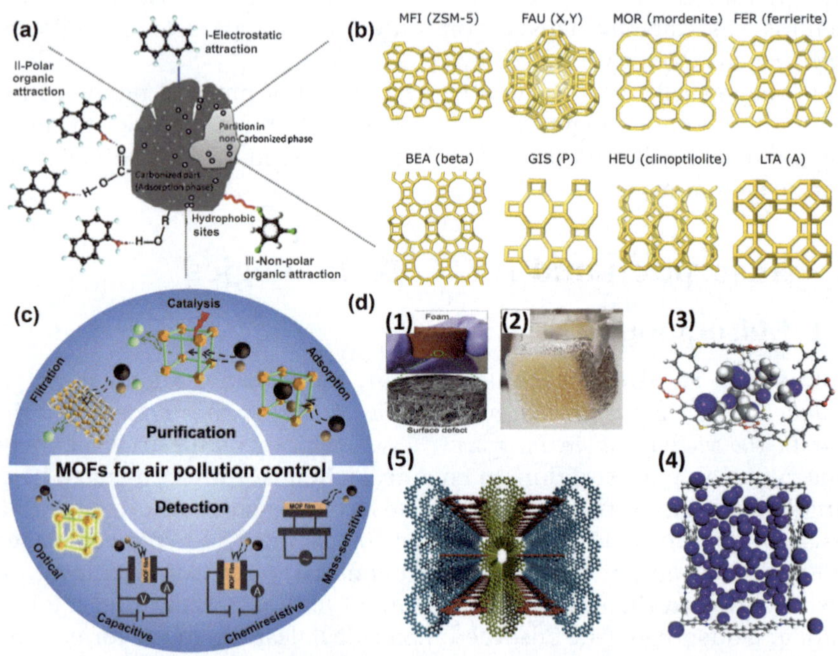

Figure 4.1 (a) VOC mitigation mechanism by biochar. Reproduced from ref. 40 with permission from Elsevier, Copyright 2017. (b) Front view of the main pores of selected zeolites. The three-letter codes above each image represent the official designations assigned to each structure by the International Zeolite Association. Adjacent to the codes, the common names of the respective zeolites are provided in parentheses. Reproduced from ref. 34, https://doi.org/10.1021/acs.chemrev.2c00140, under the terms of the CC BY 4.0 license, https://creativecommons.org/licenses/by/4.0/. (c) Schematic overview illustrating the utilisation of MOFs in mitigating air pollution. Reproduced from ref. 36 with permission from John Wiley & Sons, Copyright 2023 Wiley-VCH GmbH. (d) Images depicting several COFs: (1) TpPa-2-foam with its corresponding 3D structure; (2) COF-VF@foam photographed while submerged in water; adsorption configurations of (3) CH_3I in COF-103, and (4) I_2 in 3D-Py-COF, where atoms are colour-coded (C: grey; H: white; N: blue; Si: yellow, I: violet spheres); (5) the structure of recently developed COFs (3D-Py-COF-TANM). Reproduced from ref. 39 with permission from Elsevier, Copyright 2019.

framework. It yields activated carbons with excellent porosity and is known for its efficiency in good yield production.[30]

Zeolites are crystalline microporous aluminosilicates, and they find extensive use as catalysts, adsorbents, and ion exchangers. From a mineralogical point-of-view, they belong to the group of tectosilicates, and some are naturally occurring. With a well-defined pore size, compositional flexibility, thermal stability, and commercial availability since the 1950s, they have become integral in industrial and domestic applications.[31] The structural framework of zeolites is composed of TO_4 tetrahedra, where T commonly represents Si or Al atoms. Tetrahedrally coordinated Al atoms introduce negative charges in the framework, which are balanced by additional cationic species outside the framework.[32] The flexibility of the T–O–T angle enables various spatial arrangements of the tetrahedra, giving rise to millions of hypothetical porous structures (Figure 4.1(b)).[33] This enables zeolites to be shape- and size-selective adsorbents.[34]

Nevertheless, these commonly used adsorption materials require ongoing research and development to improve their adsorption capacity, hydrophobicity, thermal stability, and renewability. The extensive production and use of these materials present significant challenges, including cost considerations and potential environmental impacts.[35] Metal–organic frameworks (MOFs) and covalent organic frameworks (COFs) are novel materials in the field of environmental remediation. MOFs, composed of inorganic nodes and organic ligands, offer a well-defined (micro) porous structure and highly customisable properties, exhibiting significant promise for applications in air pollution control (Figure 4.1(c)).[36] They have indeed been identified as effective absorbents for air pollutants, exploiting their unique advantages of high porosity and adaptable functionality. The significant porosity of MOFs enables efficient pathways for guest molecules to reach the adsorption sites. Simultaneously, the metal nodes and organic ligands of MOFs can be strategically designed to enhance the adsorption affinity towards the target adsorbates.[37] COFs represent an emerging class of crystalline porous polymers enabling the atomically precise integration of organic units into their extended structures with periodic skeletons and ordered nanopores (Figure 4.1(d)). A key attribute of COFs is their designability, allowing for control over the geometry and dimensions of their building blocks.[38] Due to their high porosity (exceeding 4000 $m^2 g^{-1}$) and low density (approximately 0.15 $g cm^{-3}$), they have significant potential for the sequestration of air contaminants. Thus, while research on COFs is in its early stages, their considerable potential in environmental remediation is gradually gaining recognition.[39]

4.2.3 HVAC Technologies

Urbanisation, along with the growing number of people living in cities, has significantly increased indoor air pollution, a challenge further amplified by the COVID-19 pandemic. These trends emphasise the crucial need for a sustainable and healthy indoor environment. HVAC systems play essential roles in addressing this concern, providing effective ways to remove air pollutants and improve indoor air quality.[41] Before the pandemic, HVAC systems were designed to maintain acceptable indoor air quality. However, in the post-pandemic era, there has been a push towards adopting mixed-mode and natural ventilation strategies. Additionally, the deployment of ultraviolet germicidal radiation technology has been encouraged to reduce virus concentrations in built environments, thereby mitigating the risk of infection.[42] To ensure a good indoor air-sustainability, future research should focus on how to provide enough fresh air through the combination of reasonable natural ventilation and efficient air purification to achieve a low infection risk while maintaining low energy consumption during the pandemic.[43] Likewise, it is imperative to improve air filter efficiency by reducing pressure drops and operational costs. Emerging filtration technologies, such as nanofibers, Trombe walls, and botanical biofiltration, are poised to gain widespread adoption globally in the coming years. Exploring the integration of air filtration technologies with building information modelling platforms presents a promising avenue for future research. Furthermore, expanding indoor environment monitoring and refining air quality predictions are essential directions to advance research in this field.[11]

4.3 Light-induced Removal Technologies

4.3.1 Semiconductor Photocatalysis

When considering light-induced technologies for air sustainability, semiconductor photocatalysis immediately comes to mind. Serpone and co-authors maintain that the term "*photocatalysis*" first appeared in a 1910 photochemistry textbook.[44] Palmisano and co-authors, in their work, dubbed photocatalysis to be "*a promising route for 21st-century organic chemistry*".[45] While it can indeed be applied as a green synthetic route in organic chemistry – considered as one of the most innovative methods in green chemistry[46] – its primary application lies in the mineralisation of organic and inorganic pollutants. This occurs in both vapour and liquid phases, all under mild operating conditions.[47]

TiO_2 is undoubtedly the main character, as its story began in the late 1960s with photoelectrochemical solar energy conversion.[18] Its prominence as a photocatalyst is underscored by the extensive number of review papers dedicated to this oxide – a literary genre that seems to be ever-expanding, much alike the research itself.[48] While it crystallises in a large number of polymorphs, the most abundant and naturally occurring ones are: rutile (tetragonal crystal system, space group: $P4_2/mnm$), anatase (tetragonal, $I4_1/amd$), and brookite (orthorhombic, *Pbca*).[49] Their crystal structures are depicted in Figure 4.2.

A semiconductor-based photocatalyst shall ideally convert solar energy to chemical fuel from H_2O and CO_2 (Figure 4.3).[53] Thus, upon excitation,

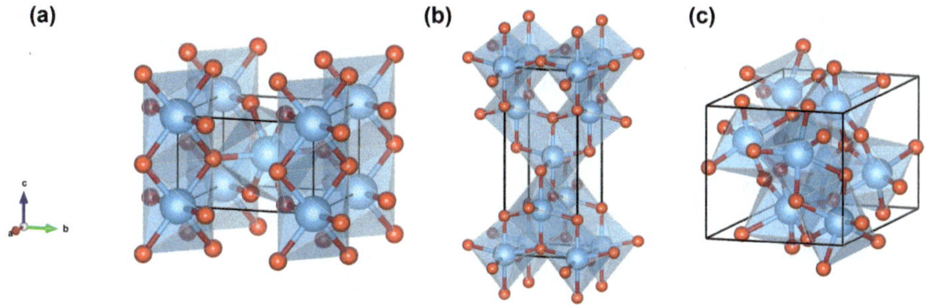

Figure 4.2 3D ball-and-stick renderings of: (a) rutile, (b) anatase, and (c) brookite crystal structures created using the VESTA software suite.[50] The axis orientation is consistent across all three structures. The light-blue spheres represent titanium atoms, and the red spheres represent oxygen. The initial crystal structures were taken from the literature.[51,52]

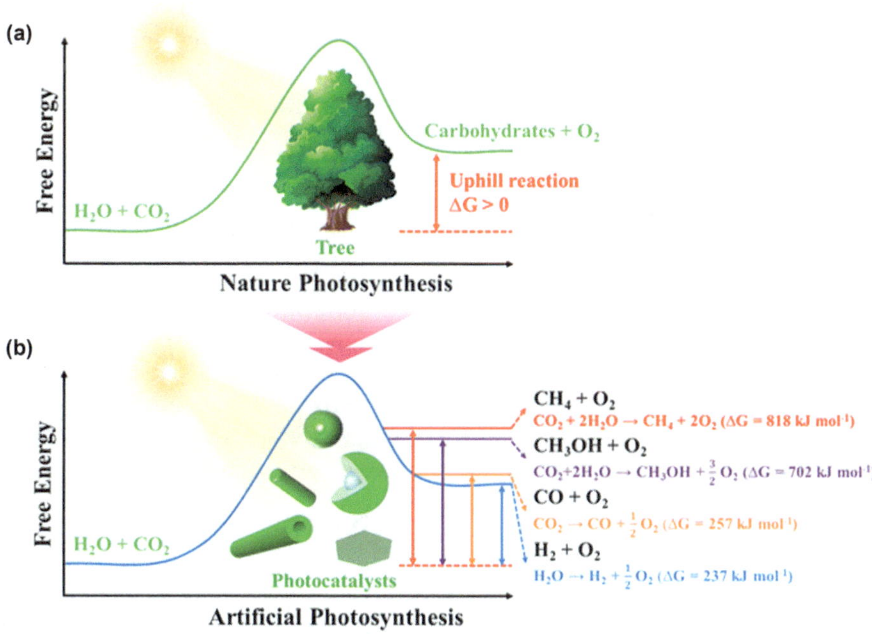

Figure 4.3 Schematic representation depicting: (a) natural and (b) artificial photosynthetic systems. Reproduced from ref. 53 with permission from the Royal Society of Chemistry.

an electron (e^-) is promoted from the valence band (VB) to the conduction band (CB), leaving a positively charged hole (h^+) behind, see eqn (4.1):

$$TiO_2 + h\nu \rightarrow e^-_{CB} + h^+_{VB} \tag{4.1}$$

The photogenerated exciton can in turn initiate the reduction/oxidation of species adsorbed on the catalyst surface, depending on the relative positions of their redox potentials. At neutral pH, the redox potential of the hole on TiO_2 is +2.53 V, while that of the excited electron is −0.52 V [both referenced to the standard hydrogen electrode (SHE)].[54] Hence, reduction can only occur when the redox potential of the e^- is negative enough to reduce the oxidant, while oxidation only takes place when the redox potential of the h^+ is more positive than that of the reductant.

Regarding the degradation of organic contaminants, the generally accepted reaction mechanism involves a radical pathway. The radical chain reaction typically commences with the generation of hydroxyl radicals ($OH^{\bullet}$) from H_2O (+2.27 V *vs.* SHE), and superoxide anion radicals ($O_2^{\bullet-}$) from O_2 (−0.28 V *vs.* SHE), facilitated by the photogenerated h^+ and e^-, respectively. These reactions are listed as eqn (4.2) and (4.3):[54]

$$H_2O + h^+_{VB} \rightarrow OH^{\bullet} + H^+ \tag{4.2}$$

$$O_2 + e^-_{CB} \rightarrow O_2^{\bullet-} \tag{4.3}$$

The generated charge carriers can also experience trapping, recombination, and interfacial transfer before initiating the redox reaction at the TiO_2 surface, thereby impeding the photocatalytic reaction,[55] thus dissipating energy,[56] *cf.* eqn (4.4):

$$e^-_{CB} + h^+_{VB} \rightarrow \text{energy} \tag{4.4}$$

4.3.1.1 *Photocatalytic Processes to Remove NO_x and VOCs*

There is a widespread consensus on identifying NO_x and VOCs as among the most hazardous indoor and outdoor air pollutants. The NO_x abatement process primarily involves the oxidation of NO by $O_2^{\bullet-}$ species, as from eqn (4.3). Photo-oxidation aims to convert NO into HNO_3 through the intermediates HNO_2 and NO_2. The NO photo-oxidation mechanism in TiO_2 photocatalysts progresses through three stages ($NO \rightarrow HNO_2 \rightarrow NO_2 \rightarrow HNO_3$), and the reactions defining these stages are presented below:[57]

$$NO + OH^{\bullet} \rightarrow HNO_2 \tag{4.5}$$

$$HNO_2 + OH^{\bullet} \rightarrow NO_2 + H_2O \tag{4.6}$$

$$NO_2 + OH^{\bullet} \rightarrow HNO_3 \tag{4.7}$$

$$NO + O_2^{\bullet-} \rightarrow NO_3^- \tag{4.8}$$

$$NO_3^- + H^+ \rightarrow HNO_3 \tag{4.9}$$

$$NO_2 + OH^{\bullet} \rightarrow HNO_3 \tag{4.10}$$

It is suggested that the ideal closure of the de-NO_x cycle involves the removal of HNO_3 through water treatment from the photocatalyst surface:[58]

$$HNO_{3(ads)} \rightarrow HNO_{3(aq)} \tag{4.11}$$

It is essential to mention an emerging indoor pollutant, nitrous acid (HONO), known for its adverse health effects, including respiratory tract irritation. Indoor HONO can originate from combustion processes, with concentrations often surpassing outdoor levels. Recent laboratory studies reveal that the photo-enhanced reaction of NO_2 on common indoor surfaces acts as a significant source of HONO. HONO exemplifies a potential downside of photocatalytic de-NO_x, as this technology may serve as a contributor to HONO levels:[59]

$$2NO_2 + H_2O \rightarrow HONO + HNO_3 \tag{4.12}$$

Hence, when evaluating light-induced NO_x removal, it is advisable to analyse the intermediates in the de-NO_x process. Indeed, as our exposure to indoor environments grows, there is an increasing need for a deeper understanding of the chemical cocktail evolving within our living and working spaces, and its potential implications for human health.[60,61]

The fundamental mechanisms governing the photocatalytic oxidation of gaseous VOCs are well-established: radicals generated through eqn (4.2) and (4.3) serve as strong oxidising agents, capable of transforming gaseous targets, such as VOC molecules, into simpler and environmentally benign compounds, ultimately resulting in carbon dioxide and water.[62] Nevertheless, the photocatalytic elimination of VOCs poses a considerably more intricate challenge, especially when attempting to identify the intermediates of their oxidation processes. Jenkin and colleagues found that the complete atmospheric oxidation of isoprene (C_5H_8) involves a staggering 1926 distinct degradation steps, resulting in the formation of 602 unique intermediate organic species before ultimately yielding CO_2 and water.[63] Therefore, recognising all intermediates in VOC photocatalytic oxidation, both qualitatively and quantitatively, is an extremely challenging task. As a general guideline, studies have indicated that oxygenated intermediate compounds, particularly formaldehyde and acetaldehyde, serve as reliable benchmarks for estimating the progression of the degradation process. These molecules can serve as effective indicators in evaluating the performance of photocatalytic oxidation-based systems.[64] As with the de-NO_x process, mineralisation of VOCs can lead to the formation of by-products/intermediates which not only might be hazardous, they can also block the active sites on the photocatalyst surface, and eventually result in their deactivation.[65] However, despite extensive studies on photocatalytic technology for VOCs removal in laboratories, limited research has addressed catalyst deactivation, by-product production, and moderate removal efficiency.[66] The generation and accumulation of by-products and intermediates on the catalyst surface may deactivate it.[67] A concern restricting full-scale photocatalytic applications in air purification systems is their moderate efficiency, especially under high flow rates resembling real-world conditions.[68] Combining adsorption and photocatalytic degradation is swiftly emerging as a promising technology for VOCs removal. It is not only environmentally friendly with low-energy consumption, it also proves to be effective and renewable.[69]

4.3.1.2 Strategies of Visible Light Absorption

With a bandgap ranging between 3.2 eV (for anatase) and 3.0 eV (for rutile),[70] equivalent to wavelengths of 388 nm and 413 nm, respectively, TiO_2 requires ultraviolet (UV) light for excitation, which comprises ~3–5% of the total solar radiation.[71] Together with the dissipation of energy, this is widely acknowledged as the Achilles' heel of TiO_2 photocatalysis.

Various strategies have been explored to create TiO_2 photocatalysts with visible light activity, as reported in a forefront review by Dionysiou.[72] To address this limitation and shift the bandgap into the visible region, the electronic structure of TiO_2 has been firstly modified through the introduction of various metallic (cationic) dopants.[73] However, these dopants have been found to act as charge carrier recombination centres, thus negatively affecting the photocatalytic efficiency.[74]

The trailblazing work of Asahi *et al.* opened the doors of research to non-metal doping.[75] They synthesised N-doped TiO_2 through sputter deposition of TiO_2 in an N_2/Ar atmosphere, followed by annealing under N_2. Since then, N-doped TiO_2 has attracted significant interest, and its sensitivity to visible light has been confirmed by numerous researchers. Typically, this is due to the formation of 2p electronic states above the VB able to produce efficient charge transfer electronic transitions to the 3d CB of TiO_2, providing in that way high photocatalytic activities. Following the success of N-doped TiO_2, various non-metal-doped TiO_2 types, including sulfur,[6] carbon, and iodine, have been extensively investigated as photocatalysts sensitive to visible light.[76] While non-metal doping of TiO_2 shows promise, the oxidising working conditions may lead to titania self-cleaning, expelling N^{3-} anions from the anionic substructure through oxidation.[77] This results in a loss of photocatalytic activity, supporting Herrmann's critique of TiO_2 non-metal doping, which suggests a cautious "*wait and see*" approach.[78]

Visible light sensitive TiO_2 can also be achieved through intrinsic doping, involving the introduction of oxygen vacancies and the formation of Ti^{3+} centres in the TiO_2 structure (Figure 4.4(a)). The synthesis of these materials, *cf.* Figure 4.4(b), typically involves high-temperature treatment of TiO_2 in various reducing atmospheres.[79] Chen was a pioneer in this area, reporting the first black TiO_2 nanomaterial for photocatalysis. Through thermal treatment at 200 °C under high pressure of H_2, stoichiometric anatase TiO_2 was transformed into defective nanocrystals with high visible light absorption, and enhanced photocatalytic efficiency.[80] The improved performance of black TiO_2 was attributed to increased light absorption resulting from the introduction of structural disorder, and H-doping, consequently narrowing the optical bandgap of black TiO_2 to 1.54 eV by introducing electronic states that formed an important VB and CB tailing.[81] Naldoni *et al.* showed that the bandgap narrowing in this material is dictated by the synergistic presence of oxygen vacancies and surface disorder.[82]

Coupling semiconductors in heterostructures has proven to be an efficient way for reducing the recombination rate of the photogenerated exciton,[83]

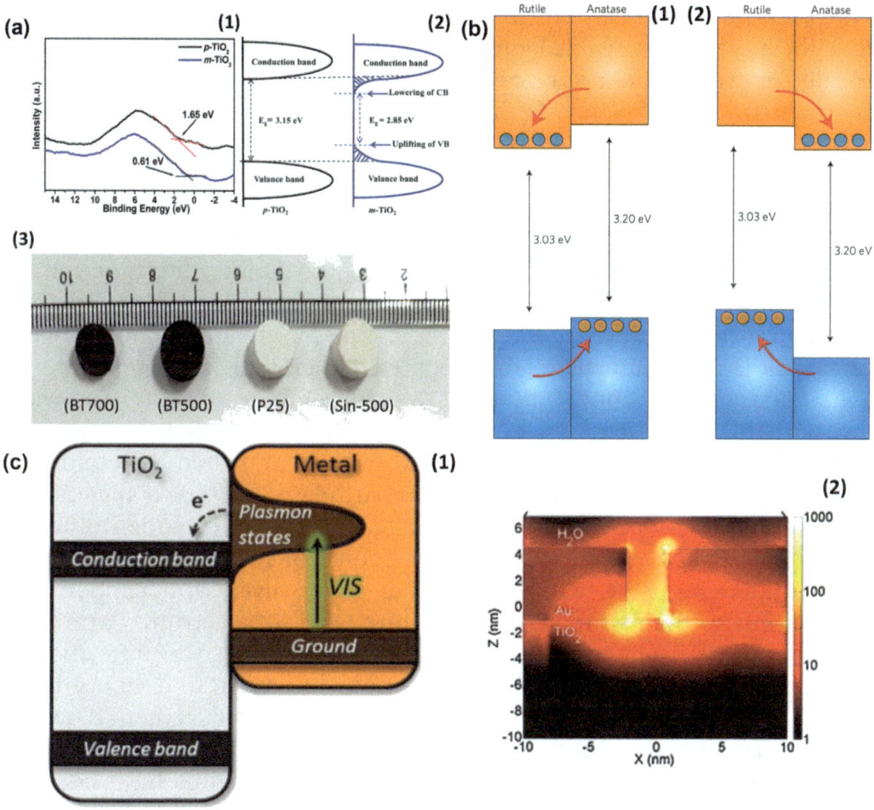

Figure 4.4 (a) (1) Valence band XPS spectra, and (2) the density of states for a commercial TiO_2, and a modified TiO_2 with Ti^{3+} centres (*p*-TiO_2 and *m*-TiO_2 respectively, in Figure 4.4(a) and (b)). Reproduced from ref. 95 with permission from the Royal Society of Chemistry. (3) Colour change in TiO_2 specimens, from off-white (commercial **P25** and **Sin-500** in the figure), to black (**BT700** and **BT500** in the figure), with the latter prepared under a hydrogen atmosphere. Reproduced from ref. 96 with permission from American Chemical Society, Copyright 2018. (b) Type-II (rutile) [panel 1], and type-II (anatase) [panel 2]. Red arrows illustrate the movement of electrons (holes) within the conduction band (valence band). Blue and orange dots symbolise electrons and holes, respectively. Reproduced from ref. 88 with permission from Springer Nature, Copyright 2013. (c) Potential mechanisms in plasmon-mediated photocatalysis. (1) Direct charge injection, where an electron from a plasmon-excited state on the metal is injected into the semiconductor's conduction band; reproduced from ref. 93 with permission from Elsevier, Copyright 2015. (2) Near-field enhancement occurring in proximity to the plasmonic metal surface; reproduced from ref. 97 with permission from American Chemical Society, Copyright 2011.

leading to the development of high efficiency photocatalysts activated by visible light.[84] Graphene, in particular, serves as an excellent companion to titania, acting as a *highway* for electron mobility and enhancing the separation of the photogenerated excitons, thereby decreasing their recombination rate.[85] While the literature on visible light-driven coupled photocatalysts is extensive,[86] the simplest yet efficient approach appears to be constructing anatase-rutile heterojunction. Proof of this is the Evonik Degussa P25, one of the most effective commercial TiO_2-based photocatalysts, a mixture of nano-anatase and nano-rutile exhibiting visible light absorption properties.[87] In their seminal work, Scanlon and co-authors demonstrated that the electron affinity of anatase is higher than that of rutile, resulting in a photogenerated alignment of approximately 0.4 eV (Figure 4.4(b)). This alignment significantly reduces the effective bandgap of composite materials, promoting efficient electron–hole separation. Such alignment is a key-factor driving the enhanced photoactivity of anatase–rutile composite materials compared to their individual counterparts.[88]

The making of plasmonic photocatalysts is yet another promising way to harvest visible light (Figure 4.4(c)).[89] The phenomenon of surface plasmon resonance (SPR) is crucial in this context.[90] It occurs when the conducting electrons on the surface of metallic nanoparticles, exposed to specific electromagnetic radiation, collectively oscillate in a plasmon mode induced by the incident light's electric field.[91] The choice of metal is a fundamental factor influencing SPR, with metals like gold, silver, and copper being common due to their pronounced SPR in the visible spectrum. Particle size and nanoparticle shape are additional parameters governing the SPR wavelength.[92] Exploring the use of plasmonic structures to enhance photocatalytic performance under visible light is an ongoing challenge. Despite extensive efforts, the precise mechanism of plasmon-mediated photocatalysis remains a subject of debate. Various mechanisms have been proposed and tentatively supported, but a consensus is yet to be reached. One common observation is that plasmonic photocatalysts exhibit peak activities when the incident light wavelength aligns with the SPR wavelength.[93] Three prevalent, non-mutually exclusive mechanisms have been highlighted in the literature: (i) direct charge injection, (ii) near-field enhancement, and (iii) extended optical path length due to plasmon-induced scattering.[94]

4.3.2 Emerging Alternative Semiconductors

Titania is definitely the most studied semiconductor material for light-induced pollutants removal. However, as highlighted in Section 4.3.1.2, despite its significant potential, titania is not without drawbacks – specifically, its reliance on UV-light activation and susceptibility to fast charge carrier trapping, recombination, and transfer during the photocatalytic process. Therefore, the scientific community faces the challenge of identifying new alternative materials capable of overcoming these limitations.

4.3.2.1 Graphitic Carbon Nitride

In 2004, Novoselov and Geim, with the discovery of graphene, drove ultrathin two-dimensional (2D) materials to the forefront of nanomaterial research, attracting significant attention.[98] Their atomic-scale thickness imparts unique characteristics, such as a large surface-to-volume ratio, robust mechanical strength, and high optical transparency. Graphitic carbon nitride (g-C_3N_4), while being a long-established material in chemical literature, has recently undergone a renaissance, emerging as a highly effective 2D photocatalyst. Its unique electronic properties offer a well-suited bandgap and band position for driving pertinent oxidation and reduction processes. Moreover, its remarkable stability against temperature variations, acids, bases, and organic solvents renders it versatile for applications beyond photocatalysis.[99] With the added advantages of straightforward synthesis, cost-effectiveness, non-toxicity, a metal-free profile, and semiconducting properties featuring a relatively narrow band gap – ranging from approximately 2.4 eV (517 nm) to 2.8 eV (443 nm) depending on preparation methods and precursors – it certainly is a promising photocatalyst activated by visible light.[100] Moreover, it exhibits the attractive properties of impeding the recombination rate of photoinduced charge carriers and augmenting the quantity of adsorbed reactant species on the photocatalyst surface.[101] Given its capacity to efficiently adsorb H_2O, O_2, and VOCs on its surface, consequently generating the respective free radicals ($OH^\bullet$, $O_2^{\bullet -}$), g-C_3N_4 proves to be a versatile material for promoting air sustainability.[102]

4.3.2.2 MXenes

MXenes, pronounced "*maxenes*", are a rapidly expanding family of 2D materials, consisting of carbides and nitrides of transition metals. In a 2D MXene flake, there are $n+1$ (where $n=1$–3) layers of early transition metals (M, elements with a green background in Figure 4.5(b)) interleaved with n layers of carbon or nitrogen (X, elements with a grey background in Figure 4.5(b)).[103] The general formula is $M_{n+1}X_nT_x$, where T_x represents surface terminations such as O, OH, F, and/or Cl (elements with a dark yellow background in Figure 4.5(b)) bonded to the outer M layers.[104] The initial discovery of MXenes occurred in 2011 with the identification of $Ti_3C_2T_x$ by Gogotsy's group.[105] Since then, over 50 MXene compositions have been recorded, with numerous others explored through computational methods.[106] The eventual count is anticipated to exceed the current number significantly, positioning MXenes to potentially become the most extensive family of known 2D materials. Their distinctive characteristics, including a metal-like electrical conductivity of approximately 20 000 S cm^{-1}, and the strong interaction with the entire electromagnetic spectrum from UV to microwaves, make them highly valuable across a broad spectrum of applications.[107] Environmental sustainability is definitely amongst those applications. Indeed, MXenes can play an important role in the elimination of pollutants from environment.[108] A recent study by Sergiienko and colleagues provides evidence that $Ti_3C_2T_x$ MXene is a robust alternative material for removing VOCs in outdoor environments, particularly under near-infrared light.[109]

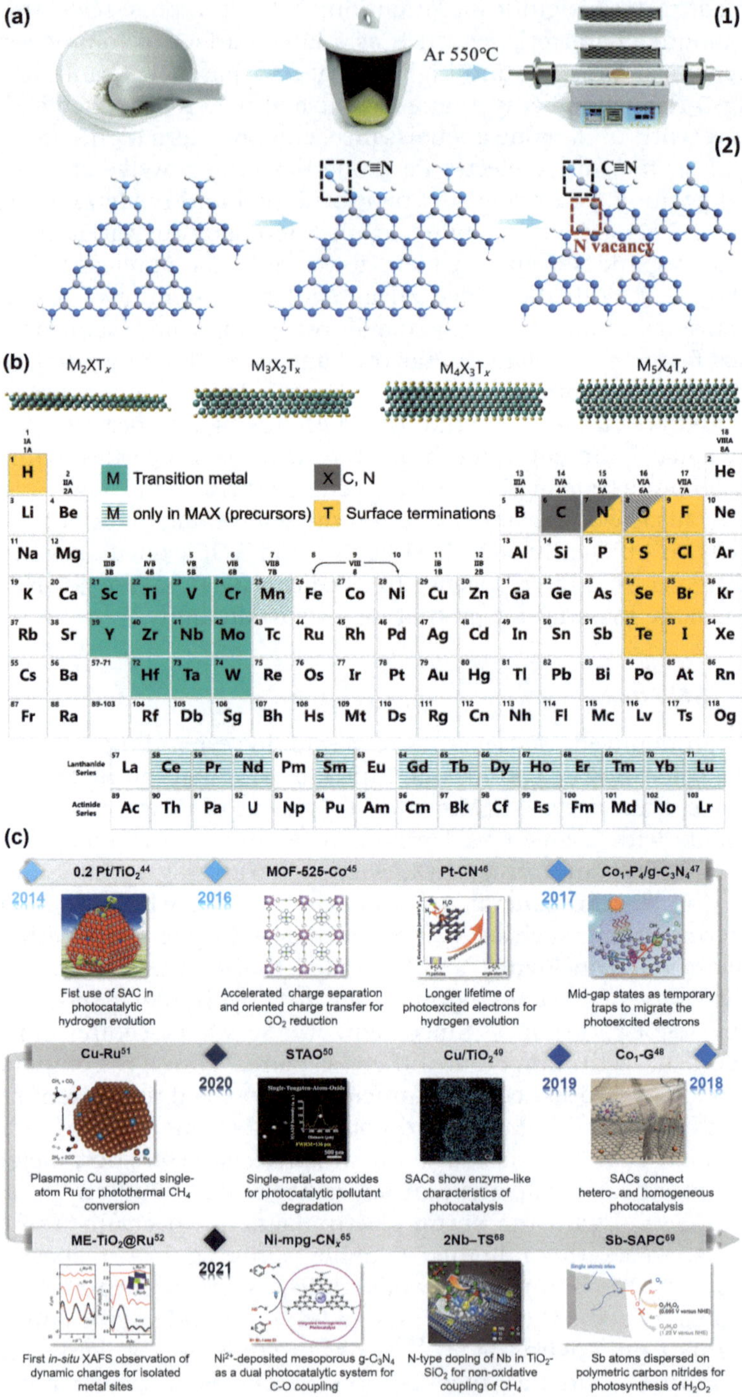
(a)
(1)
Ar 550°C
(2)
C≡N
C≡N
N vacancy
(b)
M2XTx
M3X2Tx
M4X3Tx
M5X4Tx
M Transition metal
M only in MAX (precursors)
X C, N
T Surface terminations
Lanthanide Series
Actinide Series
(c)
0.2 Pt/TiO2[44]
MOF-525-Co[45]
Pt-CN[46]
Co1-P4/g-C3N4[47]
2014
2016
2017
Fist use of SAC in photocatalytic hydrogen evolution
Accelerated charge separation and oriented charge transfer for CO2 reduction
Longer lifetime of photoexcited electrons for hydrogen evolution
Mid-gap states as temporary traps to migrate the photoexcited electrons
Cu-Ru[51]
STAO[50]
Cu/TiO2[49]
Co1-G[48]
2020
2019
2018
Plasmonic Cu supported single-atom Ru for photothermal CH4 conversion
Single-metal-atom oxides for photocatalytic pollutant degradation
SACs show enzyme-like characteristics of photocatalysis
SACs connect hetero- and homogeneous photocatalysis
ME-TiO2@Ru[52]
Ni-mpg-CNx[65]
2Nb–TS[68]
Sb-SAPC[69]
2021
First in-situ XAFS observation of dynamic changes for isolated metal sites
Ni2+-deposited mesoporous g-C3N4 as a dual photocatalytic system for C-O coupling
N-type doping of Nb in TiO2-SiO2 for non-oxidative coupling of CH4
Sb atoms dispersed on polymetric carbon nitrides for photosynthesis of H2O2

4.3.2.3 Black Phosphorus

Staying in the realm of 2D nanomaterials, black phosphorus, one of the four phosphorous polymorphs, was discovered in 1914. However, it regained widespread attention in 2014 after its exfoliation into few-layered form (phosphorene), likewise graphene, by using the scotch-tape microcleavage method owing to the van der Waals force among layers.[110] It exhibits a thickness-dependent bandgap energy, ranging from 0.3 eV for the bulk form, to 2.0 eV for monolayer phosphorene. This suggests that black phosphorus possesses a broad light absorption capability, spanning from UV light to visible light and even near-infrared light.[111] These characteristics grant black phosphorus with fascinating potential for various light-induced air remediation applications.[112]

4.3.2.4 Bismuth Oxyhalides

Bismuth oxyhalides, BiOX (with X = Cl, Br, I), crystallising in the tetragonal matlockite structure, are characterised by a layered arrangement consisting of $[Bi_2O_2]^{2+}$ slabs interleaved with double halogen atom slabs along the [001] direction. An interesting characteristic of this material class is that, as the atomic number of the halogen increases, the bandgap narrows – from 3.2 eV in BiOCl, to around 2.7 eV in BiOBr, and further to approximately 1.7 eV in BiOI. This progression implies light absorption across the UV and visible range.[113] When reduced to the nano-scale, these BiOX materials inherently exhibit a 2D nature due to their layered structure held together by van der Waals forces. Additionally, the resulting intra-electric field between $[Bi_2O_2]^{2+}$ layers and halogen atom layers can accelerate the transfer of photo-induced carriers, boosting the photocatalytic activity of BiOX.[114] Bismuth, in photocatalytic reactions, serves various roles such as electron/hole trapping, charge transfer mediation, or coordination of oxygen vacancies. In particular, BiOCl, BiOBr, and BiOI have found extensive

Figure 4.5 (a) (1) Schematic illustration of the preparation process of g-C_3N_4. (2) Engineering defects in the g-C_3N_4 crystal structure to improve light-absorption. Reproduced from ref. 121 with permission from the Royal Society of Chemistry. (b) Periodic table showing elements used to build MXenes. The green background shows metals (M elements), grey background—nonmetals (X elements) and dark yellow background—terminating elements (T). The striped background shows rare-earth elements that can be components of in-plane ordered MAX structures. The elements with diagonal stripes are only reported in MXene precursors (MAX phases). Oxygen is shown with grey diagonal stripes because of the recently discovered oxycarbide MXene. Reproduced from ref. 122 with permission from Springer Nature, Copyright 2022. (c) Timeline of the achievements and features for single atom photocatalysis. Reproduced from ref. 119 with permission from Elsevier, Copyright 2022.

applications as photocatalysts in processes such as CO_2 reduction, and environmental remediation in recent years.[115]

4.3.2.5 *Single Atom Catalysts*

An (extended) definition of a single-atom catalyst refers to an assembly comprising an atom of any element (metal, metalloid, non-metal, and halogen) that is spatially isolated from atoms of the same chemical identity (*e.g.*, adsorbed, embedded, immobilised, grafted, or confined), and directly bound to a solid carrier.[116] Formed by decorating individual metal atoms or mononuclear metal complexes onto suitable supports,[117] these catalysts have captured significant and sustained interest in the fields of chemistry and materials science, making a great breakthrough over the past decade.[118] The incorporation of isolated metal atoms can influence the band structure, amplifying the light-harvesting capacity of semiconductor supports. Additionally, these metal atoms act as electron pumps, facilitating the transfer of photoexcited electrons and thereby improving surface charge separation and transfer within single-atom photocatalytic systems.[119] This significantly enhances photocatalytic oxidation reactions, including the degradation of pollutants and the oxidation of organic airborne compounds. This emerging class of materials has marked a significant breakthrough in the field of environmental catalysis in the 21st century. It offers promising solutions, including the catalytic removal of emitted pollutants (*e.g.*, NO_x and VOC removal, CO oxidation), conversion of CO_2 into valuable chemicals, and environmentally friendly synthesis processes.[120]

4.4 Colloidal Materials

The 2023 Nobel Prize in Chemistry was awarded to MG Bawendi, LE Brus, and AI Ekimov for their ground-breaking work on quantum dots[123] – semiconductor nanocrystals with tuneable properties determined by their physical size.[124] This prestigious recognition, started with a bump,[125] has sparked renewed interest within the global colloidal chemistry community.[126] Colloidal synthesis is a widely employed technique for producing metal nanoparticles with precisely defined shapes and compositions. This method typically involves reducing a metal precursor using a chemical reducing agent, providing nanoparticle growth in the presence of a capping agent. The capping agent plays a pivotal role, ensuring that the developing particles stay dispersed in the solvent, forming a stable colloid.[127] Wet-chemistry methods are particularly recognised for their ability to systematically craft a diverse range of semiconductors, metals, and oxide nanocrystals, precisely tailored with sub-nanometre level accuracy across a wide spectrum of dimensional-morphological regimes. This is achieved through the meticulous control of thermodynamic parameters and growth kinetics in liquid environments, assisted by carefully chosen solvents, ligands, surfactants, or catalyst additives.[128] Despite the limitations of around 40 years ago, when it was challenging to systematically control the electronic properties of man-made colloids in a size-dependent manner, colloidal semiconductor particles are set to assume a crucial role in advancing nanoscience and nanotechnology in the future.[129] The methods have significantly advanced, making it now possible to synthesise, for instance, complex structures, such as quaternary chalcohalide semiconductor nanocrystals.[130]

4.4.1 Colloidal Materials for Air Sustainability

4.4.1.1 Quantum Dots

Colloidal quantum dots are ideal materials to develop highly efficient photocatalysts. Through careful control of compositions, structures, morphologies, and surface properties, they have shown significant potential in recent years.[131] With a profound comprehension of quantum confinement, one can easily tune the bandgap of quantum dots, spanning a broad spectrum of the solar spectrum.[132] Indeed, all of these factors are recognised for their substantial positive influence on the light-induced activity of semiconductor photocatalysts. A successful approach for creating an effective broad-band photocatalyst, capable of harvesting UV-A and visible light for NO_x oxidation, involves constructing a composite material architecture comprising surface-functionalised quantum dots decorated on titania. By optimising the loading of quantum dots, it has been achieved not only to convert NO, but also to exhibit high selectivity towards the storage of NO_2.[133] The utilisation of light-induced CO_2 conversion is yet another attractive strategy for easing the environmental impact of greenhouse gases. Evidence demonstrates that the incorporation of transition-metal sites into quantum dots can effectively address issues related to their low selectivity and stability in CO_2 reduction. This modification results in a system capable of facilitating highly selective photocatalytic reactions between CO_2 and H_2O, achieving 100% selectivity towards the production of CO and CH_4.[134] Colloidal nanostructures composed of *multinary* I–III–VI_2 and I_2–II–IV–VI_4 semiconductors have also been shown to be innovative nanomaterials capable of effectively acting as light-responsive catalysts for the mineralisation of several (airborne) pollutants. These inorganic compounds consist of environmentally benign elemental components, exhibit excellent light-harvesting properties, and possess band gap energies that are well-suited for solar photon absorption. Moreover, the band structures of these materials can be conveniently modified through alloying to boost their ability to harvest visible photons. These factors collectively contribute to their remarkable performance as photocatalysts in environmental remediation.[135]

4.4.1.2 Semiconductor–Metal Hybrid Nanostructures

An emerging category of colloidal materials that swiftly gained recognition and progressed rapidly is the semiconductor–metal hybrid nanoparticles.

They combine two distinct materials within a single nanosystem, unveiling distinctive synergistic effects arising from the nanoscale combination of semiconductor and metal. This results in an outcome that surpasses the mere sum of its individual components.[136] A notable feature of these nanoparticles is the light-induced charge-separation effect occurring across the semiconductor–metal nanojunction, forming the foundation for their applications in photocatalysis.[137] Literature reports indicate that these colloidal semiconductor–metal materials enhance reaction selectivity toward CO_2. Upon excitation, they overcome the high energy barrier associated with the initial step in CO_2 reduction.[138] Amongst the metals employed to engineer those semiconductor–metal hybrid nanoparticles, copper proves to be a fitting complement in various light-induced applications,[139–142] particularly for enhancing visible light absorption.[84] Colloidal Cu–TiO_2 nanocrystals have been demonstrated to effectively eliminate VOCs under simulated indoor ambient lighting conditions. This is attributed to the copper's contribution, which enhances the generation of reactive oxygen species when exposed to visible light, thus aiding the oxidation of VOCs.[143]

4.4.1.3 Lead-free Perovskite Quantum Dots

The family of metal halide perovskites has ignited significant research interest in light-induced applications, swiftly replacing their oxide-based counterparts (Figure 4.6(a)). This shift is attributed to their intriguing photophysical and structural properties, including a tuneable and suitable bandgap, excellent light-harvesting activity, high absorptivity coefficient, and a capacity for defect-tolerated optical performance.[144] Lead-based perovskite materials, in particular, have played a crucial role in advancing optoelectronic devices.[145] When in the form of colloidal nanocrystals (Figure 4.6(b)), they have attracted attention due to their unique optical versatility, high photoluminescence quantum yields and facile synthesis. For instance, the ligand assisted hot-injection synthesis method has been employed to synthesise the first $CsPbX_3$ nanocrystals at temperatures lower than 200 °C.[146] As lead halide perovskite nanocrystals possess a soft and predominantly ionic structure, the existing hot-injection synthesis methods, typically conducted at temperatures of 100–350 °C, prove ideal for producing nanocrystals of moderately ionic compounds. Moreover, their optical and electronic properties exhibit a high tolerance for structural defects and surface states,[147] making them well-suited for applications in the removal of airborne pollutants under light irradiation.

However, apprehensions regarding the safety of lead-based halides emerged. The potential leakage of Pb^{2+} from these perovskite materials raises environmental concerns due to the toxic nature of lead, posing risks to both life and the ecosystem.[148] Research directions therefore drifted to the synthesis of lead-free metal halide perovskites. In Figure 4.7(a), a selection of elements from the periodic table currently under consideration as substitutes for Pb^{2+} is presented. Lead-free halide perovskites, suitable for

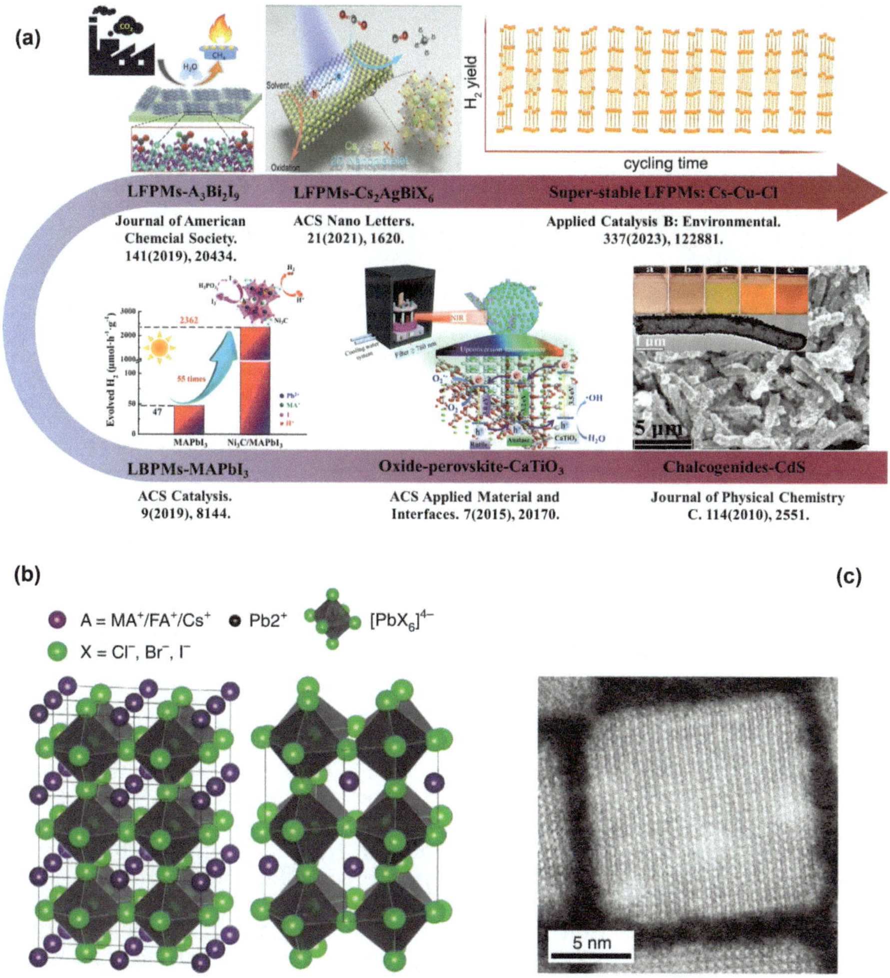

Figure 4.6 (a) Timeline evolution history for various photocatalytic applications using different photocatalysts, including the perovskite family (oxides, LBPMs, and LFPMs), and chalcogenide materials. Reproduced from ref. 144 with permission from Elsevier, Copyright 2023. Colloidal lead halide perovskite nanocrystals. (b) The $APbX_3$ perovskite structure with 3D-corner-sharing octahedra. Two typical structures are shown: cubic ($MAPbX_3$, $FAPbX_3$; two unit cells are shown) on the left, and orthorhombic ($CsPbX_3$) on the right. (c) High-angle annular dark-field scanning transmission electron micrograph (HAADF-STEM) of single, cube-shaped $CsPbBr_3$ nanocrystals, with a 15 nm edge length. Reproduced from ref. 147 with permission from Springer Nature, Copyright 2018.

photocatalysis, can be synthesised through various solution methods at low temperatures (Figure 4.7(b)). Common techniques include, but are not limited to, conventional hydrothermal (or solvothermal) methods, cooling crystallisation, antisolvent precipitation, and hot-injection methods.[149] We

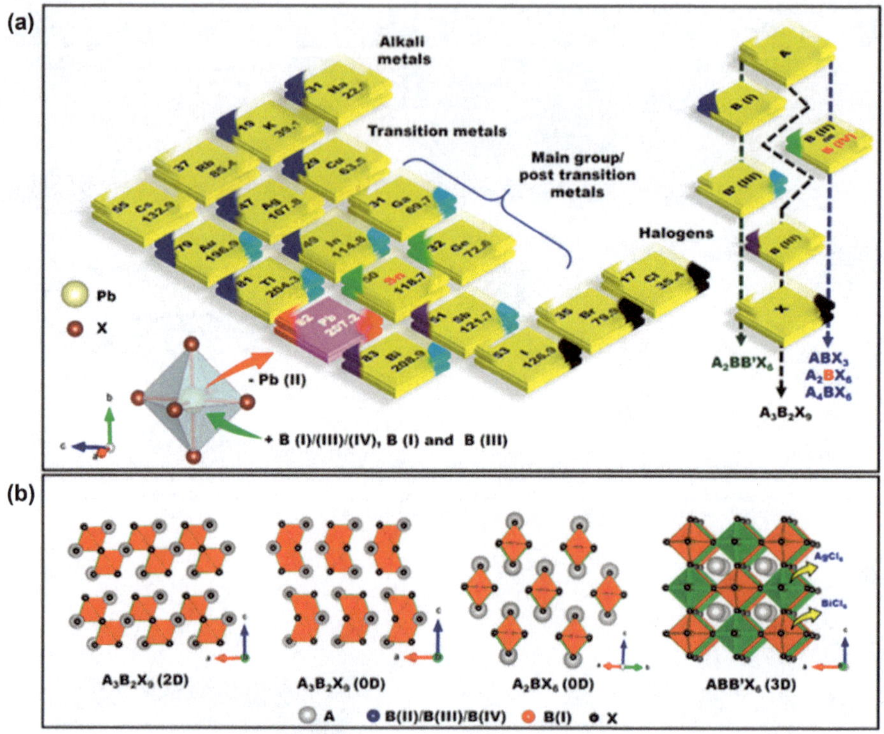

Figure 4.7 (a) Combinations of elements from part of the periodic table forming possible Pb-free metal-halide perovskites and their derivative structures. The triangular colour code specifies the site occupancy as per the compositional formula presented in the extreme right. (b) Structural presentation of Pb-free metal-halide perovskites and perovskite derivatives showing a network of octahedra in 3D, 2D, and 0D directions. Reproduced from ref. 150, https://doi.org/10.1021/acsnano.0c08903, under the terms of the CC BY 4.0 license, https://creativecommons.org/licenses/by/4.0/.

now present below examples of synthetically prepared colloidal lead-free perovskites nanocrystals with photocatalytic functionalities.

Colloidal nanocrystals of $CsSnX_3$ (with X = Cl, Br, I) have been synthesised by the hot-injection method. $CsSnCl_3$ nanocrystals exhibit a perovskite crystal structure with a cubic space group ($Pm\bar{3}m$). In contrast, $CsSnBr_3$ and $CsSnI_3$ nanocrystals have a lower symmetry, as they crystallise in the *Pnma* orthorhombic space group.[151] Bi- and Sb-halide perovskite-derivative nanocrystals have been likewise synthesised, however, charge transport is limited due to reduced structural dimensionality.[152] Colloidal, stable $Cs_2AgBiBr_6$ has also been successfully produced using a hot-injection method.[153] All these nanocrystals exhibit potential applications in photocatalytic CO_2 reduction into sustainable solar fuels (CO, CH_4), and the degradation of organic pollutants. Despite the promising advancements, the field is still in its early stages, offering ample opportunities for designing target reaction systems, improving stability and efficiency, and mitigating the toxicity of halide systems for solar-to-chemical energy conversion.[150] To address these challenges, it is imperative to develop highly stable lead-free halide perovskites with enhanced catalytic performance. Future research avenues may include, but are not limited to, exploring synthetic strategies, incorporating heterojunction engineering, and conducting in-depth investigations into photocatalytic and photoelectrochemical reaction mechanisms. To summarise, emphasis should be placed on improving fundamental properties, enhancing material stability, elucidating photophysical behaviours, and verifying interfacial dynamics.[154]

4.5 Concluding Remarks

In this chapter, we have explored the impact of indoor and outdoor air pollution on human beings. While physical removal methods have been discussed, it is evident that the complete mineralisation of pollutants offers a more viable solution. Light-induced removal technologies present a sustainable means of addressing airborne pollutants. Semiconductor photocatalysis, a simplified version of nature's photosynthesis, provides an environmentally friendly approach to eliminating airborne pollutants using solar energy. In this context, TiO_2 undeniably plays a pivotal role, albeit with its inherent drawbacks. We have examined the challenges faced by the scientific community in overcoming these drawbacks, ultimately leading to the engineering of alternative semiconductor photocatalysts. In this scenario, colloidal materials exhibit various optical and chemical properties which position them as potential alternatives to TiO_2 in numerous photocatalytic reactions. Despite possessing many desirable characteristics of both heterogeneous and homogeneous catalysts, colloidal photocatalysts come with their unique set of challenges. These challenges can be addressed through the careful selection of semiconductor core materials and surface functionalisation.[155]

Dedication

We wish to dedicate this chapter to the memory of Prof. Dionysios (Dion) D. Dionysiou, who unexpectedly and untimely passed away. This touched me very much. I had the privilege of meeting him once at a conference in Palermo, where his kindness and gentleness left a lasting impression. A true giant in science, his legacy will endure.

References

1. Editorial, Clean air for a sustainable world, *Nat. Commun.*, 2021, **12**, 5824.
2. WHO, Air pollution, https://www.who.int/health-topics/air-pollution.
3. Sustainable development goals & air pollution, https://www.who.int/teams/environment-climate-change-and-health/air-quality-and-health/policy-progress/sustainable-development-goals-air-pollution.

4. L. He, L. Rosa, D. B. Lobell, Y. Wang, Y. Yin, R. Doughty, Y. Yao, J. A. Berry and C. Frankenberg, *Proc. Natl. Acad. Sci. U. S. A.*, 2023, **120**, e2306507120.
5. W. H. O. R. O. for Europe, Air quality guidelines for Europe.
6. Editorial, So long sulphur, *Nat. Chem.*, 2009, **1**, 333.
7. WHO global air quality guidelines, https://www.who.int/publications-detail-redirect/9789240034228.
8. Household air pollution, https://www.who.int/news-room/fact-sheets/detail/household-air-pollution-and-health.
9. J. C. Nwanaji-Enwerem, J. G. Allen and P. I. Beamer, *J. Exposure Sci. Environ. Epidemiol.*, 2020, **30**, 773–775.
10. E. B. Brandt, A. F. Beck and T. B. Mersha, *J. Allergy Clin. Immunol.*, 2020, **146**, 61–63.
11. G. Liu, M. Xiao, X. Zhang, C. Gal, X. Chen, L. Liu, S. Pan, J. Wu, L. Tang and D. Clements-Croome, *Sustainable Cities Soc.*, 2017, **32**, 375–396.
12. W. Zhang, G. Li, H. Yin, K. Zhao, H. Zhao and T. An, *Environ. Sci.: Nano*, 2022, **9**, 81–104.
13. C. D. Koolen and G. Rothenberg, Air Pollution in Europe, *ChemSusChem*, 2019, **12**, 164–172.
14. *Photocatalysis: Fundamentals and Perspectives*, ed. J. Schneider, Royal Society of Chemistry, Cambridge, UK, 2016.
15. F. Arsac, D. Bianchi, J. M. Chovelon, P. Conchon, C. Ferronato, A. Lair and M. Sleiman, *Mater. Sci. Eng., C*, 2008, **28**, 722–725.
16. F. He, W. Jeon and W. Choi, *Nat. Commun.*, 2021, **12**, 2528.
17. A. Fujishima, K. Hashimoto and T. Watanabe, *TiO_2 Photocatalysis: Fundamentals and Applications*, BKC, 1999.
18. A. Fujishima and K. Honda, *Nature*, 1972, **238**, 37–38.
19. B. O'Regan and M. Grätzel, *Nature*, 1991, **353**, 737–740.
20. N. Pinna and M. Niederberger, *Angew. Chem., Int. Ed.*, 2008, **47**, 5292–5304.
21. M. Niederberger, M. H. Bartl and G. D. Stucky, *Chem. Mater.*, 2002, **14**, 4364–4370.
22. M. Karmaoui, D. M. Tobaldi, A. S. Skapin, R. C. Pullar, M. P. Seabra, J. A. Labrincha and V. S. Amaral, *RSC Adv.*, 2014, **4**, 46762–46770.
23. G. Garnweitner, H. O. Ghareeb and C. Grote, *Colloids Surf., A*, 2010, **372**, 41–47.
24. M. Thommes, K. Kaneko, A. V. Neimark, J. P. Olivier, F. Rodriguez-Reinoso, J. Rouquerol and K. S. W. Sing, *Pure Appl. Chem.*, 2015, **87**, 1051–1069.
25. P. A. M. D. Wright, *Microporous Framework Solids: Volume 4*, Royal Society of Chemistry, Cambridge, UK, 2007.
26. S. Choi, J. H. Drese and C. W. Jones, *ChemSusChem*, 2009, **2**, 796–854.
27. S. Sircar, T. C. Golden and M. B. Rao, *Carbon*, 1996, **34**, 1–12.
28. X. Yue, N. L. Ma, C. Sonne, R. Guan, S. S. Lam, Q. Van Le, X. Chen, Y. Yang, H. Gu, J. Rinklebe and W. Peng, *J. Hazard. Mater.*, 2021, **405**, 124138.

29. P. Davini, *Carbon*, 2002, **40**, 1973–1979.
30. J. A. Maciá-Agulló, B. C. Moore, D. Cazorla-Amorós and A. Linares-Solano, *Carbon*, 2004, **42**, 1367–1370.
31. J. A. Rabo and M. W. Schoonover, *Appl. Catal., A*, 2001, **222**, 261–275.
32. D. Breck, *Zeolite Molecular Sieves: Structure, Chemistry, and Use*, Krieger Pub Co, Malabar, Fla, 1984.
33. Atlas of Prospective Zeolite Structures, http://www.hypotheticalzeolites.net.
34. E. Pérez-Botella, S. Valencia and F. Rey, *Chem. Rev.*, 2022, **122**, 17647–17695.
35. R. Gonzalez-Olmos, A. Gutierrez-Ortega, J. Sempere and R. Nomen, *J. CO_2 Util.*, 2022, **55**, 101791.
36. Y. Jin, H. Liu, M. Feng, Q. Ma and B. Wang, *Adv. Funct. Mater.*, 2024, 2304773.
37. N. C. Burtch, H. Jasuja and K. S. Walton, *Chem. Rev.*, 2014, **114**, 10575–10612.
38. N. Huang, P. Wang and D. Jiang, *Nat. Rev. Mater.*, 2016, **1**, 1–19.
39. J. Wang and S. Zhuang, *Coord. Chem. Rev.*, 2019, **400**, 213046.
40. X. Zhang, B. Gao, A. E. Creamer, C. Cao and Y. Li, *J. Hazard. Mater.*, 2017, **338**, 102–123.
41. W. Zheng, J. Hu, Z. Wang, J. Li, Z. Fu, H. Li, J. Jurasz, S. K. Chou and J. Yan, *Adv. Appl. Energy*, 2021, **3**, 100040.
42. J. L. Jimenez, L. C. Marr, K. Randall, E. T. Ewing, Z. Tufekci, T. Greenhalgh, R. Tellier, J. W. Tang, Y. Li, L. Morawska, J. Mesiano-Crookston, D. Fisman, O. Hegarty, S. J. Dancer, P. M. Bluyssen, G. Buonanno, M. G. L. C. Loomans, W. P. Bahnfleth, M. Yao, C. Sekhar, P. Wargocki, A. K. Melikov and K. A. Prather, *Indoor Air*, 2022, **32**, e13070.
43. Y. Lin, J. Wang, W. Yang, L. Tian and C. Candido, *Energy Built Environ.*, 2023, S2666123323000752.
44. N. Serpone, A. V. Emeline, S. Horikoshi, V. N. Kuznetsov and V. K. Ryabchuk, *Photochem. Photobiol. Sci.*, 2012, **11**, 1121–1150.
45. G. Palmisano, V. Augugliaro, M. Pagliaro and L. Palmisano, *Chem. Commun.*, 2007, 3425–3437.
46. D. Ravelli, D. Dondi, M. Fagnoni and A. Albini, *Chem. Soc. Rev.*, 2009, **38**, 1999–2011.
47. X. Sun, S. Jiang, H. Huang, H. Li, B. Jia and T. Ma, *Angew. Chem., Int. Ed.*, 2022, **61**, e202204880.
48. X. Ruan, S. Li, C. Huang, W. Zheng, X. Cui and S. K. Ravi, *Adv. Mater.*, 2024, 2305285.
49. D. M. Tobaldi, R. C. Pullar, A. F. Gualtieri, M. P. Seabra and J. A. Labrincha, *Acta Mater.*, 2013, **61**, 5571–5585.
50. K. Momma and F. Izumi, *J. Appl. Crystallogr.*, 2008, **41**, 653–658.
51. C. J. Howard, T. M. Sabine and F. Dickson, *Acta Crystallogr., Sect. B: Struct. Sci.*, 1991, **47**, 462–468.
52. E. P. Meagher and G. A. Lager, *Can. Mineral.*, 1979, **17**, 77–85.
53. C. Wang and M. N. Ghazzal, *Energy Adv.*, 2023, **2**, 965–979.

54. A. Fujishima, T. N. Rao and D. A. Tryk, *J. Photochem. Photobiol., C*, 2000, **1**, 1–21.
55. R. Qian, H. Zong, J. Schneider, G. Zhou, T. Zhao, Y. Li, J. Yang, D. W. Bahnemann and J. H. Pan, *Catal. Today*, 2019, **335**, 78–90.
56. T. Tachikawa, M. Fujitsuka and T. Majima, *J. Phys. Chem. C*, 2007, **111**, 5259–5275.
57. J. Lasek, Y.-H. Yu and J. C. S. Wu, *J. Photochem. Photobiol., C*, 2013, **14**, 29–52.
58. J. S. Dalton, P. A. Janes, N. G. Jones, J. A. Nicholson, K. R. Hallam and G. C. Allen, *Environ. Pollut.*, 2002, **120**, 415–422.
59. S. Gligorovski, *J. Photochem. Photobiol., A*, 2016, **314**, 1–5.
60. S. Gligorovski and J. P. D. Abbatt, *Science*, 2018, **359**, 632–633.
61. D. Yang, Q. Liu, S. Wang, M. Bozorg, J. Liu, P. Nair, P. Balaguer, D. Song, H. Krause, B. Ouazia, J. P. D. Abbatt and H. Peng, *Sci. Adv.*, 2022, **8**, eabq7023.
62. K. Vikrant, C. M. Park, K.-H. Kim, S. Kumar and E.-C. Jeon, *J. Photochem. Photobiol., C*, 2019, **41**, 100316.
63. M. E. Jenkin, J. C. Young and A. R. Rickard, *Atmos. Chem. Phys.*, 2015, **15**, 11433–11459.
64. V. Héquet, C. Raillard, O. Debono, F. Thévenet, N. Locoge and L. Le Coq, *Appl. Catal., B*, 2018, **226**, 473–486.
65. Z. Xu, Y. Ren, X. Deng, M. Xu, W. Chai, X. Qian and Z. Bian, *Adv. Energy Sustain. Res.*, 2022, **3**, 2200105.
66. Z. Shayegan, C.-S. Lee and F. Haghighat, *Chem. Eng. J.*, 2018, **334**, 2408–2439.
67. M. Krichevskaya, S. Preis, A. Moiseev, N. Pronina and J. Deubener, *Catal. Today*, 2017, **280**, 93–98.
68. L. Zhong and F. Haghighat, *Build. Environ.*, 2015, **91**, 191–203.
69. W. Zou, B. Gao, Y. S. Ok and L. Dong, *Chemosphere*, 2019, **218**, 845–859.
70. J. Zhang, P. Zhou, J. Liu and J. Yu, *Phys. Chem. Chem. Phys.*, 2014, **16**, 20382–20386.
71. Z. Şen, *Solar Energy Fundamentals and Modeling Techniques: Atmosphere, Environment, Climate Change, and Renewable Energy*, Springer, London, 2008.
72. M. Pelaez, N. T. Nolan, S. C. Pillai, M. K. Seery, P. Falaras, A. G. Kontos, P. S. M. Dunlop, J. W. J. Hamilton, J. A. Byrne, K. O'Shea, M. H. Entezari and D. D. Dionysiou, *Appl. Catal., B*, 2012, **125**, 331–349.
73. M. Anpo, *Pure Appl. Chem.*, 2000, **72**, 1787–1792.
74. J.-M. Herrmann, *New J. Chem.*, 2012, **36**, 883–890.
75. R. Asahi, T. Morikawa, T. Ohwaki, K. Aoki and Y. Taga, *Science*, 2001, **293**, 269–271.
76. R. Asahi, T. Morikawa, H. Irie and T. Ohwaki, *Chem. Rev.*, 2014, **114**, 9824–9852.
77. D. M. Tobaldi, R. C. Pullar, A. F. Gualtieri, G. Otero-Irurueta, M. K. Singh, M. P. Seabra and J. A. Labrincha, *J. Solid State Chem.*, 2015, **231**, 87–100.

78. J.-M. Herrmann, *J. Photochem. Photobiol., A*, 2010, **216**, 85–93.
79. X. Chen, L. Liu and F. Huang, *Chem. Soc. Rev.*, 2015, **44**, 1861–1885.
80. X. Chen, L. Liu, P. Y. Yu and S. S. Mao, *Science*, 2011, **331**, 746–750.
81. A. Naldoni, M. Altomare, G. Zoppellaro, N. Liu, Š Kment, R. Zbořil and P. Schmuki, *ACS Catal.*, 2019, **9**, 345–364.
82. A. Naldoni, M. Allieta, S. Santangelo, M. Marelli, F. Fabbri, S. Cappelli, C. L. Bianchi, R. Psaro and V. Dal Santo, *J. Am. Chem. Soc.*, 2012, **134**, 7600–7603.
83. D. M. Tobaldi, D. Dvoranová, L. Lajaunie, N. Rozman, B. Figueiredo, M. P. Seabra, A. S. Škapin, J. J. Calvino, V. Brezová and J. A. Labrincha, *Chem. Eng. J.*, 2021, **405**, 126651.
84. D. M. Tobaldi, D. Dvoranová, L. Lajaunie, K. Czikhardtová, B. Figueiredo, J. J. Calvino, M. P. Seabra and J. A. Labrincha, *Mater. Today Energy*, 2022, **25**, 100974.
85. Q. Huang, S. Tian, D. Zeng, X. Wang, W. Song, Y. Li, W. Xiao and C. Xie, *ACS Catal.*, 2013, **3**, 1477–1485.
86. S. Wang, W. Liao, H. Su, S. Pang, C. Yang, Y. Fu and Y. Zhang, *Energy Fuels*, 2023, **37**, 1633–1656.
87. D. M. Tobaldi, L. Lajaunie, N. Rozman, A. P. F. Caetano, M. P. Seabra, A. Sever Škapin, R. Arenal and J. A. Labrincha, *J. Photochem. Photobiol., A*, 2019, **382**, 111940.
88. D. O. Scanlon, C. W. Dunnill, J. Buckeridge, S. A. Shevlin, A. J. Logsdail, S. M. Woodley, C. R. A. Catlow, M. J. Powell, R. G. Palgrave, I. P. Parkin, G. W. Watson, T. W. Keal, P. Sherwood, A. Walsh and A. A. Sokol, *Nat. Mater.*, 2013, **12**, 798–801.
89. P. K. Jain, *J. Phys. Chem. C*, 2019, **123**, 24347–24351.
90. X. Zhang, Y. L. Chen, R.-S. Liu and D. P. Tsai, *Rep. Prog. Phys.*, 2013, **76**, 046401.
91. P. Wang, B. Huang, Y. Dai and M.-H. Whangbo, *Phys. Chem. Chem. Phys.*, 2012, **14**, 9813.
92. Y. Xia and N. J. Halas, *MRS Bull.*, 2011, **30**, 338–348.
93. S. W. Verbruggen, *J. Photochem. Photobiol., C*, 2015, **24**, 64–82.
94. S. Linic, P. Christopher and D. B. Ingram, *Nat. Mater.*, 2011, **10**, 911–921.
95. M. M. Khan, S. A. Ansari, D. Pradhan, M. O. Ansari, D. H. Han, J. Lee and M. H. Cho, *J. Mater. Chem. A*, 2013, **2**, 637–644.
96. R. Katal, M. Salehi, M. H. Davood Abadi Farahani, S. Masudy-Panah, S. L. Ong and J. Hu, *ACS Appl. Mater. Interfaces*, 2018, **10**, 35316–35326.
97. Z. Liu, W. Hou, P. Pavaskar, M. Aykol and S. B. Cronin, *Nano Lett.*, 2011, **11**, 1111–1116.
98. K. S. Novoselov, A. K. Geim, I. V. Morozov, D. Jiang, Y. Zhang, S. V. Dubonos, I. V. Grigorieva and A. A. Firsov, *Science*, 2004, **306**, 666–669.
99. J. Liu, H. Wang and M. Antonietti, *Chem. Soc. Rev.*, 2016, **45**, 2308–2326.
100. W. Touati, M. Karmaoui, A. Bekka, M. F. Edelmannová, C. Furgeaud, A. Chakib, I. K. Allah, B. Figueiredo, J. A. Labrincha, R. Arenal, K. Koci and D. M. Tobaldi, *New J. Chem.*, 2022, **46**, 20679–20690.

101. W.-J. Ong, L.-L. Tan, Y. H. Ng, S.-T. Yong and S.-P. Chai, *Chem. Rev.*, 2016, **116**, 7159–7329.
102. B. Jing, Z. Ao, W. Zhao, Y. Xu, Z. Chen and T. An, *J. Mater. Chem. A*, 2020, **8**, 20363–20372.
103. Y. Gogotsi and B. Anasori, *ACS Nano*, 2019, **13**, 8491–8494.
104. B. Anasori, M. R. Lukatskaya and Y. Gogotsi, *Nat. Rev. Mater.*, 2017, **2**, 1–17.
105. M. Naguib, M. Kurtoglu, V. Presser, J. Lu, J. Niu, M. Heon, L. Hultman, Y. Gogotsi and M. W. Barsoum, *Adv. Mater.*, 2011, **23**, 4248–4253.
106. Y. Gogotsi, *Chem. Mater.*, 2023, **35**, 8767–8770.
107. M. Naguib, M. W. Barsoum and Y. Gogotsi, *Adv. Mater.*, 2021, **33**, 2103393.
108. Y. Sun and Y. Li, *Chemosphere*, 2021, **271**, 129578.
109. S. A. Sergiienko, D. M. Tobaldi, L. Lajaunie, D. V. Lopes, G. Constantinescu, A. L. Shaula, N. D. Shcherban, V. I. Shkepu, J. J. Calvino, J. R. Frade, J. A. Labrincha and A. V. Kovalevsky, *J. Mater. Chem. C*, 2022, **10**, 626–639.
110. H. Liu, A. T. Neal, Z. Zhu, Z. Luo, X. Xu, D. Tománek and P. D. Ye, *ACS Nano*, 2014, **8**, 4033–4041.
111. V. Tran, R. Soklaski, Y. Liang and L. Yang, *Phys. Rev. B: Condens. Matter Mater. Phys.*, 2014, **89**, 235319.
112. B. Li, C. Lai, G. Zeng, D. Huang, L. Qin, M. Zhang, M. Cheng, X. Liu, H. Yi, C. Zhou, F. Huang, S. Liu and Y. Fu, *Small*, 2019, **15**, 1804565.
113. H. Cheng, B. Huang and Y. Dai, *Nanoscale*, 2014, **6**, 2009.
114. J. Jiang, K. Zhao, X. Xiao and L. Zhang, *J. Am. Chem. Soc.*, 2012, **134**, 4473–4476.
115. J. He, X. Hu, Z. Liu, W. Chen and G. Longo, *Adv. Funct. Mater.*, 2023, **33**, 2306075.
116. S. K. Kaiser, Z. Chen, D. Faust Akl, S. Mitchell and J. Pérez-Ramírez, *Chem. Rev.*, 2020, **120**, 11703–11809.
117. L. Li, X. Chang, X. Lin, Z.-J. Zhao and J. Gong, *Chem. Soc. Rev.*, 2020, **49**, 8156–8178.
118. J. Zhuang and D. Wang, *Mater. Today Catal.*, 2023, **2**, 100009.
119. Z.-H. Xue, D. Luan, H. Zhang and X. W. (David) Lou, *Joule*, 2022, **6**, 92–133.
120. J. Wu, H. Shi, K. Li and X. Guo, *Curr. Opin. Chem. Eng.*, 2023, **40**, 100923.
121. X. Zhang, P. Ma, C. Wang, L. Gan, X. Chen, P. Zhang, Y. Wang, H. Li, L. Wang, X. Zhou and K. Zheng, *Energy Environ. Sci.*, 2022, **15**, 830–842.
122. B. Anasori and Y. Gogotsi, *Graphene 2D Mater.*, 2022, **7**, 75–79.
123. D. D. Sarma and P. V. Kamat, 2023 Nobel Prize in Chemistry: A Mega Recognition for Nanosized Quantum Dots, *ACS Energy Lett.*, 2023, **8**, 5149–5151.
124. Nobel Prize in Chemistry 2023, https://www.nature.com/collections/hhbaafijej.
125. A. Moscatelli, *Nat. Nanotechnol.*, 2023, **18**, 1376–1378.

126. L. Manna, *Nano Lett.*, 2023, **23**, 9673–9676.
127. M. L. Personick, in *One Hundred Years of Colloid Symposia: Looking Back and Looking Forward*, American Chemical Society, 2023, vol. 1457, pp. 247–283.
128. L. Carbone and P. D. Cozzoli, *Nano Today*, 2010, **5**, 449–493.
129. K. Yu and K. S. Schanze, *ACS Cent. Sci.*, 2023, **9**, 1989–1992.
130. D. Quarta, S. Toso, A. Fieramosca, L. Dominici, R. Caliandro, A. Moliterni, D. M. Tobaldi, G. Saleh, I. Gushchina, R. Brescia, M. Prato, I. Infante, A. Cola, C. Giannini, L. Manna, G. Gigli and C. Giansante, *Chem. Mater.*, 2023, **35**, 9900–9906.
131. X.-B. Li, C.-H. Tung and L.-Z. Wu, *Nat. Rev. Chem.*, 2018, **2**, 160–173.
132. D. V. Talapin, J.-S. Lee, M. V. Kovalenko and E. V. Shevchenko, *Chem. Rev.*, 2010, **110**, 389–458.
133. M. Balci Leinen, D. Dede, M. U. Khan, M. Çağlayan, Y. Koçak, H. V. Demir and E. Ozensoy, *ACS Appl. Mater. Interfaces*, 2019, **11**, 865–879.
134. J. Wang, T. Xia, L. Wang, X. Zheng, Z. Qi, C. Gao, J. Zhu, Z. Li, H. Xu and Y. Xiong, *Angew. Chem., Int. Ed.*, 2018, **57**, 16447–16451.
135. M. D. Regulacio and M.-Y. Han, *Acc. Chem. Res.*, 2016, **49**, 511–519.
136. T. Mokari, E. Rothenberg, I. Popov, R. Costi and U. Banin, *Science*, 2004, **304**, 1787–1790.
137. Y. Ben-Shahar, D. Stone and U. Banin, *Chem. Rev.*, 2023, **123**, 3790–3851.
138. F. Arcudi, L. Đorđević, B. Nagasing, S. I. Stupp and E. A. Weiss, *J. Am. Chem. Soc.*, 2021, **143**, 18131–18138.
139. D. M. Tobaldi, N. Rozman, M. Leoni, M. P. Seabra, A. S. Škapin, R. C. Pullar and J. A. Labrincha, *J. Phys. Chem. C*, 2015, **119**, 23658–23668.
140. D. M. Tobaldi, L. Lajaunie, D. Dvoranová, V. Brezová, B. Figueiredo, M. P. Seabra, J. J. Calvino and J. A. Labrincha, *Mater. Today Energy*, 2020, **17**, 100460.
141. D. M. Tobaldi, C. Espro, S. G. Leonardi, L. Lajaunie, M. P. Seabra, J. Calvino, S. Marini, J. A. Labrincha and G. Neri, *J. Mater. Chem. C*, 2020, **8**, 9529–9539.
142. D. M. Tobaldi, K. Kočí, M. Edelmannová, L. Lajaunie, B. Figueiredo, J. J. Calvino, M. P. Seabra and J. A. Labrincha, *Mater. Today Energy*, 2021, **19**, 100607.
143. F. Wang, S. Yang, Q. Lu, W. Liu, P. Sun, Q. Wang and W. Cao, *Colloids Surf., A*, 2023, **658**, 130741.
144. G. Getachew, A. Wibrianto, A. S. Rasal, S. Kizhepat, W. B. Dirersa, V. Gurav and J.-Y. Chang, *Prog. Mater. Sci.*, 2023, **140**, 101192.
145. J. S. Manser, J. A. Christians and P. V. Kamat, *Chem. Rev.*, 2016, **116**, 12956–13008.
146. L. Protesescu, S. Yakunin, M. I. Bodnarchuk, F. Krieg, R. Caputo, C. H. Hendon, R. X. Yang, A. Walsh and M. V. Kovalenko, *Nano Lett.*, 2015, **15**, 3692–3696.
147. Q. A. Akkerman, G. Rainò, M. V. Kovalenko and L. Manna, *Nat. Mater.*, 2018, **17**, 394–405.

148. K. S. Schanze, P. V. Kamat, P. Yang and J. Bisquert, *ACS Energy Lett.*, 2020, **5**, 2602–2604.
149. P. Cheng, K. Han and J. Chen, *ACS Mater. Lett.*, 2023, **5**, 60–78.
150. A. Dey, J. Ye, A. De, E. Debroye, S. K. Ha, E. Bladt, A. S. Kshirsagar, Z. Wang, J. Yin, Y. Wang, L. N. Quan, F. Yan, M. Gao, X. Li, J. Shamsi, T. Debnath, M. Cao, M. A. Scheel, S. Kumar, J. A. Steele, M. Gerhard, L. Chouhan, K. Xu, X. Wu, Y. Li, Y. Zhang, A. Dutta, C. Han, I. Vincon, A. L. Rogach, A. Nag, A. Samanta, B. A. Korgel, C.-J. Shih, D. R. Gamelin, D. H. Son, H. Zeng, H. Zhong, H. Sun, H. V. Demir, I. G. Scheblykin, I. Mora-Seró, J. K. Stolarczyk, J. Z. Zhang, J. Feldmann, J. Hofkens, J. M. Luther, J. Pérez-Prieto, L. Li, L. Manna, M. I. Bodnarchuk, M. V. Kovalenko, M. B. J. Roeffaers, N. Pradhan, O. F. Mohammed, O. M. Bakr, P. Yang, P. Müller-Buschbaum, P. V. Kamat, Q. Bao, Q. Zhang, R. Krahne, R. E. Galian, S. D. Stranks, S. Bals, V. Biju, W. A. Tisdale, Y. Yan, R. L. Z. Hoye and L. Polavarapu, *ACS Nano*, 2021, **15**, 10775–10981.
151. T. C. Jellicoe, J. M. Richter, H. F. J. Glass, M. Tabachnyk, R. Brady, S. E. Dutton, A. Rao, R. H. Friend, D. Credgington, N. C. Greenham and M. L. Böhm, *J. Am. Chem. Soc.*, 2016, **138**, 2941–2944.
152. Z. Ma, Z. Shi, D. Yang, F. Zhang, S. Li, L. Wang, D. Wu, Y. Zhang, G. Na, L. Zhang, X. Li, Y. Zhang and C. Shan, *ACS Energy Lett.*, 2020, **5**, 385–394.
153. L. Zhou, Y.-F. Xu, B.-X. Chen, D.-B. Kuang and C.-Y. Su, *Small*, 2018, **14**, 1703762.
154. Y. Tang, C. H. Mak, G. Jia, K.-C. Cheng, J.-J. Kai, C.-W. Hsieh, F. Meng, W. Niu, F.-F. Li, H.-H. Shen, X. Zhu, H. M. Chen and H.-Y. Hsu, *J. Mater. Chem. A*, 2022, **10**, 12296–12316.
155. M. S. Kodaimati, K. P. McClelland, C. He, S. Lian, Y. Jiang, Z. Zhang and E. A. Weiss, *Inorg. Chem.*, 2018, **57**, 3659–3670.

CHAPTER 5

Air Purification by Means of Colloid-based Air Filters

LEI ZHANG,[a,b] GUO-HONG TAO*[a] AND LING HE*[a]

[a] Sichuan University, College of Chemistry, Wangjiang Road 29, 610064, Chengdu, China; [b] Southwest Jiaotong University, School of Chemistry, Jiaoda Road 144, 610031, Chengdu, China
*Emails: taogh@scu.edu.cn; lhe@scu.edu.cn

5.1 Introduction

Air pollution is a critical environmental issue that affects the health and well-being of people, ecosystems, and the climate worldwide.[1–3] It occurs when harmful substances, including particulates, biological molecules, and gases, are introduced into the Earth's atmosphere. Three primary categories of pollutants—particulate matter (PM), viruses, and volatile organic compounds (VOCs)—have been identified as major concerns due to their prevalence and potential health risks. These pollutants originate from various sources such as industrial activities, vehicle emissions, agricultural practices, and natural events like wildfires and volcanic eruptions. As urbanization and industrialization continue to grow, the concentration of pollutants in the air has increased, leading to severe consequences. The World Health Organization has indicated that over 80 percent of people living in urban areas globally are exposed to harmful levels of air pollution, with the risk being especially pronounced in low- and middle-income countries.[4] Moreover, radioactive contaminants such as I_2 pollution are gaining increasing attention.[5–7] Air pollution is responsible for a range of health problems, from respiratory and cardiovascular diseases to premature death. It also contributes to environmental degradation, harming wildlife,

Contribution of Colloidal Materials to Air, Water and Soil Environmental Sustainability: From Basic Concepts to Advanced Applications
Edited by Izabel C. Riegel-Vidotti, Lizandra Maria Zimmermann and Heloise R. de Barros

Published by the Royal Society of Chemistry, www.rsc.org

forests, and water bodies, and exacerbating climate change through the release of greenhouse gases. Addressing air pollution requires comprehensive strategies that involve reducing emissions, adopting cleaner technologies, and developing methods for reduction.

PM refers to a mixture of tiny particles and droplets suspended in the air. These particles vary in size and composition and are classified primarily into $PM_{0.3}$ (particles with diameters of 0.3 μm or less, also called ultrafine particles), $PM_{2.5}$ (particles with diameters of 2.5 micrometers (μm) or less) and PM_{10} (particles with diameters of 10 μm or less). $PM_{0.3}$ move in erratic, zig-zag patterns because their small size and minimal mass allow them to collide with gas molecules, causing them to bounce around in multiple directions.[8–10] Due to their small size, $PM_{0.3}$ can travel over significant distances, infiltrate small airways, and serve as carriers for various bacteria and viruses. Many emerging infectious diseases (EIDs) are transmitted through these fine particulates in the air. For example, in 2019 the novel coronavirus (SARS-CoV-2 (COVID-19)) spread rapidly worldwide.[11] Although $PM_{2.5}$ and PM_{10} tend to move in predictable straight lines, some PM especially $PM_{2.5}$ can also penetrate deep into the respiratory tract, causing irritation and inflammation of the airways. This can lead to conditions such as bronchitis, asthma, and other chronic respiratory diseases. Prolonged exposure to PM is associated with an increased risk of heart attacks, strokes, and other cardiovascular conditions. $PM_{2.5}$ also have significant environmental consequences. They can reduce visibility, contributing to haze and smog, and impact ecosystems by depositing harmful substances onto soil and water bodies. Additionally, $PM_{2.5}$ can affect plant life by blocking sunlight and damaging leaf surfaces.

Viruses are microscopic pathogens that can cause a wide range of infectious diseases. Many viruses, such as influenza and SARS-CoV-2, primarily target the respiratory system, causing symptoms ranging from mild colds to severe pneumonia.[11] Viruses typically range in size from about 20 to 300 nanometers (nm) in diameter, which are roughly 25 to 40 times smaller than $PM_{2.5}$. For example, the size of SARS-CoV-2 is around 60–140 nm. Due to their small size, $PM_{2.5}$ can provide a surface for viruses to attach to, allowing them to remain airborne for extended periods and travel greater distances. This increases the likelihood of human exposure and infection.

Radioactive iodine especially I-131 is primarily produced through the fission of uranium and plutonium in nuclear reactors and during nuclear weapon detonations. It can be released into the environment during nuclear accidents, such as the Chernobyl disaster in 1986 and the Fukushima Daiichi nuclear disaster in 2011. I-131 poses significant health risks due to its radioactive properties and its affinity for the thyroid gland. When inhaled or ingested, I-131 is rapidly absorbed by the thyroid, where it emits beta particles and gamma rays, causing damage to thyroid tissues.[6] This can lead to thyroid cancer, hypothyroidism, and other cancers.

Air purification is a critical aspect of maintaining healthy indoor environments, especially in the face of increasing air pollution and the prevalence of airborne contaminants. Air filters play a central role in this process by trapping and removing PM, viruses, and iodine vapor from the air. The effectiveness of air filters in purifying air is determined by the materials and technologies used in their construction. The earliest air filters used simple materials like woven fabrics and fiberglass to capture large dust particles and debris. These basic filters were suitable for rudimentary applications but were insufficient for capturing viruses and iodine vapor. As industrialization and urbanization increased air pollution, the demand for more advanced filtration materials became apparent.

With the development of materials science, various materials including activated carbon, metal organic frameworks (MOFs), covalent organic frameworks (COFs), ionic liquids and their composites were introduced for their ability to increase purification efficiency and/or adsorb gases.[12–15] For example, activated carbon filters that trap chemical pollutants and VOCs are often used alongside particulate filters. MOFs and COFs offer potential for advanced air filtration, capable of selectively capturing specific pollutants at the molecular level. Compared with these materials mentioned above, materials derived from ionic liquid colloids offer benefits including cost-effectiveness, a variety of structures, and high efficiency.[16] This chapter introduces the various types of ionic liquids and their solid analogues used in air filters, discussing mechanisms of action and applications in different settings.

5.2 Performance Evaluation

The performance evaluation for removing PM with air filters is a critical process that ensures these devices effectively remove contaminants and provide clean air in various environments. Several metrics are used to evaluate the performance of air filters.

(i) Filtration efficiency (η) is defined as,

$$\eta = 1 - \frac{c_{\text{out}}}{c_{\text{in}}} \tag{5.1}$$

where c_{out} and c_{in} denote the concentration ($\mu g\, m^{-3}$) of PM at the outlet and the concentration ($\mu g\, m^{-3}$) of PM at the inlet, respectively (Figure 5.1). η is a key performance parameter for air filters. High-efficiency particulate air (HEPA) filters can theoretically remove at least 99.97% of dust, pollen, mold, bacteria, and any airborne particles with a size of 300 nm.

(ii) Pressure drop (ΔP) is expressed as,

$$\Delta P = P_{\text{in}} - P_{\text{out}} \tag{5.2}$$

where P_{in} (Pa) and P_{out} (Pa) are the pressures before and after the PM filtration, respectively. It reflects the resistance to airflow caused by the filter; a lower pressure drop indicates less resistance and more

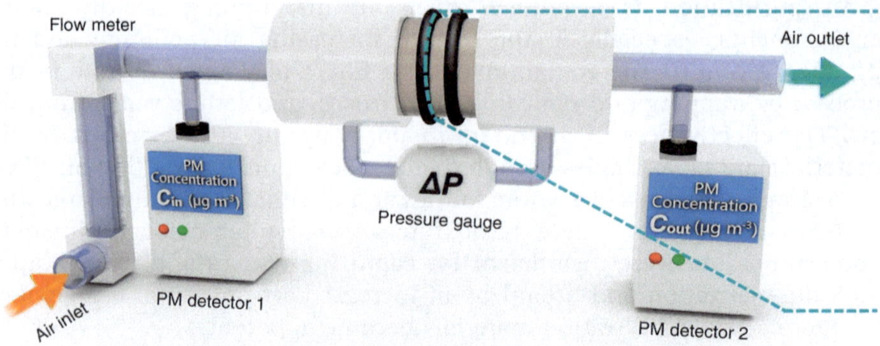

Figure 5.1 Schematic diagram of the device for the removal experiment and the measurement of air flow rate, pressure drop, and PM concentrations. Reproduced from ref. 28, https://doi.org/10.1038/s41467-020-15502-7, under the terms of the CC BY 4.0 license, http://creativecommons.org/licenses/by/4.0/.

efficient airflow. According to the US Department of Energy, a fine pressure drop is below *ca.* 325 Pa at an air face velocity of 5 $\mathrm{cm\,s^{-1}}$.[17]

(iii) Quality factor (QF) can be expressed as follows,

$$\mathrm{QF} = -\ln\left(\frac{\ln(1-\eta)}{\Delta P}\right) \tag{5.3}$$

QF is a metric used in the performance evaluation of air filters to assess their overall effectiveness. It takes into account both the filter's efficiency in removing particles and the resistance it adds to airflow. A high QF value suggests an optimal balance between pressure drop and filtration efficiency. This means that the air filter effectively removes particles while maintaining low resistance to airflow. Achieving a high QF value in air filters is a challenging goal. However, η and ΔP are typically mutually restrictive—increasing filtration efficiency usually results in a higher pressure drop, and *vice versa*. Consequently, attaining a high QF value places significant demands on the design and materials of the filter.

5.3 Colloids Based on Ionic Liquids

Colloids are mixtures where one substance is dispersed evenly throughout another at a microscopic level, typically in the range of 1 to 1000 nm. Colloidal systems can be classified based on the phases of the dispersed substance and the dispersion medium, such as aerosols (solid or liquid particles in gas), emulsions (liquid in liquid), and sols (solid in liquid).

Colloids based on ionic liquids (ILs) represent a fascinating area of research in material science, leveraging the unique properties of ionic liquids to create stable, versatile colloidal systems.[16] ILs are salts in a liquid state, typically composed of organic cations and inorganic or organic anions, which remain liquid at relatively low temperatures (often below 100 °C). Their distinctive characteristics, such as low volatility, high thermal stability, and tunable solubility, make them ideal for various applications in colloidal science.[18–20] IL based colloid materials exhibit many excellent properties and advantages.

i. Tunability: the physicochemical properties of ILs can be easily modified by altering the combination of cations and anions, allowing for precise control over the colloidal system's characteristics.
ii. Stability: ILs provide an excellent medium for colloids due to their ability to stabilize dispersed materials, preventing aggregation and sedimentation over extended periods.
iii. Conductivity: many ILs exhibit high ionic conductivity, which can be beneficial for applications requiring conductive colloidal suspensions.
iv. Compatibility: ILs can dissolve a wide range of substances, including organic, inorganic, and polymeric materials, making them highly versatile for creating various colloidal systems.

IL colloids incorporate the unique properties of ILs into colloidal systems. These colloids combine the stability and tunability of ILs with the versatility of colloidal materials, theoretically offering several advantages for air filtration applications. The ionic nature of these materials could enhance the electrostatic interactions between the ILs and PM to promote the capture and retention of pollutants. ILs in colloidal form provide excellent adsorption capabilities for gaseous pollutants. These colloids can be tailored to target specific chemicals, such as VOCs and iodine vapor, making them highly effective in purifying air from harmful gases.

The fabrication of colloid-based air filters involves several steps (Figure 5.2):

1. Synthesis of colloids: colloids are synthesized using methods such as sol–gel processes, precipitation, and chemical reduction. The choice of method depends on the desired properties of the colloid.
2. Incorporation into filters: colloids are incorporated into filter media, which can be fibers, foams, or membranes. Techniques like electrospinning, dip-coating, and layer-by-layer assembly are employed to achieve uniform distribution.
3. Optimization: the filter's performance is optimized by adjusting parameters like colloid concentration, filter thickness, and airflow rate.

5.4 Mechanisms of Colloid-based Air Filtration

Conventional fibers employed for air filters are primarily derived from petroleum-derived polymers such as polyethylene (PE) and polypropylene (PP). Due to the inert surface, such air filters decontaminate particles mainly through four size-based mechanisms: sieving, inertial impaction,

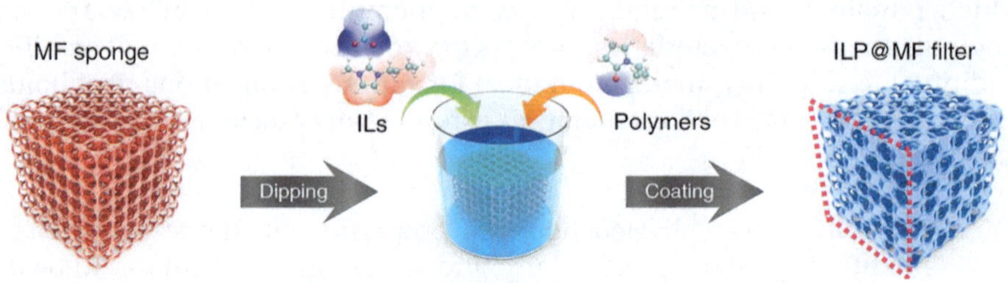

Figure 5.2 Schematic representation of the dipping-coating progress for the fabrication of the IL colloid based-air filter. Reproduced from ref. 28, https://doi.org/10.1038/s41467-020-15502-7, under the terms of the CC BY 4.0 license, http://creativecommons.org/licenses/by/4.0/.

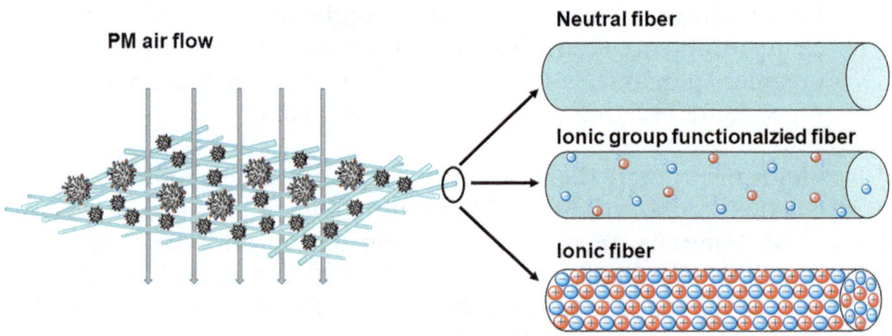

Figure 5.3 Schematic of fibrous air filters for PM removal based on different fibers. Adapted from ref. 26 with permission from the Royal Society of Chemistry.

interception, and diffusion. These mechanisms work together to trap particles of various sizes. However, it is extremely difficult for these air filters to reach a high filtration efficiency while enjoying a low pressure drop, as a result, they always possess an exceedingly low QF value. For instance, the QF value of commercial polypropylene-based HEPA filters is about 0.02 Pa^{-1}.[21] To improve the QF, many charged materials which can attract oppositely charged pollutants are developed and used in air filters (Figure 5.3). These materials provide an inborn advantage in particulate matter removal owing to their in-built ionizable groups, since electrostatic force can influence the motion of particles and give a capability for effective particle removal, which enhances filtration efficiency. Protein air filters with ionic groups carboxyl (–COOH) and amine ($-NH_2$ and $-NH_3^+$) exhibit excellent removal efficiencies and increased QF values (~0.07 Pa^{-1}) that can be more than three times higher than those of HEPA filters.[22,23] Moreover, the porous structures of conventional plastic micrometer-sized fibers are often ineffective at removing chemical gases due to their inert surfaces and limited surface areas. Porous materials with a high surface area to adsorb chemicals into their pores and functional materials with the ability to interact with various chemicals are also developed to enhance the removal efficiencies for VOCs and iodine vapor.

5.5 Development of Colloid-based Air Filters

5.5.1 Small Molecule-based Colloids for Air Filters

Self-assembly of small molecules and ions into supramolecular nanofibers in organic liquids or water is a bottom-up and easy approach and is utilized to efficiently prepare organo- and hydrogels. This method does not require specialized equipment, and the structure of nanofibers can be regulated by adjusting self-assembly conditions, such as solvents, molecular structures, temperature, and time. The first example of employing colloidal self-assembled nanofibers for air filtration was reported by the H.-W. Schmidt group.[24] Supramolecular nanofiber webs based on *N*,*N*′,*N*″-tris(2-ethylhexyl)-1,3,5-benzenetricarboxamide (BTA) (Scheme 5.1) are *in situ* formed inside nonwoven scaffolds. Immersion solutions of BTA in 2-butanone were prepared with concentrations of 0.4, 0.6, 0.8, and 1.0 wt% by suspending the corresponding amount of BTA in the solvent and heating to 50 °C until a clear solution was obtained. This solution was then transferred to a custom-made immersion bath, which was maintained at 50 °C using a thermostat. The cut nonwoven fabric was clamped into a supporting metal frame with an open quadratic area of 8.5 × 8.5 cm. The frame with the microfiber fabric was dipped into the immersion solution for 30 seconds. After soaking, the fabric was removed and dried in the frame under ambient conditions for 10 minutes and then for 30 minutes in a vacuum. The filtration efficiency of the nonwoven fabric is poor, with an efficiency ranging from 5% to 6%. After incorporating BTA nanofibers, the removal efficiency improves and continues to increase with higher BTA content. In the case of the composites with the highest nanofiber content, a significant performance improvement is observed. This substantial improvement is attributed to the high coverage with supramolecular nanofibers. The nanofiber composite achieves a

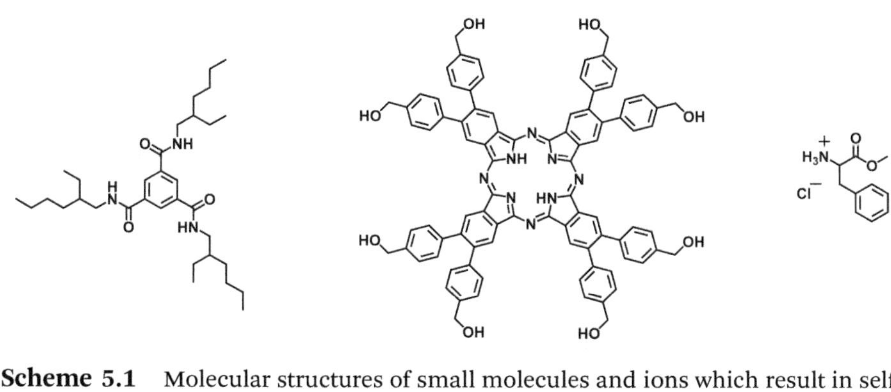

Scheme 5.1 Molecular structures of small molecules and ions which result in self-assembly into nanofibers.

filtration efficiency of 67% for 0.2 μm particles and 87% for 1.0 μm particles. This difference is due to the denser nanofiber web of the composite and is also reflected in the substantial increase in ΔP from 9 Pa for the nonwoven fabric to 429 Pa for the composite. Moreover, during the entire 24-hour airstream test, no fragments were detected. Scanning electron microscopy (SEM) images showed no significant difference in the morphology of the composite before and after the airstream test. This result demonstrates that the supramolecular nanofibers are robust and do not break up during the experiment.

Multifunctions of air filters can be achieved by rational design of molecule structures used in self-assembly. S. C. Tan reported colloidal self-assembly of 4,4″-bis(hydroxymethyl)-[1,1′:2′,1″-terphenyl]-4′,5′-dicarbonitrile (Scheme 5.1) in 1-pentanol.[25] The molecular building block naturally has the ability to shield against ultraviolet (UV) rays, while the nanofibrous mesh constructed from it effectively captures particulate matter while allowing for good airflow. Similar to previous research, increasing the content of self-assembled fibers results in a higher ΔP for such filters. The air filter with a basis weight of 1.6 mg cm^{-2} was able to filter 89.69% of $PM_{2.5}$ and 93.08% of PM_{10} with a ΔP of 374 Pa. As a result, the QF value is about 6.5×10^{-3} Pa^{-1}, which is higher than those of commonly used commercial respirators under the same test conditions. However, the QF values of these nanofibrous air filters remain limited due to the inert surface.

Ionic group functionalized materials such as proteins were employed to manufacture air filters. Compared to neutral nanofibers, protein-based nanofibers possess an inherent advantage in particulate matter removal due to the presence of ionizable groups derived from amino acids.[22] These groups can generate electrostatic forces that influence particle motion, thereby enhancing the effectiveness of particle removal. These protein air filters demonstrate excellent removal efficiencies and increased QF values (~0.07 Pa^{-1}), which can be more than three times greater than that of HEPA filters. Nevertheless, there is still room for improvement in the QF value of protein air filters. Motivated by the preceding research, G.-H. Tao and L. He utilized colloidal self-assembly of amino acid salts for air purification.[26] L-Phenylalanine methyl ester hydrochloride ([L-PheC_1]Cl) (Scheme 5.1) assembled into a dense network of nano- and micro-fibrils in a mixed solution of acetonitrile and ethanol. The self-assembled ionic fibrils are in-site incorporated into four commercially available and cheap porous materials including melamine-formaldehyde (MF) resin foam, polyurethane (PU) foam, polypropylene (PP) nonwoven and Nylon nonwoven, and the resulting composites are named LIF@MF, LIF@PU, LIF@PP, and LIF@Nylon, respectively. SEM micrographs of the general and corresponding magnified sections of ionic fiber composites based on the different scaffolds are shown in Figure 5.4. The average fiber diameters of the [L-PheC_1]Cl ionic fibers in the composites are as follows: 2.7 ± 1.0 μm for LIF@MF, 2.5 ± 1.0 μm for LIF@PU, 2.5 ± 1.2 μm for LIF@PP, and 2.6 ± 1.2 μm for LIF@Nylon. These values are larger than those of the [L-PheC_1]Cl ionic fibers formed without a porous skeleton, and these differences may be attributed to the presence of

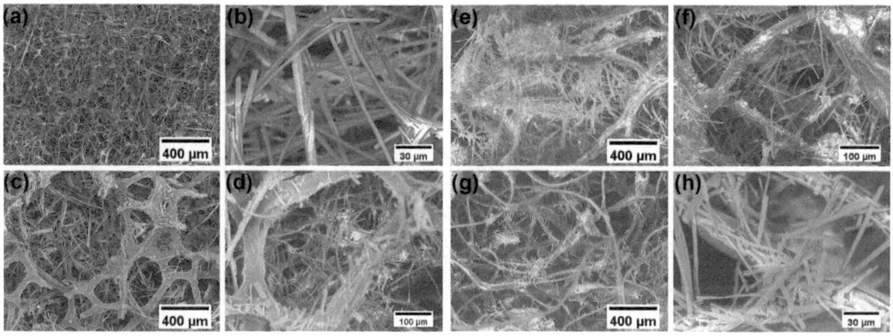

Figure 5.4 SEM micrographs of composites containing [L-PheC_1]Cl supramolecular ionic fibers. (a) and (b) LIF@MF; (c) and (d) LIF@PU; (e) and (f) LIF@Nylon; (g) and (h) LIF@PP. Reproduced from ref. 26 with permission from the Royal Society of Chemistry.

the porous skeleton in the solvent environment. The LIF@MF, LIF@PU, and LIF@Nylon demonstrate homogeneous filling of the porous skeletons with supramolecular fibers, akin to the nonwoven composites made from non-ionic fibers. For LIF@MF, LIF@PU, and LIF@Nylon, the removal efficiencies for $PM_{2.5}$ and PM_{10} consistently exceed 98% for at least 16 hours without notable decline. The average $PM_{2.5}$ removal efficiencies for these composites are as follows: LIF@MF at 99.60% (±0.06%), LIF@PU at 99.71% (±0.18%), LIF@Nylon at 99.82% (±0.05%), and LIF@PP at 88.00% (±6.50%). In PM_{10} filtration, the efficiencies are 99.73% (±0.07%) for LIF@MF, 99.75% (±0.24%) for LIF@PU, 99.84% (±0.05%) for LIF@Nylon, and 89.75% (±5.48%) for LIF@PP. The relatively low filling fraction of LIF@PP may contribute to its lower removal efficiency. LIF@MF, LIF@PU, and LIF@Nylon function as highly efficient filters, achieving $PM_{2.5}$ removal efficiencies of 95–100%, comparable to top filter materials created through electrospinning, such as silk fibroin nanofabrics (98.8%) and zein nanofabrics (99.6%). However, the removal efficiency significantly decreased as the RH increased to 85%, with average $PM_{2.5}$ removal efficiencies dropping to 65.01% (±4.14%) for LIF@MF, 65.65% (±3.19%) for LIF@PU, and 63.34% (±7.75%) for LIF@Nylon. All composites exhibit a commendable average ΔP of 5 Pa, comparable to that of high-performance triboelectric nanogenerator-enhanced nanofiber air filters. QF values for these composites range from 0.4241 to 1.2640. LIF@PP exhibits a lower QF value of 0.4241, which may limit its practical applications. In contrast, LIF@MF, LIF@PU, and LIF@Nylon show exceptionally high QF values (>1.0000 Pa^{-1}), significantly surpassing those of ionic group functionalized fiber-based air filters, including silk fibroin nanofabric (0.027 Pa^{-1}) and zein nanofabric (~0.07 Pa^{-1}).

To gain insight into the factors contributing to the exceptional performance of these materials, the electrostatic potential (ESP) on the molecular van der Waals (vdW) surface is studied. High absolute ESP values may significantly influence the movement of PM. The ESP-mapped vdW surface of

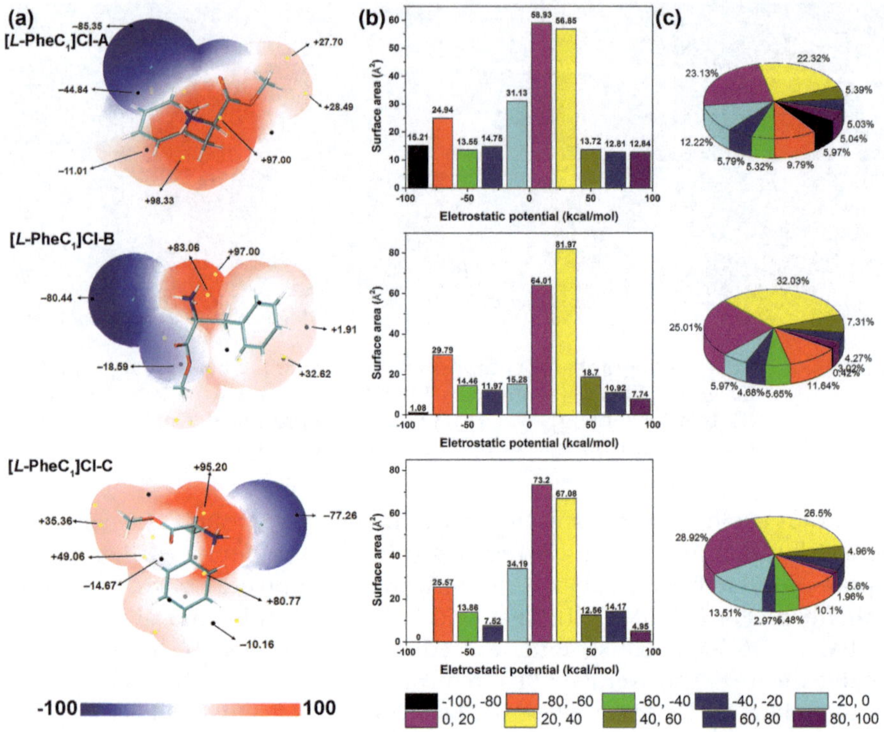

Figure 5.5 (a) ESP-mapped molecular vdW surface of the cluster conformers in the [L-PheC$_1$]Cl crystal. The unit is in kcal mol^{-1}. Surface local minima and maxima of ESP are represented as black and yellow points, respectively. (b) Surface area in each ESP range on the vdW surface. (c) Area percent in each ESP range on the vdW surface. Reproduced from ref. 26 with permission from the Royal Society of Chemistry.

cluster conformers in [L-PheC$_1$]Cl crystals is illustrated in Figure 5.5(a). Figure 5.5(b) shows how the entire molecular surface was partitioned, and the distribution of different ESP ranges is presented in Figure 5.5(c). All conformers exhibit a considerable vdW surface area with very high absolute ESP values (*i.e.*, <−40 kcal mol^{-1} and >+40 kcal mol^{-1}). A specific case study of [L-PheC$_1$]Cl-A reveals that it has a global minimum ESP of −85.35 kcal mol^{-1} on its surface, significantly lower than that of PP (−4.79 kcal mol^{-1}), polyacrylonitrile (PAN) (−38.06 kcal mol^{-1}), and neutral phenylalanine methyl ester (L-PheC1) (−34.04 kcal mol^{-1}). In contrast, the global maximum ESP on the surface is +98.33 kcal mol^{-1}, exceeding those of PP (+9.10 kcal mol^{-1}), PAN (+26.70 kcal mol^{-1}), and neutral L-PheC$_1$ (+19.70 kcal mol^{-1}). These significant differences indicate that the ESP distribution on the vdW surface fluctuates more dramatically in [L-PheC$_1$]Cl-A compared to neutral phenylalanine methyl ester, PAN, and PP. Notably, 20.90% of the vdW surface exhibits a large negative ESP, attributed to areas close to the chloride ion. Additionally, 25.65% of the vdW surface shows ESP values exceeding +40 kcal mol^{-1}, primarily associated with the hydrogens in the protonated amino group. Furthermore, 35.35% of the vdW surface of [L-PheC_1]Cl-A has lower ESP values (ranging from −20 to +20 kcal mol^{-1}), predominantly located in the benzyl unit. The extensive vdW surface characterized by significant negative or high positive ESP values in these self-assembled ionic fibers may interfere with PM movement, thus enhancing PM removal efficiency. These results suggest that colloidal ionic materials have considerable potential for air filtration applications.

5.5.2 Polymer-based Colloids for Air Filters

Compared with self-assembled colloidal materials based on small molecules and ions, polymer-based colloidal materials exhibit higher mechanical stability. G.-H. Tao and coworkers first reported colloidal materials where polymers are dispersed rare earth (Ln) ionic liquids for PM removal.[13] 1-Methyl-3-octylimidazolium hexanitratolanthanate ($[C_8mim]_3La(NO_3)_6$) and 1-methyl-3-octylimidazolium pentanitratoeuropate ($[C_8mim]_2Eu(NO_3)_5$) were employed as the disperse phase. Three polymers including polyvinyl pyrrolidone (PVP), polyvinyl alcohol (PVA), and polyacrylamide (PAM) with low toxicity, physiological inertia, and excellent biological compatibility were used. Typically, ionic liquids were added to an aqueous solution of PVP, PVA, or PAM in a mole ratio of 1 : 10. The solution was heated to 120 °C to remove water, resulting in a yellow solid raw product. This solid was rinsed thoroughly three times with deionized water and then dried in a vacuum. The polymer composite materials, designated as La-10PVP, Eu-10PVP, La-10PVA, and La-10PAM, were all obtained as yellow solids. Fourier transform infrared (FT-IR) spectra show that a redshift from 1655 cm^{-1} to 1643 cm^{-1} ascribed to the stretching vibration of C=O of PVP occurs, which indicates the oxygen atoms of carbonyl in PVP may coordinate with Ln^{3+} or Eu^{3+} (Figure 5.6(a)). The surface of these materials naturally adsorbs nanoparticles under ambient conditions. Under high PM concentration conditions, the adsorption efficiency for $PM_{2.5}$ and PM_{10} exceeds 90% within 15 hours for both La-10PVP and Eu-10PVP. The photo images of Eu-10PVP are displayed under a UV lamp before (Figure 5.6(b)) and after (Figure 5.6(c)) the PM absorption. With the increase in PM absorption time, the photoluminescent performance of the Eu-10PVP film significantly weakened. A notable decrease in fluorescence intensity was also observed in the fluorescence spectra of Eu-10PVP (Figure 5.6(d)). The asymmetry coefficient changed noticeably as Eu-10PVP absorbed PM, indicating that the absorbed PM alters the coordination environment of Eu^{3+}. Interestingly, these colloidal materials can be reused, which is beneficial to reduce the cost. The adhered PM on the surface was removed after sonication in ethyl acetate for 30 minutes at ambient temperature. After 5 cycles, no obvious reduction of the adsorption efficiency was observed (Figure 5.6(e)).

To reduce costs associated with expensive rare metals and enhance functionality, biopolymers were utilized. Alginate is a fully ionized polymer,

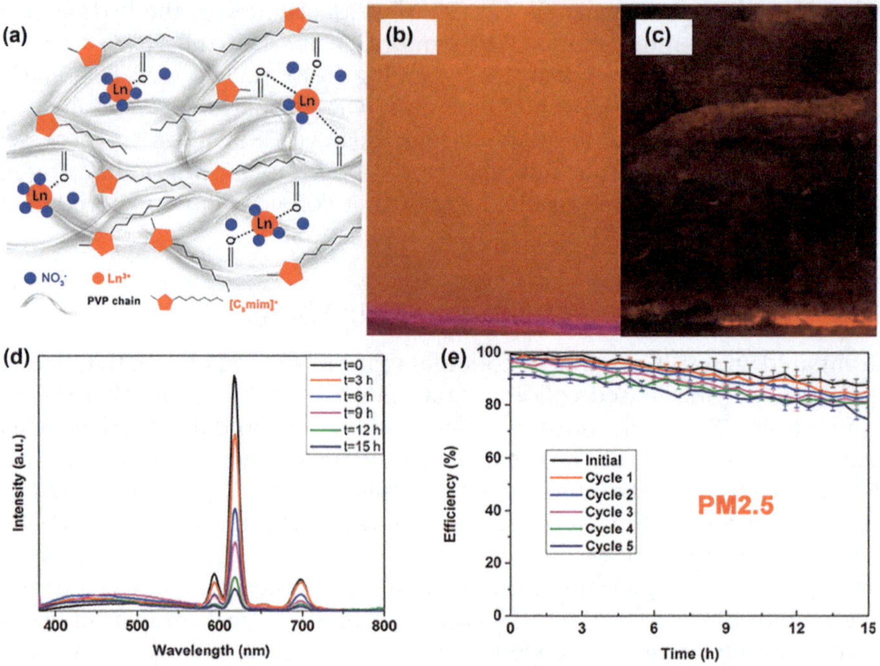

Figure 5.6 (a) Schematic structure of Ln-10PVP. (b) The photo images of the Eu-10PVP film. (c) The photo images of the Eu-10PVP film after adsorbing PM under a UV lamp. (d) The fluorescence spectra of the Eu-10PVP film over absorption time. (e) The recycling experiment curves of capturing $PM_{2.5}$. Reproduced from ref. 13 with permission from John Wiley & Sons, Copyright 2017 Wiley-VCH Verlag GmbH & Co. KGaA, Weinheim.

with one carboxylate group present in each building block (Figure 5.7).[27] The antimicrobial properties of alginate are enhanced by incorporating suitable cationic components. Three petroleum-derived organic cations including 1-ethyl-3-methylimidazolium $[Emim]^+$, 1-butyl-3-methylimidazolium $[Bmim]^+$, tetraethylammonium $[N_{2222}]^+$, and a natural cation (cholinium $[Ch]^+$) were introduced into alginate *via* ion exchange with sodium alginate (Na[Alg]), and the corresponding materials are named [Emim][Alg], [Bmim][Alg], $[N_{2222}]$[Alg], and [Ch][Alg], respectively. These alginate-based ionic polymers are then loaded into porous melamine formaldehyde (MF) foam. The resulting composites [Emim][Alg]@MF, [Bmim][Alg]@MF, $[N_{2222}]$[Alg]@MF, and [Ch][Alg]@MF show average $PM_{2.5}$ removal efficiencies of 97.65% (±0.52%), 96.18% (±0.26%), 99.22% (±0.21%), and 99.69% (±0.27%), respectively. The corresponding PM_{10} removal efficiencies of these four composites are 99.30% (±0.16%), 98.78% (±0.10%), 99.74% (±0.03%), and 99.89% (±0.09%), respectively. The average ΔP for all composites are below 2 Pa, resulting in QF values of 1.8754, 1.6325, 2.4268, and 2.8882 for [Emim][Alg]@MF, [Bmim][Alg]@MF, $[N_{2222}]$[Alg]@MF, and [Ch][Alg]@MF, respectively, which surpasses commercial air filters by approximately two orders of magnitude.

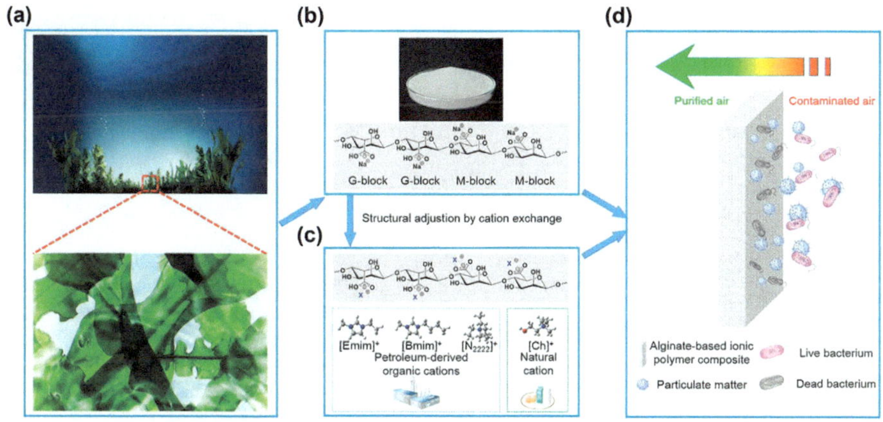

Figure 5.7 A schematic of alginate-based ionic polymer composites for air purification. (a) Left top: Digital photograph. Left bottom: Local magnified image of *Laminaria hyperborea*. (b) Middle top: Real product image and structure of sodium alginate: heterogeneous blocks of α-L-guluronate (G-block) and β-D-mannuronate (M-block). (c) Middle bottom: Structures of conventional organic cations and the natural cation used in this work. (d) Right: Illustration of PM removal and sterilization of alginate-based ionic polymer composites. Reproduced from ref. 27 with permission from John Wiley & Sons, Copyright 2020 Wiley-VCH Verlag GmbH & Co. KGaA, Weinheim.

Airborne Gram-negative bacteria, such as *E. coli*, typically measure around $0.5 \times (1–3)$ μm, which is comparable in scale to $PM_{2.5}$. Research indicates that Gram-negative bacteria associated with $PM_{2.5}$ can contribute to pulmonary infections, highlighting the need for air filter materials with antimicrobial properties. The antimicrobial activity of alginate-based ionic polymers against *E. coli* was evaluated using optical density (OD) measurements and the inhibition zone method. The bacteriostatic rates for Na[Alg], [Emim][Alg], [Bmim][Alg], [N_{2222}][Alg], and [Ch][Alg] are 28.96% (±1.55%), 29.07% (±2.69%), 48.80% (±2.83%), 69.66% (±2.81%), and 92.70% (±0.84%), respectively (Figure 5.8(a)). These results suggest that the antimicrobial activity of alginate polymers is influenced by cation structure, with [Ch][Alg] demonstrating the highest antimicrobial effect. Figure 5.8(b) illustrates the diameters of the inhibition zones for Na[Alg], [Emim][Alg], [Bmim][Alg], [N_{2222}][Alg], and [Ch][Alg], measuring 11.0, 12.5, 15.0, 17.0, and 20.5 mm, respectively, which align well with the OD values obtained. [Ch][Alg] shows the highest positive values of ESP (+60.70 kcal mol^{-1}). The antibacterial mechanism was proposed according to ESP values on the vdW surface. Cholinium can attract the negatively charged membranes of bacteria through electrostatic forces. As a result, the external surface of the cell membrane comes into contact with cholinium. This interaction alters the permeability of the membranes due to the combination of cholinium and the cell membrane, leading to changes in the superficial structure and

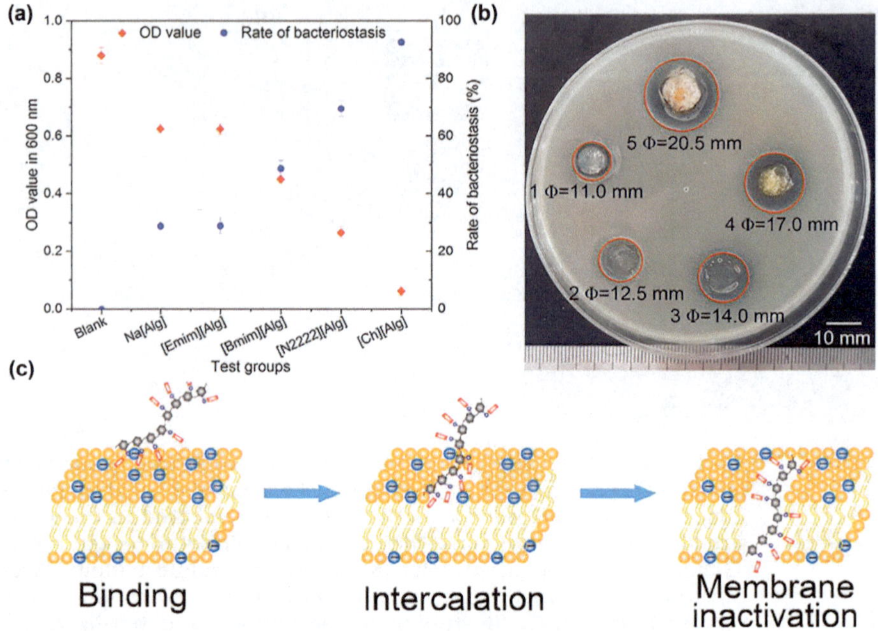

Figure 5.8 (a) and (b) The optical density tests and inhibition zone tests of the alginate-based ionic polymers against *E. coli*. Labeled: (1) Na[Alg], (2) [Emim][Alg], (3) [Bmim][Alg], (4) $[N_{2222}]$[Alg], and (5) [Ch][Alg]. (c) Proposed mechanism of the antimicrobial activity of [Ch][Alg] *vs. E. coli*. Reproduced from ref. 27 with permission from John Wiley & Sons, Copyright 2020 Wiley-VCH Verlag GmbH & Co. KGaA, Weinheim.

facilitating the intercalation of [Ch][Alg] into the membrane. The flexible alginate backbone allows the remaining long chains to easily adhere to the membranes. Consequently, permeation into the bacterial membrane continues until complete insertion occurs, which ultimately disrupts the membrane structure and inactivates its function, resulting in bacterial cell death (Figure 5.8(c)). These results suggest that the rational design of molecular structures can enhance the functionality of colloidal materials, thereby improving the effectiveness of air purification.

Capturing $PM_{0.3}$ with conventional purification technologies, which rely on size-dependent mechanisms, is challenging due to their ultra-low mass and small particle size. While traditional filters with standard pore sizes effectively intercept larger particles, they struggle with $PM_{0.1}$. Some of these tiny particles may continue to follow the gas streamlines and pass through the filtration material, reducing filtration efficiency. Most existing filters demonstrate low efficiency in intercepting $PM_{0.3}$ particles and face an inevitable trade-off between particulate matter removal and air permeability. To resolve this limitation, self-powered colloidal materials consisting of ionic liquids and polymer were prepared. In theory, the ionic liquids significantly enhance conductivity, enabling the colloid materials to generate a strong electric field at low voltage. The colloidal materials named ILP are prepared using the metal- and halogen-free ionic liquid 1-alkyl-3-methylimidazolium acetate ([$C_n$mim][OAc]) combined with hydrophilic polymers including poly(acrylamide) (PAM), poly(vinyl alcohol) (PVA), and poly(vinylpyrrolidone) (PVP).[28] To achieve a desirable partially porous structure, these ILP colloids are irregularly coated onto the skeleton of a melamine-formaldehyde (MF) resin sponge, forming the ILP@MF filter. Under ambient conditions, the average $PM_{2.5}$ removal efficiencies of the ILP@MF filters containing [C_4mim][OAc], [C_6mim][OAc] and [C_8mim][OAc] are 98.78 ± 0.42%, 97.53 ± 0.51% and 96.21 ± 0.53%, respectively. For PM_{10}, the removal efficiencies are 99.02 ± 0.48% for [C_4mim][OAc], 98.64 ± 0.41% for [C_6mim][OAc] and 97.47 ± 0.44% for [C_8mim][OAc].

Pure PVP exhibits low conductivity with an electron conductivity (σ) value lower than 10^{-8} mS m^{-1}. In contrast, the [C_4mim][OAc]–PVP composite demonstrates a significantly higher conductivity of 1.03×10^2 mS m^{-1}, enabling the generation of a strong electric field for capturing $PM_{0.3}$. Figure 5.9(a) illustrates a sketch of the filtration test with an applied voltage. The variable applied voltage and electric field distribution are provided by a numerical

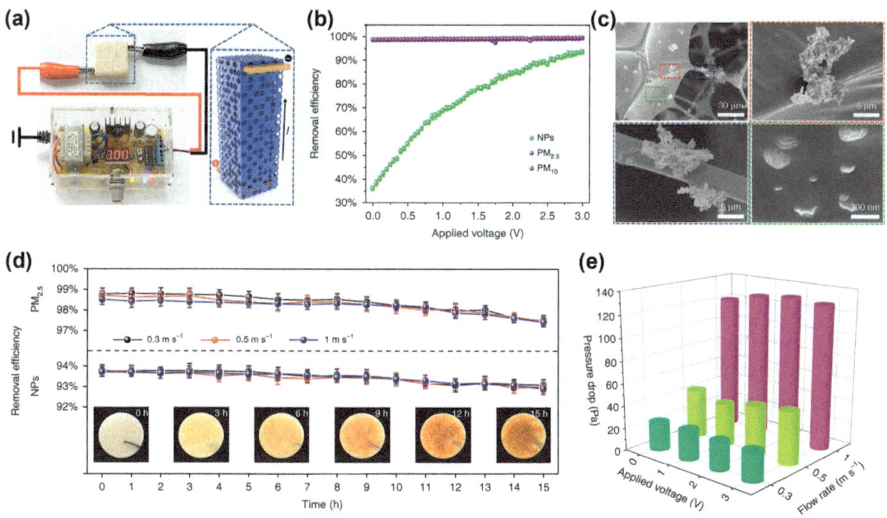

Figure 5.9 (a) Schematic of the charged [C_4mim][OAc]–PVP@MF filter. (b) Efficiency points of the [C_4mim][OAc]–PVP@MF filter for $PM_{0.3}$, $PM_{2.5}$ and PM_{10} under different voltages. (c) SEMs of the [C_4mim][OAc]–PVP@MF filter after a 15 h removal experiment. Error bar represents the standard deviation of three replicate measurements. (d) 15 h test of the removal efficiency for $PM_{2.5}$ and $PM_{0.3}$ of the [C_4mim][OAc]–PVP@MF filter. The inset images are photos of the [C_4mim][OAc]–PVP@MF filter during the purification test. (e) Pressure drop of the [C_4mim][OAc]–PVP@MF filter under various applied voltage (0, 1, 2, and 3 V) and at different air flow rates (0.3, 0.5, and 1 m s^{-1}). Reproduced from ref. 28, https://doi.org/10.1038/s41467-020-15502-7, under the terms of the CC BY 4.0 license, http://creativecommons.org/licenses/by/4.0/.

control transformer. Two pieces of copper electrodes were inserted at the ends of the $[C_4mim][OAc]$–PVP@MF filter, allowing the $[C_4mim][OAc]$–PVP composites on the sponge to contact the electrode surface and form a closed loop. Compared to the uncharged $[C_4mim][OAc]$–PVP@MF filter, the removal efficiencies of $PM_{2.5}$ increased as the applied voltage increased to 3 V ($99.59 \pm 0.31\%$). Similar results were observed for PM_{10} removal, with the removal efficiency increasing from $99.01 \pm 0.23\%$ (0 V) to $99.75 \pm 0.22\%$ (3 V). An obvious improvement in $PM_{0.3}$ removal is observed (Figure 5.9(b)). At 0 V, the removal efficiency of the uncharged $[C_4mim][OAc]$–PVP@MF filter for $PM_{0.3}$ is $35.12 \pm 0.69\%$, attributed to the great hydrophilicity and ESP values of the $[C_4mim][OAc]$–PVP composite.

With increased voltage, the removal efficiencies exhibit significant enhancement. The removal efficiency of $PM_{0.3}$ reaches up to $93.77 \pm 1.05\%$ when the voltage rises to 3 V. Furthermore, the removal results of PM were confirmed by SEM and dynamic light scattering (DLS). SEM images show that the $[C_4mim][OAc]$–PVP@MF filter changes significantly compared to the original, with agglomerated $PM_{2.5}$ and PM_{10} deposited on the filter surface (Figure 5.9(c)). Upon closer inspection, nano-sized particles with non-uniform diameters ranging from dozens to hundreds of nanometers adhere to the filter surface (Figure 5.9(c)).

The removal performance of the $[C_4mim][OAc]$–PVP@MF filter for 15 hours under an applied voltage of 3 V is shown in Figure 5.9(d). The removal efficiencies for $PM_{2.5}$ and $PM_{0.3}$ remained high, over 97% and 92%, respectively. Moreover, the removal efficiencies of the $[C_4mim][OAc]$–PVP@MF filter showed no significant change as the airflow rate increased from 0.3 to 1 $m\,s^{-1}$. After 15 hours, the color of the $[C_4mim][OAc]$–PVP@MF filter changed from white to brown (Figure 5.9(d)), indicating particle uptake from polluted air. The average ΔP of the $[C_4mim][OAc]$–PVP@MF filter at 3 V was only 26 Pa at a flow rate of 0.3 $m\,s^{-1}$, which is far lower than the U.S. Department of Energy's standard (approximately 325 Pa at an airflow rate of 5 $cm\,s^{-1}$). As a result, a button lithium cell or a silicon-based solar panel equipped with a switch can serve as a portable power supply platform. These results indicate that ion colloid materials can effectively remove not only $PM_{2.5}$ but also $PM_{0.3}$ under applied voltage conditions.

5.5.3 Polymer-based Colloid for Virus Capture

Hazardous biological pathogens in the air present a significant public health concern, as infected individuals release virus-laden aerosols (VLAs) during everyday respiratory activities. Colloidal materials named GILP comprising glycine ionic liquids (GILs) and polyvinylpyrrolidone (PVP) have been developed for the simultaneous filtration and collection of VLAs, taking advantage of the inherent electrostatic attraction of the filtration materials.[29] GILs possess the smallest ion volume among amino acid cations, resulting in high polarity and a dense array of functional groups. Additionally, GILP exhibits significant hydrophilicity, strong polarity, and unique biocompatibility, enhancing its effectiveness in capturing virus aerosol particles in high-humidity environments. GILP is wrapped around a melamine-formaldehyde

(MF) resin skeleton, utilizing the plasticity of PVP to create a conductive and porous air filter (GILP@MF). By applying a low voltage across the filter, the GILP composite is polarized by the electric field, allowing it to capture both VLAs and non-virus-laden aerosols. The high porosity of the skeleton ensures good air permeability and minimizes pressure drop.

The virus concentration before and after capture by the GILP@MF filter was measured using the Brilliant Blue G-250 staining method (Figure 5.10(a)). The time-dependent UV-vis absorption spectra of VLAs adsorbed by GILP@MF were compared to the spectra of the VLA starting solution without any filtration. The ionic composite exhibits a distinctly increased trend in absorbance at 595 nm over time, gradually approaching the initial VLA levels (Figure 5.10(d)). This indicates a gradual increase in the concentration of captured virus aerosols. SEM images captured at different time points show that the microscopic surface of the pore skeleton in the filter transitions from smooth and clean to having many tiny whitish spots (Figure 5.10(b)). As the concentration of VLAs increases, these whitish spots accumulate and form a strip-like appearance at 30 seconds, eventually becoming wrinkles that adhere to the material surface. Microscope imaging confirms that the

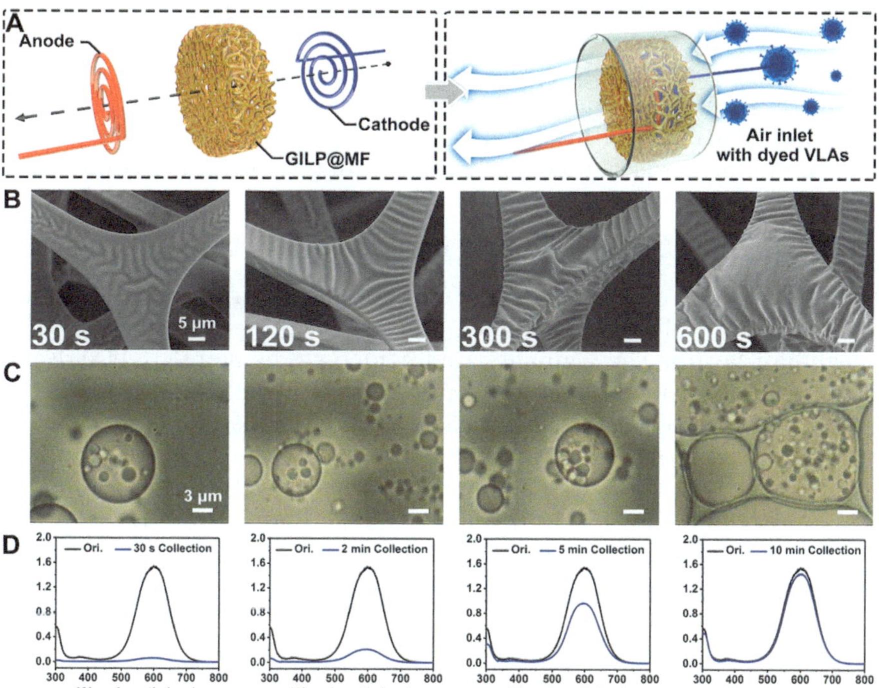

Figure 5.10 (A) Composition of the GILP@MF filter and adsorption of dyed VLAs. (B) SEM images of the adsorbed VLAs on the surface of GILP@MF at different times. (C) Microscope photos of the adsorbed VLAs. (D) UV-vis spectral changes of VLAs captured by GILP@MF at different times. Reproduced from ref. 29 with permission from Elsevier, Copyright 2024.

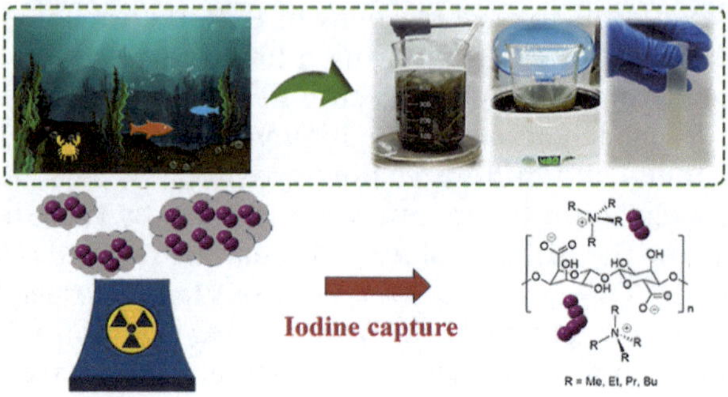

Figure 5.11 Fabrication of alginate-based ionic polymer composites for iodine capture. Reproduced from ref. 30 with permission from American Chemical Society, Copyright 2024.

adsorbed viral aerosols are in a microsphere state (Figure 5.10(c)). These microspheres are encapsulated in many vesicle structures, likely due to the agglomeration of virus aerosols. Moreover, due to the biocompatibility and water solubility of GIL and PVP, the composite can detach from the MF after capturing VLAs, releasing the adsorbed viruses into a buffer solution. Thus, the GILP@MF filter can be employed as a self-powered mask that does not require an external power supply while effectively protecting against VLAs in the environment and collecting aerosol samples generated by the wearer. The proposed GILP@MF mask could offer a non-invasive method for sampling respiratory viruses, potentially contributing to medical diagnosis.

5.5.4 Polymer-based Colloids for Iodine Capture

Materials that can effectively adsorb iodine vapor from off-gases generated during nuclear fuel reprocessing are crucial for both environmental safety and human health. The alginate-based composite named [N_{1111}][Alg]@MF demonstrates considerable effectiveness in iodine capture (Figure 5.11).[30] The starting materials, [N_{1111}][Alg]@MF, are both accessible and cost-effective. Thanks to the abundant functional groups present in alginate and the high porosity provided by melamine-formaldehyde (MF), the iodine adsorption capacity of [N_{1111}][Alg]@MF reached an impressive 5.82 $g\,g^{-1}$, with adsorption equilibrium achieved within 9 hours. The numerous hydroxyl and carboxyl functional groups in $[Alg]^-$ play a pivotal role in iodine accumulation. Moreover, the composite exhibits exceptional physicochemical stability, allowing for excellent recyclability in iodine adsorption, underscoring its potential for practical applications.

5.6 Challenges and Future Directions

Despite the significant progress made with ionic liquid-based colloidal materials in the field of air treatment, several challenges remain that need to be addressed in future research and applications. Firstly, improving the efficiency of filters in capturing ultrafine particles and viruses continues to be an important goal. Current filtration technologies do not perform optimally when processing tiny particulate matter, highlighting the urgent need to develop more effective filtering materials to enhance their capture capabilities. Secondly, the miniaturization of filtering devices especially for devices that require applied voltage is a challenge that cannot be overlooked. To enhance wearing comfort, the development of compact filtration systems is essential. This not only meets the spatial requirements of modern environmental control equipment but also improves treatment efficiency within a smaller volume. Concurrently, the development of durable and cost-effective materials is also a vital aspect of enhancing filter performance. Filters operating over extended periods must possess good chemical and physical stability to withstand various complex environments, while also reducing maintenance costs and improving economic viability. Lastly, the environmental impact of used filters is an emerging concern. If these contaminated filters are not disposed of properly, they may lead to secondary pollution, posing threats to ecosystems and human health. Therefore, the research and development of recyclable or biodegradable filter materials, along with the exploration of scientifically sound disposal methods, become an indispensable part of future studies. Furthermore, effectively integrating new filtration technologies with existing systems adds complexity to their application. This necessitates deeper technical research and collaboration across various sectors to ensure compatibility and ease of operation among different systems.

In conclusion, tackling these challenges is essential not only for enhancing air quality and advancing environmental protection but also for ensuring public health. We believe that the development of advanced colloidal materials will help overcome these obstacles. Interdisciplinary innovation and collaboration are crucial for developing more efficient, cost-effective, and environmentally sustainable air treatment solutions.

References

1. O. Raaschou-Nielsen, Z. J. Andersen, R. Beelen, E. Samoli, M. Stafoggia and G. Weinmayr, *et al.*, *Lancet Oncol.*, 2013, **14**, 813.
2. J. Lelieveld, J. S. Evans, M. Fnais, D. Giannadaki and A. Pozzer, *Nature*, 2015, **525**, 367.
3. W. J. Guan, X. Y. Zheng, K. F. Chung and N. S. Zhong, *Lancet*, 2016, **388**, 1939.
4. A. M. Fiore, V. Naik, D. V. Spracklen, A. Steiner, N. Unger and M. Prather, *et al.*, *Chem. Soc. Rev.*, 2012, **41**, 6663.
5. J. Fu, J.-Y. Liu, Y.-R. Zhou, L. Zhang, S.-L. Wang, S. Qin, M. Fan, G.-H. Tao and L. He, *Chem. Eng. J.*, 2024, **488**, 150913.
6. X.-L. Yang, Q.-H. Zhu, G.-H. Zhang, J. Fu, S.-L. Wang, L. Ma, S. Qin, G.-H. Tao and L. He, *Chem. Commun.*, 2024, **60**, 1168.

7. G.-H. Zhang, Q.-H. Zhu, S.-J. Guo, L. Zhang, C. Yu, S. Qin, L. He and G.-H. Tao, *Adv. Mater.*, 2024, **36**, 2311990.
8. G. Q. Gu, C. B. Han, C. X. Lu, C. He, T. Jiang, Z. L. Gao, C. J. Li and Z. L. Wang, *ACS Nano*, 2017, **11**, 6211.
9. S. Zhang, H. Liu, N. Tang, S. Zhou, J. Yu, B. Li and B. Ding, *Adv. Mater.*, 2020, **32**, 2006930.
10. M. Yang, X. Li, N. Yao, J. Yu, X. Yin, S. Zhang and B. Ding, *ACS Nano*, 2024, **18**, 16895.
11. Z. Tang, N. Kong, X. Zhang, Y. Liu, P. Hu and S. Mou, *et al.*, *Nat. Rev. Mater.*, 2020, **5**, 847.
12. Y. Chen, S. Zhang, S. Cao, S. Li, F. Chen and S. Yuan, *et al.*, *Adv. Mater.*, 2017, **29**, 1606221.
13. K. X. Song, L. He, L. Zhang and G. H. Tao, *Adv. Mater. Interfaces*, 2018, **5**, 1700448.
14. P. Li, J. Li, X. Feng, J. Li, Y. Hao and J. Zhang, *et al.*, *Nat. Commun.*, 2019, **10**, 2177.
15. J. Wang and S. Zhuang, *Coordin. Chem. Rev.*, 2019, **400**, 213046.
16. Q. Li, F. Yan and J. Texter, *Chem. Rev.*, 2024, **124**, 3813.
17. US Department of Energy, Specification for HEPA filters used by DOE contractors (USDOE, 2005).
18. K. Ueno, S. Imaizumi, K. Hata and M. Watanabe, *Langmuir*, 2009, **25**, 825.
19. J. Yuan, S. Soll, M. Drechsler, A. H. E. Müller and M. Antonietti, *J. Am. Chem. Soc.*, 2011, **133**, 17556.
20. B. Oh, S. Baek, K. S. Nam, C. Sung, C. Yang and Y.-S. Lim, *et al.*, *Nat. Commun.*, 2024, **15**, 5839.
21. C. Liu, P. C. Hsu, H. W. Lee, M. Ye, G. Y. Zheng, N. A. Liu, W. Y. Li and Y. Cui, *Nat. Commun.*, 2015, **6**, 6205.
22. H. Souzandeh, K. S. Johnson, Y. Wang, K. Bhamidipaty and W. H. Zhong, *ACS Appl. Mater. Interfaces*, 2016, **8**, 20023.
23. H. Tian, X. Fu, M. Zheng, Y. Wang, Y. Li, A. Xiang and W.-H. Zhong, *Nano Res.*, 2018, **11**, 4265.
24. H. Misslitz, K. Kreger and H. W. Schmidt, *Small*, 2013, **9**, 2053.
25. V. K. Singh, S. K. Ravi, W. X. Sun and S. C. Tan, *Small*, 2017, **13**, 1601924.
26. L. Zhang, W. L. Yuan, Z. Zhang, G. H. Zhang, H. Chen, N. R. Zhao, L. He and G. H. Tao, *J. Mater. Chem. A*, 2019, **7**, 4619.
27. Y. Wang, W. L. Yuan, L. Zhang, Z. Zhang, G. H. Zhang, S. L. Wang, L. He and G. H. Tao, *Adv. Sustain. Syst.*, 2020, **4**, 2000046.
28. G. H. Zhang, Q. H. Zhu, L. Zhang, F. Yong, Z. Zhang, S. L. Wang, Y. Wang, L. He and G. H. Tao, *Nat. Commun.*, 2020, **11**, 1653.
29. G.-H. Zhang, Q.-H. Zhu, L. Zhang, L. Li, J. Fu, S.-L. Wang, W.-L. Yuan, L. He and G.-H. Tao, *J. Hazard. Mater.*, 2024, **465**, 133480.
30. Q. Huang, J. Fu, Y.-H. Wang, S.-L. Wang, J.-Y. Liu, S.-J. Guo, S. Qin, G.-H. Tao and L. He, *ACS Sustainable Chem. Eng.*, 2024, **12**, 3089.

CHAPTER 6

CO_2 Adsorption: Leveraging Colloidal Systems as a Strategic Approach

WESLEY F. MONTEIRO,[a] ALINE S. AQUINO,[a] FRANCINE BERTELLA,[b] VINICIUS DEMETRIO DA SILVA,[c] ANDERSON J. SCHWANKE,[a] CHRISTIAN W. LOPES[b] AND KATIA BERNARDO-GUSMÃO*[a]

[a] Laboratório de Reatividade e Catálise, Instituto de Química, Universidade Federal do Rio Grande do Sul (UFRGS), Porto Alegre, RS, Brazil; [b] Laboratório de Catálise e Caracterização de Materiais, Departamento de Química, Universidade Federal do Paraná (UFPR), Curitiba, PR, Brazil; [c] Programa de Pós-Graduação em Engenharia e Tecnologia de Materiais, Escola Politécnica – Pontifícia Universidade Católica do Rio Grande do Sul (PUCRS), Brazil
*Email: katia.gusmao@ufrgs.br

6.1 Introduction

The elevated consumption of fossil fuels over the past decades has resulted in a significant increase in the amount of CO_2 present in the atmosphere, which is one of the main contributors to the greenhouse effect. Nevertheless, the International Energy Agency (IEA) reported that global CO_2 emissions fell by 5.1% in 2020 (or almost 2 Gt CO_2) due to the drastic reduction in global economic activity and mobility caused by the coronavirus pandemic.[1] Similarly, British Petroleum (BP) reported a 6.3% reduction in global CO_2 emissions in 2020.[2] This decline is the largest since the end of World War II

Contribution of Colloidal Materials to Air, Water and Soil Environmental Sustainability: From Basic Concepts to Advanced Applications
Edited by Izabel C. Riegel-Vidotti, Lizandra Maria Zimmermann and Heloise R. de Barros

Published by the Royal Society of Chemistry, www.rsc.org

and almost five times greater than the decrease in 2009 following the global financial crisis. Despite the decline in 2020, global energy-related CO_2 emissions remained at 31.5 Gt, contributing to CO_2 reaching its highest average annual concentration in the atmosphere of 412.5 parts per million in 2020 – about 50% higher than when the industrial revolution began.[1] However, the world has experienced an extremely rapid economic recovery since then, driven by fiscal changes and monetary stimulus, and a rapid – although unequal – distribution of vaccines.[1] The recovery in energy demand in 2021 was exacerbated by adverse weather conditions and energy market conditions, despite renewable energy generation recording its highest annual growth.[1] More recently, CO_2 emissions reached a new record of 37.4 Gt in 2023.[3]

Figure 6.1 shows global CO_2 emissions from the energy sector between 1900 and 2023 (top) and their change relative to the previous year (bottom).

Carbon capture and storage (CCS) refers to the process of capturing carbon dioxide emissions from industrial and power generation sources, and transporting them to a storage location.[4] In recent years, the acronym CCS has been expanded to CCSU, carbon capture, storage, and utilization, which includes the use of captured CO_2 or its conversion into higher value-added products, such as fuels and chemicals.[5]

Among the CCS processes, the most relevant is the chemical absorption of CO_2 from natural gas using an aqueous amine solution

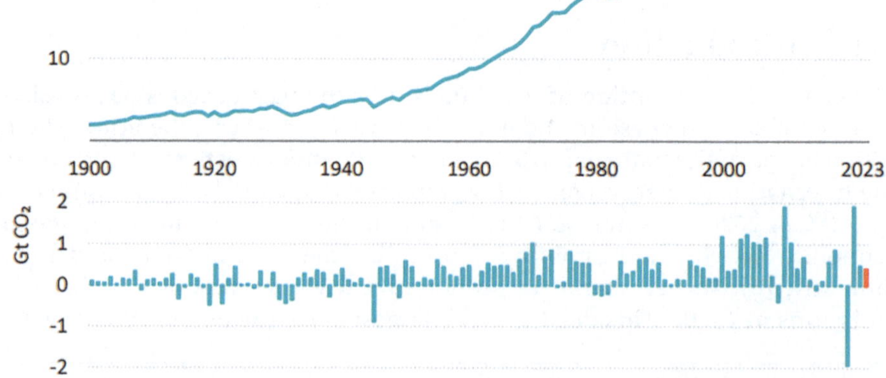

Figure 6.1 Total CO_2 emissions from energy combustion and industrial processes and their annual variation (1900–2023).[3] Reproduced from ref. 3, https://www.iea.org/reports/co2-emissions-in-2023, under the terms of the CC BY 4.0 license, https://creativecommons.org/licenses/by/4.0/.

(20–30% by mass of monoethanolamine), which has a technology readiness level (TRL) of 9, demonstrating the maturity of the process.[5] However, amines can induce equipment and pipeline corrosion, are volatile, and can be decomposed, which drives the search for alternatives for the chemical absorption of CO_2 and is one of the imminent technological trends.[5] The use of ionic liquids as substitutes for amines is an interesting alternative due to their high stability, low volatility, and reasonable CO_2 absorption.[6] These characteristics will be further addressed throughout this chapter.

Another alternative to the CO_2 chemisorption processes with amines is adsorption processes involving porous solids or colloids, which were initially considered in the 1990s and have gained significant attention from both industry and academia over the years.[5] The advantages of the adsorption process lie in the fact that adsorption columns can be adapted and optimized in plants that generate CO_2. Moreover, adsorbents can be used over a wide range of pressures and temperatures, making it possible to use them in pre- and post-combustion processes, although the use of adsorbents is more common in processes that involve air capture and contain a low concentration of carbon dioxide.[5] Efforts in this area are mainly focused on increasing the working capacity for CO_2, enhancing selectivity relative to other gases, and improving the adsorbent's tolerance to impurities.[5]

Traditional methods, although somewhat effective, often fall short in terms of efficiency, scalability, and environmental impact. In this context, colloidal systems emerge as a promising frontier in the search for advanced CO_2 sorption techniques. This chapter investigates the multifaceted world of colloidal systems, offering an exploration of their potential in CO_2 capture. At the center of this exploration are emulsion systems, whose unique properties facilitate enhanced CO_2 capture. Emulsions, with their large interfacial areas and adjustable properties, present a versatile platform for developing new CO_2 capture solutions. Building on this foundation, we introduce emulsions and ionic liquid-based emulsions, demonstrating their ability to combine the high solubility of CO_2 with the advantageous characteristics of emulsions.

Further expanding the scope, we examine the role of polymers synthesized through emulsion methods. These polymers not only offer large surface areas and customized functionalities but also bring a level of structural versatility that is crucial for effective CO_2 adsorption. Additionally, aerogel polymers and aerogel-based solvents represent a cutting-edge approach, combining the lightweight, porous nature of aerogels with the interactive capabilities of polymers to create superior adsorption materials.

In this chapter, we aim to show the strategic potential of colloidal systems in CO_2 adsorption. By leveraging the unique properties of emulsions, ionic liquids, polymers, and aerogels, we can enable more efficient, sustainable, and scalable carbon capture technologies.

6.2 Emulsion Systems for CO_2 Adsorption

Emulsions are liquid–liquid dispersions typically stabilized by incorporating molecular surfactants or amphiphilic polymers that adsorb at the oil/water interface, preventing the coalescence of emulsified droplets through electrostatic and steric repulsive forces.[7] These characteristics enable a broader range of applications in industry and everyday products, including food, cosmetics, paints, and more.[8–10] Beyond these applications, the use of emulsions for CO_2 capture has emerged as a promising approach due to their unique properties and advantages. These colloidal systems provide high surface area contact between the gas and liquid phases, significantly enhancing the efficiency of CO_2 capture processes. Moreover, emulsions can be customized to optimize their properties for CO_2 sorption by adjusting the type and concentration of surfactants, droplet size, and the nature of the dispersed and continuous phases. These adjustments influence the mass transfer rate of CO_2 and the overall stability of the emulsion. Additionally, the versatility of emulsions allows for the incorporation of various additives that can enhance CO_2 solubility and reactivity, further improving the effectiveness of this innovative approach.

In this scenario, a series of formulations are being explored to obtain an emulsion that can be applied as a solvent for CO_2 sorption. As an example, Sridevi *et al.*[11] prepared an emulsion based on water, kerosene, sodium lauryl sulfate, sodium hydroxide, triethylamine and amine-functional ion exchange resin applied to CO_2 capture. Optical microscopy analysis of the emulsion revealed that the emulsion appeared completely homogeneous before the introduction of CO_2. After CO_2 was passed through the emulsion, a series of bubbles were observed in the sample, which also exhibited dark regions. These dark regions are likely due to the formation of bicarbonate ions, indicating that CO_2 interacted with the emulsion, leading to chemical changes within the sample. The results obtained by the group indicated that the emulsion system exhibited a CO_2 capture capacity of above 95%.[11]

A high internal phase emulsion (HIPE), another class of emulsion that can be applied to CO_2 capture, is a type of emulsion used to prepare a porous CO_2 adsorbing foam with high throughput, by stabilizing a more dispersed phase with a less continuous phase. A surfactant-free emulsion method can be used to prepare porous foams using an oil-in-water emulsion template, as demonstrated by Zhang *et al.*[12]

To obtain the porous foams, the initially prepared emulsion was subjected to photopolymerization by light irradiation. This process caused the emulsion to solidify into a foam structure. Following photopolymerization, the toluene solvent was completely removed under low vacuum conditions to ensure the formation of a stable and porous foam. The authors showed that increasing the fraction of the oil phase and shearing rate in the emulsion enriches PEI (polyethyleneimine) on pore surfaces, enhancing CO_2 capture capability due to increased interfacial areas and porosity.

Beyond these conventional approaches, emulsions utilizing ionic liquids represent a groundbreaking advancement. As a result, emulsions with ionic liquids open new horizons in the development of advanced materials and technologies for effective CO_2 capture and sequestration.

6.3 Ionic Liquid-based Emulsions

The evolution of ionic liquids (ILs) has been substantial since the initial publication of nitrates of alkylammonium bases in 1911 by Rây and Rakshit.[13] Ionic liquids present numerous advantages when used in CO_2 sorption processes. Firstly, ILs exhibit high selectivity, as they can be specifically designed to interact with CO_2. This targeted interaction leads to a high selectivity for CO_2 over other gases, thereby enhancing the efficiency of CO_2 capture, particularly in gas mixtures. One of the most significant benefits of ILs is their customizability. By selecting specific cations and anions, ILs can be tailored for specific applications. This ability to fine-tune IL properties allows them to meet the particular requirements of the sorption process, thereby enhancing overall performance.[14,15]

Furthermore, ILs have low volatility, with negligible vaporization preventing solvent loss into the gas stream during sorption processes. Their low vapor pressure ensures that the IL remains within the system, making them highly advantageous for gas separation applications. Additionally, IL exhibits high thermal and chemical stability, as well as excellent solvation capacity for a wide range of substances, which further supports their efficacy in CO_2 sorption.[16,17]

Lastly, ILs possess a high loading capacity, making them efficient sorbents for CO_2 sorption. Their capability to hold a substantial amount of CO_2 contributes to effective carbon capture from gas streams. Additionally, IL can be synthesized from renewable and non-toxic raw materials, aligning with sustainable practices for CO_2 sorption and utilization. In summary, ILs offer a promising and sustainable approach to CO_2 capture.[14]

Ionic liquids, frequently referred to as designer solvents, have garnered significant interest across various disciplines, including synthesis and catalysis, materials chemistry, polymer chemistry, electrochemistry, and biotechnology.[18,19]

It can be stated that the increasing interest in ILs is largely due to their ability to engage in a diverse array of intermolecular interactions, including strong and weak ionic interactions, hydrogen bonding, van der Waals forces, dispersive forces, and n–π and π–π interactions. These interactions confer exceptional solvation properties on ILs, setting them apart from traditional organic solvents.[20] However, the use of pure ILs presents certain drawbacks that should be taken into account. One primary concern is their high viscosity, which can hinder mass transfer rates and diffusion processes, ultimately impacting reaction kinetics.[21]

Additionally, the cost of ILs is relatively high compared to traditional organic solvents, making them less cost-effective for industrial applications.

Furthermore, limited solubility and potential environmental impact should be considered when using ILs. Despite being touted as greener alternatives, some ILs may still pose toxicity or disposal concerns. Therefore, careful evaluation of the environmental implications of IL use is essential.[22]

To address these drawbacks, researchers have explored integrating ILs into emulsion systems. This strategy offers the potential to alleviate some of the limitations associated with pure ILs and improve system performance overall. Leveraging the unique properties of ILs within emulsions may lead to better mass transfer rates, enhanced solubility, and potentially reduced costs, expanding the applicability of ILs in various chemical processes.[22,23]

In the past decade, a subclass of ionic liquids has seen a significant increase in the number of publications. These are known as surface-active ionic liquids (SAILs). These compounds feature long hydrophobic chains paired with hydrophilic segments, which can be present in either the cation or the anion.[24–26]

Within this context, the synthesis of emulsions using ionic liquids involves significant features, such as the presence of hydrophobic or hydrophilic nanodomains and a micelle core based on an IL. These characteristics enhance their potential applications in areas such as compound separation or extraction.[27]

The utilization of ionic liquids also facilitates the production of emulsions without the presence of water. This feature proves advantageous, especially when dealing with compounds that exhibit poor solubility in water. In such cases, specific ILs serve as the polar phase within these emulsions, allowing for formulations that are entirely free of water. These particular emulsions are commonly referred to as non-aqueous IL microemulsions. To prepare these microemulsions, an IL, a non-polar organic solvent, and a surfactant are necessary.[22]

In this context, researchers are investigating various precursors to develop an ionic liquid-based emulsion. However, emulsions with ionic liquids as the dispersed phase, though promising and supported by relevant studies on synthesis and characterization, have few studies regarding potential applications, particularly in relation to CO_2 adsorption.

Lad and Murthy (2014)[28] evaluated the average droplet size formed by the dispersed phase using the ionic liquid 1-butyl-3-methylimidazolium hexafluorophosphate [Bmim][PF_6] and the surfactant Tween-80. Among the results, the group demonstrated that the ratio between the concentrations of the IL and surfactant allows for the control of droplet size, with the smallest size being 38 μm when using 6% IL.

Li *et al.*[29] conducted a study evaluating water–IL–water emulsions with two ionic liquids, 1-octyl-3-methylimidazolium hexafluorophosphate [Omim][PF_6] and 1-butyl-3-methylimidazolium hexafluorophosphate [Bmim][PF_6], in the presence of a surfactant. The results presented by the group indicate two behaviors dependent on the water volume fraction in the system.[29] The behavior occurs within the range of 0.68 to 0.75, enabling the formation of multiple emulsions, meaning droplets encapsulated by external droplets.

Recently, Birrer *et al.*[30] studied the effects of a range of different surfactants on the stability, interfacial tensions, and solubilization of various imidazolium and phosphonium IL-in-water emulsions. Their research revealed that the stability of IL-in-water emulsions in surfactant solutions is heavily influenced by the surfactant's type and charge. They observed that even at surfactant concentrations significantly higher than the critical micelle concentration, emulsions were generally unstable beyond a few minutes. However, some emulsions with stability longer than one month were reported.[30]

Hasib-ur-Rahman and Larachi (2013)[31] investigated the kinetic behavior of CO_2 absorption at room temperature in emulsions comprising diethanolamine (DEA) and an ionic liquid, 1-hexyl-3-methylimidazolium bis(trifluoromethylsulfonyl)imide [Hmim][Tf_2N], using a laboratory-scale stirred-cell reactor. The study aimed to evaluate the impact of amine concentration, CO_2 partial pressure, stirring rate, and temperature variations on the absorption process.

Their findings emphasize the critical role of dispersing amine within the ionic liquid phase to optimize absorption capacities in these immiscible systems. Stirring rate was identified as a significant factor influencing the rate of CO_2 capture, with higher speeds enhancing absorption efficiency.

Furthermore, the research investigated the effects of varying amine concentration, CO_2-to-gas ratio, and temperature on absorption dynamics. Overall, the study highlights the unique advantages of amine–ionic liquid blends, including the potential for precipitating CO_2-captured products, which could inform the design of efficient absorption columns for CO_2 capture applications.[31]

Iliuta *et al.*[32] also explored the efficiency of CO_2 capture using a blend of diethanolamine (DEA) and the ionic liquid [Hmim][Tf_2N]. They employed a gas–liquid/liquid stirred-cell reactor to investigate the effects of various parameters such as amine concentration, CO_2 partial pressure, agitation speed, and temperature on the kinetics of CO_2 absorption. The absorption mechanism is explained through the zwitterion mechanism, which results in the formation of solid carbamate crystals within the ionic liquid phase. This approach demonstrates the potential for efficient CO_2 capture with lower energy requirements for regeneration and minimal environmental impact compared to traditional aqueous amine solutions.[32]

Liu *et al.*[33] conducted a study on the development of CO_2-responsive microemulsions utilizing an ionic liquid, 1,1,3,3-tetramethylguanidine-oleic acid (TMG-OA), as a surfactant to mitigate the volatilization of amines effectively. The synthesis and characterization of TMG-OA highlight its impressive switchability, stability, and surface-active properties, positioning it as a promising candidate for sustainable applications. The reversible transformations of microstructures within the microemulsions are governed by the interconversion between TMG-OA and oleic acid constituents, facilitating controlled microstructural adjustments triggered by CO_2. While the study primarily focuses on the reversible changes in microstructure induced by CO_2, it does not extensively explore the CO_2 adsorption characteristics of TMG-OA.

In conclusion, the high costs associated with ionic liquids represent a primary concern in the development of CO_2 capture technologies. Consequently, there is a need for a system that ensures high capture efficiency at minimal expense. One potential approach involves enhancing the CO_2–ionic liquid contact area,[34] which can be achieved through emulsion synthesis. Emulsions may exhibit greater gas sorption capacity compared to single-phase solutions because the dispersed phase can contain nanoscale droplets, thereby significantly increasing the gas contact area and consequently enhancing mass transfer rates.[33]

Building on this, recent advancements have expanded the use of emulsions beyond CO_2 sorption to include the preparation of polymers specifically designed for this purpose. By utilizing the unique properties of emulsions in the synthesis of these polymers, a new class of materials is being developed with higher efficiency and performance in CO_2 capture systems.

6.4 Polymers Through Emulsion Methods

Emulsion-formed polymers are produced through free radical polymerization, where an initiator and a surfactant are present in an aqueous phase while the monomer remains insoluble (organic phase), resulting in a heterogeneous system, see Figure 6.2. A key feature of this method is the formation of polymers with high molecular weight, facilitated by the compartmentalization of growing polymer chains within micelles, which serve as reaction centers.[35]

Sattari *et al.*[36] published a review focusing on various classes of polymers utilized for CO_2 adsorption, with particular emphasis on porous polymers. These polymers are noted for their high adsorption capacity and physicochemical stability. The classes discussed include porous organic polymers (POPs), porous coordination polymers (PCPs), microporous organic polymers (MOPs) and amine-functionalized porous polymers.[36]

Porous polymers are best produced using emulsion polymerization techniques, such as the HIPE (high internal phase emulsion) method, enabling

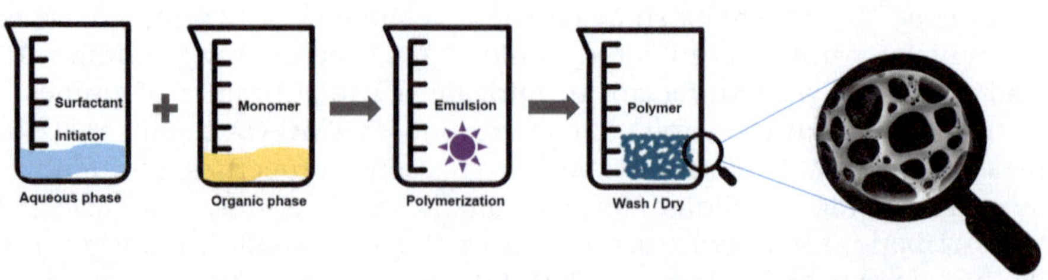

Figure 6.2 Illustration of the emulsion polymerization procedure for porous polymers.

precise control over the size of pores within the polymeric network. Furthermore, this approach can lead to interconnected pores, creating a three-dimensional macroporous structure that enhances CO_2 adsorption efficiency, as shown in some studies.[37–39]

The monomers described for emulsion polymerization usually have acrylic or vinyl functions, the polymer base most applied to paints and varnishes, for example. The emulsion polymers used in the synthesis reported in the studies evaluated are mostly *n*-butyl acrylate (*n*-BA), methyl methacrylate (MMA), methacrylic acid (MAA), styrene (S), divinylbenzene (DVB), and 4-vinylbenzyl chloride (VBC). It is important to note that in emulsion polymerization there is a predominant combination of two monomers, a process known as co-polymerization.[35]

In addition to the monomers, which are in the organic phase (oil), an initiator and a surfactant also take part in this polymerization in the aqueous phase. The use of persulphates (mainly ammonium and potassium) as initiators is well-popularized due to their low cost and action in the reaction temperature range of 60–95 °C, for example.

Given that the initiator generates a negatively charged species, it is crucial to use tensoactives (surfactants) to ensure more effective colloidal stability, based on electrostatic repulsion between the particles. Among the most commonly used are: sorbitan oleate (Span 80), sodium dodecyl sulfate (SDS), cetyltrimethylammonium bromide (CTAB), polyoxyethylene–polyoxypropylene block copolymer (Pluronic F-68) and others.

Starting with the structured polymer, the proposed functionalization to improve the CO_2 adsorption capacity of these materials is implemented.

Depending on the polymer structure, post-synthesis functionalization can be carried out by modifying the surface to increase reactivity and graft the amine groups onto the polymer.[39] An example is the incorporation of silanes such as vinyltriethoxysilane (VTEO) and tetraethyl orthosilicate (TEOS), as proposed by Wang *et al.*,[40] which enhances solubility in the organic phase and creates active sites without damaging the structure, as stated by the author.

Amine-functionalized porous polymers are widely used in CO_2 capture processes due to the common use of amine solutions as solvents. Thus, the amine groups dispersed in the polymers significantly increase their CO_2 adsorption performance. Many types of amines are used to modify adsorbents, such as ethylenediamine (EDA), triethylenetetramine (TETA), tetraethylenepentamine (TEPA), imidazole (IM), aminoalkoxysilane (AAS), aziridine (AZI), piperazine, polyethyleneimine (PEI), and others.[40] Furthermore, studies on CO_2 capture solvents have always reported better gas selectivity in fluorinated compounds. In ionic liquids, anions such as $[BF_4]^-$, $[PF_6]^-$ and $[NTf_2]^-$, especially the latter, showed good CO_2 sorption capacity.[41]

In this regard, the work of Stiernet, who developed amino/imidazolium-based bifunctional poly(HIPE) *via* a one-pot procedure, is noteworthy. This is because, in addition to using functionalized monomers, it performed anion exchange for

Table 6.1 Overview of CO_2 adsorption data on emulsion polymers.

Polymer poly (HIPE)	Monomers	Functionalization	CO_2 adsorption ($mmol\,g^{-1}$)	Ref.
Poly(VBC–DVB)	Vinylbenzylchloride Divinylbenzene	Ethylenediamine Triethylene Tetramine	Poly(VBC–DVB)/ EDA = 0.32 Poly(VBC–DVB)/ TETA = 0.33	37
Poly(EHMA–DVB)	Ethylhexyl methacrylate Divinylbenzene	Polyethylenimine	Poly(EHMA–DVB)/ PEI = 4.0	43
Poly(S–DVB)	Styrene Divinylbenzene	Polyethylenimine Tetraethylene Pentamine	Poly(S–DVB)/ PEI = 0.82 Poly(S–DVB)/ TEPA = 0.72	44
Poly(S–DVB)	Styrene Divinylbenzene	Aziridine	Poly(S–DVB)/ AZI = 1.71	40
Poly(AEAm–IBAm)	Aminoethyl acrylamide Imidazolium bis-acrylamide	Anions exchange, NaOAc and LiTFSI	Poly(AEAm–IBAm)/ OAc = 0.44 Poly(AEAm–IBAm)/ TFSI = 1.79	42

acetate (OAc) or $[CH_3COO]^-$ and bis-trifluoromethanesulfonimidate (TFSI) or $[NTf_2]^-$, and again, the latter fluorinated exhibited superior CO_2 adsorption results.[42]

Specifically for emulsion polymers, there is recent research on CO_2 adsorption that warrants closer examination, as outlined in Table 6.1.

The CO_2 adsorption results of Muchan and co-authors for poly(VBC–DVB) show that functionalization with TETA was slightly more effective, obviously because its structure has twice as many amine groups as EDA. The tests were carried out at room temperature, atmospheric pressure, and with a gas mixture of 4% CO_2/N_2.[37]

The proposed functionalization for poly(EHMA–DVB) was the impregnation of polyethyleneimine in different loadings (20 to 70 wt%). In addition to the amount of PEI, the influence of temperature, moisture (vol% H_2O) and regenerability after 10 CO_2 adsorption/desorption cycles with the 10% CO_2/N_2 gas mixture was evaluated. The most promising result was 4 $mmol\,g^{-1}$ of CO_2 adsorbent for poly(EHMA–DVB) with 60% PEI, 70 °C and 3 vol% H_2O.[43]

Azman *et al.*[44] prepared poly(S–DVB) with three different ratios of S–DVB monomers (0 : 100, 20 : 80, 80 : 20), as well as surface modification using PEI and TEPA. The CO_2 adsorption tests were conducted with a gas mixture of 15% CO_2/N_2 at 2 bar and room temperature. According to the results presented by the authors, all the S–DVB combinations had better results when modified with PEI, with the 20 : 80 ratio exhibiting the results shown in Table 6.1, again the amount of amine groups is greater than TEPA. Some differences in properties such as specific surface area and degradation temperature between the poly(HIPE) were reported in the study.[44]

Following a similar approach with poly(S–DVB), Wang employed aziridine already in the synthesis stage of poly(HIPE), termed copolymerization with *in situ* anchoring. Additionally, hydroxyl groups were introduced through silanols derived from silicon alkoxides (VTEO and TEOS). These modifications led to the production of porous polymers with a high specific surface area (513 $m^2 g^{-1}$) and achieving a CO_2 adsorption capacity of 1.71 mmol g^{-1} in tests conducted with 10% CO_2/N_2 at 25 °C.[40]

An interesting proposal was the use of amino-functionalized monomers.[42] The bifunctional amino/imidazolium porous polymers poly(AEAm–IBAm) were synthesized in a one-step HIPE polymerization using various monomer combinations. The values presented for CO_2 adsorption refer to the composition of 67% amino (AEAm) and 33% imidazolium (IBAm). Additionally, anion exchange (specifically OAc and TFSI, as previously mentioned) was conducted, effectively enhancing the CO_2 adsorption capacities of the materials.

Finally, a recent paper shows the prospects for the manufacture and application of environmentally friendly colloid systems, in which CO_2-switchable surfactants are used in emulsions, commonly long-chain tertiary amines.[45] In one way or another, the amines must be incorporated into the polymer for CO_2 adsorption, either through post-synthesis impregnation, as a functionalized monomer, or even as the surfactant responsible for the emulsion. The applications of these reactive colloids include emulsion polymerization, catalysis, controlled drug release, and enhanced oil recovery, the latter a very established field of activity for CO_2 capture.

6.5 Aerogel-based Polymer Systems

Another class of materials that has been widely studied for CO_2 capture is aerogels. An aerogel is a porous material synthesized through a sol–gel approach, followed by the replacement of the liquid in the gel with a gas, resulting in a highly porous structure.[46,47] These materials possess unique properties, such as adjustable density, which can vary from 1000 kg m^{-3} to 1 kg m^{-3} (above the solid density in the former and lower than the density of air in the latter).[47] Additionally, they can exhibit ultralow thermal conductivity and dielectric constant, high surface area, among other properties.[47,48] Aerogels can be classified according to their shape (as powders, films, or monoliths), porous structures (microporous, mesoporous, macroporous, or with mixed porosity), or composition (organic, inorganic, or hybrid).[49,50]

The control of the synthesis procedures is key to adjusting the final properties of the aerogel. Commonly, aerogels are prepared following three crucial steps using the sol–gel technique: (i) sol–gel preparation, (ii) aging, and (iii) drying.[49,51] Figure 6.3 exhibits the general preparation steps of aerogels. In short, the precursors and solvents are mixed, and after hydrolysis, they react to form a colloidal solution of small polymeric particles and/or dissolved polymer clusters. These particles then grow and agglomerate, ultimately forming a three-dimensional solid network. During the

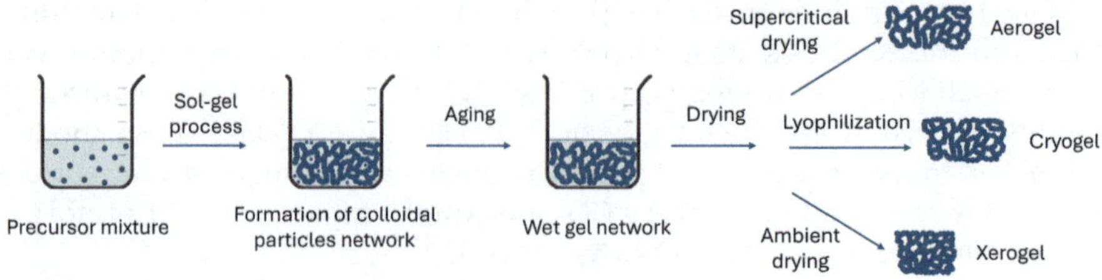

Figure 6.3 General preparation steps of aerogels.

aging step, the unreacted species present in the liquid react with the hydrolyzed surface of the gel, strengthening the solid structure. Finally, during the drying step, the solvent inside the wet gel is replaced with air without seriously damaging the microstructure network.[46,51] The liquid from the gel can be extracted using different techniques, such as supercritical drying, lyophilization, and ambient drying (Figure 6.3). Each of these procedures will result in a solid network with different degrees of pore structure shrinkage.[49] After obtaining the aerogel, further modifications can be performed, such as surface functionalization or carbonization, depending on the intended application.[51,52]

Polymer aerogels can be synthesized using different precursors, such as resorcinol/formaldehyde (RF),[53] melamine/formaldehyde (MF),[54] polyimide,[55] and poly(vinyl alcohol) (PVA),[56] among others. Additionally, organic aerogels can be prepared from bio-based polymers, such as cellulose[50] and chitosan.[57] Due to their properties (high porosity, low density, and relatively higher mechanical strength compared to silica-based aerogels), polymer aerogels are promising candidates for CO_2 adsorption.[49,51]

The CO_2 adsorption capacity of aerogels depends on several factors, such as the texture and surface chemistry of the aerogel, the CO_2 flow pressure, and temperature, among others.[50] Anas *et al.*[58] evaluated the CO_2 adsorption capacity of silica, resorcinol–formaldehyde, carbon, and wheat starch aerogels. They found that the excess CO_2 uptake was proportional to the mesopore surface area (calculated by subtracting the micropore area from the BET specific surface area) of the aerogels at various pressures. Silica aerogel ($S_{BET} = 1158\ m^2\ g^{-1}$) had a maximum absolute CO_2 adsorption uptake of 14 $mmol\ g^{-1}$, and RFA-19 (resorcinol–formaldehyde aerogel, $S_{BET} = 620\ m^2\ g^{-1}$) had a maximum uptake of 13 $mmol\ g^{-1}$ at 35 °C and 35 bar.

Table 6.2 shows a comparison of the CO_2 adsorption capacities for different types of polymer-based aerogels reported in the literature.

Another strategy to increase CO_2 uptake is to functionalize the surface of the aerogel with amine groups.[49,59,64] Zhu *et al.*[59] synthesized two nanocellulose aerogels, cellulose nanofibrils (CNFs) and cellulose nanocrystals (CNCs), and functionalized their surfaces with amino groups (*N*-(2-aminoethyl)-3-aminopropylmethyldimethoxysilane, AEAPDMS). The amino-functionalized aerogels (AEAPDMS-CNC and AEAPDMS-CNF) exhibited CO_2 uptakes of 1.81 and 2.44 $mmol\ g^{-1}$, respectively (measured at 25 °C and 1 bar of CO_2), which

Table 6.2 Summary of the CO_2 adsorption capacity of various polymer-based aerogels.

Aerogel	Analysis conditions	CO_2 adsorption capacity ($mmol\,g^{-1}$)	Ref.
Epoxy-functionalized polyethyleneimine modified epichlorohydrin-cross-linked cellulose aerogel (EBPCa)	25 °C	6.45	60
Amino-functionalized nanocellulose aerogels (AEAPDMS-CNF and AEAPDMS-CNC)	25 °C, 1 bar	1.81 (CNC) 2.44 (CNF)	59
Amine hybrid resorcinol–formaldehyde/silsesquioxane composite (AH-RFSA)	30 °C, 1% CO_2	3.57 (dry) 4.43 (humid)	61
Resorcinol–formaldehyde aerogel carbonized at 1050 °C	0 °C	6.43	62
Carbon aerogel derived from lignin and cellulose nanofibers	0 °C	5.23	63

were considerably higher than the CO_2 uptakes of the pristine aerogels (0.13 and 0.17 for CNCs and CNFs, respectively).

For further details on factors influencing CO_2 adsorption by aerogels, readers are referred to the works of Baraka and Labidi[50] and Keshavarz *et al.*[49]

6.6 Aerogel-based Inorganic Systems

As mentioned, aerogels can be classified according to their chemical composition. Inorganic aerogels are an important class of materials, and, compared to polymer-based aerogels, their uniqueness lies in their high thermal stability, an important property of most inorganic substances. This property, combined with their high specific area, chemical stability, and suitable porous structure, makes the use of inorganic aerogels attractive for the selective capture of CO_2 in fossil fuel power plants. The inorganic aerogels include silica aerogels, oxide aerogels, and mixed inorganic aerogels.

Silica aerogels are the most studied, however, their ability to adsorb CO_2 is restricted to physical adsorption due to the inactive surface silanol groups (≡Si–OH). Cui *et al.*[65] reported that modification of silica aerogel with APTES (3-(triethoxysilyl)propan-1-amine) resulted in a decrease in specific area and pore volume, however, the CO_2 capture capacity increased by 200% compared to the unmodified aerogel.[65] Figure 6.4 presents a general preparation scheme of pure silica aerogels (PSAs) and amino-modified silica aerogels (AMSAs) by a grafting methodology. The preparation involves the dripping of HCl and $NH_3{\cdot}H_2O$, respectively, into a solution containing the silica precursor (TEOS) and the solvent (ethanol, EtOH) to promote the hydrolysis and condensation reactions, thus forming the silica gels. The formed gel is aged for 1 day at room temperature and the EtOH and hexane solutions are replaced three times and subsequently, the material is dried at

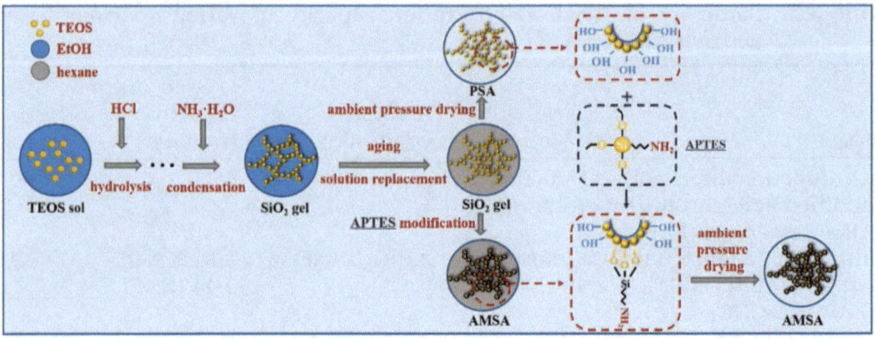

Figure 6.4 General scheme of the preparation of pure silica aerogels (PSA) and amine-modified silica aerogels (AMSA). Reproduced from ref. 66 with permission from Elsevier, Copyright 2023.

ambient pressure. The AMSA material is obtained after the replacement step, adding the APTES, which diffuses into the gels and grafts onto the surface of the aerogel.

Thus, attention has been directed to the study and modification of the silica aerogel surface with amino groups, which have been shown to play an important role in the adsorptive performance. Wörmeyer *et al.*[67] and Kong *et al.*[68] demonstrated that the number of amino groups significantly influences the adsorption capacity. The CO_2 adsorption of the silica aerogel functionalized with tri-amino groups was superior (0.87 mmol g^{-1} at 250 Pa) to the aerogel functionalized with mono-amino groups. Regarding amino modification methodology, the two main ones are grafting (shown in Figure 6.4) and impregnation by pore volume considering weak molecular forces. The grafting method demonstrated better stability after several cycles of adsorption and desorption, while impregnation by pore volume presents higher adsorption capacity.[69]

Silica aerogels can be synthesized from different silica precursors, including tetramethoxysilane (TMOS), tetraethoxysilane (TEOS), and methyltriethoxysilane (MTES). TMOS exhibits a higher hydrolysis rate than TEOS, and results in narrow and uniform pores, and a higher specific area. Trimethoxymethylsilane (MTMS) can be added to TMOS to increase hydrophobicity, and TMOS can be added to TEOS to increase hydrophobicity.[70]

From an economic perspective, the cost associated with the production of aerogels is still an obstacle to be overcome. Supercritical drying processes, despite producing better quality aerogels, require high pressure and temperature conditions. Another obstacle lies in the cost and amount of waste to obtain the raw materials. Although silica is relatively cheap, the commonly used silica precursor alkoxides (TMOS, TEOS, TPOS, MTES, MTMS, and DMDMS) to produce silica aerogels are expensive for large-scale implementation. Some authors have used waste and/or low-cost raw materials containing silica. Fan *et al.*[71] used fly ash silica to obtain aerogels modified with APTES, achieving CO_2 adsorption of 4.95 mmol g^{-1} at 70 °C.

Rani *et al.*[72] used rice husk silica to produce sodium silicate and then silica aerogels impregnated with APTES. The material exhibited CO_2 adsorption of 0.88 mmol g^{-1}. In general, there is a gain from an environmental point of view, while the adsorption performance is reduced.

References

1. *International Energy Agency – IEA Global Energy Review 2021*, 2021, https://www.iea.org/reports/global-energy-review-co2-emissions-in-2021-2.
2. *British Petroleum – BP Statistical Review of World Energy 2021*, 2021, https://www.bp.com/content/dam/bp/business-sites/en/global/corporate/pdfs/energy-economics/statistical-review/bp-stats-review-2021-full-report.pdf.
3. *International Energy Agency – IEA Global Energy Review 2023*, 2023, https://www.iea.org/reports/co2-emissions-in-2023.
4. N. MacDowell, N. Florin, A. Buchard, J. Hallett, A. Galindo, G. Jackson, C. S. Adjiman, C. K. Williams, N. Shah and P. Fennell, *Energy Environ. Sci.*, 2010, **3**, 1645.
5. M. Bui, C. S. Adjiman, A. Bardow, E. J. Anthony, A. Boston, S. Brown, P. S. Fennell, S. Fuss, A. Galindo, L. A. Hackett, J. P. Hallett, H. J. Herzog, G. Jackson, J. Kemper, S. Krevor, G. C. Maitland, M. Matuszewski, I. S. Metcalfe, C. Petit, G. Puxty, J. Reimer, D. M. Reiner, E. S. Rubin, S. A. Scott, N. Shah, B. Smit, J. P. M. Trusler, P. Webley, J. Wilcox and N. Mac Dowell, *Energy Environ. Sci.*, 2018, **11**, 1062.
6. M. E. Boot-Handford, J. C. Abanades, E. J. Anthony, M. J. Blunt, S. Brandani, N. M. ac Dowell, J. R. Fernández, M.-C. Ferrari, R. Gross, J. P. Hallett, R. S. Haszeldine, P. Heptonstall, A. Lyngfelt, Z. Makuch, E. Mangano, R. T. J. Porter, M. Pourkashanian, G. T. Rochelle, N. Shah, J. G. Yao and P. S. Fennell, *Energy Environ. Sci.*, 2014, **7**, 130.
7. J. Tang, P. J. Quinlan and K. C. Tam, *Soft Matter*, 2015, **11**, 3512.
8. W. Meng, H. Sun, T. Mu and M. Garcia-Vaquero, *Trends Food Sci. Technol.*, 2024, **150**, 104610.
9. D. Venkataramani, A. Tsulaia and S. Amin, *Adv. Colloid Interface Sci.*, 2020, **283**, 102234.
10. C. Thillaye du Boullay, M. Jaber, M. Le Denic, F. Gerony, R. Bordes, G. Mériguet, A.-L. Rollet, P. Walter and L. de Viguerie, *Colloids Surf., A*, 2023, **673**, 131816.
11. M. Sridevi, K. Muralirajan and S. Angayarkanny, *Mater. Lett.*, 2024, **364**, 136335.
12. S. Zhang, D. Wang, Q. Pan, Q. Gui, S. Liao and Y. Wang, *ACS Appl. Mater. Interfaces*, 2017, **9**, 34497.
13. P. C. Rây and J. N. Rakshit, *J. Chem. Soc., Trans.*, 1911, **99**, 1470.
14. Z.-Z. Yang, Y.-N. Zhao and L.-N. He, *RSC Adv.*, 2011, **1**, 545.
15. M. M. Dharman, H.-J. Choi, D.-W. Kim and D.-W. Park, *Catal. Today*, 2011, **164**, 544.

16. N. H. Solangi, A. Anjum, F. A. Tanjung, S. A. Mazari and N. M. Mubarak, *J. Environ. Chem. Eng.*, 2021, **9**, 105860.
17. Z. Lei, C. Dai, J. Hallett and M. Shiflett, *Chem. Rev.*, 2024, **124**, 7533.
18. N. V. Plechkova and K. R. Seddon, *Chem. Soc. Rev.*, 2008, **37**, 123.
19. J. P. Hallett and T. Welton, *Chem. Rev.*, 2011, **111**, 3508.
20. H. Olivier-Bourbigou, L. Magna and D. Morvan, *Appl. Catal., A*, 2010, **373**, 1.
21. M. A. Martins, C. M. Neves, K. A. Kurnia, P. J. Carvalho, M. A. Rocha, L. M. Santos, S. P. Pinho and M. G. Freire, *Fluid Phase Equilib.*, 2016, **407**, 188.
22. M. Hejazifar, O. Lanaridi and K. Bica-Schröder, *J. Mol. Liq.*, 2020, **303**, 112264.
23. O. Zech, A. Harrar and W. Kunz, in *Ionic Liquids: Theory, Properties, New Approaches*, ed. K. Alexander, IntechOpen, 2011, p. 245.
24. I. A. Sedov and B. N. Solomonov, *Fluid Phase Equilib.*, 2016, **425**, 9.
25. R. L. Vekariya, A. Dhar and J. Lunagariya, *Compos. Interfaces*, 2017, **24**, 801.
26. R. L. Vekariya and N. S. Kumar, *Colloids Surf., A*, 2017, **529**, 203.
27. Y. Gao, J. Zhang, H. Xu, X. Zhao, L. Zheng, X. Li and L. Yu, *Chem. Phys. Chem.*, 2006, **7**, 1554.
28. V. N. Lad and Z. V. P. Murthy, *J. Ind. Eng. Chem.*, 2014, **20**, 274.
29. J. Li, J. Zhang, B. Han, Y. Zhao and G. Yang, *J. Colloid Interface Sci.*, 2012, **368**, 395.
30. S. G. Birrer, P. Quinnan and L. D. Zarzar, *Langmuir*, 2023, **39**, 10795.
31. M. Hasib-ur-Rahman and F. Larachi, *Sep. Purif. Technol.*, 2013, **118**, 757.
32. I. Iliuta, M. Hasib-ur-Rahman and F. Larachi, *Chem. Eng. J.*, 2014, **240**, 16.
33. H. Liu, S. Zhan, R. Li, Y. Liu, P. Guo, Z. Wang, J. Du, Y. Wen, P. Dai and H. Liao, *Chem. Eng. J.*, 2020, **400**, 125369.
34. I. S. Elizarova and P. F. Luckham, *J. Colloid Interface Sci.*, 2017, **491**, 286.
35. P. A. Lovell and F. J. Schork, *Biomacromolecules*, 2020, **21**, 4396.
36. A. Sattari, A. Ramazani, H. Aghahosseini and M. K. Aroua, *J. CO_2 Util.*, 2021, **48**, 101526.
37. P. Muchan, C. Saiwan, D. deMontigny and P. Tontiwachwuthikul, *Energy Procedia*, 2013, **37**, 151.
38. H. He, W. Li, M. Lamson, M. Zhong, D. Konkolewicz, C. M. Hui, K. Yaccato, T. Rappold, G. Sugar, N. E. David, K. Damodaran, S. Natesakhawat, H. Nulwala and K. Matyjaszewski, *Polymer*, 2014, **55**, 385.
39. T. Zhang, R. A. Sanguramath, S. Israel and M. S. Silverstein, *Macromolecules*, 2019, **52**, 5445.
40. S. Wang, X. Qiu, Y. Chen and S. Chen, *Microporous Mesoporous Mater.*, 2022, **330**, 111585.
41. A. S. Aquino, F. L. Bernard, M. O. Vieira, J. V. Borges, M. F. Rojas, F. D. Vecchia, R. A. Ligabue, M. Seferin, S. Menezesd and S. Einloft, *J. Braz. Chem. Soc.*, 2014, 25.

42. P. Stiernet, M. Mazaj, S. Kovačič and A. Debuigne, *Chem. Eng. J.*, 2022, **446**, 137012.
43. M. Irani, A. T. Jacobson, K. A. M. Gasem and M. Fan, *Energy*, 2018, **157**, 1.
44. M. I. Azman, J. Chungprempree, J. Preechawong, P. Sapsrithong and M. Nithitanakul, *Polymers*, 2021, **13**, 2247.
45. S. Dowlati, R. Mokhtari, L. Hohl, R. Miller and M. Kraume, *Adv. Colloid Interface Sci.*, 2023, **315**, 102907.
46. L. Zuo, Y. Zhang, L. Zhang, Y. E. Miao, W. Fan and T. Liu, *Materials*, 2015, **8**, 6806.
47. A. Du, B. Zhou, Z. Zhang and J. Shen, *Materials*, 2013, **6**, 941.
48. A. C. Pierre and G. M. Pajonk, *Chem. Rev.*, 2002, **102**, 4243.
49. L. Keshavarz, M. R. Ghaani, J. M. D. MacElroy and N. J. English, *Chem. Eng. J.*, 2021, **412**, 128604.
50. F. Baraka and J. Labidi, *Sci. Total Environ.*, 2024, **912**, 169093.
51. S. Montes and H. Maleki, in *Colloidal Metal Oxide Nanoparticles*, ed. S. Thomas, A. Tresa Sunny and P. Velayudhan, Elsevier, 2020, p. 337.
52. A. S. Jatoi, Z. Hashmi, S. A. Mazari, R. Abro and N. Sabzoi, *J. Porous Mater.*, 2021, **28**, 1919.
53. S. A. Al-Muhtaseb and J. A. Ritter, *Adv. Mater.*, 2003, **15**, 101.
54. E. Masika and R. Mokaya, *RSC Adv.*, 2013, **3**, 17677.
55. Y. Gao, C. Dong, F. Zhang, H. Ma and Y. Li, *J. Appl. Polym. Sci.*, 2022, **139**, 1.
56. H.-B. Chen, E. Hollinger, Y.-Z. Wang and D. A. Schiraldi, *Polymer*, 2014, **55**, 380.
57. A. A. Alhwaige, T. Agag, H. Ishida and S. Qutubuddin, *RSC Adv.*, 2013, **3**, 16011.
58. M. Anas, A. G. Gönel, S. E. Bozbag and C. Erkey, *J.* CO_2 *Util.*, 2017, **21**, 82.
59. W. Zhu, M. Chen, J. Jang, M. Han, Y. Moon, J. Kim, J. You, S. Li, T. Park and J. Kim, *Carbohydr. Polym.*, 2024, **323**, 121393.
60. X. Chen, J. Lin, H. Wang, Y. Yang, C. Wang, Q. Sun, X. Shen and Y. Li, *Carbohydr. Polym.*, 2023, **302**, 120389.
61. Y. Kong, X. Shen, S. Cui and M. Fan, *Appl. Energy*, 2015, **147**, 308.
62. J. R. Guarín Romero, J. C. Moreno-Piraján and L. Giraldo Gutierrez, *C*, 2018, **4**, 52.
63. S. Geng, J. Wei, S. Jonasson, J. Hedlund and K. Oksman, *ACS Appl. Mater. Interfaces*, 2020, **12**, 7432.
64. A. Sami, N. E. Rabat, K. Johari, M. R. Shamsuddin, S. M. M. Ghani, A. S. M. Ghumman and M. S. Khan, *Colloids Surf., A*, 2024, **690**, 133778.
65. S. Cui, W. Cheng, X. Shen, M. Fan, A. Russell, Z. Wu and X. Yi, *Energy Environ. Sci.*, 2011, **4**, 2070.
66. J. Feng, L. Fan, M. Zhang and M. Guo, *Colloids Surf., A*, 2023, **656**, 130510.
67. K. Wörmeyer and I. Smirnova, *Chem. Eng. J.*, 2013, **225**, 350.

68. Y. Kong, G. Jiang, M. Fan, X. Shen, S. Cui and A. G. Russell, *Chem. Commun.*, 2014, **50**, 12158.
69. B. Cui, X. Ju, H. Ma, S. Meng, Y. Liu, J. Wang, D. Wang and Z. Yang, *Sep. Purif. Technol.*, 2025, **354**, 128794.
70. L. Keshavarz, M. R. Ghaani and N. J. English, *Molecules*, 2021, **26**, 5023.
71. L. Fan, Y. Mu, J. Feng, F. Cheng, M. Zhang and M. Guo, *Chem. Eng. J.*, 2023, **452**, 138945.
72. H. A. N. Rani, F. N. Mohammad, I. S. S. Jamaludin, F. M. K. N. Teo, S. M. N. Sah and J. M. Jalil, *Recent Innovations Chem. Eng.*, 2020, **13**, 407.

CHAPTER 7

Colloids and Solar Energy Interactions

MAIARA DE JESUS BASSI, JÚLIA KETZER MAJEWSKI AND LUCIMARA STOLZ ROMAN*

Federal University of Paraná, Physics Department, 81531-980, Curitiba, PR, Brazil
*Email: lucimara.roman@ufpr.br

7.1 Introduction

Energy is one of the basic principles vital to human society and its maintenance on Earth. Furthermore, energy consumption has increased every year to pursue technological and industrial development across the world.

Energy based on fossil fuels currently meets most energy requirements, however, these non-renewable sources are responsible for the production of carbon dioxide (CO_2) and other greenhouse gases, which are the main components of climate change and relevant environmental impacts.

In search of new environmentally friendly alternatives, great attention is being devoted to photovoltaic technology that captures energy from sunlight providing a supply of clean, safe, and sustainable energy.

The predominant technology for converting solar energy into electrical energy is currently based on inorganic materials and has been used as sources of renewable electrical energy in applications ranging from satellites to residential roofs. However, the manufacturing processes for these devices are complex, involving several steps, using very high temperatures, in addition to being rigid and difficult to integrate into some surfaces.

Contribution of Colloidal Materials to Air, Water and Soil Environmental Sustainability: From Basic Concepts to Advanced Applications
Edited by Izabel C. Riegel-Vidotti, Lizandra Maria Zimmermann and Heloise R. de Barros

Published by the Royal Society of Chemistry, www.rsc.org

Organic photovoltaic devices (OPVs), consisting of organic semiconductor materials, have attracted notable interest as a possible alternative to conventional photovoltaic technologies. These have great potential for manufacturing light flexible substrates over large areas, through solution processing, enabling deposition by roll-to-roll printing, in addition to the low-cost perspective.[1]

Despite advances, the active layer of OPVs is mainly processed from toxic halogenated or aromatic solvents such as chloroform (CB), chlorobenzene (CB) and dichlorobenzene (DCB), as shown in Table 7.1, which are harmful to the environment, human health and limit the large-scale production and expansion of OPVs.[2] Therefore, replacing harmful solvents with green solvents, such as water, is an environmentally friendly alternative that can be accepted by the industry.

Due to the water insolubility of the polymers/molecules used as the active layer in OPVs, a strategy for the correct and ecological processing of these materials is through the synthesis of organic semiconductor nanoparticles (NPs), with colloidal dispersion, through a miniemulsion process. The original concept of this miniemulsion technique was developed by Landfester,[3] and since then several studies have been carried out using this method.[4]

In addition to this technique presenting the advantage of using environmentally friendly solvents, it has also demonstrated a level of control over the morphology of the active layer unattainable by conventional processing methods, preforming the desired nanoscale phase structure of the particles prior to device fabrication and facilitating the transfer of charge and energy between materials that are limited to the diameter of the NPs.[5] These characteristics are extremely important

Table 7.1 Chemical structure and associated chemical risks for the commonly used solvents for OPV fabrication.

Solvent	Chloroform	Chlorobenzene	Dichlorobenzene	Water
Chemical structure	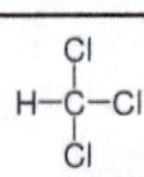	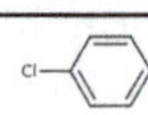	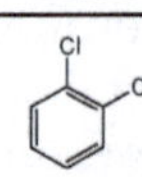	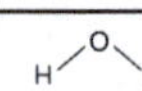
Pictograms				

for photovoltaic performance. Colloids can range from nanospherical shapes, as shown in Figure 7.1(a), to nanodonuts.[6] The structure of the nanodonuts shown in Figure 7.1(b) may bring new insights into the engineering of active layers for solar cells.

The design of organic solar cells using aqueous nanoparticles in the active layer is still reduced to the laboratory scale due to the difficulties of deposition and homogeneity of the films, as shown in Figure 7.2. But with advances in technology, the deposition of aqueous NPs is now possible using the slot-die printing method, which can be adapted into a roll-to-roll process, which is an important factor for industrialization.[7]

This chapter was prepared with the aim of providing a roadmap on the formation of aqueous dispersions of semiconductor nanoparticles, presenting its optoelectrical characteristics and application as an active layer of OPVs.

Figure 7.3 shows the chemical structure of the materials used in the devices discussed here. The electrical, optical, and morphological aspects of nanostructured active layers for devices in bilayer and bulk heterojunction architectures will be addressed. The conjugated polymer PSiF-DBT was studied using halogenated solution processing and in aqueous NP through stability, steady-state and non-equilibrium photoexcitation dynamics. The results show that there is a significant difference in the electron transition time due to the morphology of the material. F8T2:MDMO-PPV and Zn(5BrTTP):PSiF-DBT nanoparticles exhibited Förster energy transfer between the materials, contributing to improved device performance. Furthermore, devices using aqueous P(TBT-DPP): ICBA NPs exhibited better performance when compared to devices using chlorinated solvent, higher short circuit current density (J_{sc}) and higher fill factor (FF).

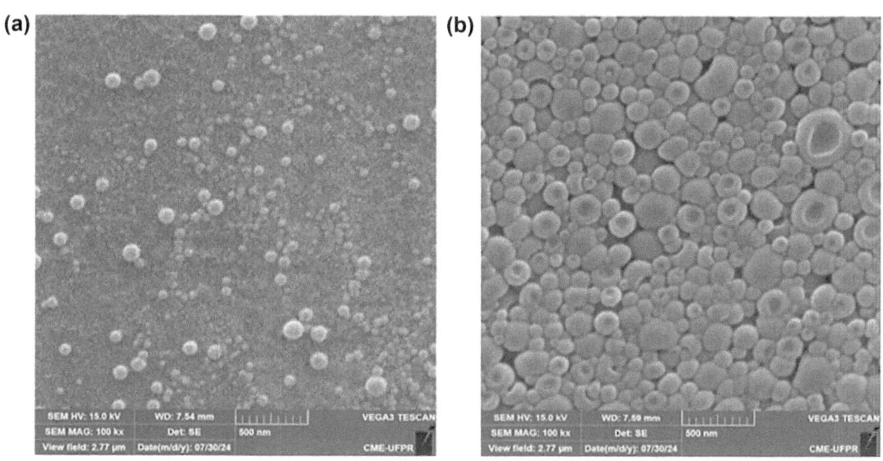

Figure 7.1 Scanning electron microscopy (SEM) images of nanospherical (a) and nanodonut (b) shaped colloids.

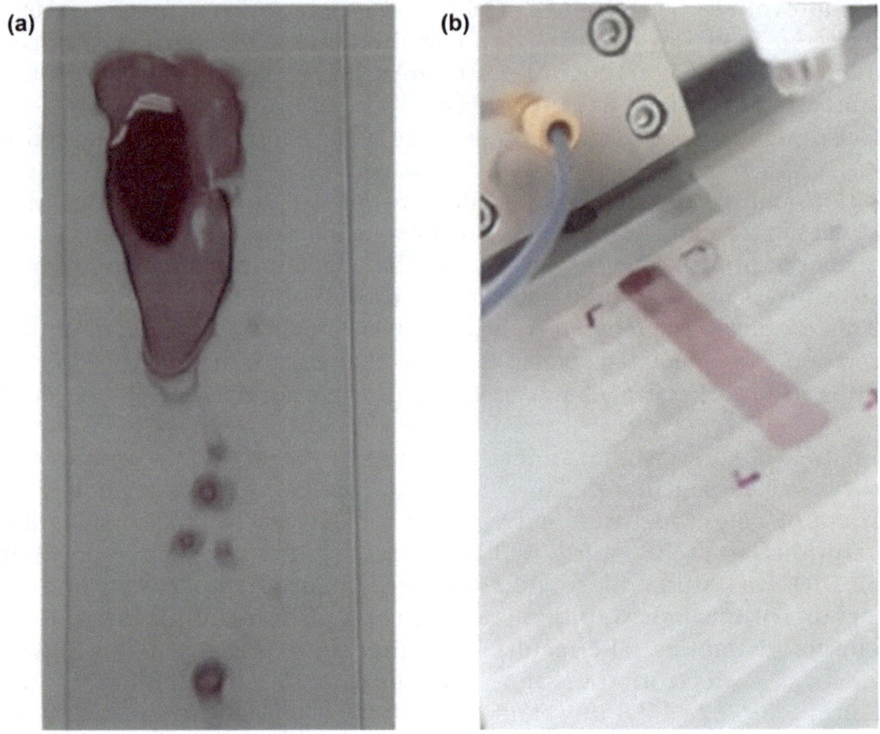

Figure 7.2 Printing of the aqueous and nanostructured OPV active layer on: (a) substrate without surface treatment and (b) substrate with surface treatment.

7.2 Colloidal Dispersion of Semiconducting Polymers by the Miniemulsion Process

The miniemulsion process is one of the most environmentally friendly methods used in the development of organic photovoltaic devices. This process is schematized in Figure 7.4 and basically consists of dissolving semiconductor materials in a suitable organic solvent which is added to an aqueous solution containing surfactant (a). This mixture is subjected to magnetic stirring (b). An ultrasonication process is then carried out, with the aim of considerably reducing the dimensions of the spheres (c). Therefore, the dispersions containing the nanoparticles and surfactant are heated to evaporate the organic solvent (d), and a centrifugal dialysis process is carried out with the aim of concentrating the dispersions and eliminating excess surfactant (e), thus forming nanoparticles (NPs) of polymers, fullerenes or other organic molecules, in addition to the mixture between materials.

In this context, the nanostructured PSiF-DBT active layer used as an electron donor material in OPVs was studied and compared with the conventional halogenated morphology.

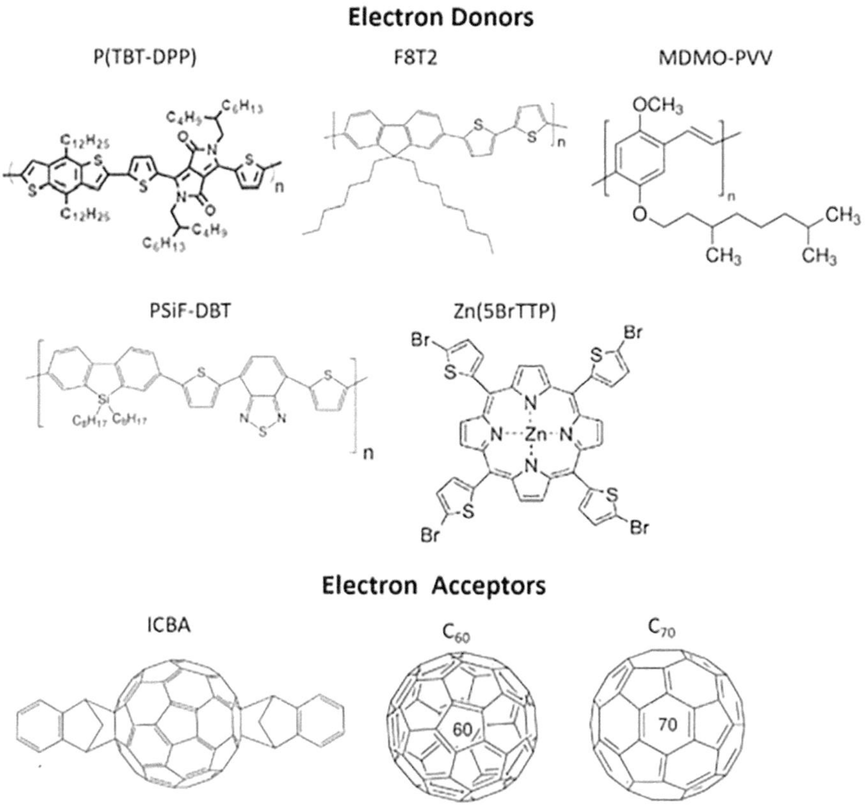

Figure 7.3 Chemical structure of the materials used in this chapter along with their chemical acronym.

Figure 7.5(a) shows the surface images of a nanoparticle film obtained by SEM. The image shows that the PSIF-DBT NP presented a compact morphology, and the polymer is spherical-like shaped.[8] The scanning microscopy measurements were complemented with chemical species identification by X-ray fluorescence (Figure 7.5(b)). The composition obtained by the EDS analysis showed the presence of the constituents of the PSIF-DBT and the surfactant SDS as well as the presence of aluminum, the constituent material of the sample holder. The high intensity of the silicon is related to the presence of Si atoms in the copolymer units and in the substrate used for these measurements.

Figure 7.5(c) exhibits the TEM image of the PSiF-DBT nanoparticles. The PSiF-DBT nanoparticles presented a spherical shape, as confirmed by the SEM image. The diameter of 60 nm for the nanoparticles was obtained by the Hough transform. The image (Figure 7.5(c)) shows both isolated nanoparticles and agglomerated nanoparticles forming structures like small-scale grape clusters and larger clusters.

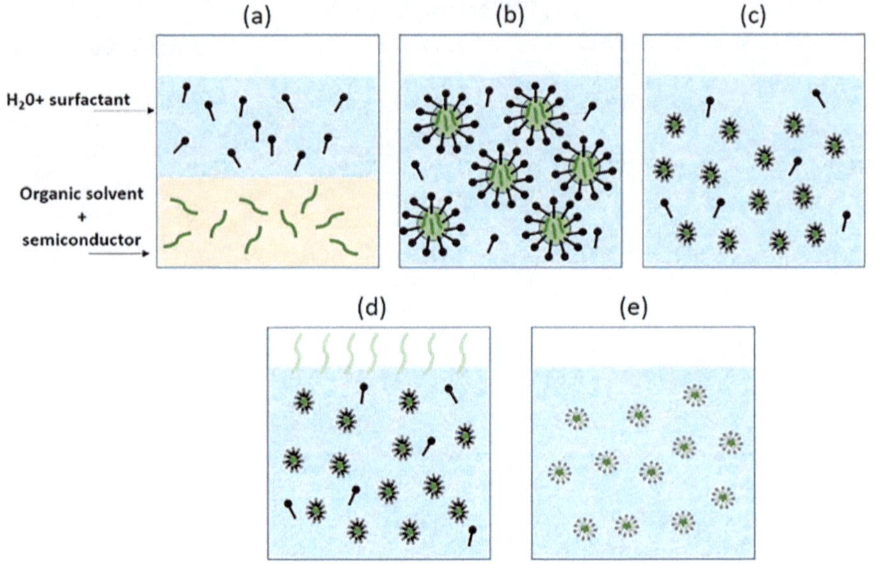

Figure 7.4 Schematic of the miniemulsion process.

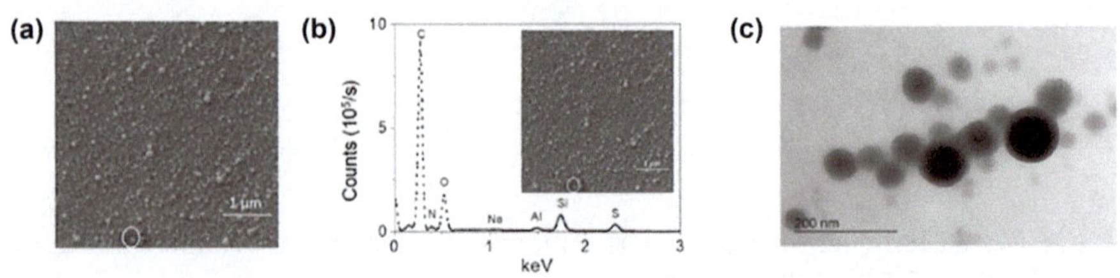

Figure 7.5 SEM image of nanoparticles in (a) showing the film pattern of drop cast deposition and in (b) EDS measurement presenting the components of polymer PSiF-DBT (carbon, nitrogen, sulfur, silicon) and surfactant SDS (carbon, sodium, oxygen) and (c) TEM image presenting the spherical form of the nanoparticles in clusters and line structures.

The exciton decay curves of PSiF-DBT films at different morphology are presented in Figure 7.6(a). From the results it is possible to observe that the decay of the PSiF-DBT polymer depends on the morphology, exhibiting lower time to the exciton investigated in the NP films. For the emission at 690 nm acquired for the PSiF-DBT copolymer film in the conventional morphology, the average lifetime of 378 ps was obtained. On the other hand, it is observed that the average decay time decreases to 112 ps in the NP films. Thereby, it is possible to conclude that the interaction between the copolymer's moieties becomes stronger in the NP films, enabling the exciton migration to be performed faster.

In order to investigate the stability of PSiF-DBT semiconducting behavior in films with these two morphologies films, we performed experiments of

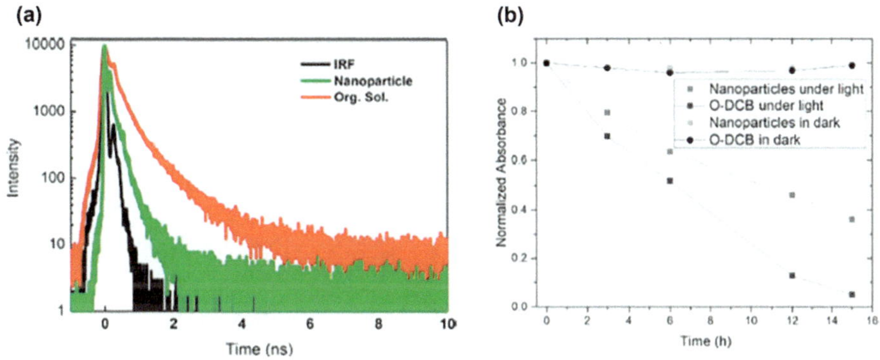

Figure 7.6 (a) Time resolved photoluminescence in nanoparticles and organic solution of PSiF-DBT polymer and (b) PSiF-DBT stability measurements solubilized in O-DCB and nanoparticles by absorbance spectra.

light soaking and temperature and compared them with their stability in dark. We tested the films for fifteen hours continuously in a controlled environment. To do that, we developed equipment to control environmental conditions like humidity, temperature, and illuminance. The spectroscopic measurements were intercalated with the exposition of the conditions of the environmental chamber. The first absorption spectrum was measured right after the film's deposition. The sequential measurements were taken after every 3 hours in the environmental chamber. This procedure was repeated for 15 hours. The absorbance variation intensity of the 580 nm peak for the PSiF-DBT solubilized in O-DCB and nanoparticles is shown in Figure 7.6(b). It can be noticed that the NP film presented better stability, maintaining the light absorbance longer. This behavior may be associated with the surfactant micelles capping the tied polymer chains.

To further characterize the PSiF-DBT nanoparticle film regarding its electronic structure, degree of ordering and transport properties, synchrotron-based spectroscopic techniques were employed. Sulfur K-edge NEXAFS measurements of the PSiF-DBT nanoparticle film were performed using the surface-sensitive TEY mode (Figure 7.7(a)). Two groups of electronic transitions characterize the sulfur K-edge NEXAFS spectra of PSiF-DBT: T1 and T2 transitions related to the thiophene unit and B1 and B2 to the benzothiadiazole unit. They are assigned as the S1s transition to π^* and σ^* orbitals, respectively. These assignments have been extensively studied in the literature and corroborated by theoretical calculations.[9,10] If we compare the NEXAFS spectrum of thiophene with that of PSiF-DBT it is remarkable the difference between them, since the latter has sulfur in two different chemical environments, which could be clearly identified in this deep core-level spectrum. Other structures appearing above the ionization potential of sulfur are considered as transitions to higher unoccupied states.

The NEXAFS spectra were measured by varying the angle of the incidence of the synchrotron light. The intensity of the electronic transition reaches a

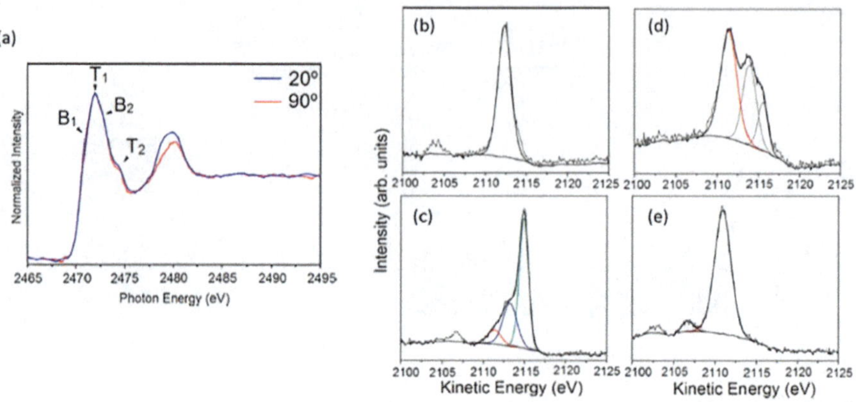

Figure 7.7 (a) Angular dependent sulfur K-edge NEXAFS spectra of nanoparticle PSiF-DBT film at grazing (20°) and normal (90°) photon incidence. Deconvoluted sulfur KL2,3L2,3 Auger decay spectra obtained for PSiF-DBT at different photon energies of the NEXAFS spectrum: (b) 2471.9 eV, (c) 2472.6 eV, (d) 2474.3 eV and (e) 2479.7 eV.

maximum when the electric field vector of the linearly polarized synchrotron light is parallel to the maximum amplitude of the molecular orbital that characterizes it. Because of this, the preferential molecular orientation of the films can be determined through the angular variation of the NEXAFS spectra. However, no dichroism was observed between the intensities of the resonances as a function of grazing and normal incident angles, suggesting that the PSiF-DBT film in NP format is disordered.

The core-hole clock (CHC) spectroscopy has proven to be an important tool for pursuing ultrafast electron dynamics in semiconducting materials.[11] In this spectroscopy the relaxation of core holes is monitored by the resonant Auger decay spectra. By taking into consideration the core-hole lifetime, which ranges from the femtosecond to attosecond interval, charge transfer times can be calculated from the spectroscopic data. Auger decay spectra were measured for photon energies covering the sulfur K-edge and some results are presented in Figure 7.7(b)–(e). The features observed in the spectra change according to the transitions in the NEXAFS spectrum. A peak with constant kinetic energy of 2112 eV could be fitted in all spectra, which was considered as the contribution of the normal Auger decay, also called charge transfer channel, being the predominant peak at photon energies higher than the ionization threshold of thiophene and benzothiadiazole,[9] as shown in Figure 7.7(e). Additionally, two more structures are easily identified in the spectrum of Figure 7.7(c): the peak appearing at 2113.3 eV is considered the spectator Auger decay channel related to the S1s→π^* transition, while the contribution at 2115 eV is associated with S1s→σ^*(S-C) spectator Auger. For higher excitation energies (Figure 7.7(d)), a new σ^* spectator arises. For the spectrum recorded at 2471.9 eV photon energy, it is possible to observe a single peak. This energy is related to the B1 transition, so this peak is considered as a π* spectator. This result is not surprising since the benzothiadiazole unit is the acceptor unit in the PSiF-DBT copolymer, expected to contribute less to the charge transfer process probed by the CHC method. The charge transfer times were calculated through the ratio between the areas of the deconvoluted spectator and normal peaks presented in the Auger decay spectra for the energies 2472.6 eV and 2474.3 eV.[12] Sulfur core-hole lifetime (1.27 fs) was used as an internal reference.[13] The charge transfer times calculated for resonances T1 and T2 were 10.7 fs and 1.1 fs, respectively. Compared with a previous publication,[10] it is possible to notice that the charge transfer time was reduced from 12.6 fs to 10.7 fs by using the nanoparticle films, which may reflect the agglomerated nanoparticles structure as verified by the TEM image.

7.3 Architecture, Operation, and Electrical Characterization of OPVs

The monolayer structure is the simplest structure in organic photovoltaic devices. In general, they consist of an organic light-absorbing semiconductor layer, also called an active layer, arranged between two electrodes with asymmetric work functions, as shown in Figure 7.8(a). Indium tin oxide (ITO) and doped tin oxide with fluorine (FTO) are generally used as one of the electrodes, which are (semi)-transparent so that a large amount of light reaches the active layer. These are called hole collectors, *i.e.* anode. The other electrode can be composed of aluminum, calcium, magnesium, silver or gold, with the function of collecting electrons, therefore, they are the cathodes.[14]

The first step for the photovoltaic process to occur is the absorption of light. When light is absorbed by the active layer, an electron is promoted from the HOMO (highest occupied molecular orbital) to the LUMO (lowest

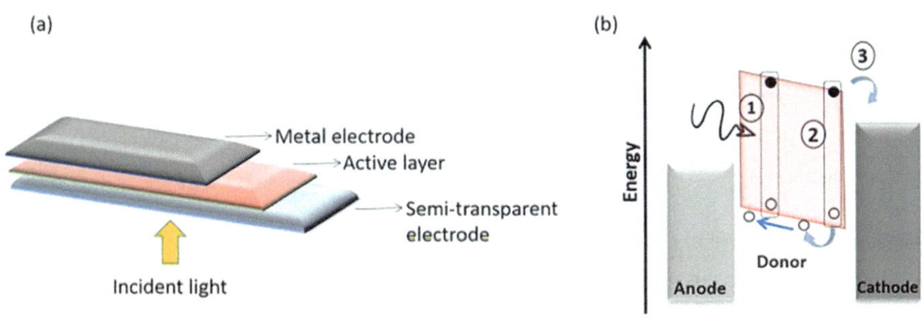

Figure 7.8 (a) Structure schematic of an organic photovoltaic device composed of a conjugated polymer as the active layer and (b) energy diagram of a device in the monolayer structure. The excitons formed by optical excitation in the active layer (1) diffuse until they reach the polymer/metal interface (2). Exciton dissociation occurs; thus, electrons and holes are transported to the respective electrodes (3).

unoccupied molecular orbital) forming an electron–hole pair, also called an exciton, which is bound by a Coulombic force. This only occurs when the semiconductor material absorbs photons with energy greater than or equal to the material's band gap (E_g). In general, the difference in the work function of the electrodes provides an electric field. This electric field is rarely sufficient to disrupt the photogenerated exciton; instead, the exciton diffuses into the organic layer, where it can be dissociated into free charges.[15]

In order to achieve charge separation to produce electrical current, a high electric field is required, found near the polymer/metal interfaces, which is provided by the ionization energy/work function asymmetries of the electrodes.[15,16] These charges then need to travel to the device's respective electrodes, holes to the anode and electrons to the cathode, as shown in Figure 7.8(b).

Because photons generate tightly bound excitons, which in turn do not dissociate into free charges easily, the first devices, which were built in a monolayer format, had power conversion efficiencies well below 1%.[17] In order to improve the performance of devices, other structures were created, such as heterojunctions, which are structures that have two or more layers of organic semiconductors. Heterojunctions are subdivided into bilayer and bulk junctions.

In 1986, Tang demonstrated an organic solar cell with a power conversion efficiency of 1%,[18] this being 1–2 orders of magnitude larger than the devices of that time. The main innovation was the introduction of an interface between two organic semiconductors with different electro affinity values: one of these materials is called an electron donor and the other, an electron acceptor, as shown in Figure 7.9(a).

Figure 7.9 (a) Structure schematic of a bilayer heterojunction device where the photosensitive layer is composed of an electron donor and an electron acceptor and (b) energy diagram illustrating the charge transfer process that occurs at the D/A interface of an organic solar cell in a bilayer structure. The excitons formed by optical excitation (1) in the donor material diffuse until they meet the D/A interface (2). The electron is transferred to the acceptor material (3), while the hole remains in the donor material. The electron is transported between molecules of the acceptor material, just as the holes are transported in the donor material, until they are collected by the respective electrodes (4).

As the excited states in organic molecules (excitons) are characterized by large binding energies, typically on the order of 0.2–1 eV, the use of a donor/acceptor (D/A) interface provides a driving force, to overcome the exciton binding energy. Thus, excitons can be efficiently dissociated into electrons and holes, contributing to the photocurrent.

Figure 7.9(b) illustrates the charge transfer process. In bilayer devices, photogenerated excitons can diffuse within the donor toward the planar interface with the second material, the acceptor, which is typically chosen to be strongly electronegative. The acceptor material provides the necessary condition for the exciton to be separated, so the electron can go to a much lower energy state inside the acceptor. This charge transfer dissociates the exciton, the electrons move to the acceptor material, while the hole remains in the donor. This charge transfer, or electron transfer, is very fast. Indeed, it has been found to be faster than 100 fs in donor/acceptor systems, and very efficient, as alternative loss mechanisms are much lower.

The limiting factor of devices built on the bilayer heterojunction structure is that, for complete absorption of light for energy conversion, the film thickness must be greater than 100 nm. This is much more than the exciton diffusion length, about 10 nm in disordered and semi-crystalline polymers and small molecules. With such a large thickness, only a small fraction of the excitons can reach the heterojunction interface.

This problem can be overcome through the concept of bulk heterojunction (BHJ), where the electron donor and acceptor materials are intimately mixed throughout the entire volume of the OPV. This structure offers a distributed junction between the donor and acceptor materials: both components interpenetrate each other, so that the interface between them is no longer flat as in the bilayer structure, but in this case, spatially distributed so that the donor/acceptor interface (D/A) is within a distance smaller than the exciton diffusion length, as illustrated in Figure 7.10(a). As the interface is dispersed throughout the volume, this structure contributes to the excitons finding the D/A interface, these can be dissociated within their lifetime, as shown in Figure 7.10(b).

Another improvement can be made with the use of ternary blends in BHJ. The active layer of OPVs usually contains two organic semiconductors: an electron donor and an electron acceptor. However, this binary system often shows insufficient coverage of the solar irradiation spectrum due to the narrow optical absorption of organic compounds. The use of three complementary absorption materials (for example, two donors and one acceptor, or one donor and two acceptors) has shown to be very promising, as it presents better absorption of sunlight, with the prospect of presenting better performance compared to the binary structure.[19]

To determine the performance and study the electrical characteristics of organic solar cells, several parameters can be defined. The parameters used to characterize and compare the performance of different solar cells are described in this section.

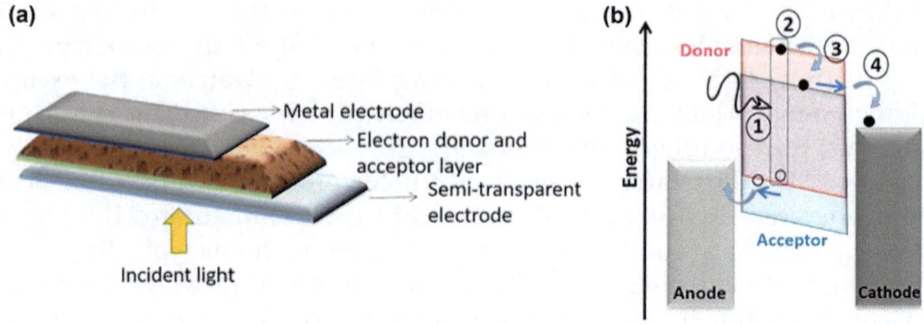

Figure 7.10 (a) Schematic of the structure of a bulk heterojunction device where the active layer is made up of a mixture of electron donor and acceptor materials. Energy diagram of a volume heterojunction device and (b) excitons formed by optical excitation in the donor material (1) diffuse until they meet the D/A interface (2). The dissociation of excitons occurs by transferring electrons to the acceptor molecules (3). The electron is transported between acceptor molecules, just as holes are transported in the donor material to the respective electrodes (4).

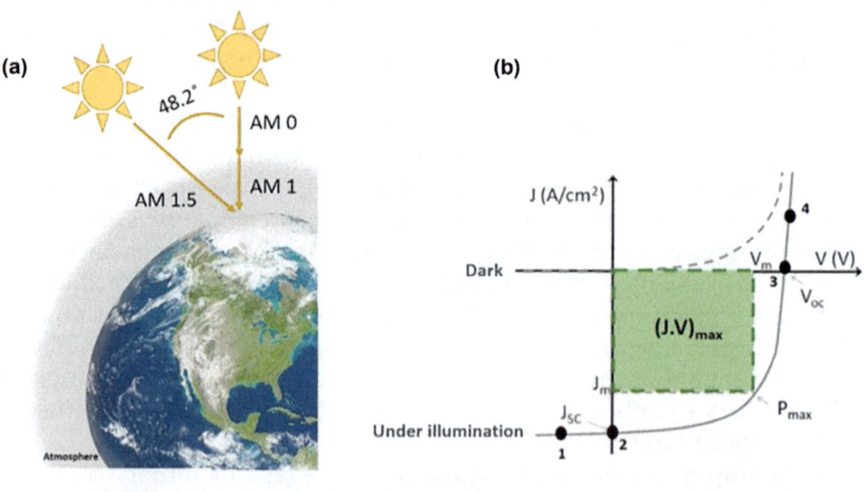

Figure 7.11 (a) Definition of the optical spectrum AM 0, AM 1 and AM 1.5 and (b) characteristic curve of an ideal photovoltaic device, in the dark and under illumination, indicating the parameters V_{oc}, J_{sc} and FF.

The solar spectrum is an environmental factor that affects the performance of photovoltaic devices. Because external solar spectral distributions vary with atmospheric conditions and sun angle, a representative solar spectral irradiance distribution is standardized so that solar cells can be compared on a common basis.

The optical path taken by sunlight through the atmosphere is known as air mass (AM). As shown in Figure 7.11(a), the AM 0 spectrum corresponds to radiation outside the Earth's atmosphere and the AM 1 spectrum is when sunlight reaches the Earth's surface with direct incidence. To compare and characterize different types of solar cells, the AM 1.5 spectrum is used as a standard condition. The measuring surface is defined in this case as a plane inclined at 37° towards the equator, pointing towards the Sun, which is at an angle of 48.2° with respect to normal incidence. Total sunlight irradiance is defined as 100 mW cm^{-2} for the standard '1 Sun' measurement.[20]

The electrical response of each photovoltaic device is obtained by the characteristic curve of current density *versus* applied voltage (J–V). Figure 7.11(b) shows a hypothetical characteristic curve for a device in the dark and under illumination.

The curve in the dark represents the response for a rectifier diode where there is electric current only for the direct voltage, that is, the electrode with the highest work function is positively polarized and the electrode with the lowest work function is negatively polarized. The curve under illumination represents the open circuit voltage (V_{oc}) and the short circuit current density (J_{sc}). This curve is shifted to the third and fourth quadrants where $J<0$, as the absorption of photons from radiation promotes the production of electron–hole pairs that generate a current contrary to the dark current, and the contribution of these pairs to the current is important and is responsible for the displacement of the curve. In addition to the V_{oc} and J_{sc} parameters, through the J–V curve under lighting it is possible to obtain other photovoltaic parameters such as the fill factor (FF) and power conversion efficiency (η). The equations that define and relate these quantities will be presented below.

The fill factor is defined as the ratio between the maximum power $(JV)_{max}$ provided by the device and its nominal power ($J_{sc}V_{oc}$), as indicated in Figure 7.11(b), that is:

$$\mathrm{FF} = \frac{(JV)_{max}}{J_{sc}V_{oc}} \tag{7.1}$$

Power conversion efficiency (η) is an essential parameter to describe the performance of a solar cell and is expressed by the maximum electrical power generated by the cell (P) and the incident light power (P_{in}) through the expression:

$$\eta = \frac{P}{P_{in}} = \frac{(JV)_{max}}{I_0} = \mathrm{FF}\,\frac{J_{sc}V_{oc}}{I_0}, \tag{7.2}$$

where I_0 is the intensity of the light incident on the device.

7.4 Förster Non-radiative Energy Transfer in Colloidal Systems Applied in OPVs

FRET was discovered by Theodor Förster in 1948.[21] FRET is a process by which an excited state of a donor molecule transfers energy to a proximal acceptor molecule in the ground state, through space non-radiative dipole–dipole coupling.[22–25] The overlap between the fluorescence spectrum of the donor molecule and the absorption spectrum of the acceptor molecule is essential for FRET type energy transfer to occur. The greater the spectral overlap, the more efficiently the transfer happens.[26] This property can be used in organic photovoltaics to improve the total photoconversion of the blend device as using Förster energy transfer within the donor layer allows the absorbed light to be channeled to the C_{60} acceptor layer, as shown in ref. 27

Förster determined that the energy transfer rate, K_{FRET}, from the donor to the acceptor is given by:

$$K_{FRET} = \frac{1}{\tau_0}\left(\frac{R_0}{r}\right)^6, \tag{7.3}$$

where τ_0 is the fluorescence lifetime of the exciton, r is the distance between donor and acceptor and R_0 is the Förster radius.[21] The Förster radius is the critical distance in ängstroms between the donor and acceptor molecules where the probability of transferring resonant energy between the molecules is 50%. The Förster radius can be written as:

$$R_0 = 9.78 \times 10^2\left[\frac{k^2 Q_D}{n^4}\int F_D(\lambda)\varepsilon_A(\lambda)\lambda^4 d\lambda\right]^{\frac{1}{6}}, \tag{7.4}$$

where k^2 is the orientation factor between the donor and acceptor, Q_D is the quantum efficiency of the donor's photoluminescence, n is the refractive index, ε_A is the acceptor's molar extinction coefficient and F_D is the donor's emission spectrum.

In this context, we synthesize water-dispersed polymeric nanoparticles from the F8T2 copolymer, the MDMO-PPV homopolymer and a mixture of both, with the aim of investigating the interaction between these materials.[28,29] Polymeric nanoparticles were synthesized as described in Figure 7.4.

Two different sets of nanoparticles were prepared: two-component nanoparticles and one-component nanoparticles, as shown in Figure 7.12(a) and (b).

The two-component nanoparticles were synthesized in such a way that the F8T2 and MDMO-PPV materials were mixed within the same nanoparticle (Figure 7.12(a)), making the interaction between them limited by their diameter, which was obtained by dynamic light scattering (DLS) presented in Table 7.2.

For the second set of samples, nanoparticle syntheses were carried out separately (Figure 7.12(b)). Thus, the synthesis of F8T2 nanoparticles was mixed with the synthesis of MDMO-PPV nanoparticles, under the same conditions as the previous case.

As shown in Figure 7.12(c), the use of these two materials to act together was chosen due to the overlap of the photoluminescence spectrum of F8T2 and the absorption spectrum of MDMO-PPV. In this way, F8T2 can act as a donor material and MDMO-PPV as an acceptor material in a Förster-type non-radiative energy transfer system. This implies that F8T2 can transfer all of its energy to MDMO-PPV in the overlap region *via* FRET.

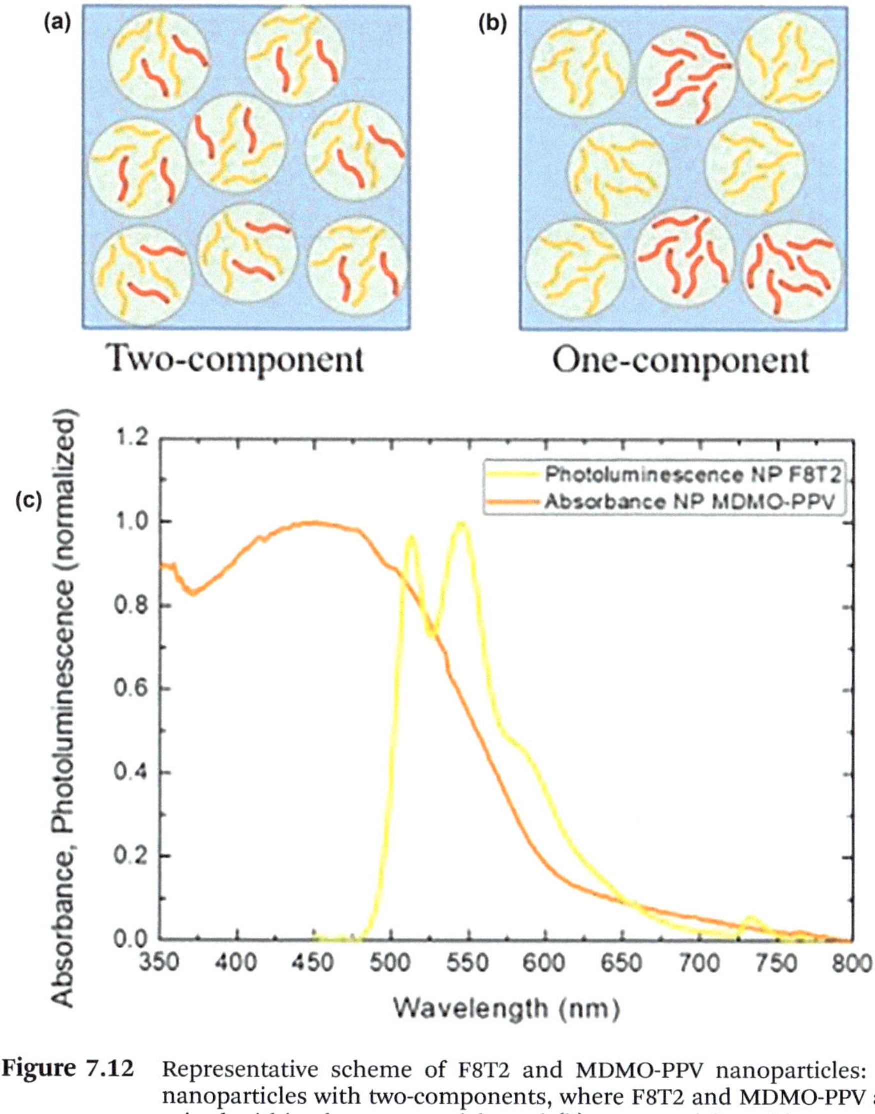

Figure 7.12 Representative scheme of F8T2 and MDMO-PPV nanoparticles: (a) nanoparticles with two-components, where F8T2 and MDMO-PPV are mixed within the nanoparticle and (b) nanoparticles with one-component, where the materials F8T2 and MDMO-PPV are in many different nanoparticles. (c) Overlapping of the absorption and emission spectra of the MDMO-PPV and F8T2 nanoparticles respectively.

Figure 7.13(a) and (b) display the photoluminescence spectra of F8T2 nanoparticle films, MDMO-PPV and their mixtures in the one- and two-component configuration. The photoluminescence spectrum of F8T2 presents two well-defined peaks at 510 nm and 545 nm and a more discrete peak at 590 nm, and for MDMO-PPV an emission maximum is shown at 560 nm.

Table 7.2 Relationship of the diameter of the nanoparticles determined by the measure of DLS – dynamic light scattering.

Nanoparticles	Average diameter (nm)
F8T2	52.6
MDMO-PPV	42.7
5% de MDMO-PPV	34.0
10% de MDMO-PPV	31.5

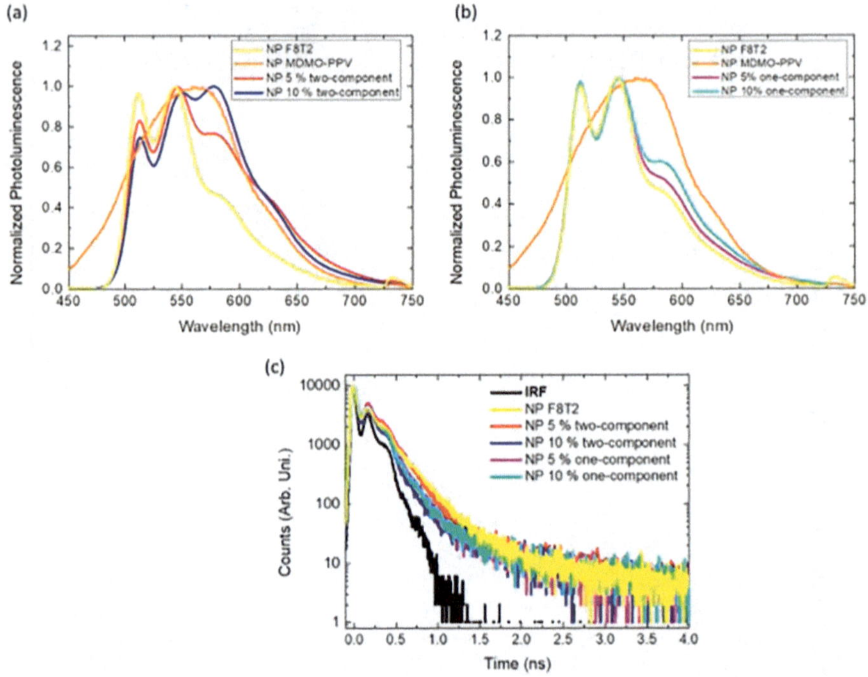

Figure 7.13 Optical properties of nanoparticle films: (a) normalized photoluminescence spectrum of two-component nanoparticles (400 nm excitation); (b) normalized photoluminescence spectrum of one-component nanoparticles (400 nm excitation). Photoluminescence spectra of films with F8T2 and MDMO-PPV nanoparticles were plotted for comparison and (c) time resolved photoluminescence spectra of nanoparticle films in one and two-component configurations.

Analyzing the photoluminescence spectra of two-component nanoparticles shown in Figure 7.13(a), the film with 5% MDMO-PPV nanoparticles shows a small emission corresponding to MDMO-PPV, at the same time that there is a decrease in the emission related to F8T2. The emission from the film with 10% MDMO-PPV shows a greater extinction in the F8T2 region and an increase in emission in the MDMO-PPV region. This indicates that two-component nanoparticle films can exhibit Förster-type nonradiative energy transfer, where the exciton created in the F8T2 donor can be transferred to the MDMO-PPV acceptor.

The photoluminescence spectra of films with one-component nanoparticles and 5% and 10% MDMO-PPV did not show a decrease in emission in the region corresponding to F8T2 with the increase in the proportion of MDMO-PPV, instead, it exhibits the emission of the two materials (Figure 7.13(b)). Förster's theory explains that since energy must be conserved, the energy acceptor material MDMO-PPV must have available energy states such that the energy gain from MDMO-PPV is equal to the energy loss from F8T2, therefore films using one-component nanoparticles may not show effective energy transfer.

According to Förster's theory, the energy transfer rate, K_{FRET}, from the donor to the acceptor is inversely proportional to the sixth power of the distance between the materials, as shown in eqn (7.3). Thus, in the system corresponding to two-component nanoparticles, a greater energy transfer between F8T2 and MDMO-PPV may be occurring, as they are confined within the same nanoparticle, unlike one-component nanoparticle films, where each material is in a different nanosphere, therefore, these materials may be at a distance greater than the Förster radius, which may contribute to their radiative recombination, resulting in a less efficient energy transfer.[30]

FRET theory predicts a decrease in the excited state lifetime of the energy donor with increasing acceptor concentration, since FRET offers an alternative non-radiative decay pathway from the donor to the acceptor.[31,32] Therefore, measurements to evaluate the excited state lifetime of one- and two-component nanoparticle films were carried out. The energy transfer efficiency was calculated as follows:

$$E = 1 - \frac{\tau_{DA}}{\tau_D}, \tag{7.5}$$

where τ_{DA} is the emission decay time of the donor material in the presence of the acceptor, and τ_D is the emission decay time of the donor material alone.

Lifetime measurements confirm the shorter lifetime for two-component nanoparticle films. As the proportion of MDMO-PPV was increased, there was a decrease in times, from 0.147 ns (NP 5% two components) to 0.091 ns (NP 10% two components), indicating a non-radiative energy transfer efficiency of the type Förster of 49.0% and 68.4% respectively. As expected, for one-component nanoparticle films, the times were very similar to the F8T2 NP film, exhibiting a time of 0.280 ns (NP 5% one component) and 0.269 ns (NP 10% one component), with a transfer of non-radiative Förster energy of 2.8% and 6.6%, respectively. These data are summarized in Figure 7.13(c) and Table 7.3. Therefore, the lifetime measurements corroborate the photoluminescence measurements presented previously.

The photovoltaic properties of nanoparticle syntheses with one and two components were investigated as an active layer in organic devices in the bilayer heterojunction structure (ITO/PEDOT:PSS/NP/C_{60}/Al), as shown in Figure 7.14(a).

Table 7.3 Lifetime relationship for one and two component nanoparticle films, along with their percentage of energy transfer.

	τ (ns)	% Energy transfer – calculated through decay times
NP F8T2	0.288 ± 0.006	
NP 5% two-component	0.147 ± 0.007	49.0 ± 3.5%
NP 10% two-component	0.091 ± 0.003	68.4 ± 1.7%.
NP 5% one-component	0.280 ± 0.006	2.8 ± 4.1%
NP 10% one-component	0.269 ± 0.006	6.6 ± 3.9%

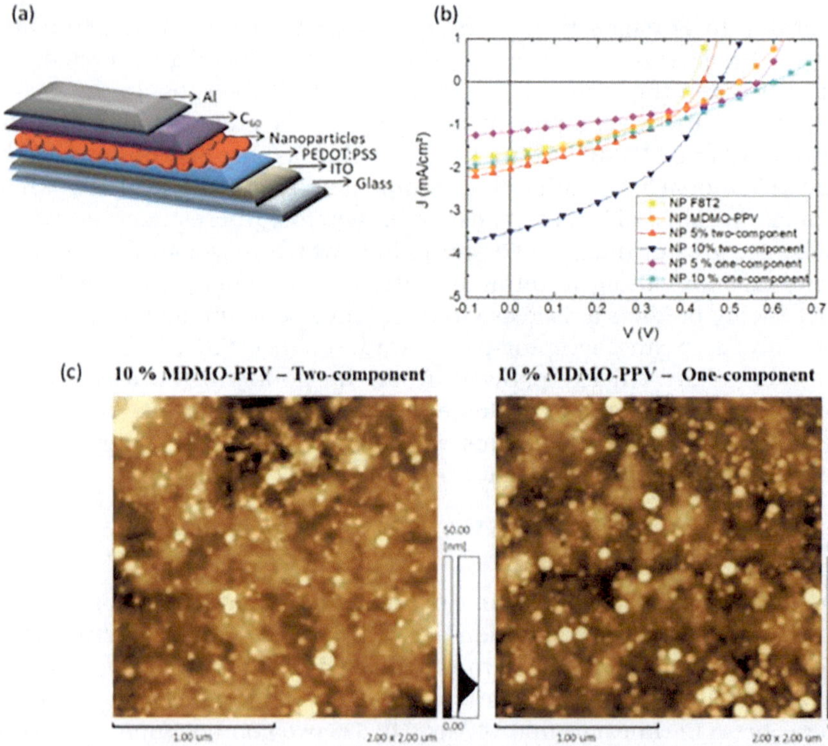

Figure 7.14 (a) Device structure using nanoparticles as the active layer; (b) curves (*J–V*) of devices in the bilayer heterojunction structure with the active layer composed of nanoparticles in the configuration of one and two-components; (c) morphology of nanoparticle films with 10% MDMO-PPV of one and two components, deposited on glass substrate/ITO/PEDOT-PSS obtained by atomic force microscopy (*Z* scale: 50 nm).

Figure 7.14(b) presents the *J–V* curves of the devices using nanoparticles with F8T2 and MDMO-PPV as the active layer, with binary devices used as the reference, and the mixture of both materials in one- and two-component situations.

Table 7.4 Comparison of the photovoltaic parameters of the bilayer heterojunction devices with the active layer composed of nanoparticles in the one- and two-component configurations.

	V_{OC} (V)	J_{SC} (mA cm^{-2})	FF (%)	η (%)
NP F8T2	0.41	1.6	46	0.30
NP MDMO-PPV	0.51	1.8	31	0.29
NP 5% two-component	0.43	2.0	40	0.35
NP 10% two-component	0.48	3.5	41	0.69
NP 5% one-component	0.56	1.1	38	0.25
NP 10% one-component	0.59	1.8	32	0.34

It is possible to observe that devices with two-component nanoparticles and 10% MDMO-PPV showed a higher J_{SC} short circuit current density compared to all other devices, according to Table 7.4.

Comparing the AFM images of the devices with 10% MDM-PPV nanoparticles with one and two components (Figure 7.14(c)) it was possible to observe that both films presented nanospheres distributed throughout the sample, but as the nanoparticles with two components present the materials F8T2 and MDMO-PPV confined within the same nanoparticle, they are more likely to have a greater energy transfer compared to single-component nanoparticles. This was confirmed by time-resolved photoluminescence and photoluminescence spectra. Furthermore, the two-component nanoparticles with 10% MDMO-PPV showed higher FRET and smaller diameter (Table 7.2) compared to the nanoparticles with 5% MDMO-PPV, which may be contributing to an improvement in short circuit current density (J_{sc}) and consequently the power conversion efficiency (η).

In this same context, we studied one- and two-component nanoparticles using materials with complementary absorption spectra based on a Zn metalloporphyrin (5BrTTP) and a copolymer (PSiF-DBT).[30] When NPs were applied as an active layer following the ITO/PEDOT:PSS/NP/C_{70}/Al structure, an increase in the short circuit current density (J_{sc}) of two-component NPs was also observed when compared to one-component NPs, as shown in Figure 7.15(a). The films are shown in Figure 7.15(b) and (c). This result is related to the emissions shown in Figure 7.15(d). One-component NPs exhibit the emission of both materials and in the case of two components there is a high probability of lossless exciton transfer by radiative recombination, indicating a higher energy transfer rate than in films with individual nanoparticles.

Measurements to evaluate the excited state lifetime of one- and two-component nanoparticle films and conventional morphology were also studied, as shown in Figure 7.16(a) and (c). These results also confirm a shorter time for nanoparticles with closer materials (0.44 and 0.32 ns for one and two components, respectively). This shows us the high probability of lossless exciton transfer by radiative recombination, indicating a higher

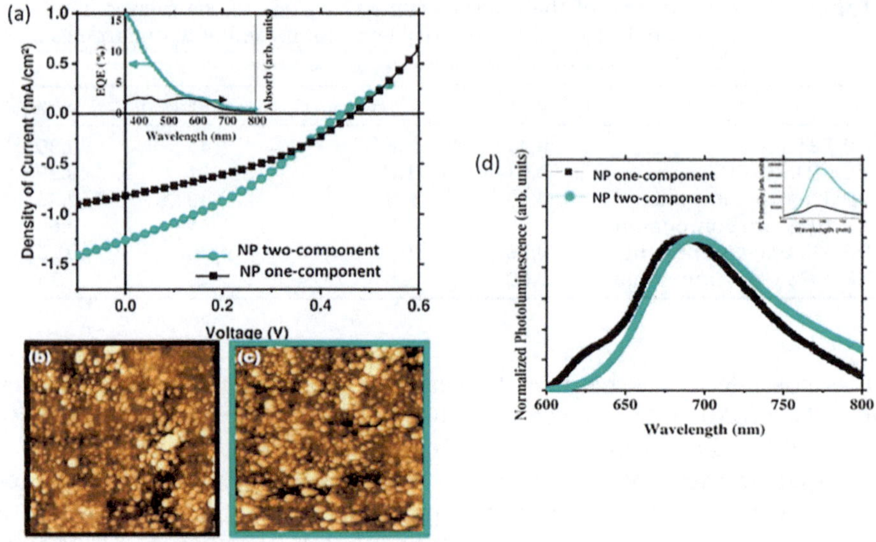

Figure 7.15 (a) J–V curves under illumination (100 mW cm^{-2}) for glass/ITO/PEDOT:PSS/nanoparticle film/C70/Al two-component and one-component. It is inside the same nanoparticle. The inset shows the external quantum efficiency of the two-component nanoparticle. (b) and (c) Topographic images of one-component and two-component nanostructures, respectively. (d) Normalized photoluminescence measurement of the nanoparticle films with one component and two components. The inset shows the intensity emission of photoluminescence.

energy transfer rate than in films with individual nanoparticles. Furthermore, the exciton lifetime in nanoparticle systems becomes shorter when compared with conventional morphology (0.32 and 0.46 ns, respectively), as shown in Figure 7.16(b) and (c). This indicates that the interaction between Zn (5BrTTP) and PSiF-DBT becomes stronger in the nanoparticle system, allowing exciton hopping to be easily performed.

7.5 Colloidal Nanoparticles as Active Layers in Organic Photovoltaic Devices: Theoretical and Experimental Comparison

The phase segregation of the materials that make up the blend is a determining factor in the performance of the devices. In conventional blend films, the phase segregation of materials depends on the surface properties of the substrate, the solvent used, its evaporation rate and thermal treatments. The nanoparticles prepared by the miniemulsion technique improve the nanostructure not allowing great phase segregation, within a mixture in each individual particle obtained in the synthesis regardless of the solvent

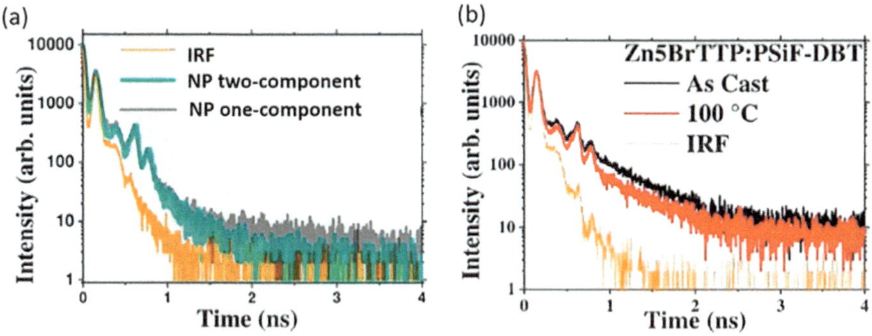

(c)

	$T_{625\ nm}$(ns)
Conventional morphology – as cast	0.54
Conventional morphology – 100 °C	0.48
Nanoparticle one-component	0.44
Nanoparticle two-component	0.32

Figure 7.16 Optical properties of films: (a) excited-state decay curve of Zn(5BrTTP):PSiF-DBT for nanoparticle films with one and two components. (b) Excited-state decay of Zn(5BrTTP):PSiF-DBT with and without thermal treatment. Measured at 625 nm and (c) lifetime relationship for one and two component nanoparticle films.

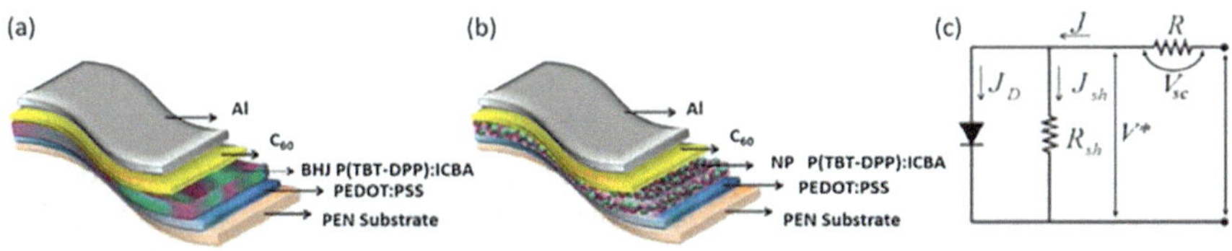

Figure 7.17 Device structures: (a) conventional blend of PEN/PEDOT:PSS/BHJ P(TBT-DPP):ICBA/C_{60}/Al and (b) nanoparticles PEN/PEDOT:PSS/NP P(TBT-DPP):ICBA/C_{60}/Al. (c) Equivalent circuit used to model the dark *J*–*V* response of the devices: *R* is the series resistance, R_{sh} is the shunt resistance, V_{sc} is the potential drop across *R*, J_D is the diode current, J_{sh} is the shunt current and *J* is the total current of the system.

evaporation time or the deposition method. It can be an efficient method to control the morphology from the solution, preparing nanostructured inks that form nanostructured films.

The active layer of the flexible photovoltaic devices reported as an example here is composed of a nanoparticle bulk heterojunction of the polymer P(TBT-DPP) with the fullerene ICBA and conventional BHJ as the control. The structure of conventional BHJ devices is schematized in Figure 7.17(a) and with NPs in Figure 7.17(b). In all cases, three donor:acceptor weight ratios (1 : 1, 3 : 7 and 1 : 3) were tested. The aqueous nanoparticles were synthesized following the miniemulsion technique, illustrated in Figure 7.4,

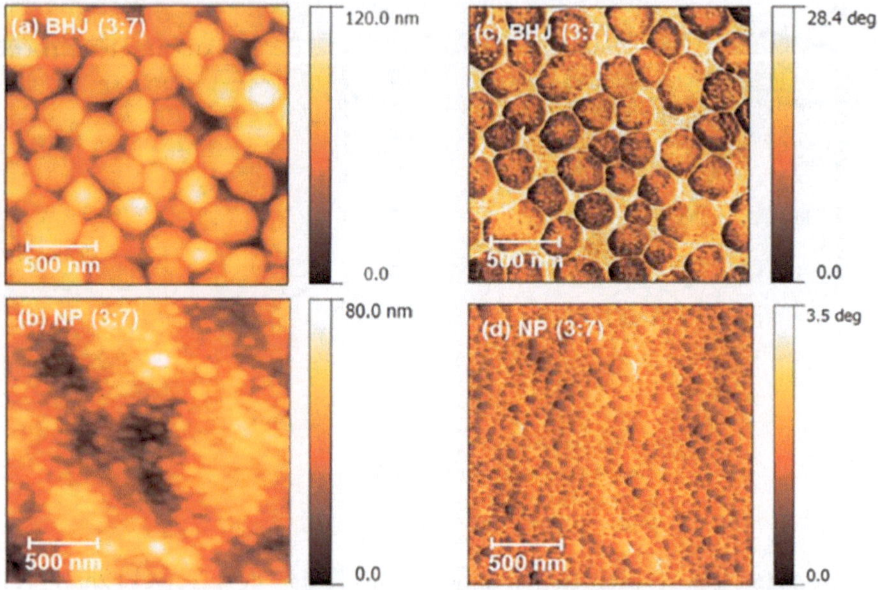

Figure 7.18 Atomic force microscopy (AFM) topography map of (a) conventional P(TBT-DPP):ICBA BHJ and (b) NP films blade coated onto PEN/PEDOT:PSS substrates and phase contrast images simultaneously acquired for (c) conventional P(TBT-DPP):ICBA BHJ and (d) NP films. The scan size is 2 μm × 2 μm.

and all devices were printed by doctor blade on a PEN/PEDOT:PSS substrate.[33]

Film morphologies for 3 : 7 compositions for BHJ chloroform devices and aqueous NPs were performed *via* atomic force microscopy (AFM) and the images are displayed in Figure 7.18(a) and (b), respectively. The simultaneously acquired phase contrast images are shown in Figure 7.18(c) and (d). The conventional BHJ film coated with chloroform solutions resulted in large domains on the order of 500 nm in diameter. This phase segregation in these dimensions does not favor the dissociation of excitons. In contrast, the P(TBT-DPP):ICBA film on aqueous nanoparticles exhibited smaller phase segregation, on the order of 100 nm. Nanostructuring materials on a smaller scale results in devices with higher current density and superior power conversion efficiency.

The characteristic curves (*J*–*V*) under 100 mW cm^{-2} illumination for the conventional blend (BHJ) and nanoparticle devices with the blend ratios of P(TBT-DPP):ICBA equal to 1 : 1, 3 : 7 and 1 : 3 are presented in Figure 7.19. For all proportions of the P(TBT-DPP):ICBA blend, devices based on nanoparticles demonstrated superior photovoltaic performance when compared to devices based on the conventional blend. According to the morphology indicated in Figure 7.18, the miniemulsion technique is an efficient method for controlling the structuring of materials on a smaller scale, resulting in

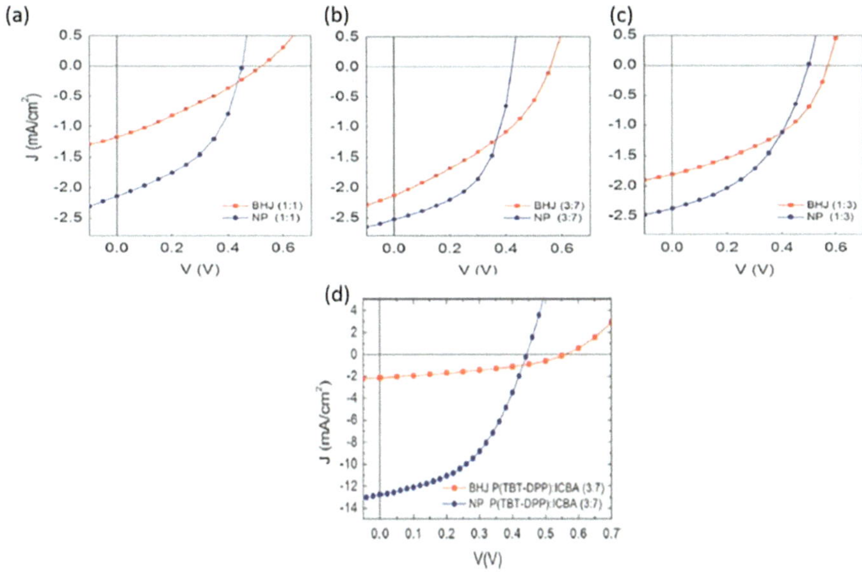

Figure 7.19 *J–V* curves under AM 1.5 illumination of 100 mW cm^{-2} from conventional BHJ and NP devices based on the blend P(TBT-DPP):ICBA in the weight ratios: (a) 1 : 1, (b) 3 : 7, (c) 1 : 3 and (d) conventional BHJ and optimized NP (addition of 20% of EtOH) devices based on the blend P(TBT-DPP):ICBA in the weight ratio of 3 : 7.

higher short circuit current density (J_{sc}) for nanoparticle devices in the three blend proportions tested. The improvement in J_{sc} and FF values is consistent with the decrease in series resistance (R_s) for the nanoparticle devices. It is also observed that the open circuit voltage (V_{oc}) values are lower for all nanoparticle devices when compared to conventional blend devices. The most efficient device was obtained for the blend of aqueous nanoparticles in a 3 : 7 ratio, presenting an open circuit voltage (V_{oc}) of 0.42 V, short circuit current density (J_{sc}) of 2.53 mA cm^{-2}, factor filling capacity (FF) equal to 52% and power conversion efficiency equal to 0.56%.

Given the good performance shown by devices with a 3 : 7 aspect ratio, we further optimized the ink formulation to improve wettability and promote better film formation. Therefore, 20% ethanol by volume was added at 3 : 7 to decrease the surface tension of the aqueous nanoparticle dispersions. Figure 7.19(d) shows the *J–V* curve under illumination for the optimized NP device and the conventional BHJ under the same parameters and conditions. The NP device exhibited a sixfold increase in Jsc increasing the PCE from 0.44% to 2.63%.

With the aim of better understanding the electrical and photovoltaic behaviors described above, the characteristic curves (*J–V*) in the absence of light for these devices were analyzed using the phenomenological model proposed by Koehler *et al.*[34]

The model assumes that these curves can be obtained using an equivalent circuit composed of a nonlinear series resistance (R) limited by space charge (SCL) that depends on J. This resistance is connected to a diode (D), which in turn, is connected in parallel to a resistance R_{sh}, as shown in Figure 7.17(c). The system current is given by:

$$J = J_{sh} + J_D, \tag{7.6}$$

where J_{sh} is the shunt current and J_D is the diode current and are given by the following expressions: $J_{sh} = (V - V_{SC})/(R_{sh} + A)$ e $J_D = J_0(\exp[q(V - V_{SC})/KT)] - 1)$, where A is the active area of the device, J_0 is the diode saturation current and V_{SC} is the potential drop across a space charge limited series resistance (R), given by:

$$V_{SC} = \frac{1}{3}\sqrt{8JL^3/(\mu_{eff}\varepsilon)}, \tag{7.7}$$

where L is the film thickness and ε is the dielectric constant of the active layer ($\varepsilon \sim 3\varepsilon_0$).

Therefore, the system current can be expressed by:

$$J = J_{sh} + J_0 \exp\left[\frac{q(V - V_{SC})}{kT}\right] - 1. \tag{7.8}$$

The J–V characteristics in the dark of this system can be obtained from the solution of eqn (7.8), with J_0 and R_{sh} being adjustment parameters. From the Mott–Gurney law, R_{sh} is given by

$$R_{sh} = 8L^3/9\varepsilon\mu_{eff}^{low}A(V - V_{sc}), \tag{7.9}$$

where L is the sample thickness and μ_{eff}^{low} is the fitting parameter that represents the effective mobility at low V.

Fitting this SCL-D model to the experimental curves provides the effective mobility (μ_{eff}) of the charge carriers. μ_{eff} is also a tuning parameter that determines the space charge limited (SCL) current for high values of applied voltage. R_{sh} is responsible for the ohmic behavior at low values of applied voltage and J_0 is given by the abrupt increase in current between the ohmic and SCL regimes. J_0 and μ_{eff} depend on the intrinsic morphology of the film and can be directly related to the photovoltaic response of the device. For example, low values of J_0 are associated with higher values of V_{oc} by the expression:

$$V_{oc}^{calc} = \left(\frac{kT}{e}\right)\ln\left(\frac{J_{sc}}{J_0}\right). \tag{7.10}$$

Furthermore, the diode current can be written as:

$$J_0 = J_{mat}\exp\left(-\frac{E_{DA}}{2kT}\right), \tag{7.11}$$

where E_{DA} is the effective bandgap energy at the donor/acceptor interface and J_{mat} is the factor that determines the carrier recombination rate, but which does not depend on the energy barrier at the heterojunction. J_{mat} quantifies different material properties, such as the intermolecular overlap of donor and acceptor molecules and the effective area of the donor/acceptor interface. Effective mobility (μ_{eff}) is also affected by film morphology at the molecular level. Large domains of donor (acceptor) material generate high barriers for the transport of holes (electrons) towards the collector electrodes. This effect increases the chance of bimolecular recombination after exciton dissociation. A higher bimolecular recombination rate decreases the value of μ_{eff}, resulting in lower short circuit currents (J_{sc}).

Figure 7.20 shows that there is a good agreement between the model proposed here and the experimental data obtained for both types of morphology. In Table 7.5 we report the values of μ_{eff}, J_0 and R_{sh} obtained from the model at $V = 0.1$ V. As shown in Table 7.5, μ_{eff} and J_0 increase for devices made with higher ICBA density nanoparticle solutions. In the BHJ film they produce higher fluctuations in the local density of ICBA molecules. These fluctuations mask the dependence of μ_{eff} and J_0 on the ICBA bulk density that is only apparent in NP

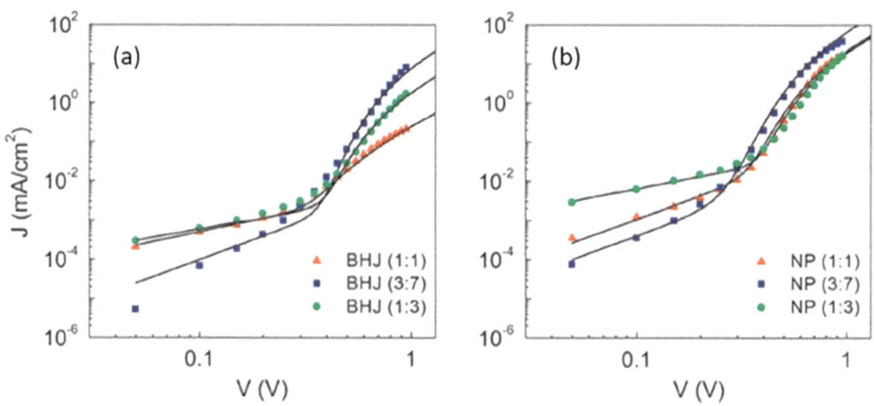

Figure 7.20 *J–V* curves in the dark from conventional (a) BHJ and (b) NP devices based on the blend P(TBT-DPP):ICBA in the weight ratios: 1 : 1 (triangle), 3 : 7 (square) and 1 : 3 (circles). Solid lines are obtained from the fitting of the experimental curves using the proposed model.

Table 7.5 Fitting parameters of the experimental dark *J–V* curves from Figure 7.20 taking shunt resistance values given by eqn (7.9) at $V = 0.1$ V.

Blend weight ratio	J_0 [$mA\,cm^{-2}$]	μ_{eff} [$cm^2\,V^{-1}\,s^{-1}$]	R_{sh} [$\Omega\,cm^2$]	V_{oc}^{exp} [V]	V_{oc}^{calc} [V]
BHJ (1 : 1)	3×10^{-8}	6.5×10^{-14}	1.43×10^{5}	0.52 ± 0.09	0.45
NP (1 : 1)	4×10^{-8}	7.0×10^{-12}	7.90×10^{5}	0.45 ± 0.02	0.47
BHJ (3 : 7)	1×10^{-9}	5.0×10^{-12}	5.66×10^{5}	0.55 ± 0.01	0.56
NP (3 : 7)	2×10^{-7}	1.0×10^{-11}	3.34×10^{5}	0.42 ± 0.01	0.42
BHJ (1 : 3)	1×10^{-9}	1.0×10^{-12}	1.66×10^{5}	0.57 ± 0.01	0.55
NP (1 : 3)	2×10^{-8}	8.0×10^{-12}	1.43×10^{4}	0.49 ± 0.03	0.48

devices where the mixture of acceptor and donor materials is finer and more uniformly distributed throughout the film. The values in Table 7.5 clearly show that NP devices always present higher μ_{eff} and J_0 when compared to the respective BHJ cells for the same blend proportion.

7.6 Conclusion

In this chapter, we provide an overview of colloidal aqueous dispersions of nanoparticles of organic semiconductor polymers and molecules, from preparation techniques to device formation. The nanoparticle-based organic solar cells can bring new possibilities for nanostructuring for the blends of electron-donating and electron-accepting materials, which requires fine control of thin film phase separation on the nanometer scale for increasing the interfaces where exciton dissociation and charge generation occurs. Nevertheless, the use of aqueous nanoparticles in active layers eliminates the use of chlorinated, aromatic and halogenated solvents in the device manufacturing process, which makes conditions more sustainable and presents promising results compared to conventional devices.

Acknowledgements

The studies reported here were performed by many students and coworkers. We would like to acknowledge M. Ramoni, L. Wouk, K. R. M. Pacheco, A. B. Fernández, W. Renzi, J. L. Duarte, M. L. M. Rocco, C. K. Oliveira, I. A. Heisler, C. F. do N. Marchiori, F. Von Kieseritzky, N. A. D. Yamamoto, A. C. Arias, M. E. Payne, M. Koehler, and A. Facchetti. The funding through the CNPq, CAPES, INCT Nanocarbon and INCT Nanovida is gratefully acknowledged.

References

1. K. Fukuda, K. Yu and T. Someya, *Adv. Energy Mater.*, 2020, **10**, 1–10.
2. S. Luo, C. Li, J. Zhang, X. Zou, H. Zhao, K. Ding, H. Huang, J. Song, J. Yi, H. Yu, K. S. Wong, G. Zhang, H. Ade, W. Ma, H. Hu, Y. Sun and H. Yan, *Nat. Commun.*, 2023, **14**, 1–11.
3. B. K. Landfester, R. Montenegro, U. Scherf, R. Güntner, U. Asawapirom, S. Patil, D. Neher and T. Kietzke, *Adv. Mater.*, 2002, **14**, 651–655.
4. P. Marlow, F. Manger, K. Fischer, C. Sprau and A. Colsmann, *Nanoscale*, 2022, **14**, 5569–5578.
5. D. Darwis, N. Holmes, D. Elkington, A. L. D. Kilcoyne, G. Bryant, X. Zhou, P. Dastoor and W. Belcher, *Sol. Energy Mater. Sol. Cells*, 2014, **121**, 99–107.
6. X. Zhou, C. Gao, Y. Gao, Y. Fan and H. Sun, *ACS Macro Lett.*, 2023, **12**, 821–827.
7. R. E. Ginesi, M. R. Niazi, G. C. Welch and E. R. Draper, *RSC Appl. Interfaces*, 2024, **1**, 323–328.
8. M. Ramoni, M. de Jesus Bassi, L. Wouk, K. R. M. Pacheco, A. B. Fernández, W. Renzi, J. L. Duarte, M. L. M. Rocco and L. S. Roman, *Thin Solid Films*, 2021, **721**, 138536.

9. U. Aygu, D. Batchelor, U. Dettinger, S. Yilmaz, S. Allard, U. Scherf, H. Peisert and T. Chasse, *J. Phys. Chem. C*, 2012, **116**, 4870–4874.
10. Y. Garcia-Basabe, C. F. N. Marchiori, B. G. A. L. Borges, N. A. D. Yamamoto, A. G. Macedo, M. Koehler, L. S. Roman and M. L. M. Rocco, *J. Appl. Phys.*, 2014, **115**, 134901.
11. B. G. A. L. Borges, C. F. N. Marchiori, M. Glaser, Y. Garcia-Basabe, C. E. V. de Moura, A. B. Rocha, L. S. Roman, T. Chassé, M. B. Casu and M. L. M. Rocco, *Thin Solid Films*, 2016, **615**, 165–170.
12. Y. Garcia-basabe, C. F. N. Marchiori, C. E. V. De Moura, A. B. Rocha, L. S. Roman and M. L. M. Rocco, *J. Phys. Chem. C*, 2014, **118**, 23863–23873.
13. J. L. Campbell and T. Papp, *At. Data Nucl. Data Tables*, 2001, **77**, 1–56.
14. H. Spanggaard and F. C. Krebs, *Sol. Energy Mater. Sol. Cells*, 2004, **83**, 125–146.
15. J. Nelson, *Curr. Opin. Solid State Mater. Sci.*, 2002, **6**, 87–95.
16. S. Günes, H. Neugebauer and N. S. Sariciftci, *Chem. Rev.*, 2007, **107**, 1324–1338.
17. G. A. Chamberlain, *Sol. Cells*, 1983, **8**, 47–83.
18. C. W. Tang, *Appl. Phys. Lett.*, 1986, **48**, 183–185.
19. B. Kan, Y. Yi, X. Wan, H. Feng, X. Ke, Y. Wang, C. Li and Y. Chen, *Adv. Energy Mater.*, 2018, **8**, 1800424.
20. H. J. Snaith, *Nat. Photonics*, 2012, **6**, 337–340.
21. T. Forster, *Ann. Phys.*, 1948, **437**, 55–75.
22. H. Sahoo, *J. Photochem. Photobiol., C*, 2011, **12**, 20–30.
23. R. M. Clegg, *Curr. Opin. Biotechnol*, 1995, **6**, 103–110.
24. N. Mastour and H. Bouchriha, *Phys. Lett. A: Gen. At. Solid State Phys.*, 2016, **380**, 3866–3869.
25. N. Mastour, Z. Ben Hamed, A. Benchaabane, M. A. Sanhoury and F. Kouki, *Org. Electron.*, 2013, **14**, 2093–2100.
26. I. B. Berlman, *Energy transfer parameters of aromatic compounds*, 1973.
27. L. Chen, L. S. Roman, D. M. Johansson, M. Svensson, M. R. Andersson, R. A. J. Janssen and O. Inganäs, *Adv. Mater.*, 2000, **12**, 1110–1114.
28. M. de Jesus Bassi, L. Wouk, W. Renzi, C. K. Oliveira, J. L. Duarte, I. A. Heisler and L. S. Roman, *Synth. Met.*, 2021, **275**, 116740.
29. A. Roozbeh, M. de Jesus Bassi, A. B. Pereira, L. S. Roman, T. Buckup and I. A. Heisler, *J. Phys. Chem. C*, 2020, **124**, 27946–27953.
30. L. C. W. de Menezes, W. Renzi, C. F. D. N. Marchiori, C. K. B. Q. M. de Oliveira, F. Von Kieseritzky, J. L. Duarte and L. S. Roman, *J. Phys. Chem. C*, 2018, **122**, 5796–5804.
31. A. A. Mohapatra, V. Kim, B. Puttaraju, A. Sadhanala, X. Jiao, C. R. Mcneill, R. H. Friend and S. Patil, *ACS Appl. Energy Mater.*, 2018, **1**, 4874–4882.
32. V. Gupta, V. Bharti, M. Kumar, S. Chand and A. J. Heeger, *Adv. Mater.*, 2015, **27**, 4398–4404.
33. N. A. D. Yamamoto, M. E. Payne, M. Koehler, A. Facchetti, L. S. Roman and A. C. Arias, *Sol. Energy Mater. Sol. Cells*, 2015, **141**, 171–177.
34. M. Koehler, N. A. D. Yamamoto, A. G. Macedo, D. Z. Grodniski, L. S. Roman and M. G. E. da Luz, *Appl. Phys. Lett.*, 2013, **103**, 1–6.

Section II: Water

CHAPTER 8

Origin and Composition of Main Water Contaminants

FATEMEH ASADI ZEIDABADI,[a] EHSAN BANAYAN ESFAHANI,*[a] RAPHAELL MOREIRA[b] AND MADJID MOHSENI*[a]

[a] Department of Chemical and Biological Engineering, University of British Columbia, 2360 East Mall, Vancouver, Canada; [b] Institute of Applied and Physical Chemistry, Universität Bremen, Leobener Str. 6, D-28359 Bremen, Germany
*Emails: ebanayan@mail.ubc.ca; madjid.mohseni@ubc.ca

8.1 Introduction

Water is essential for all life on Earth, forming the foundation of ecosystems, human health, and industrial processes. However, the quality of water resources is constantly threatened by contaminants introduced through human activities and natural processes. Human activities, such as industrial discharges, agricultural runoff, and improper waste disposal, significantly contribute to the introduction of harmful substances into water bodies. Natural processes, including geological weathering and microbial activities, also play a role in contaminating water supplies.

Water contaminants can be classified into several categories, each with distinct origins, compositions, and impacts: organic, inorganic, microbial, and radiological. Each category poses unique challenges for water quality management. Organic and inorganic contaminants can be toxic and persistent, making them difficult to remove from water sources. Microbial contaminants are particularly concerning due to their potential to cause waterborne diseases. Radiological contaminants pose serious health risks due to their radioactive nature.

Contribution of Colloidal Materials to Air, Water and Soil Environmental Sustainability: From Basic Concepts to Advanced Applications
Edited by Izabel C. Riegel-Vidotti, Lizandra Maria Zimmermann and Heloise R. de Barros

Published by the Royal Society of Chemistry, www.rsc.org

By exploring the sources, mechanisms of introduction, and impacts of these contaminants, we can better protect water quality, safeguard public health, and ensure sustainable water resources. This can be achieved through the development of targeted regulations, advanced monitoring techniques, and effective remediation strategies.

In this chapter, we explore the origins and composition of major water contaminants. The diverse sources of these contaminants, along with their pathways into aquatic systems and their potential impacts on both environmental and human health are examined. Through this exploration, we highlight the regulatory/advisory levels for water contaminants, the importance of proactive water management practices, and the need for continuous research and innovation in the field of water quality protection and treatment.

8.2 Types of Water Contaminants

8.2.1 Organic Contaminants

Organic contaminants are chemical compounds containing carbon atoms, primarily derived from human activities. Commonly used products contain chemical components that can persist through waste and sewage treatment plants, eventually entering aquatic environments as parent compounds, metabolites, or transformation products. Pharmaceuticals and personal care products (PPCPs), pesticides, industrial chemicals, and other emerging contaminants (ECs) have been detected in water (typically in $ng\,L^{-1}$) and found bound to sediment and plastic particles (typically in $ng\,g^{-1}$). Despite advancements in knowledge since the late 1990s, the fate of these contaminants and their transformation products in the aquatic environment remains relatively unresolved. Additionally, naturally occurring compounds can transform into hazardous substances during treatment, particularly through halogenation. Even with improved wastewater treatment practices, organic contamination is unlikely to decrease due to the continued and perhaps expanded use of synthetic chemicals and the inherent difficulties of rigorous source control. The following provides a comprehensive overview of the sources, types, properties, classifications, occurrence, effects on humans and biota, regulations, and treatment technologies of these organic contaminants.

8.2.1.1 Sources of Organic Contaminants

Organic contaminants in the environment originate from various sources, which can be broadly categorized into point and non-point sources:

8.2.1.1.1 Point Sources

Wastewater treatment plant (WWTP) effluents: WWTPs treat domestic and industrial wastewaters but often fail to remove all organic contaminants like pharmaceuticals, personal care products, and industrial chemicals.

Artificial recharge: Intentional infiltration of surface water into groundwater aquifers to replenish levels can introduce organic contaminants, such as pesticides and industrial chemicals, from surface water into the groundwater.

Landfills: Landfill leachate, formed as waste decomposes, may contain a variety of organic contaminants, including solvents, pesticides, and pharmaceuticals. If not properly managed, leachate can seep into the surrounding soil and groundwater, causing contamination.

Leakage of septic tanks: Malfunctioning or poorly maintained septic tanks can release organic contaminants, such as household chemicals and pharmaceuticals, into the surrounding soil and groundwater.

8.2.1.1.2 Non-point Sources. Non-point sources are diffuse, not confined to a single location, and contribute to the widespread distribution of organic contaminants in the environment, often carried into water bodies by precipitation, runoff, and atmospheric deposition. Key non-point sources include:

Agricultural runoffs: Pesticides and fertilizers from agricultural fields can wash into nearby water bodies, contaminating surface water and groundwater. Irrigation practices can also introduce these contaminants into soil and underlying aquifers.

Animal farms: Waste from animal farms often contains antibiotics, hormones, and other organic contaminants. Runoff from animal farms can transport these into surface water, while leaching affects groundwater quality.

Urban runoff: Runoff from urban areas, including lawns, gardens, and roads, can carry organic contaminants such as lawn chemicals, pet waste, oil, and grease into water bodies. Urban runoff is a major source of 6PPD (*N*-(1,3-dimethylbutyl)-*N*′-phenyl-*p*-phenylenediamine), an antioxidant widely used in the rubber industry, particularly automobile tires. As tires wear down, tiny particles containing 6PPD are deposited onto the road and are washed off into water bodies.[1]

Atmospheric deposition: Organic contaminants released into the atmosphere from sources like vehicle emissions and industrial processes can then settle onto land or water surfaces through precipitation or dry deposition. Recent studies from Stockholm University found that the levels of selected perfluoroalkyl acids (PFAAs) in rainwater now exceed levels deemed safe by health and environment advisory agencies.[2]

8.2.1.2 Types, Properties, and Classification of Organic Contaminants

Organic contaminants can be categorized by their sources, chemical structures, and impacts on human health and the environment. Below are key categories and examples, though this list is not exhaustive, and many other organic contaminants also pose significant risks.

8.2.1.2.1 Persistent Organic Pollutants (POPs). POPs are synthetic chemicals known for their environmental persistence, bioaccumulation potential, and long-range transport. While there is no universally agreed-upon half-life for a substance to be deemed "persistent", POPs typically have half-lives spanning years to decades in soil and sediment, and several days in the atmosphere.[3] Their stable chemical structures make them resistant to metabolic breakdown, allowing them to persist in the environment and bioaccumulate in food chains.[3] POPs can volatilize and exist in both gas and particle phases, enabling long-range transport through air and water, which often leads to contamination far from their original source. POPs encompass several chemical families, primarily chlorinated and brominated aromatic compounds. Key classes of POPs include:

i. ***Polychlorinated biphenyls (PCBs):*** PCBs encompass 209 congeners, each with distinct physicochemical properties and toxicity levels based on their chlorine atom arrangement. Initially synthesized in 1929, PCBs were widely used in numerous industrial applications due to their chemical stability and heat resistance. They were employed as additives in heat exchange fluids, components in transformers and capacitors, plasticizers for plastics, additives in cement, and in products such as printing inks and paints. However, their persistence in the environment and associated toxicity prompted regulatory examination, leading to their prohibition in manufacturing, processing, and distribution in the US by the late 1970s.[4] Despite the global bans on manufacturing and usage of PCBs, these chemicals persistently and gradually leak into the environment from aging equipment and waste disposal sites.[5] PCBs are notably more prevalent in marine environments, including water, sediments, and aquatic organisms, as these areas serve as their final repository.[6] PCBs exhibit environmental stability, undergo long-distance atmospheric transport, and accumulate in the food chain, posing risks to both ecosystems and human health. In aquatic environments, PCBs bind to particulate matter, eventually accumulating in sediments, where environmental disturbances like ship movement or flooding can resuspend them, perpetuating contamination.[7] The sorption of PCBs onto sediments depends on the characteristics of both the congeners and the sediments themselves; more chlorinated congeners tend to bind more strongly.[8] However, less chlorinated congeners can migrate further from their original emission sources and sometimes appear at higher concentrations.[9] Specific congeners such as PCB28, 52, 101, 118, 138, 153, and 180 are designated as indicators of environmental pollution (ind-PCBs),[10] while coplanar PCBs (co-PCBs) like PCB77, 81, 105, 114, 118, 123, 126, 156, 157, 167, 169, and 189 are employed for evaluating PCB toxicity.[11]

ii. ***Polychlorinated dibenzo-p-dioxins (PCDDs) and dibenzofurans (PCDFs):*** PCDDs and PCDFs collectively referred to as dioxins are a group of persistent environmental pollutants. While they can be released through natural phenomena like volcanic eruptions and forest fires, they primarily result from human activities, such as the incineration of hazardous, municipal, and medical waste, chlorine bleaching of paper pulp, cement production, and metal smelting.[12] The primary route of human exposure to PCDD/Fs is through the diet. PCDD/Fs are highly hydrophobic, leading to their accumulation in the fatty tissues of living organisms. The toxicity of PCDD/Fs varies depending on the number and position of chlorine atoms in the molecules, with 2,3,7,8-tetrachlorodibenzo-*p*-dioxin (TCDD) being the most toxic congener, and is currently classified as a known human carcinogen.[13]

iii. ***Polybrominated diphenyl ethers (PBDEs):*** PBDEs are brominated flame retardants (BFRs) composed of two phenyl rings bonded to varying numbers of bromine atoms. Less brominated PBDEs, like tetra-, penta-, and hexa-BDEs, have a high affinity for lipids, allowing them to bioaccumulate in animals and humans.[14] Widely used in consumer goods, including flame retardants, electronics, construction materials, coatings, textiles, and polyurethane foam, PBDEs have been extensively utilized, with an estimated 46 000 tons of commercial penta-BDE, 25 000 tons of octa-BDE, and 380 000 tons of deca-BDE used in the US and Canada from 1970 to 2020.[15] Structurally similar to PCBs, PBDEs resist environmental degradation, leading to their persistence and widespread contamination. They can enter the environment during production, application, disposal, and recycling[15] and have been detected in water, air, sediment, and human blood.[16] Exposure to PBDEs is linked to endocrine disruption, neurodevelopmental deficits, DNA damage, and potential carcinogenicity.[17] As a result, many countries have implemented regulations to phase out or restrict the use of certain PBDEs, promoting the development of safer alternatives and more sustainable flame retardant solutions.

8.2.1.2.2 Pesticides. The development of synthetic pesticides accelerated during World War II to boost food production, leading to the widespread use of chemicals like dichlorodiphenyltrichloroethane (DDT), aldrin, dieldrin, endrin, parathion, and 2,4-dichlorophenoxyacetic acid (2,4-D) often without consideration of environmental or health risks.[18] The publication of Rachel Carson's "Silent Spring" in 1962[19] and subsequent studies raised awareness of the dangers associated with pesticides, leading to the US ban on DDT in 1972.[20] Since then, more selective and safer pesticides were developed, along with Integrated Pest Management (IPM) systems that aim to reduce pesticide use by incorporating natural pest control methods.[19] Despite these advancements, pesticides remain essential for crop protection and disease control. However, their environmental and health impacts are significant, with approximately 5.2 billion pounds used globally each year.[21] Many pesticides are stable, bioaccumulative, and toxic, posing risks through environmental contamination and exposure. Less than 1% of applied pesticides reach their target pests, with the rest lost through spray drift, runoff, and other means, causing unintended harm on ecosystems and human health.[19] Balancing pest control needs with environmental and health considerations remains a critical challenge.

8.2.1.2.3 Pharmaceuticals and Personal Care Products (PPCPs). PPCPs include a wide range of chemicals used in healthcare and cosmetics, such as medications, veterinary drugs, fragrances, sunscreens, and other personal care products. While essential for health and hygiene, their presence in the environment has become a growing concern. PPCPs enter the ecosystem through excretion, improper disposal, and manufacturing waste. Many contain synthetic and natural ingredients like surfactants, preservatives, and fragrances that can harm environmental and human health. The Environment Agency (EA) of England and Wales ranked the top 10 PPCPs of concern based on their persistence, bioaccumulation, and toxicity (PBT) as lofepramine, dextropropoxyphene, procyclidine, tramadol, paracetamol, clotrimazole, thioridazine, mebeverine, aminophylline, and tamoxifen.[22]

Once in the environment, PPCPs may undergo transformation processes like biodegradation, but conventional wastewater treatments often fail to remove them completely. This leads to their accumulation in water, sediments, and even in organisms within these ecosystems.[23] Chronic exposure to low levels of PPCPs can disrupt human health, potentially causing endocrine issues, antibiotic resistance, and allergic reactions.[24] In the environment, PPCPs can harm aquatic ecosystems, leading to reproductive issues in wildlife and broader ecological disturbances.

8.2.1.2.4 Industrial Chemicals

i. ***Bisphenol A (BPA):*** Bisphenols are synthetic chemicals with two phenolic hydroxyl groups. Notable analogs like BPA, BPS, BPF, and BPAF have been widely used in polycarbonate plastics and epoxy resins since the 1960s. These chemicals are commonly found in everyday items like food containers, baby bottles, dental sealants, and paper products. BPA, the most prevalent bisphenol, has an annual demand growth of 6% to 10%.[25] BPA leaches from containers into food, especially when heated, making dietary intake the primary source of BPA exposure for over 90% of people.[26] It can also enter the body through dental sealants and cross the placenta into fetal tissues.[27] Over 90% of urine samples from populations in the US show detectable BPA levels.[28] Levels of BPA in maternal plasma have been reported in the range of 0.3 to 18.9 $ng\,mL^{-1}$, with lower concentrations found in fetal plasma and breast milk.[27] As an endocrine-disrupting chemical (EDC), BPA interferes with hormone function, binding to estrogen receptors and other nonclassical receptors.[29] This disruption is linked to increased risks of cardiovascular diseases, obesity, diabetes, and reproductive problems.[29] In response to these concerns, regulatory bodies worldwide have restricted BPA use, particularly in products for infants and children. The industry has also developed BPA-free alternatives to reduce human exposure.

ii. ***Microplastics:*** Since 1950, global plastic production has surged, reaching approximately 368 million tons in 2019, with further growth expected. However, this exponential growth has also led to widespread plastic pollution, particularly in the form of microplastics – particles smaller than 5 mm. Microplastics are classified into two categories: primary microplastics, intentionally manufactured to be small and are found in commercial products such as personal care items and cosmetics, and secondary microplastics, which result from the breakdown of larger plastic debris. Secondary microplastics are the most prevalent in marine environments due to their persistence and low degradation rates. Once in the environment, microplastics can be transported by wind and water currents, leading to widespread contamination in different environmental matrices and even in remote locations such as the Arctic.[30] In aquatic systems, microplastics can be ingested by marine organisms, accumulate in sediments, or float on the water surface. Microplastics also serve as carriers for other pollutants, such as heavy metals and persistent organic pollutants, which adhere to their surfaces.[31] The potential health effects of microplastics on humans and wildlife are under investigation, with concerns including chemical toxicity from absorbed pollutants, physical harm like blockages and abrasions in the digestive tract, and disruption of cellular processes.[32]

   Detection and quantifying micro and nanoplastics (1–1000 nm) in water is challenging due to their diverse properties, widespread distribution, and the complexities of environmental matrices. Sampling and collection are hindered by heterogeneity and contamination, while separation and isolation are complicated by varying densities and small sizes. Advances in analytical techniques, such as enhanced Raman and FTIR spectroscopy, have improved the chemical identification and imaging of smaller particles, although challenges remain.[33]

iii. ***Per- and poly-fluoroalkyl substances (PFAS):*** PFAS are a group of >14 000 man-made chemicals, widely used since the 1940s due to their unique chemical properties. PFAS have in common a carbon chain fully or partially saturated with fluorine atoms attached to a functional group (*e.g.*, carboxylic or sulfonic). Characterized by their strong carbon–fluorine bonds, PFAS are highly resistant to heat, water, and oil, leading to their persistence in the environment and thus being referred to as "forever chemicals". PFAS are used in non-stick cookware, water-repellent clothing, stain-resistant fabrics, food packaging, firefighting foams, and various industrial processes.[34–37] The two most studied and widely used PFAS are perfluorooctanoic acid (PFOA) and perfluorooctane sulfonate (PFOS), which have been phased out in many countries due to health concerns but are still prevalent in the environment due to their persistence.[38,39]

PFAS enter the environment through industrial discharges, wastewater treatment plants, landfill leachates, and the use of PFAS-containing products such as firefighting foams. Once in the environment, PFAS are highly mobile, contaminating soil, water, and air. They can travel long distances through groundwater and surface water, leading to widespread environmental distribution.[40,41] PFAS are extremely persistent and bioaccumulate in living organisms, including humans, raising concerns about the long-term exposure risks. Studies have linked PFAS to various health issues, including developmental problems in fetuses and infants, liver damage, thyroid disease, decreased fertility, high cholesterol, and increased risk of certain cancers.[42] The widespread presence of PFAS in the environment and human blood samples indicates that exposure is nearly ubiquitous.[43] In response, regulatory agencies have taken steps to limit the use of PFAS, setting limits for PFAS in drinking water, restricting their use in consumer products, and promoting the development of safer alternatives. Ongoing research is crucial to understand the health impacts of PFAS and develop effective remediation strategies.

iv. ***p-Phenylenediamines (PPDs):*** PPDs are a class of chemicals with aromatic amine groups known for their antioxidant properties, making them essential in enhancing the durability and lifespan of rubber products, particularly tires. PPDs are commonly added to rubber compounds during tire manufacturing to prevent cracking, aging, and deterioration caused by oxidative stress. PPDs are also utilized in various rubber products where protection against environmental degradation is essential, such as conveyor belts, hoses, and seals. One of the most used PPDs in various industries is 6PPD, with high production volume. In 2020, China's annual industrial output of 6PPD reached 0.2 million tons.[44] The primary source of 6PPD release into the environment is through tire wear,[1] which can leach into aquatic ecosystems through stormwater runoff. In the environment, 6PPD can transform into harmful by-products such as 6PPD-quinone (6PPD-Q) formed *via* the reaction of 6PPD with O_3.[45]

PPDs have been detected in various environmental contexts such as snow, dust, surface water, and sediment. A study analyzing 150 urine samples in South China detected 6PPD and 6PPD-Q with frequencies between 60% and 100%.[44] The metabolism of PPDs in humans is not well understood. Studies on mice demonstrated that ingesting 6PPD led to enlarged liver sizes and affected liver metabolism.[46] A recent study suggested that 6-PPDQ was identified as the cause of acute mortality of coho salmon.[47] The prevalence of 6PPD and 6PPDQ in the environment and their potential harmful impacts have gained increasing attention in recent years.

8.2.1.2.5 Disinfection By-products (DBPs). Water disinfection is essential for preventing waterborne diseases, with chlorine-based disinfectants being the most common due to their low cost, efficiency, and persistence. However, these disinfectants can lead to the formation of disinfection by-products (DBPs), which pose potential health risks. DBP formation is influenced by factors such as natural and anthropogenic contaminants, treatment parameters (*e.g.*, disinfectant type, pH, and contact time), and the presence of brominated and iodinated compounds. Among the various types of disinfection by-products (DBPs), trihalomethanes (THMs) and haloacetic acids (HAAs) are the most commonly detected in chlorinated water.[48] Of particular concern are non-regulated nitrogenous DBPs (N-DBPs), like nitrosamines, which are more cytotoxic and genotoxic than many regulated DBPs, highlighting the need for further research and potential regulation.[49]

Chlorine residual, crucial for preventing the regrowth of microbial pathogens, can lead to the continuous formation of DBPs, exposing humans to health risks through ingestion, inhalation, and dermal absorption.[50] Health effects associated with DBP exposure include an increased risk of cancer (particularly bladder cancer), reproductive and developmental issues, liver and kidney damage, and respiratory problems.[48] The level of risk depends on the type and concentration of DBPs, duration of exposure, and individual susceptibility.

8.2.1.2.6 Volatile Organic Compounds (VOCs). Organic compounds with a boiling point $\leq$100 °C and/or a vapor pressure >1 mm Hg at 25 °C are generally classified as volatile organic compounds (VOCs). These compounds are of significant concern for several key reasons. First, their volatility enables easy transition to the vapor state, increasing environmental dispersion. Second, some VOCs in the atmosphere can pose serious public health risks. Third, VOCs contribute to the formation of harmful photochemical oxidants in the atmosphere.[51] VOCs encompass a diverse group of chemicals, with some of the most common types being aromatic hydrocarbons such as benzene, toluene, ethylbenzene, and xylene (BTEX), as well as halogenated hydrocarbons like chloroethylene and trichloroethylene. Carcinogenic VOCs (cVOCs) are particularly concerning due to their cancer-causing potential.[52] While VOCs originate from both natural and anthropogenic sources, human activities like industrial processes, fuel burning, and chemical manufacturing are the primary contributors.[52] In water, VOCs can enter through direct discharge, runoff, and leaching from contaminated sites. Exposure to VOC-contaminated water occurs through drinking, bathing, swimming, and other activities, posing risks of both cancer and non-cancer health effects.[53] Once in water, VOCs may dissolve, volatilize, or adsorb onto sediments, with their fate influenced by their physical and chemical properties.

Emissions of VOCs pose significant environmental and health challenges, contributing to ozone depletion, ground-level ozone formation,[54] and global warming.[55] Microbial VOCs (mVOCs), formed in the metabolism of fungi and bacteria, can cause health issues such as mucosal irritation and lethargy.[53] Even at concentrations below the maximum contamination level (MCL), VOCs can have long-term, synergistic effects, leading to reproductive and birth defects.[56]

Figure 8.1 shows the chemical structures of different types of organic contaminants found in water sources.

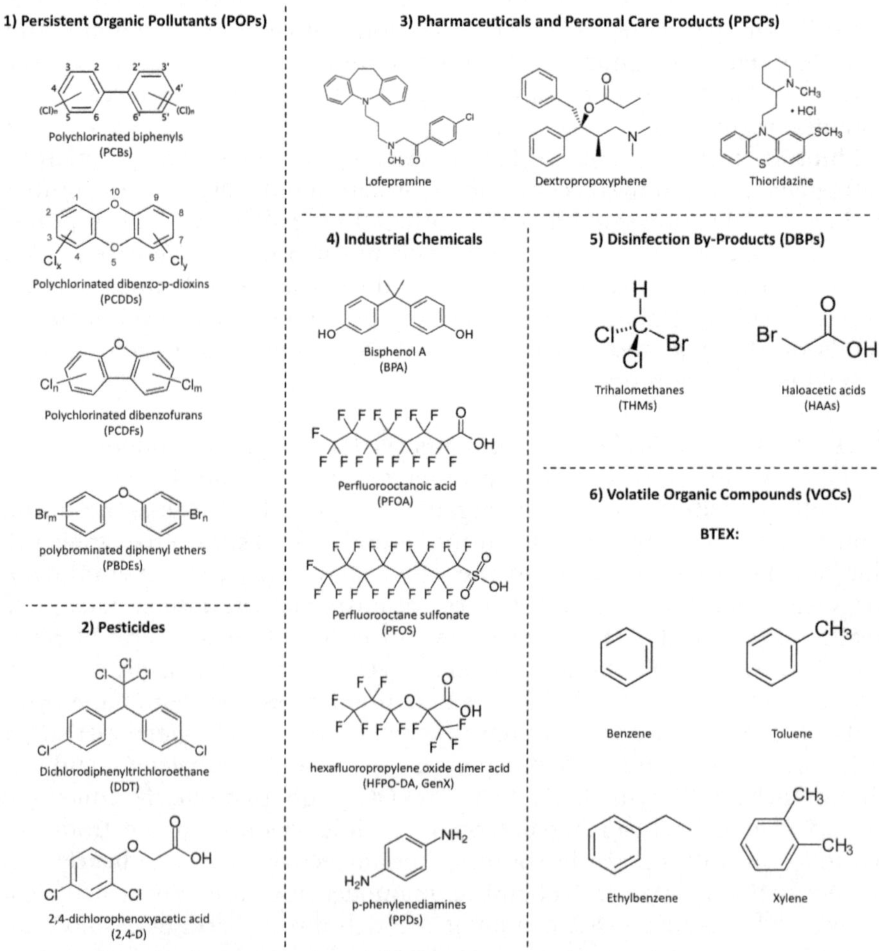

Figure 8.1 Chemical structures of organic contaminants in water.

8.2.1.3 Regulation and Treatment Technologies – Organic Contaminants

Effective management of organic contaminants is essential for protecting environmental integrity and public health. These contaminants in water pose significant risks by harming human health and disrupting aquatic ecosystems. Regulatory frameworks play a vital role in establishing stringent standards to control contamination in air, water, and soil, with variations reflecting regional environmental priorities and health considerations. Table 8.1 summarizes key contaminants, their permissible levels in water, health impacts, and sources of contamination, as outlined by authoritative bodies such as the United States Environmental Protection Agency (USEPA), Health Canada, and the World Health Organization (WHO).[57–59]

To meet regulatory standards for water safety, effective remediation strategies are essential. Various treatment technologies, including biological processes, adsorption methods, high-pressure membranes, and destruction techniques, are employed to remove organic contaminants. Each method has specific advantages and disadvantages, detailed in Table 8.2, guiding the selection of appropriate technologies based on contaminant types, concentrations, and environmental conditions. *Biological processes* are cost-effective for water treatment but face challenges with certain micropollutants.[60] *Adsorption methods*, like activated carbons, are versatile but may struggle with hydrophilic contaminants.[61] *High-pressure membranes* are efficient but costly and prone to fouling.[62] *Destruction techniques*, such as photochemical and electrochemical processes, effectively decompose persistent pollutants but can be expensive and may generate toxic by-products.[63,64] Integrating these technologies can enhance overall treatment efficiency and sustainability.[65]

8.2.2 Inorganic Contaminants

8.2.2.1 Origin of Inorganic Contaminants

Inorganic contaminants, including metals, salts, and minerals are widespread in water, originating from both natural and human activities:

Natural sources include:

Geological weathering releases inorganic substances like calcium, sulfate, sodium, and chloride during the breakdown of rocks and minerals into water bodies.

Volcanic activity introduces metals such as arsenic, mercury, and selenium into the atmosphere and subsequently into water sources through precipitation.

Soil erosion mobilizes inorganic elements, increasing contaminant levels in water, particularly in regions with high rainfall.

Seawater intrusion elevates salt and inorganic constituent levels in freshwater aquifers, especially in coastal areas.

Table 8.1 Organic contaminants – permissible levels, health impacts, and sources of contamination.

Contaminant category	Substance	USEPA, MCL[a] ($\mu g\,L^{-1}$)[57]	WHO, GV[b] ($\mu g\,L^{-1}$)[59]	Health Canada, MAC[c] ($\mu g\,L^{-1}$)[58]	Health impact[57,58]	Sources of contamination in drinking water[57,58]
Persistent organic pollutants (POPs)	PCB	0.5	—	—	Skin changes; thymus gland problems; immune deficiencies; reproductive or nervous system difficulties; increased risk of cancer	Runoff from landfills; discharge of waste chemicals
	Carbofuran	40	7	90	Problems with blood, nervous system, or reproductive system	Leaching of soil fumigant used on rice and alfalfa
	Dioxin (2,3,7,8-TCDD)	0.00003	—	—	Reproductive difficulties; increased risk of cancer	Emissions from waste incineration and other combustion; discharge from chemical factories
	Pentachlorophenol (PCP)	1	9	60	Liver or kidney problems; increased cancer risk	Discharge from wood preserving factories
Pesticides	Atrazine	3	100	5	Cardiovascular system or reproductive problems	Runoff from herbicide used on row crops
	2,4-D	70	30	100	Kidney, liver, or adrenal gland problems	Runoff from herbicide used on row crop
	Endrin	2	0.6	—	Liver problems	Residue of banned insecticide
	Toxaphene	3	—	—	Kidney, liver, or thyroid problems; increased risk of cancer	Runoff/leaching from insecticide used on cotton and cattle

	Alachlor	2	20	—	Eye, liver, kidney or spleen problems; anemia; increased risk of cancer	Runoff from herbicide used on row crops
	Lindane	0.2	2	—	Liver or kidney problems	Runoff/leaching from insecticide used on cattle, lumber, gardens
	Methoxychlor	40	20	—	Reproductive difficulties	Runoff/leaching from insecticide used on fruits, vegetables, alfalfa, livestock
	Chlorpyrifos	—	30	90	Nervous system effects (cholinesterase inhibition)	Leaching and/or runoff from agricultural or other uses
	Diquat	20	—	50	Cataracts	Runoff from herbicide use
	Glyphosate	700	—	280	Kidney problems; reproductive difficulties	Runoff from herbicide use
Industrial chemicals	HFPO-DA PFBS PFNA PFHxS	Hazard index $= \frac{\text{HFPO}-\text{DA}_{\text{ng L}^{-1}}}{10\ \text{ng L}^{-1}} + \frac{\text{PFBS}_{\text{ng L}^{-1}}}{2000\ \text{ng L}^{-1}} + \frac{\text{PFNA}_{\text{ng L}^{-1}}}{10\ \text{ng L}^{-1}} + \frac{\text{PFHxS}_{\text{ng L}^{-1}}}{10\ \text{ng L}^{-1}} = 1$	—	—	Liver, immune, thyroid, and developmental effects following exposure during pregnancy and/or childhood.	Discharge from manufacturing and industrial chemical facilities, use of certain consumer products, occupational exposures, and certain firefighting activities.
	PFOA	0.004	0.1^d	0.2	Cardiovascular, liver, immune, thyroid, and developmental effects following exposure during pregnancy and/or childhood	Discharge from manufacturing and industrial chemical facilities, use of certain consumer products,

Table 8.1 (*Continued*)

Contaminant category	Substance	USEPA, MCL[a] ($\mu g\,L^{-1}$)[57]	WHO, GV[b] ($\mu g\,L^{-1}$)[59]	Health Canada, MAC[c] ($\mu g\,L^{-1}$)[58]	Health impact[57,58]	Sources of contamination in drinking water[57,58]
					increased incidence of certain types of cancers including kidney and testicular.	occupational exposures, and certain firefighting activities.
	PFOS	0.004	0.1[d]	0.6	Cardiovascular, liver, immune, thyroid, and developmental effects following exposure during pregnancy and/or childhood increased incidence of certain types of cancers including liver	Discharge from manufacturing and industrial chemical facilities, use of certain consumer products, occupational exposures, and certain firefighting activities.
	Di(2-ethylhexyl)-phthalate	6	8		Liver cancer (classified as probable carcinogen)	By-product of drinking water disinfection with chlorine; industrial effluents and spills
	2,4,6-Trichlorophenol	—	200	5	Liver cancer (classified as probable carcinogen)	By-product of drinking water disinfection with chlorine; industrial effluents and spills
Disinfection by-products (DBP)	Bromodichloromethane (BDCM)	0	60	—	Liver, kidney or central nervous system problems; increased risk of cancer	By-product of drinking water disinfection
	Dibromochloromethane (DBCM)	60	100	—		
	Chloroform	70	300	—		
	Bromoform	0	100	—		

	Total trihalomethanes (THMs)	80	$\frac{C_{BDCM}}{GV_{BDCM}} + \frac{C_{DBCM}}{GV_{DBCM}} + \frac{C_{Chloroform}}{GV_{Chloroform}} + \frac{C_{Bromoform}}{GV_{Bromoform}} \leq 1$	100	Liver, kidney or central nervous system problems; increased risk of cancer	By-product of drinking water disinfection
	Haloacetic acids – total (HAAs)	60	—	80	Increased risk of cancer	By-product of drinking water disinfection with chlorine
Volatile organic compounds (VOCs)	Benzene	5	10	5	Anemia; decrease in blood platelets; increased risk of cancer	Discharge from factories; leaching from gas storage tanks and landfills
	Carbon tetrachloride	5	4	2	Liver problems; increased risk of cancer	Industrial effluents and leaching from hazardous waste sites
	1,4-Diaxane	—	50	50	liver effects that occur before the development of cancer	Contaminations of drinking water supplies near landfills and industrial sites
	Toluene	1000	700	60	Nervous system, kidney, or liver problems	Discharge from petroleum factories
	Xylenes	10 000	500	90	Nervous system damage	Discharge from petroleum factories; discharge from chemical factories
	1,4-Dichlorobenzene	—	300	5	Benign liver tumours and adrenal gland tumours (classified as probable carcinogen)	Releases or spills from industrial effluents; use of urinal deodorants

Table 8.1 (*Continued*)

Contaminant category	Substance	USEPA, MCL[a] ($\mu g\,L^{-1}$)[57]	WHO, GV[b] ($\mu g\,L^{-1}$)[59]	Health Canada, MAC[c] ($\mu g\,L^{-1}$)[58]	Health impact[57,58]	Sources of contamination in drinking water[57,58]
	1,2-Dichloroethane	5	30	5	Increased risk of cancer	Discharge from industrial chemical factories
	Dichloromethane	5	20	50	Liver problems; increased risk of cancer	Discharge from drug and chemical factories
	Ethylbenzene	700	300	140	Liver or kidneys problems	Discharge from petroleum refineries
	Tetrachloroethylene	5	—	10	Liver problems; increased risk of cancer	Discharge from factories and dry cleaners
	Trichloroethylene	5	—	5	Liver problems; increased risk of cancer	Discharge from metal degreasing sites and other factories

[a] MCL = maximum concentration level.
[b] GV = guideline value.
[c] MAC = maximum acceptable concentration.
[d] Provisional guideline value.

Table 8.2 Advantages and disadvantages/challenges of treatment technologies for organic contaminants remediation.

Category	Technology	Advantages	Disadvantages/ challenges
Biological processes	Activated sludge Biofiltration Bioreactors Anaerobic digestion Constructed wetland Phytoremediation	• Utilizes natural processes and minimizes chemical usage, reducing secondary pollution. • Generally lower operating costs compared to physical and chemical treatments. • Capable of degrading a wide range of biodegradable organic compounds, including industrial chemicals.	• Less effective for non-biodegradable or highly persistent pollutants like PFAS. • Requires regular maintenance and monitoring, with complexity in design and operation. • Generates waste sludge and can lead to contaminant accumulation in biomass, necessitating further treatment and careful disposal. • Requires long reaction times for specific contaminants.
Adsorption	GAC	• Effective at removing a wide range of organic pollutants, including POPs, pesticides, PPCPs, and VOCs. • High surface area provides extensive adsorption capacity.	• Requires periodic replacement or reactivation, which can be costly. • Limited effectiveness for certain highly hydrophilic or low molecular weight compounds such as short-chain PFAS.
	IX resin	• High specificity for certain pollutants, leading to efficient removal. • Can be tailored for selective adsorption of specific contaminants. • Can be regenerated and reused, saving on the cost.	• Regeneration can be complex and expensive. • Produces concentrated brine waste that requires appropriate disposal. • Some types of resins are single use.
	Metal–organic frameworks (MOFs)	• Extremely high surface area and tunable pore sizes. • Potential for high selectivity and capacity for various pollutants.	• High production costs and potential instability in aqueous environments. • Limited large-scale application and commercialization.

Table 8.2 (*Continued*)

Category	Technology	Advantages	Disadvantages/challenges
		• Versatile and can be designed for specific applications.	• Potential of leaching metals to the aqueous solutions.
High pressure membrane	Nanofiltration (NF) Reverse osmosis (RO)	• Effective for removal of industrial chemicals, pesticides, PPCPs, and certain DBPs. • Can selectively remove smaller organic molecules while allowing certain ions to pass.	• High energy consumption, especially for RO. • Membrane fouling and scaling can reduce efficiency and require frequent maintenance. • Produces concentrated brine waste that requires appropriate disposal.
Advanced destruction techniques	Electrochemical processes	• Capable of degrading a wide range of organic pollutants. • Does not require external chemical additives, reducing the risk of secondary pollution. • Can achieve complete mineralization of contaminants, breaking them down into harmless end products.	• Requires significant electrical energy, leading to high operational costs. • Potential for the formation of toxic by-products that may require further treatment. • Requires specialized and costly equipment, making it less accessible for small-scale applications.
	Photochemical processes	• Effective for degrading various organic pollutants using UV light and catalysts like TiO_2. • Do not leave chemical residues in treated water, enhancing safety. • Can be combined with other AOPs (*e.g.*, H_2O_2) to enhance effectiveness and efficiency.	• UV light generation requires significant energy, increasing operational costs. • Potential for incomplete degradation and formation of harmful by-products. • Efficiency decreases in turbid or colored waters due to light scattering and absorption.
	Sonochemical processes	• Effective at breaking down refractory and complex organic pollutants that are resistant to other treatments.	• Ultrasound generation requires substantial energy, leading to high operational costs.

Table 8.2 (*Continued*)

Category	Technology	Advantages	Disadvantages/ challenges
		• Ultrasound waves can enhance reaction rates, leading to faster degradation. • Use physical energy (ultrasound) instead of chemical additives, reducing secondary pollution risk.	• Requires specialized equipment that can be costly to purchase and maintain. • Currently less developed and applied on a large scale compared to other AOPs, limiting widespread use.

Anthropogenic sources include:

Agricultural practices involve the use of fertilizers and pesticides, leading to the leaching of nitrates, phosphates, and heavy metals into water, causing eutrophication and health risks.

Industrial discharges release heavy metals like lead, cadmium, and mercury through mining, manufacturing, and energy production.

Urban runoff introduces road salts, metals, and construction materials to water bodies, posing toxic risks.

Waste disposal, when done improperly, can release heavy metals and metalloids into groundwater. Treated sewage still contains residual inorganic substances like phosphates, nitrates, and heavy metals, impacting water quality. Additionally, landfills produce leachate – a liquid rich in ammonium, heavy metals, and salts – that can contaminate both groundwater and surface water.

8.2.2.2 Types, Properties, and Impact of Inorganic Contaminants on Human Health

Inorganic contaminants in water can be broadly categorized into several groups based on their chemical nature and sources:

Heavy metals: Heavy metals are dense, high atomic weight metals and metalloids, often defined by a density over 5 $g\,cm^{-3}$. Their contamination in water is a growing global concern due to widespread industrial, domestic, agricultural, and technological use. In the latter half of the 20th century alone, significant quantities of heavy metals, including 22 000 tons of cadmium, 783 000 tons of lead, and 1 350 000 tons of zinc, were released into the environment.[66] These metals pose serious risks to human health, depending on factors like dose, exposure route, and individual characteristics.[67] Key heavy metals include:

Lead (Pb): Lead is commonly released from industrial processes, old plumbing, and lead-based paints.[68] Lead contamination, especially from corroding pipes, is a major health risk, particularly affecting children's nervous systems and cognitive development.[68]

Mercury (Hg): Mercury gets into drinking water from coal-fired power plants, mining activities, agricultural runoff, as well as seepage from landfills.[69] It bioaccumulates in aquatic organisms,[69] leading to toxic effects in the food chain. Mercury can cause neurological disorders, immunological and reproductive effects, abortion, and disruption of the endocrine system.[70]

Cadmium (Cd): Cadmium is found in products like polyvinyl chloride (PVC), dyes, and Ni–Cd batteries. It is also linked to processes such as copper and nickel smelting, fossil fuel burning, and the application of phosphate fertilizers.[71] Cadmium exposure is linked to cardiovascular, renal and hepatic dysfunction, pulmonary edema, and reproductive issues.[71]

Arsenic (As): Arsenic is a naturally occurring element found in both organic and inorganic forms, with the latter being far more toxic. Arsenic enters the environment through natural processes such as volcanic activity and erosion, as well as human activities like mining, metal smelting, and the use of arsenic-based pesticides. Notably, arsenic contamination in drinking water has caused massive poisoning incidents, particularly in India and Bangladesh.[72,73] Chronic exposure can lead to severe health issues including skin thickening and discoloration, developmental problems, cardiovascular disease, neurotoxicity, and increased cancer risk. Acute arsenic poisoning, though less common, can cause vomiting, abdominal pain, diarrhea, and even death. The toxicity of arsenic varies significantly with its chemical form, with arsenite (As III) being ten times more toxic than arsenate (As v).[74]

Zinc (Zn): Zinc is an essential trace element vital for biological functions but becomes toxic at high concentrations, causing gastrointestinal distress and potential immune and neurological issues.[75] Commonly used in coatings, batteries, wood preservatives, and other products, zinc contamination in water mainly arises from industrial discharges, agricultural runoff, and urban sources.[67] Elevated zinc levels in water can be hazardous, leading the WHO and EPA to set guidelines to limit its concentrations in drinking water. However, compared to several other heavy metals, zinc is relatively harmless, resulting in higher regulatory limits. Notably, zinc contamination poses a greater concern for the environment and aquatic species, leading to more stringent discharge limits in environmental regulations.

Chromium (Cr): Chromium is a highly toxic heavy metal, naturally occurring and widely used in various industries. It exists primarily as trivalent chromium [Cr(III)], essential for metabolism, and hexavalent chromium [Cr(VI)], which is highly toxic.[76] Cr(VI) contamination, often from industrial processes like metal plating and leather tanning, can lead to severe health issues, including respiratory problems and lung and gastric cancer.[76] High levels of Cr(VI) in drinking water and soil have been linked to a 60-fold increase in gastric cancer rates, in terms of loss of life expectancy (LLE) and medical expenditure values of US$1337 and US$1162 million year^{-1}, respectively.[76]

Iron (Fe): Iron is a common contaminant in drinking water due to its abundance in the Earth's crust. It leaches from rocks and minerals into groundwater, existing mainly as ferric (Fe(III)) and ferrous (Fe(II)) ions. Iron contamination affects both surface- and groundwaters, with surface water primarily contaminated by industrial effluents and groundwater by rainwater infiltration through iron-rich soils. In severe cases, iron levels in surface waters can range even up to 15 mg L^{-1},[77] leading to reduced water quality, a reddish color and metallic taste. The USEPA and Health Canada guidelines established an aesthetic limit of 0.3 mg L^{-1} for iron in drinking water, though, levels above 0.05 mg L^{-1} may still give rise to aesthetic problems.[78]

Manganese (Mn): Manganese is an essential nutrient involved in the metabolism of amino acids, proteins, and lipids, but it becomes neurotoxic in excess. It exists as a soluble divalent ion (Mn^{2+}) and often co-occurs with iron in surface and groundwater due to the weathering and leaching of metal-bearing minerals. Groundwater typically contains higher manganese levels. Recently, the potential neurotoxicity of manganese in drinking water has gained attention, with studies linking exposure to impaired neurological development in children, including lower intelligence quotient (IQ) and poor neurobehavioral function, even at low concentrations.[79] In Quebec, studies found a 6.2-point difference in IQ scores between children exposed to low and high manganese levels in water.[79] Consequently, the USEPA and Health Canada guidelines set an aesthetic objective limit of 0.05 mg L^{-1} for manganese in drinking water, while Health Canada proposed a health-based maximum acceptable concentration of 0.1 mg L^{-1} based on studies showing neurotoxic effects in rats.[80]

Figure 8.2 shows the average concentration (μg L^{-1}) of 8 heavy metals – Cd, Pb, Cr, Zn, Cu, Ni, Mn, and Fe – in river and lake waters across five continents from 1970 to 2017.[81] The results can be analyzed both geographically and temporally:

i. *Geographical trend:* Heavy metal sources vary by continent: waste discharge and rock weathering are dominant in Africa; mining, manufacturing, and rock weathering in Asia and South America; fertilizer and pesticide use, and rock weathering in North America; and mining, manufacturing, waste discharge, and rock weathering in Europe.[81] Europe shows the lowest mean concentrations, indicating effective pollution control measures, such as strict emission standards and

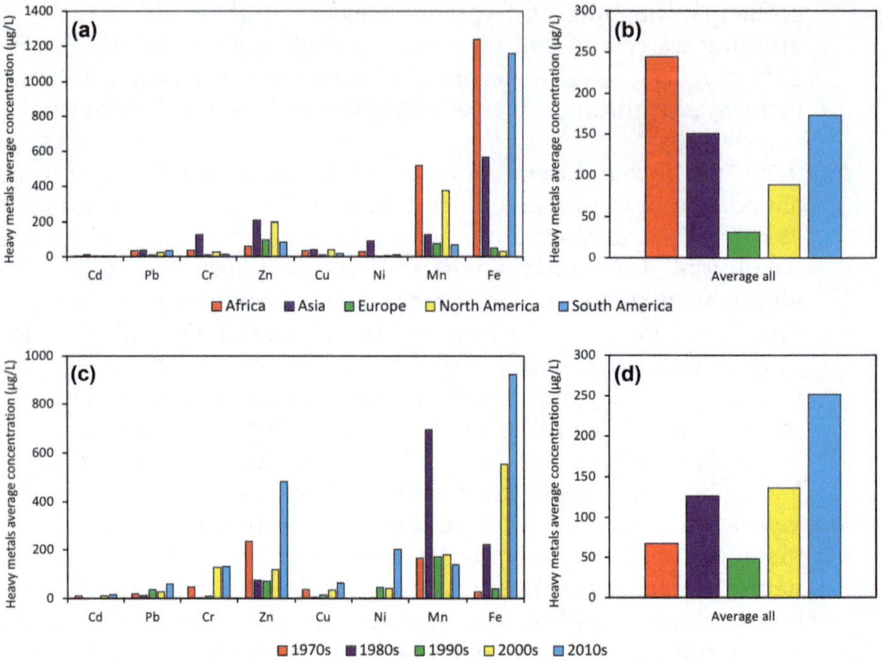

Figure 8.2 Average concentrations ($\mu g\,L^{-1}$) of dissolved heavy metals in global river and lake waters, categorized by continent and decade: (a) average concentrations of each heavy metal for five continents from 1970 to 2017; (b) overall average concentration of all studied heavy metals per continent; (c) average concentrations of each heavy metal for five decades from the 1970s to the 2010s globally; (d) overall average concentration of all studied heavy metals per decade (adapted from ref. 81 with permission from Springer Nature, Copyright 2019).

rigorous waste treatment. Despite these efforts, cancer risks associated with Cr were above hazardous levels in Africa, Asia, and North America.[81]

ii. *Temporal trend:* The primary sources of heavy metals in water have shifted over time. In the 1970s to the 1980s, mining, manufacturing, and agricultural were predominant. By the 1990s to 2010s, rock weathering also became a key source.[81] The global concentrations of heavy metals in rivers and lakes have generally increased, indicating rising pollution levels over the decades.

Inorganic acids:

Inorganic acids, widely used in industrial applications, can become significant contaminants in water sources. While typically considered low-hazard due to their ability to be neutralized in the environment, excessive levels can have severe adverse effects on both the environment and human health.[67] Industrialization and human activities have increased emissions of nitrogen oxides (NO and NO_2) and sulfur dioxide (SO_2) from vehicles and power plants. These gases react in the atmosphere to form nitric acid (HNO_3)

and sulfuric acid (H_2SO_4), which precipitate as acid rain or snow, significantly altering the pH of natural water bodies.[67]

Nutrients:

Nitrogen and phosphorus are essential nutrients for microorganism growth and plant photosynthesis, but excessive levels in water can cause eutrophication. This leads to algal blooms, which block sunlight, causing underwater plants to die and decompose, consuming oxygen in the process and resulting in hypoxia. This lack of oxygen can kill fishes and other aquatic organisms, reducing biodiversity and harming water quality.[82] Human activities, particularly agricultural runoff, industrial waste, and sewage disposal, accelerate eutrophication, causing economic losses and environmental degradation. Effective management, including reducing nutrient inputs through better agricultural practices and wastewater treatment, and internal methods like sediment dredging and aeration, are critical in mitigating these impacts.

Excess nitrogen, particularly in the forms of nitrate (NO_3^-) and nitrite (NO_2^-), poses significant health risks. Nitrate can convert to nitrite in the body, leading to methemoglobinemia or "blue baby syndrome" in infants, a serious condition that impairs the ability of blood to carry oxygen. Additionally, nitrite can react with certain compounds in the body to form *N*-nitroso compounds, which are known carcinogens.[83] High levels of nitrate in drinking water have been associated with an increased risk of gastric and esophageal cancers, although studies show mixed results for other cancer types.[84] Furthermore, nitrates can disrupt thyroid function by inhibiting iodine uptake.[85] To address these risks, the USEPA has set an MCL for total nitrate/nitrite at 10 $mg\,L^{-1}$ and for nitrite alone at 1.0 $mg\,L^{-1}$. Similarly, high phosphorus concentrations can cause kidney damage and osteoporosis, prompting regulations like those from the Wisconsin Department of Natural Resources (WDNR) to limit phosphorus to 1.0 $mg\,L^{-1}$ for point sources.[82]

Other Inorganic Compounds:

Selenium (Se): Selenium is a metalloid with properties between metals and nonmetals. Once considered toxic, recent studies have highlighted its essential role in human health. Selenium is crucial for antioxidant defense, thyroid hormone metabolism, and immune support.[86] However, both deficiency and excessive intake of selenium can lead to serious health issues: deficiency may lead to Keshan disease and hypothyroidism,[87] while excessive intake, known as selenosis, can cause gastrointestinal issues, hair loss, and nerve damage.[88]

Fluoride (F^-): Fluoride comes from both natural sources, such as volcanic eruptions and the weathering of fluoride-rich rocks, and human activities like fertilizer production and industrial processes. In moderate amounts (0.5–1 $mg\,L^{-1}$), fluoride helps prevent tooth decay and strengthens dental enamel. However, excessive fluoride exposure can cause dental fluorosis, characterized by tooth discoloration and mottling, and long-term high exposure can lead to skeletal fluorosis, a severe bone disease.[89] Elevated fluoride levels are reported in groundwater in over 20 countries, including India, where 19 states face severe fluorosis issues.[90]

Cyanides (CN^-): Cyanides can enter natural water sources from metal finishing industries, iron and steel mills, landfill runoff, pesticide application, and the use of cyanide-containing road salts. Cyanide poses serious environmental and health risks due to its high toxicity. Even at low concentrations, cyanides are lethal to aquatic life.[91] In humans, cyanide exposure can lead to acute health effects such as respiratory failure, neurological damage, and, in extreme cases, death.[92] The persistence and toxicity of cyanides in water necessitate stringent monitoring and regulation to prevent contamination and protect public health.

Perchlorates (ClO_4^-): Perchlorates are inorganic compounds found in rocket propellants, explosives, fireworks, and certain fertilizers. They pose a significant risk as they can disrupt iodine uptake in the thyroid gland, leading to thyroid dysfunction, which is particularly harmful to pregnant women, infants, and individuals with existing thyroid issues.[93] Perchlorates are highly soluble in water and can persist in the environment, making their detection and removal from water sources a challenging task.[64]

Other cationic/anionic water constituents: In addition to the above-mentioned inorganic contaminants, water may contain other cationic and anionic substances that impact both human health and the environment. Cationic contaminants such as calcium (Ca^{2+}), magnesium (Mg^{2+}), sodium (Na^+), and potassium (K^+) are essential for biological functions but can cause issues such as water hardness, reduced soap effectiveness, scale buildup, and increased heating costs when present in excess. High levels of sodium in drinking water can worsen health conditions like hypertension,[94] especially in individuals with salt-sensitive conditions. Anionic contaminants like chloride (Cl^-), sulfate (SO_4^{2-}), bicarbonate (HCO_3^-), and carbonate (CO_3^{2-}), are also often introduced into water systems through natural processes, such as mineral dissolution, or anthropogenic sources like industrial discharge, agricultural runoff, and wastewater. Chloride and sulfate can affect water taste and pipe corrosion, while high sulfate levels may cause gastrointestinal problems and impact livestock.[95] Bicarbonates and carbonates influence the pH of water, affecting the solubility and toxicity of other contaminants. They may also reduce the efficacy of treatment processes for removing other water contaminants.

8.2.2.3 *Regulatory Framework and Remediation Strategies – Inorganic Contaminants*

Various countries have developed regulatory frameworks to manage inorganic contaminants in drinking and natural water. The USEPA has established the lead MCL of 15 $\mu g\ L^{-1}$ in drinking water, mandating actions to control corrosion if exceeded. Notably, the Maximum Contaminant Level Goal (MCLG) for lead is set at zero, as no safe exposure level is known. The WHO has a stricter limit of 10 $\mu g\,L^{-1}$ for lead.

For arsenic, the USEPA reduced the MCL from 50 $\mu g\,L^{-1}$ to 10 $\mu g\,L^{-1}$ in 2001, aligning with WHO and Health Canada limits to reduce health risks such as cancer. The MCLG for arsenic is also zero due to its high toxicity.

The differences between MCLs and MCLGs reflect a balance between health protection and technical and economic feasibility. While MCLGs represent levels with no expected adverse health effects, MCLs may be higher due to the challenges and costs associated with achieving such low levels. This approach is also seen with organic contaminants like PFOA and PFOS, which have MCLGs of zero and enforceable MCLs of 4.0 ppt. Table 8.3 summarizes MCLs and guidelines for various inorganic substances, highlighting the global efforts to ensure safe and sustainable water resources.

Effective regulatory frameworks and remediation strategies are essential for maintaining water quality and protecting public health from inorganic contaminants. Key factors in selecting treatment processes include contaminant speciation, influent and effluent concentrations, and flow variability, while minimizing costs is also crucial. *Adsorption*, using materials like activated carbon and advanced adsorbents (*e.g.*, biochar, zeolite, and MOFs), effectively removes contaminants such as lead and arsenic. *Membrane filtration techniques*, including reverse osmosis (RO) and nanofiltration (NF), can efficiently treat contaminants and are particularly used for desalination but may require pre-treatment to handle co-existing components. *Ion exchange processes* are effective for removing both cationic and anionic contaminants with various selectivity (refer to Figure 8.3). Cation exchange

Table 8.3 Inorganic contaminant guidelines ($mg\,L^{-1}$) for drinking water quality.

Inorganic substance	EPA (MCL)[a]	WHO (guideline value)[b]	Health Canada (MAC)[c]
Lead (Pb)	0.015	0.01	0.005
Arsenic (As)	0.01	0.01	0.01
Zinc (Zn)	5.0	3.0	5.0
Mercury (Hg)	0.002	0.006	0.001
Chromium (Cr)	0.1	0.05	0.05
Cadmium (Cd)	0.005	0.003	0.007
Copper (Cu)	1.3	2	2
Iron (Fe)	0.3	—	0.3
Manganese (Mn)	0.05	0.08	0.12 (AO:[d] 0.02)
Selenium (Se)	0.05	0.04	0.05
Antimony (Sb)	0.006	0.02	0.006
Nitrates (NO_3^-)	10 as N	50 as NO_3	45 as NO_3, 10 as N
Nitrates (NO_2^-)	1 as N	3 as NO_2	3 as NO_2; 1 as N
Fluoride (F^-)	4.0	1.5	1.5
Cyanide (CN^-)	0.2	—	0.2

[a] USEPA, National Primary Drinking Water Regulations.[57]
[b] WHO Guideline for Drinking Water, 4th Edition.[96]
[c] Health Canada, Guidelines for Canadian Drinking Water Quality.[58]
[d] Aesthetic Objective.

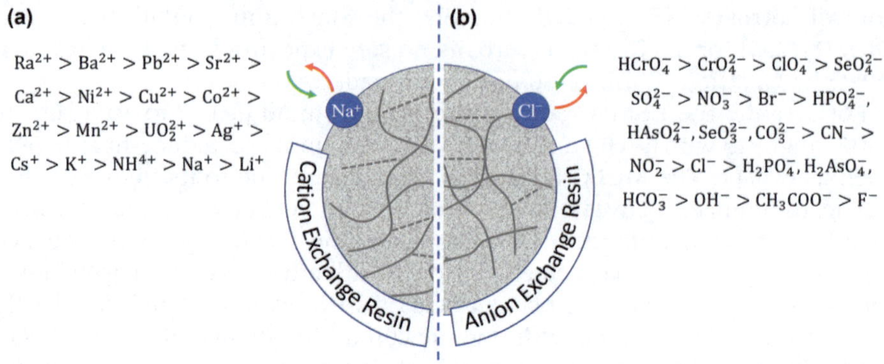

Figure 8.3 Selectivity of ions during (a) cation and (b) anion exchange processes.[97]

uses resins to preferentially capture divalent cations, while anion exchange uses strong-base or weak-base resins to remove anions. *Chemical precipitation* converts dissolved contaminants into solid particles for removal, effectively addressing metals like arsenic and lead. *Bioremediation* leverages microorganisms and plants to detoxify contaminants, including nitrates and cyanides. *Advanced oxidation processes* (AOPs), which use strong oxidizers like hydroxyl radicals and ozone, degrade contaminants efficiently, with technologies such as photocatalysis demonstrating high removal efficiencies for heavy metals. These diverse technologies highlight the comprehensive approaches available for addressing inorganic contamination and ensuring safe water resources.

8.2.3 Microbial Contaminants

Microbial contamination in water encompasses various microorganisms, including bacteria, viruses, and parasites, each with distinct characteristics and health implications. Pathogens are microorganisms that cause disease and can lead to infection in their host. Microbial contaminants can cause both acute and chronic effects and thrive in various environments. These microorganisms often adhere to suspended solids in water, and their concentrations can fluctuate, making infection risk difficult to assess based solely on average levels. Factors like pathogen dose, invasiveness, virulence, and the individual's immune system influence disease risk. Once inside the host, pathogens can multiply, potentially leading to illness. Waterborne pathogens can also multiply in food, beverages, or warm water systems, increasing the infection risk. Understanding microbial contamination is key to developing effective water treatment and public health strategies.

8.2.3.1 Types and Properties of Microbial Contaminants

Infectious diseases caused by pathogenic bacteria, viruses, and parasites such as protozoa and helminths are the most widespread health risks associated with drinking water. Some waterborne pathogens lead to severe and sometimes life-threatening diseases such as typhoid, cholera, infectious hepatitis (caused by hepatitis A or hepatitis E virus), and disease caused by *Shigella* spp. and *E. coli* O157. Others are typically associated with less severe outcomes, such as self-limiting diarrheal disease (*e.g.* noroviruses and *Cryptosporidium*). Breakdown in water supply safety (source, treatment, and distribution) may lead to large-scale contamination and disease outbreaks. These pathogens vary in health impact, infectivity, and resistance (refer to Figure 8.4).

Bacteria are generally the most sensitive group of pathogens to inactivation by disinfection. Some free-living pathogens, such as Legionella and non-tuberculous mycobacteria, can grow in water environments, but enteric bacteria typically do not grow in water and survive for shorter periods than viruses or protozoa. Many bacterial species that are infective to humans are carried by animals. Pathogenic bacteria in water including *Escherichia coli*, *Vibrio cholerae*, *Salmonella typhi*, and *Shigella* are known to cause waterborne diseases like diarrhea, cholera, and typhoid.

- ***Escherichia***, a Gram-negative, rod-shaped, and often antibiotic-resistant bacterium,[98] makes infections like diarrhea and gastroenteritis difficult

	Health significance	Relative infectivity	Resistance to chlorine
Bacteria			
Burkholderia	High	Low	Low
Campylobacter	High	Moderate	Low
Escherichia coli – Diarrhoeagenic	High	Low	Low
E. coli – Enterohaemorrhagic	High	High	Low
Francisella	High	High	Moderate
Legionella	High	Moderate	Low
Mycobacteria (nontuberculous)	Low	Low	High
Salmonella typhi	High	Low	Low
Other salmonellae	High	Low	Low
Shigella	High	High	Low
Vibrio	High	Low	Low
Viruses			
Adenoviridae	Moderate	High	Moderate
Astroviridae	Moderate	High	Moderate
Caliciviridae	High	High	Moderate
Hepeviridae	High	High	Moderate
Picornaviridae	High	High	Moderate
Reoviridae	High	High	Moderate
Protozoa			
Acanthamoeba	High	High	High
Cryptosporidium	High	High	High
Cyclospora	High	High	High
Entamoeba	High	High	High
Giardia	High	High	High
Naegleria	High	Moderate	Low
Helminths			
Dracunculus	High	High	Moderate

High ■ Moderate ■ Low ■

Figure 8.4 Pathogens in drinking water (adapted from ref. 96 with permission from the WHO, Copyright 2022).

to treat. *Escherichia* species include *E. coli* (EC), *E. albertii*, *E. fergusonii*, and *E. hermannii*, with *E. coli* being the most common in drinking water.[99] While many *E. coli* strains are harmless, harmful types like enterotoxigenic *E. coli* (ETEC), enteropathogenic *E. coli* (EPEC), enterohemorrhagic *E. coli* (EHEC), and enteroinvasive *E. coli* (EIEC) can cause severe diseases. ETEC, found in cattle and human feces, contaminates water sources, especially in developing countries with poor sanitation, causing diarrhea in children and "traveler's diarrhea" in visitors to developing regions.[100]

- ***Vibrio***, a curved-shaped, Gram-negative bacterium, includes species like *V. cholerae* and *V. parahaemolyticus*. *V. cholerae* O1 and O139 cause cholera, affecting 1.3 to 4 million people annually, with up to 143 000 deaths, mostly in areas lacking safe water and sanitation.[101] *V. parahaemolyticus* thrives in warm, low-salinity waters and mainly causes gastroenteritis, especially in Asian countries like Japan.[102] Gastroenteritis affects 3 to 5 billion people annually, with nearly 2 million deaths among children under five.[103]
- ***Salmonella***, a rod-shaped Gram-negative bacterium, causes typhoid, paratyphoid fever, and gastroenteritis. It includes two species, *S. enterica* and *S. bongori*, with *S. enterica* divided into six subspecies. *Salmonella* can be transmitted through contaminated water and animals.[104]
- ***Shigella***, another Gram-negative, rod-shaped bacterium, causes bacillary dysentery (shigellosis) and includes four species: *S. dysenteriae*, *S. flexneri*, *S. boydii*, and *S. sonnei*. *S. flexneri* is prevalent in low- and middle-income countries, while *S. sonnei* is found in high-income regions. *Shigella* spreads primarily through the fecal–oral route *via* contaminated water.[105]

Viruses are the smallest pathogens and are harder to remove by physical processes such as filtration. They are often less sensitive to disinfection than bacteria and parasites [*e.g.* adenovirus is less sensitive to ultraviolet (UV) light] and can persist in water for long periods. Waterborne viruses, such as adenovirus, astrovirus, hepatitis A and E viruses, rotavirus, norovirus, and enteroviruses like coxsackieviruses and polioviruses, primarily cause gastroenteritis, leading to diarrhea, abdominal cramps, vomiting, and fever but can also result in more severe conditions like encephalitis, meningitis, myocarditis, cancer, and hepatitis.[96] Contaminated water can transmit viruses through ingestion, inhalation (*e.g.*, during showering), and contact with skin and eyes.

Hepatitis A and E viruses, spread through contaminated water, cause liver disease, with hepatitis E causing more severe damage. The WHO estimated 7134 deaths from hepatitis A and 44 000 from hepatitis E in 2015–2016. Hepatitis A is prevalent in low- and middle-income countries with poor sanitation, while hepatitis E is more common in East and South Asia.[106]

Rotavirus is a leading cause of severe childhood gastroenteritis, accounting for about one-third of diarrhea hospitalizations. It spreads through contaminated water and causes the majority of deaths in developing countries due to inadequate healthcare.[107] Norovirus, responsible for 18% of diarrheal diseases worldwide, leading to 64 000 hospitalizations, 900 000 clinic visits among children in developed countries, and around 200 000 deaths of children under five in developing countries each year. Like other viruses, norovirus spreads primarily through the fecal–oral route, often *via* contaminated drinking water.[108]

Parasites including protozoa and helminths can infect humans through contaminated water, causing about 842 000 deaths annually.[109] *Giardia intestinalis*, responsible for giardiasis, affects around 2% of adults and 8% of children in developed nations and about 33% in developing countries, primarily through the fecal–oral route.[110] *Entamoeba histolytica*, causing amoebic dysentery, is the third deadliest parasitic infection, with nearly 100 000 annual deaths, mainly in areas with poor sanitation.[111]

Cyclospora cayetanensis causes cyclosporiasis, a gastroenteric disease linked to diarrhea. It is a significant health concern in developed countries due to the consumption of imported food from developing regions, where the parasite is endemic due to water and sanitation issues.[112] *Cryptosporidium*, causing cryptosporidiosis, includes over 30 species, but *Cryptosporidium parvum* and *Cryptosporidium hominis* primarily infect humans. This parasite is prevalent in developing countries due to inadequate clean water and poor sanitation.

Parasites are the group of pathogens that is least sensitive to inactivation by chemical disinfection. UV light is effective against *Cryptosporidium*, which is highly resistant to oxidizing disinfectants such as chlorine. Protozoa, which are larger than 2 μm, can be removed by physical processes. They can survive for long periods in water. They are moderately species specific, with livestock and humans as sources of protozoa such as *Cryptosporidium* and *Balantidium*, while pathogenic *Cyclospora* and *Entamoeba* are only found in humans. Infective doses are typically low.

8.2.3.2 Impacts of Microbial Contaminants on Human Health

The well-known pathogens of concern, their related major diseases, and primary sources are shown in Table 8.4. All these infectious agents are spread by the fecal–oral route, often involving water (refer to Figure 8.5). Environmental factors, water quality, food, and poor hygiene contribute to the spread of these infectious agents. In 2001, infectious diseases accounted for an estimated 26% of global deaths.[113] Moreover, ongoing social and environmental changes continue to bring new or re-emerging waterborne pathogens. For instance, in 2000, climate change was linked to 2.4% of worldwide diarrhea cases, 6% of malaria cases in some middle-income countries, and 7% of dengue fever cases in industrialized nations, leading to an estimated 154 000 deaths (0.3% of total mortality) and a burden of

Table 8.4 Pathogens of concern – their related diseases and primary sources (reproduced from ref. 115 with permission from Elsevier, Copyright 2004).

Pathogens	Major diseases	Primary sources
Bacteria		
Campylobacter jejuni	Gastroenteritis	Human and animal feces
Enteropathogenic *E. coli*	Gastroenteritis	Human and animal feces
Legionella pneumophila	Acute respiratory illness (legionellosis)	Thermally enriched water
Leptospira spp.	Leptospirosis	Animal and human urine
Salmonella typhi	Typhoid fever	Human feces
Salmonella paratyphi	Paratyphoid fever	Human feces
Other Salmonella	Salmonellosis	Human and animal feces
Shigella spp.	Bacillary dysentery	Human feces
Vibrio cholera	Cholera	Human feces and freshwater zooplankton
Yersinia enterocolitica	Gastroenteritis	Human and animal feces
Viruses		
Adenoviruses	Upper respiratory and gastrointestinal illness	Human feces
Coxsackie viruses A	Aseptic meningitis	Human feces
Coxsackie viruses B	Aseptic meningitis	Human feces
Echo viruses	Aseptic meningitis	Human feces
Hepatitis A virus	Infectious hepatitis	Human feces
Hepatitis E virus	Infectious hepatitis; miscarriage and death	Human feces
Norovirus	Gastroenteritis	Fomites and water
Rotaviruses	Gastroenteritis	Human and animal feces
Polio viruses	Poliomyelitis	Human feces
Protozoa		
Balantidium coli	Balantidosis (dysentery)	Human and animal feces
Cryptosporidium homonis, C. parvum	Cryptosporidiosis (gastroenteritis)	Water, human and other mammal feces
Entamoeba histolytica	Amoebic dysentery	Human and animal feces
Giardia lamblia	Giardiasis (gastroenteritis)	Water and animal feces
Naegleria fowleri	Primary amoebic meningoencephalitis	Warm water
Helminths		
Ascaris lumbricoides	Ascariosis	Animal and human feces

5.5 million disability adjusted life years (DALYs) (0.4%). Approximately 46% of this burden was concentrated in WHO-designated SEAR-D regions (with 23% in AFR-E) and an additional 14% in EMR-D regions.[114]

Health impacts and signs. Most diseases resulting from microbial pollutants commonly cause diarrhea. This includes illnesses such as cholera, gastroenteritis, salmonellosis, shigellosis, giardiasis, cyclosporiasis, and cryptosporidiosis.[116] Annually, around 1.7 billion children are diagnosed

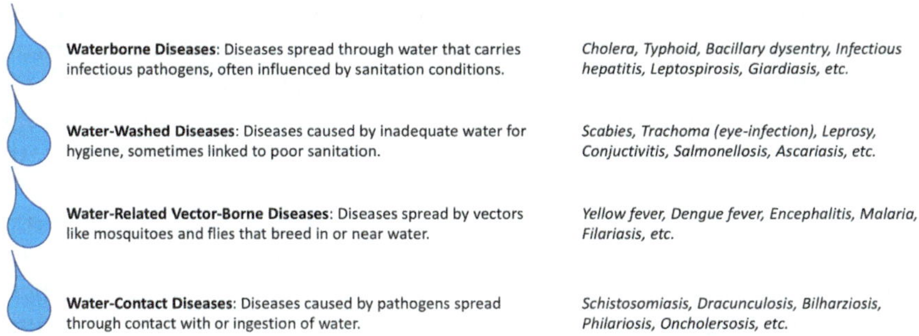

Figure 8.5 Water supply related diseases and intermediate roles of water (adapted from ref. 115 with permission from Elsevier, Copyright 2004).

with diarrhea, with approximately 525 000 fatalities. Additionally, in 2017, diarrhea claimed the lives of about 991 265 adults. Diarrhea, if untreated, can lead to severe dehydration and subsequent death due to the loss of essential water and salts from the body. There are three clinical types of diarrhea: acute watery diarrhea (lasting several hours or days, often caused by cholera), acute bloody diarrhea (caused by shigellosis), and persistent diarrhea (lasting 14 days or more). Acute watery diarrhea is particularly dangerous, potentially fatal within hours without treatment. Treatment options include oral rehydration salts, nutrient-rich food, and zinc supplements.[117]

Additionally, many diseases caused by microbial pollutants also result in fever, such as gastroenteritis, salmonellosis, shigellosis, cyclosporiasis, and cryptosporidiosis. Fever is characterized by a body temperature above 36.5–37.5 °C, leading to sweating, shivering, headaches, muscle aches, poor appetite, rash, restlessness, and general weakness. Fevers can vary in severity; low-grade fevers are associated with diseases like gastroenteritis, cyclosporiasis, and cryptosporidiosis, while high-grade fevers (40.1–41.1 °C) from salmonellosis and shigellosis, may cause confusion, excessive sleepiness, irritability, and seizures.[118,119] Treatments include staying hydrated, consuming light, easily digestible foods, and taking medications like ibuprofen, acetaminophen, or aspirin.[118,119] Moreover, many illnesses from microbial pollutants can also cause body and muscle aches. These diseases included gastroenteritis, salmonellosis, giardiasis, amoebiasis, shigellosis, cyclosporiasis, and cryptosporidiosis. Gastroenteritis can cause headaches, muscle aches, abdominal pain, or joint aches lasting anywhere from 1 to 10 days, while the rest of the diseases can cause abdominal pain, which could last for several days.[120]

8.2.3.3 Worldwide Detection of Microbial Contaminants

Microbial contamination and the resulting diseases are more prevalent in countries with inadequate water treatment, poor sanitation, and limited access to clean drinking water. These conditions are typically found in developing countries. These regions often have large populations living in poverty, which worsens the issues related to water contamination and disease spread. Efforts to improve water quality, sanitation infrastructure, and public health education are crucial in reducing the prevalence of these diseases.

Despite advanced water treatment and sanitation, industrialized countries are still vulnerable to microbial contamination and disease. Localized outbreaks can occur due to lapses in water treatment, aging infrastructure, and natural disasters, as seen in the 1993 Milwaukee cryptosporidiosis outbreak. Imported foods and international travel can introduce pathogens like *Cyclospora* and norovirus, respectively. Recreational water use of lakes, rivers, and swimming pools also poses a risk. Emerging pathogens, including antibiotic-resistant bacteria, add ongoing challenges. However, industrialized nations are better equipped to detect, respond to, and mitigate these impacts due to strong healthcare systems and regulatory frameworks.

8.2.3.4 Regulatory Frameworks and Remediation Strategies – Microbial Contaminants

Regulatory frameworks: Across different countries and organizations, the standard for *E. coli* in drinking water is set at zero detectable organisms per 100 mL due to the severe health risks of fecal contamination. Standards also typically cover total coliforms and other microbial pathogens to ensure comprehensive water safety.

USEPA:
- ***E. coli***: MCLG is zero, indicating fecal contamination.
- **Total coliforms:** Indicator of microbial contamination; triggers further testing for *E. coli* if positive.
- ***Giardia lamblia***: MCLG is zero, requiring a treatment technique (TT) of 99.9% removal/inactivation.
- **Legionella:** MCLG is zero, with a specific TT for controlling Legionella.
- Enteric viruses: MCLG is zero, requiring a TT of 99.99% removal/inactivation.
- ***Cryptosporidium***: MCLG is zero, with a TT requirement of 99–99.7% removal.

WHO:
- ***E. coli***: Zero detectable organisms per 100 mL, as a primary indicator of fecal contamination.
- **Other pathogens**: Zero detectable enteric viruses and protozoa (*e.g.*, *Giardia* and *Cryptosporidium*) per 100 mL. Emphasis on comprehensive, multi-barrier water treatment for safety.

European Union (EU):

- ***E. coli***: Zero colony-forming units (CFU) per 100 mL.
- **Enterococci**: Zero CFU per 100 mL.
- ***Clostridium perfringens* (including spores)**: Zero CFU per 100 mL in water influenced by surface water.

Canada:

- ***E. coli***: Zero detectable organisms per 100 mL; considered an immediate risk requiring prompt action.
- **Total coliforms**: Zero per 100 mL.
- **Enteric viruses**: Aim for at least 99.99% inactivation or removal through effective treatment.
- **Protozoa (*e.g.*, Giardia, Cryptosporidium)**: Aim for at least 99.9% removal or inactivation through treatment.

Australia:

- ***E. coli***: Zero detectable organisms per 100 mL, indicating fecal contamination.
- **Enteric viruses**: Absence of detectable viruses in a 100 mL sample.
- **Protozoa (*e.g.*, Giardia, Cryptosporidium)**: Absence of detectable protozoa in a 100 mL sample, with a focus on effective treatment.
- **Legionella**: No specific MCLG, but routine monitoring and effective disinfection are recommended.

Remediation strategies: Effective remediation strategies for microbial contaminants in water include chemical, physical, irradiation, and thermal treatments. *Chemical disinfection* uses chlorine-based technologies or ozone to eliminate microbes, with chlorine being widely used in developing countries due to its affordability and ease of use. *Physical methods* include coagulation, sedimentation, filtration (*e.g.*, membrane, ceramic, and composite filters), and granular media filters, which physically remove and retain microbes. *Irradiation-based processes* use solar or UV light to inactivate microbes, though UV treatments require reliable electricity and are less effective in highly turbid water. *Thermal technologies* involve boiling or distillation, effectively killing microbes but requiring significant energy and careful maintenance.

8.2.4 Radiological Contaminants

8.2.4.1 Types and Properties of Radiological Contaminants

Background radiation has been a constant presence since life began on Earth 4.5 billion years ago. Radioactive substances, or radionuclides, are found in sand, soil, sediments, water, and biota. The global average annual radiation dose per person is about 3.0 millisieverts (mSv), with 80% (2.4 mSv) from natural sources, 19.6% (~0.6 mSv) from medical diagnostics, and 0.4% (~0.01 mSv) from other human-made sources.[121] Individual exposure varies by location, diet, lifestyle, medical treatments, and occupation. Radiological water contaminants, emitting radiation, include over 3000 known radionuclides, both natural and artificial.[122]

Natural radiological water contaminants include isotopes like uranium (U), radium (Ra), radon (Rn), thorium (Th), and potassium-40 (K-40), naturally found in rocks, soil, and groundwater. These elements can dissolve into water through geological processes. For example, uranium and thorium decay chains produce radium and radon, which can leach into groundwater from surrounding rock formations. Potassium-40, a naturally occurring isotope, is also commonly found in water sources. The presence of these natural radionuclides can vary depending on the geological characteristics of the region and the composition of underlying rock formations.[123]

Artificial (human-made) radiological water contaminants result from activities like nuclear power generation, medical treatments, industrial processes, and nuclear weapons testing. Key isotopes include tritium, cesium-137, iodine-131, strontium-90, and technetium-99, which can enter water through spills, leaks, or improper disposal of radioactive materials. Tritium is produced in nuclear reactors and weapons testing; iodine-131 is used in medical diagnostics and treatments. Cesium-137 and strontium-90, by-products of nuclear fission, are released during accidents or improper waste disposal, while technetium-99 is used in medical imaging.[123,124]

Unit of measurement: In the context of measuring radionuclide concentration, becquerel (Bq) and picocuries (pCi) are key units. The becquerel, part of the International System of Units (SI), quantifies radioactivity as one disintegration per second, reflecting the rate at which radioactive material decays. It is often expressed as $\mathrm{Bq\,L^{-1}}$ when measuring radioactivity in water. Picocuries per liter ($\mathrm{pCi\,L^{-1}}$) is another unit used, where one picocurie equals one trillionth (10^{-12}) of a curie. The conversion between these units is: 1 $\mathrm{Bq\,L^{-1}}$ equals 27.027 $\mathrm{pCi\,L^{-1}}$.

8.2.4.2 Transport and Fate of Radiological Contaminants

The transport and fate of radionuclides in water have significant environmental and health implications. Contaminated water harms aquatic ecosystems, reduces biodiversity, and poses risks to humans through drinking water, seafood consumption, and recreational activities. Some radionuclides persist for decades or centuries, requiring long-term monitoring. The transport of radionuclides in water depends on their source, type, and physicochemical properties (*e.g.*, solubility and sorption) and environmental factors (*e.g.*, pH, redox conditions, presence of organic matter, and sediment type).[125] Radionuclides, like technetium-99m from medical waste, can travel long distances in water, interacting with particles through sorption and desorption.[121] Uranium exists in different oxidation states, with U(VI) being the dominant form in oxidized conditions. The mobility of U(VI) in water is contingent upon pH levels and the presence of substances like iron oxides, phosphate minerals, and organic matter, which can affect its sorption behavior.[126,127] Under oxic conditions, U(VI), remains stable in aqueous solution; however, its mobility is regulated by factors such as pH and the concentrations of inorganic and organic carbon. In oxic water environments with a pH below 5, the predominant dissolved form of uranium is uranyl $U(VI)O_2^{2+}$. At neutral pH levels, the mobility of U(VI) may decrease due to its adsorption onto substances such as iron oxides, phosphate minerals, organic matter, and clays.[126] Poorly-structured iron oxides exhibit particularly strong affinity for U(VI).[127] Sorption of U(VI) tends to be weaker at higher pH levels, typically above 8.[125]

Sedimentation causes radionuclides attached to particles to settle and accumulate in sediments, reducing their availability in the water column. However, disturbances like strong currents or human activities can resuspend these sediments, reintroducing radionuclides into the water and increasing their potential for transport and biological uptake. Contaminated sediments can also be used as fertilizers, raising soil radioactivity. Freiler *et al.*[128] conducted a yearly study on radon concentrations in a spring located in the Sopron Mountains, Hungary, aiming to identify the likely source of the elevated radon levels in the water. Their analysis revealed that the maximum radon concentration in the sedimentary rocks ($c_{pot} = 2020$ Bq g^{-1}) was approximately eight times higher than the measured radon concentration in the water. In contrast, the maximum potential radon concentration for soils was significantly lower (41.3 Bq L^{-1}), implying that the sedimentary rocks surrounding the spring are likely the primary source of radon release into the water, as opposed to the soil in the area.[128]

Biological uptake plays a crucial role in the migration and fate of radionuclides, impacting human health by affecting redox status, which in turn influences the speciation and fate of redox-sensitive radionuclides like U and Pu in the environment.[129] Radioactive decay is a fundamental process that shapes the fate of radionuclides, transforming them into different elements over time. The uranium, thorium, and actinium decay series serve as prime examples, demonstrating a sequence of decay steps with unique half-lives for each radionuclide involved.[130] For instance, ^{210}Bi undergoes rapid decay with a short half-life of approximately five days, while ^{238}U persists for billions of years (refer to Figure 8.6). This variation in decay rates illustrates the complex behavior of radionuclides in the environment. Bioaccumulation further complicates this, as radionuclides build up in aquatic organisms like fish, leading to increased exposure for organisms higher up the food chain, including humans.[130]

8.2.4.3 Impacts on Human Health and the Environment

Radionuclide contamination of water bodies is a major environmental concern, stemming from sources like nuclear accidents, industrial discharges, and improper waste disposal. Historical incidents like the Chernobyl and Fukushima disasters provide stark examples of how such events can lead to

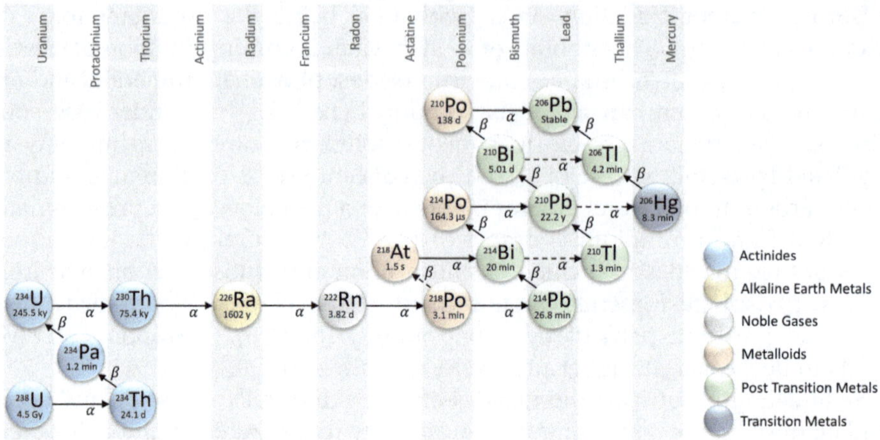

Figure 8.6 Uranium-238 decay series of steps to become a stable form of lead. Each step in the illustration indicates a different nuclide. The numbers below each label indicate the length of the radionuclide's half-life. The arrows show the mode of decay; horizontal arrows: alpha particle emission and 45°-angle arrow: beta particle emission (adapted from ref. 130 with permission from Elsevier, Copyright 2007).

severe and widespread radioactive contamination. After Chernobyl, groundwater near the plant became contaminated with radionuclides such as ^{137}Cs, ^{90}Sr, ^{239}Pu, and ^{240}Pu, with levels of ^{137}Cs initially at 50 MBq L^{-1}, decreasing to 200 Bq L^{-1} by 2001.[131] Similarly, the Fukushima disaster released around 18 000 teraBq of ^{137}Cs into the Pacific Ocean, highlighting the severe impact on water sources. This release not only affected the immediate area but also had far-reaching implications for marine life and human health, given the extensive reach of ocean currents.[123] Beyond these high-profile incidents, radionuclide contamination can result from routine operations and accidents in medical facilities, industrial processes involving uranium and thorium, and energy production. Improper disposal of radioactive waste in each of these industries poses a long-term risk to groundwater and surface water quality. Exposure to radiological contaminants in water poses significant health risks. Ingesting or inhaling radionuclides can increase the risk of cancer, kidney damage, and other health issues.[132,133] For example, radon and radium exposure is linked to bladder, lung, and bone cancer. Environmental impacts include bioaccumulation in aquatic organisms, affecting biodiversity and ecosystem health. Long-term exposure to low radiation levels can cause chronic health issues for both humans and wildlife.[134]

8.2.4.4 Regulatory Framework and Remediation Strategies – Radiological Contaminants

Regulations controlling radionuclides in water are essential for safeguarding public health, with national and international agencies setting strict limits

Table 8.5 MCLs of radionuclide substances in drinking water.

Health authority	Radionuclide elements	MCL	Ref.
USEPA	Combined ^{226}R and ^{228}R	5 pCi L^{-1}	135
	Rn	4000 pCi L^{-1}	
	U	30 μg L^{-1}	
WHO	^{210}Po, ^{228}Ra	0.1 Bq L^{-1}	96
	^{232}Th, ^{235}U, ^{226}Ra	1 Bq L^{-1}	
	^{238}U, ^{137}Cs	10 Bq L^{-1}	
Japan	Cs, Sr, Pu, Ru	10 Bq kg^{-1}	136
Health Canada	^{3}H	7000 Bq L^{-1}	58
	^{226}Ra	0.5 Bq L^{-1}	
	^{210}Pb	0.2 Bq L^{-1}	
	^{90}Sr	5 Bq L^{-1}	
	^{131}I	6 Bq L^{-1}	
	^{137}Cs	10 Bq L^{-1}	

to ensure safe levels of radioactive substances. Table 8.5 provides an overview of MCLs set by these authorities, highlighting the stringent standards put in place to protect water quality and human well-being.

The growing volume of nuclear waste and stricter radionuclide regulations have intensified the need for effective water treatment and pollutant recovery. Before 1990, methods like evaporation, volume reduction, chemical precipitation, coagulation, filtration, and ion exchange were widely used but often proved inefficient, costly, and time-consuming.[123,137] Post-1990, advanced techniques such as reverse osmosis,[138] micro- and nano-filtration,[139] as well as meso- and micro-porous adsorbents[140] improved removal efficiency. Over the past two decades, researchers have focused on developing innovative methods like nano-level composites,[141] forward osmosis,[142] bioremediation,[143] and electro-remediation[144] methods to enhance treatment processes. A recent review by Adeola *et al.*,[145] listed the removal efficiencies of different remediation technologies, with high-pressure membranes leading at 97% average efficiency, followed by electro-remediation (89%), ion exchange (86%), adsorption (84%), and bioremediation (78%).

8.3 Conclusion

The complex and varied nature of water contaminants – including organic, inorganic, microbial, and radiological categories – emphasizes the urgent need for robust water quality management. These contaminants, whether introduced from natural processes or human activities, pose substantial challenges due to their diverse origins, properties, and far-reaching impacts on both environmental and human health. Effectively addressing these issues demands a comprehensive approach, integrating stringent regulatory frameworks, advanced monitoring techniques, and innovative remediation strategies. This chapter explored the origins and composition of major water contaminants, examining their sources, pathways into aquatic systems, and their potential consequences. By highlighting regulatory and advisory standards, the importance of proactive water management, and the necessity for ongoing research and innovation, we highlight the critical efforts required to safeguard water quality and ensure the sustainability of this essential resource.

Abbreviations

2,4-D	2,4-Dichlorophenoxyacetic acid
6PPD	*N*-(1,3-Dimethylbutyl)-*N*′-phenyl-*p*-phenylenediamine
6-PPDQ	6PPD-quinone
ADWG	Australian drinking water guidelines
AOM	Algal organic matter
AOPs	Advanced oxidation processes
BFRs	Brominated flame retardants
BPA	Bisphenol A
BPAF	Bisphenol AF
BPF	Bisphenol F
BPS	Bisphenol S
CAFOs	Concentrated animal feeding operations
CFU	Colony-forming units
ChNNP	Chernobyl nuclear power plant
CPPD	*N*-Phenyl-*N*′-cyclohexyl-*p*-phenylenediamine
cVOCs	Carcinogenic volatile organic compounds
DALYs	Disability adjusted life years
DBPs	Disinfection by-products
DDT	Dichloro-diphenyl-trichloroethane
DPPD	*N*,*N*′-Diphenyl-*p*-phenylenediamine
EA	Environment agency
ECs	Emerging contaminants
EDC	Endocrine-disrupting chemical
EHEC	Enterohemorrhagic *E. coli*
EIEC	Enteroinvasive *E. coli*
EPEC	Enteropathogenic *E. coli*
ER	Estrogen receptor
ERR	Estrogen-related receptor
ETEC	Enterotoxigenic *E. coli*
EU	European Union
GAC	Granular activated carbon
GV	Guideline value
HAAs	Haloacetic acids
IPM	Integrated pest management
IPPD	*N*-Isopropyl-*N*′-phenyl-*p*-phenylenediamine

IX	Ion exchange
LCR	Lead and copper rule
LLE	Loss of life expectancy
LOD	Limit of detection
MAC	Maximum acceptable concentration
MCL	Maximum contamination level
MOF	Metal–organic framework
mVOCs	Microbial volatile organic compounds
NF	Nanofiltration
NHMRC	National health and medical research council
NOM	Natural organic matter
OSPAR	Oslo and Paris conventions
PBDEs	Polybrominated diphenyl ethers
PBT	Persistence, bioaccumulation, and toxicity
PCBs	Polychlorinated biphenyls
PCDD/Fs	Polychlorinated dibenzo-*p*-dioxins/furans
PCDDs	Polychlorinated dibenzo-*p*-dioxins
PCDFs	Polychlorinated dibenzofurans
PFAS	Per- and polyfluoroalkyl substances
PFOA	Perfluorooctanoic acid
PFOS	Perfluorooctane sulfonate
POPs	Persistent organic pollutants
PPDs	*p*-Phenylenediamines
PVC	Polyvinyl chloride
RO	Reverse osmosis
SHE	Standard hydrogen electrode
SI	International System of Units
SODIS	Solar disinfection
TCDD	2,3,7,8-Tetrachlorodibenzo-*p*-dioxin
THMs	Trihalomethanes
UF	Ultrafiltration
USEPA	United States Environmental Protection Agency
UV	Ultraviolet
VOCs	Volatile organic compounds
WWTP	Water and wastewater treatment plant

References

1. X. Chen, T. He, X. Yang, Y. Gan, X. Qing, J. Wang and Y. Huang, *J. Hazard. Mater.*, 2023, **452**, 131245.
2. I. T. Cousins, J. H. Johansson, M. E. Salter, B. Sha and M. Scheringer, *Environ. Sci. Technol.*, 2022, **56**, 11172–11179.
3. K. Jones and P. De Voogt, Persistent organic pollutants (POPs): state of the science, 1999, **100**, 209–221.
4. N. L. Howell, M. P. Suarez, H. S. Rifai and L. Koenig, *Chemosphere*, 2008, **70**, 593–606.

5. Y. Peng, J. Wu, X. Luo, X. Zhang, J. P. Giesy and B. Mai, *Environ. Int.*, 2019, **130**, 104952.
6. A. V. B. Reddy, M. Moniruzzaman and T. M. Aminabhavi, *Chem. Eng. J.*, 2019, **358**, 1186–1207.
7. S. Zhou, H. Yang, A. Zhang, Y. F. Li and W. Liu, *Chemosphere*, 2014, **114**, 26–34.
8. A. Ranjbar Jafarabadi, A. Riyahi Bakhtiari, S. Mitra, M. Maisano, T. Cappello and C. Jadot, *Environ. Pollut.*, 2019, **253**, 78–88.
9. S. Gao, J. Chen, Z. Shen, H. Liu and Y. Chen, *Chemosphere*, 2013, **91**, 809–816.
10. EFSA, *EFSA J.*, 2010, **8**, 1–36.
11. F. Ge, X. Wang, K. Zhang, X. Jin, R. Guo, Y. Liu, X. Qiao, X. Zhao, B. Zheng and X. Zheng, *Chemosphere*, 2019, **234**, 763–768.
12. P. Fernández-Navarro, J. García-Pérez, R. Ramis, E. Boldo and G. López-Abente, *Environ. Res.*, 2017, **159**, 555–563.
13. M. Marquès and J. L. Domingo, *Int. J. Environ. Res. Public Health*, 2019, **16**, 3566.
14. M. A. Siddiqi, R. H. Laessig and K. D. Reed, *Clin. Med. Res.*, 2003, **1**, 281–290.
15. G. Abbasi, A. M. Buser, A. Soehl, M. W. Murray and M. L. Diamond, *Environ. Sci. Technol.*, 2015, **49**, 1521–1528.
16. H. S. Wang, G. M. Jiang, Z. J. Chen, J. Du, Y. B. Man, J. P. Giesy, C. K. C. Wong and M. H. Wong, *J. Hazard. Mater.*, 2013, **261**, 253–259.
17. K. Ji, K. Choi, J. P. Giesy, J. Musarrat and S. Takeda, *Environ. Sci. Technol.*, 2011, **45**, 5003–5008.
18. P. K. Gupta, *Toxicity of herbicides*, Elsevier Inc., 3rd edn, 2018.
19. M. F. F. Bernardes, M. Pazin, L. C. Perira and D. J. Dorta, in *Toxicology Studies – Cells, Drugs and Environment*, ed. A. C. Andreazza and G. Scola, 2015, p. 195.
20. V. R. Beasley and J. M. Levengood, *Principles of ecotoxicology*, Elsevier Inc., 2012.
21. I. Mahmood, S. R. Imadi, K. Shazadi, A. Gul and K. R. Hakeem, Effects of Pesticides on Environment, in *Plant, soil and microbes*, 2016, vol. 1.
22. D. Ashton, M. Hilton and K. V. Thomas, *Sci. Total Environ.*, 2004, **333**, 167–184.
23. A. Ziylan-Yavas, D. Santos, E. M. M. Flores and N. H. Ince, *Environ. Prog. Sustainable Energy*, 2022, **41**, e13821.
24. S. Keerthanan, C. Jayasinghe, J. K. Biswas and M. Vithanage, *Crit. Rev. Environ. Sci. Technol.*, 2021, **51**, 1221–1258.
25. E. Burridge, *Eur. Chem. News*, 2003, **17**, 14–20.
26. R. A. Rudel, J. M. Gray, C. L. Engel, T. W. Rawsthorne, R. E. Dodson, J. M. Ackerman, J. Rizzo, J. L. Nudelman and J. G. Brody, *Environ. Health Perspect.*, 2011, **119**, 914–920.
27. G. Schönfelder, W. Wittfoht, H. Hopp, C. E. Talsness, M. Paul and I. Chahoud, *Environ. Health Perspect.*, 2002, **110**, A703–A707.

28. A. M. Calafat, X. Ye, L.-Y. Wong, J. A. Reidy and L. L. Needham, *Environ. Health Perspect.*, 2008, **116**, 39–44.
29. B. S. Rubin, *J. Steroid Biochem. Mol. Biol.*, 2011, **127**, 27–34.
30. A. L. Lusher, V. Tirelli, I. O'Connor and R. Officer, *Sci. Rep.*, 2015, **5**, 1–9.
31. E. Tumwesigye, C. Felicitas Nnadozie, F. C Akamagwuna, X. Siwe Noundou, G. William Nyakairu and O. N. Odume, *Environ. Pollut.*, 2023, **330**, 121829.
32. M. Hu and D. Palić, *Redox Biol.*, 2020, **37**, 101620.
33. H. Ye, E. B. Esfahani, I. Chiu, M. Mohseni, G. Gao and T. Yang, *J. Hazard. Mater.*, 2024, **470**, 134194.
34. S. T. McBeath, A. Serrano Mora, F. Asadi Zeidabadi, B. K. Mayer, P. McNamara, M. Mohseni, M. R. Hoffmann and N. J. D. Graham, *Curr. Opin. Electrochem.*, 2021, **30**, 100865.
35. E. B. Esfahani, F. Asadi Zeidabadi and M. Mohseni, *ACS ES&T Water*, 2023, **3**, 3614–3625.
36. J. Glüge, M. Scheringer, I. T. Cousins, J. C. Dewitt, G. Goldenman, D. Herzke, R. Lohmann, C. A. Ng, X. Trier and Z. Wang, *Environ. Sci.: Processes Impacts*, 2020, **22**, 2345–2373.
37. F. A. Zeidabadi, E. B. Esfahani and M. Mohseni, *J. Hazard. Mater. Adv.*, 2023, **10**, 100322.
38. F. Asadi Zeidabadi, E. Banayan Esfahani, R. Moreira, S. T. McBeath, J. Foster and M. Mohseni, *Environ. Res.*, 2024, **246**, 118103.
39. E. Banayan Esfahani, F. Asadi Zeidabadi, M. Jafarikojour and M. Mohseni, *J. Water Process Eng.*, 2024, **61**, 105332.
40. F. Asadi Zeidabadi, E. B. Esfahani, S. T. McBeath, K. L. Dubrawski and M. Mohseni, *Chemosphere*, 2023, 137743.
41. E. Banayan Esfahani, F. Asadi Zeidabadi, S. Zhang and M. Mohseni, *Environ. Sci.: Water Res. Technol.*, 2022, **8**, 698–728.
42. K. Steenland and A. Winquist, *Environ. Res.*, 2021, **194**, 110690.
43. S. Poothong, C. Thomsen, J. A. Padilla-Sanchez, E. Papadopoulou and L. S. Haug, *Environ. Sci. Technol.*, 2017, **51**, 13388–13396.
44. B. Du, B. Liang, Y. Li, M. Shen, L. Y. Liu and L. Zeng, *Environ. Sci. Technol. Lett.*, 2022, **9**, 1056–1062.
45. X. Hua and D. Wang, *J. Hazard. Mater.*, 2023, **459**, 132265.
46. L. Fang, C. Fang, S. Di, Y. Yu, C. Wang, X. Wang and Y. Jin, *Sci. Total Environ.*, 2023, **869**, 161836.
47. Z. Tian, H. Zhao, K. T. Peter, M. Gonzalez, J. Wetzel, C. Wu, X. Hu, J. Prat, E. Mudrock, R. Hettinger, A. E. Cortina, R. G. Biswas, F. V. C. Kock, R. Soong, A. Jenne, B. Du, F. Hou, H. He, R. Lundeen, A. Gilbreath, R. Sutton, N. L. Scholz, J. W. Davis, M. C. Dodd, A. Simpson, J. K. McIntyre and E. P. Kolodziej, *Science*, 2021, **371**, 185–189.
48. R. Sinha, A. K. Gupta and P. S. Ghosal, *J. Environ. Chem. Eng.*, 2021, **9**, 106511.
49. D. Liew, K. L. Linge, A. Heitz and C. A. Joll, Nitrogenous DBPs in drinking water: toxicity, regulation, analysis, occurrence and control, in *Disinfection by-products and human health*, 2012.

50. B. Chen, J. Jiang, X. Yang, X. Zhang and P. Westerhoff, *Water Res.*, 2021, **200**, 117265.
51. S. K. Golfinopoulos, T. D. Lekkas and A. D. Nikolaou, *Chemosphere*, 2001, **45**, 275–284.
52. E. David and V.-C. Niculescu, *Int. J. Environ. Res. Public Health*, 2021, **18**, 13147.
53. P. Pandey and R. Yadav, *Int. J. Plant Environ.*, 2018, **4**, 14–26.
54. J. L. Domingo and M. Nadal, *Environ. Int.*, 2009, **35**, 382–389.
55. R. E. Hester, R. M. Harrison and R. G. Derwent, in *Volatile Organic Compounds in the Atmosphere*, The Royal Society of Chemistry, 1995.
56. P. J. Squillace, J. C. Scott, M. J. Moran, B. T. Nolan and D. W. Kolpin, *Environ. Sci. Technol.*, 2002, **36**, 1923–1930.
57. USEPA, National Primary Drinking Water Regulations, https://www.epa.gov/ground-water-and-drinking-water/national-primary-drinking-water-regulations.
58. Health Canada, Guidelines for Canadian Drinking Water Quality – Summary Tables.
59. WHO, Water quality for drinking: WHO guidelines.
60. J. Benner, D. E. Helbling, H. P. E. Kohler, J. Wittebol, E. Kaiser, C. Prasse, T. A. Ternes, C. N. Albers, J. Aamand, B. Horemans, D. Springael, E. Walravens and N. Boon, *Water Res.*, 2013, **47**, 5955–5976.
61. E. Banayan Esfahani, F. Dixit, F. Asadi Zeidabadi, M. R. Johnson, N. Mayilswamy, B. Kandasubramanian and M. Mohseni, *Curr. Opin. Chem. Eng.*, 2023, **42**, 100953.
62. N. H. Barbhuiya, U. Misra and S. P. Singh, *Chemosphere*, 2022, **286**, 131757.
63. E. B. Esfahani and M. Mohseni, *J. Environ. Chem. Eng.*, 2022, **10**, 107050.
64. F. Asadi Zeidabadi, E. Banayan Esfahani, S. T. McBeath and M. Mohseni, *Water Res.*, 2024, 121529.
65. E. Banayan Esfahani, F. Asadi Zeidabadi, L. Rajesh, S. T. Mcbeath and M. Mohseni, *Curr. Opin. Chem. Eng.*, 2024, **44**, 101018.
66. O. V. Singh, S. Labana, G. Pandey, R. Budhiraja and R. K. Jain, *Appl. Microbiol. Biotechnol.*, 2003, **61**, 405–412.
67. P. Borah, M. Kumar and P. Devi, Types of inorganic pollutants: metals/metalloids, acids, and organic forms, in *Inorganic pollutants in water*, 2020.
68. D. M. Dave and M. Yang, *J. Health Econ.*, 2022, **84**, 102644.
69. K. G. Pavithra, P. SundarRajan, P. S. Kumar and G. Rangasamy, *Chemosphere*, 2023, **312**, 137314.
70. S. A. Counter and L. H. Buchanan, *Toxicol. Appl. Pharmacol.*, 2004, **198**, 209–230.
71. G. Genchi, S. S. L. Graziantono, A. Carocci and A. Catalano, *Int. J. Environ. Res. Public Health*, 2020, **17**, 1–24.
72. S. Sharma and A. Bhattacharya, *Appl. Water Sci.*, 2017, **7**, 1043–1067.
73. D. Das, A. Chatterjee, B. K. Mandal, G. Samanta, D. Chakraborti and B. Chanda, *Analyst*, 1995, **120**, 917–924.

74. F. W. Pontius, K. G. Brown and C. J. Chen, *J. Am. Water Works Assoc.*, 1994, **86**, 52–63.
75. B. Wang, T. Fang and H. Chen, *Nutrients*, 2023, **15**, 1–27.
76. S. Prasad, K. K. Yadav, S. Kumar, N. Gupta, M. M. S. Cabral-Pinto, S. Rezania, N. Radwan and J. Alam, *J. Environ. Manage.*, 2021, **285**, 112174.
77. V. Kumar, P. K. Bharti, M. Talwar, A. K. Tyagi and P. Kumar, *Water Sci.*, 2017, **31**, 44–51.
78. WHO, Who/Sde/Wsh/03.04/08, 2003, vol. 2, p. 4.
79. M. F. Bouchard, S. Sauvé, B. Barbeau, M. Legrand, M. È. Brodeur, T. Bouffard, E. Limoges, D. C. Bellinger and D. Mergler, *Environ. Health Perspect.*, 2011, **119**, 138–143.
80. Health Canada, Manganese Drink. Water Doc. Public Consult.
81. Y. Li, Q. Zhou, B. Ren, J. Luo, J. Yuan, X. Ding, H. Bian and X. Yao, *Rev. Environ. Contam. Toxicol.*, 2020, **251**, 1–24.
82. E. Banayan Esfahani, F. Asadi Zeidabadi, A. Bazargan and G. McKay, *The Modified Bardenpho Process*, 2019.
83. J. D. Brender, in *Just Enough Nitrogen: Perspectives on how to get there for regions with too much and too little nitrogen*, ed. M. A. Sutton, K. E. Mason, A. Bleeker, W. K. Hicks, C. Masso, N. Raghuram, S. Reis and M. Bekunda, Springer International Publishing, Cham, 2020, pp. 283–294.
84. R. Picetti, M. Deeney, S. Pastorino, M. R. Miller, A. Shah, D. A. Leon, A. D. Dangour and R. Green, *Environ. Res.*, 2022, **210**, 112988.
85. E. García Torres, R. Pérez Morales, A. González Zamora, E. Ríos Sánchez, E. H. Olivas Calderón, J. de Alba Romero and E. Y. Calleros Rincón, *Int. J. Environ. Health Res.*, 2022, **32**, 984–1001.
86. J. Zembrzuska, B. Karbowska and I. Gołębiewska, in *Analytical Methods in the Determination of Bioactive Compounds and Elements in Food*, ed. M. Jeszka-Skowron, A. Zgoła-Grześkowiak, T. Grześkowiak and A. Ramakrishna, Springer International Publishing, Cham, 2021, pp. 345–369.
87. J. Köhrle and R. Gärtner, *Best Pract. Res., Clin. Endocrinol. Metab.*, 2009, **23**, 815–827.
88. M. E. Reid, M. S. Stratton, A. J. Lillico, M. Fakih, R. Natarajan, L. C. Clark and J. R. Marshall, *J. Trace Elem. Med. Biol.*, 2004, **18**, 69–74.
89. F. V. Zohoori and R. M. Duckworth, *Fluoride: Intake and Metabolism, Therapeutic and Toxicological Consequences*, Elsevier Inc., 2016.
90. P. Eswar and C. G. Devaraj, *Indian J. Dent. Adv.*, 2011, **3**, 526–533.
91. D. Madeira, J. Andrade, M. C. Leal, V. Ferreira, R. J. M. Rocha, R. Rosa and R. Calado, *Front. Mar. Sci.*, 2020, **7**, 1–11.
92. M. Hijosa-Valsero, R. Molina, H. Schikora, M. Müller and J. M. Bayona, *Water Res.*, 2013, **47**, 1701–1707.
93. M. Borjan, S. Marcella, B. Blount, M. Greenberg, J. Zhang, E. Murphy, L. Valentin-Blasini and M. Robson, *Sci. Total Environ.*, 2011, **409**, 460–464.
94. L. Pilic, C. R. Pedlar and Y. Mavrommatis, *Nutr. Rev.*, 2016, **74**, 645–658.
95. M. K. Sharma and M. Kumar, *Environ. Monit. Assess.*, 2020, **192**, 74.

96. WHO, *Guidelines for Drinking-Water Quality*, World Health Organization, Geneva, 4th edn, 2022.
97. D. Clifford, S. Subramonian and T. J. Sorg, *Environ. Sci. Technol.*, 1986, **20**, 1072–1080.
98. G. M. Rossolini, F. Arena and T. Giani, *Infectious Diseases, 2-Volume Set*, 2017, pp. 1181–1196.e1.
99. R. A. Haasdijk and J. van Ingen, *New Microbes New Infect.*, 2018, **21**, 100–101.
100. G. Garrity, *Bergey's manual® of systematic bacteriology: volume 2: the Proteobacteria, part B: the Gammaproteobacteria*, Springer Science and Business Media, 2007.
101. M. Ali, A. R. Nelson, A. L. Lopez and D. A. Sack, *PLoS Neglected Trop. Dis.*, 2015, **9**, 0003832.
102. A. Rincé, C. Balière, D. Hervio-Heath, J. Cozien, S. Lozach, S. Parnaudeau, F. S. Le Guyader, S. Le Hello, J. C. Giard, N. Sauvageot, A. Benachour, S. Strubbia and M. Gourmelon, *Front. Microbiol.*, 2018, **9**, 1–17.
103. B. R. Rezny and D. S. Evans, *StatPearls*, 2020.
104. J. A. Crump and J. Wain, *International Encyclopedia of Public Health*, 2017, pp. 425–433.
105. C. N. Thompson, P. T. Duy and S. Baker, *PLoS Neglected Trop. Dis.*, 2015, **9**, 1–13.
106. G. W. Webb, S. Kelly and H. R. Dalton, *Clin. Microbiol. Newsl.*, 2020, **42**, 171–179.
107. H. F. Clark, P. A. Offit, U. D. Parashar and R. L. Ward, *Rotavirus vaccines*, Elsevier Inc., 2nd edn, 2008.
108. B. A. Lopman, D. Steele, C. D. Kirkwood and U. D. Parashar, *PLoS Med.*, 2016, **13**, 1001999.
109. A. Omarova, K. Tussupova, R. Berndtsson, M. Kalishev and K. Sharapatova, *Int. J. Environ. Res. Public Health*, 2018, **15**, 495.
110. N. Dunn and A. L. Juergens, *StatPearls*, 2024.
111. R. Mahmud, J. Ibrahim, N. Moktar and T. S. Anuar, *Entamoeba histolytica in Southeast Asia*, 2013.
112. R. M. Karanja, W. Gatei and N. Wamae, *Afr. Health Sci.*, 2007, **7**, 62–67.
113. World Health Organization, *The world health report 2003: shaping the future*, World Health Organization, 2003.
114. World Health Organization, *WHO guideline on self-care interventions for health and well-being, 2022 revision*, World Health Organization, 2022.
115. N. J. Ashbolt, *Toxicology*, 2004, **198**, 229–238.
116. C. M. Chow, A. K. C. Leung and K. L. Hon, *Clin. Exp. Gastroenterol.*, 2010, **3**, 97.
117. WHO, *Diarrhoeal Disease*, World Health Organization, 2017.
118. M. McConnaughey, Life Cycle of Parasites, in *xPharm: The Comprehensive Pharmacology Reference*, Elsevier, 2007, pp. 1–15.
119. N. T. Desai, R. Sarkar and G. Kang, *Tropenmed. Parasitol.*, 2012, **2**, 91–98.
120. S. L. Percival, R. M. Chalmers, M. Embrey, P. R. Hunter, J. Sellwood and P. Wyn-Jones, *Microbiology of Waterborne Diseases*, Elsevier, 2004.
121. UNSCEAR, *United Nations*, 2008, vol. I, p. 66.

122. A. Berthiaume, *Heliyon*, 2023, **9**, e16602.
123. F. Hossain, *J. Environ. Radioact.*, 2020, **225**, 106423.
124. WHO, *Radiological Examination of Drinking Water*, 1978.
125. B. Santhanabharathi, K. P. Pradhoshini, M. Suhail Ahmed, M. Priyadharshini, M. H. Shafeeka Parveen, L. Alam, I. M. Mofizur Rahman, V. H. Duong, M. Ud Din War and M. Saiyad Musthafa, *Int. J. Environ. Anal. Chem.*, 2023, 1–27.
126. M. Villalobos, M. A. Trotz and J. O. Leckie, *Environ. Sci. Technol.*, 2001, **35**, 3849–3856.
127. M. Kohler, G. P. Curtis, D. E. Meece and J. A. Davis, *Environ. Sci. Technol.*, 2004, **38**, 240–247.
128. Á. Freiler, Á. Horváth, K. Török and T. Földes, *J. Environ. Radioact.*, 2016, **151**, 174–184.
129. C. Xu, P. H. Santschi, J. Y. Zhong, P. G. Hatcher, A. J. Francis, C. J. Dodge, K. A. Roberts, C. C. Hung and B. D. Honeyman, *Environ. Sci. Technol.*, 2008, **42**, 8211–8217.
130. E. R. Landa, *Radioact. Environ.*, 2007, **10**, 211–237.
131. D. A. Bugai, IAEA TM on Groundwater Contamination Following Fukushima Accident, Vienna, Austria, DOI: 10.13140/RG.2.1.1259.6248.
132. G. Jia, G. Torri and L. Magro, *J. Environ. Radioact.*, 2009, **100**, 941–949.
133. J. A. Galhardi, K. Luko-Sulato, L. N. M. Yabuki, L. M. Santos, Y. J. A. B. da Silva and Y. J. A. B. da Silva, *Rare earth elements and radionuclides*, Elsevier Inc., 2022.
134. National Research Council, *Health risks of radon and other internally deposited alpha-emitters: BEIR IV.*, National Academies Press (US), Washington (DC), 1988.
135. USEPA, 2012 Edition of the Drinking Water Standards and Health Advisories, Washington.
136. JAPAN, *New Standard limits for Radionuclides in Foods. Department of Food Safety Pharmaceutical & Food Safety Bureau. Ministry of Health, Labour and Welfare* (*translated document*), 2012.
137. V. M. Efremenkov, *Iaea Bull.*, 1989, **31**, 37–42.
138. S. Ding, Y. Yang, C. Li, H. Huang and L. A. Hou, *Water Res.*, 2016, **95**, 174–184.
139. V. O. Kaptakov and V. V. Milyutin, *Radiochemistry*, 2023, **65**, 554–559.
140. K. Patra and A. Sengupta, *Mater. Today Sustainability*, 2024, **25**, 100703.
141. C. Ding, W. Cheng, Y. Sun and X. Wang, *J. Hazard. Mater.*, 2015, **295**, 127–137.
142. M. 'Adli Nor Azman, P. S. Goh, A. F. Ismail, K. Jamaluddin, K. Y. Wong and A. S. Sahril, *J. Environ. Chem. Eng.*, 2024, **12**, 112542.
143. D. Prakash, P. Gabani, A. K. Chandel, Z. Ronen and O. V. Singh, *Microb. Biotechnol.*, 2013, **6**, 349–360.
144. S. Yang, G. Wu, J. Song and B. Hu, *Sep. Purif. Technol.*, 2022, **297**, 121568.
145. A. O. Adeola, K. O. Iwuozor, K. G. Akpomie, K. A. Adegoke, K. O. Oyedotun, J. O. Ighalo, J. F. Amaku, C. Olisah and J. Conradie, *Environ. Geochem. Health*, 2023, **45**, 2663–2689.

CHAPTER 9

Self-assembled Structures for Water Remediation

S. H. YE,[a] R. MOREIRA,[b] Y. LU,[c] H. W. ZHOU[a] AND Z. X. CHEN*[a]

[a] Future Technology School, Shenzhen Technology University, Shenzhen 518118, P. R. China; [b] Institute of Applied and Physical Chemistry, University of Bremen, Bremen, Germany; [c] Bioproducts Institute, Department of Chemical & Biological Engineering, The University of British Columbia, Vancouver, BC V6T 1Z3, Canada
*Email: chenzhixiang@sztu.edu.cn

9.1 Introduction

Water remediation has been a long-lasting problem in modern civilization for water safety, disease control, and environmental protection. It will also continue to be a hot topic within the foreseeable future, which is critical to balance with economic developments. In particular, self-assembled structures are prominent in this research area, owing to the capability of forming spontaneous complex functional structures without heavy chemistry reactions. The facileness, versatility, and multi-functionality of self-assembly systems have been demonstrated to offer new approaches and solutions to solve major water pollution challenges.

Intermolecular forces are the keys to constructing robust self-assembled structures, and include electrostatic forces, van der Waals forces, and hydrogen bonds. These interactions collectively determine the spatial arrangement of molecules and, thereby, the assembled architectures. Attributed to the non-covalent bonding natures in most self-assemblies, their structures are usually highly dynamic and adjustable, depending on their chemical origins and

Contribution of Colloidal Materials to Air, Water and Soil Environmental Sustainability: From Basic Concepts to Advanced Applications
Edited by Izabel C. Riegel-Vidotti, Lizandra Maria Zimmermann and Heloise R. de Barros

Published by the Royal Society of Chemistry, www.rsc.org

environmental conditions. Moreover, self-assembly can be endowed with well-defined structures for specific functionalities and properties. For example, self-assembled membranes can be constructed with tunable surface wettability and charge distribution for pollutant separation and filtration. Similarly, controlling molecular arrangement and spacing can enable the self-assembly of nanoparticles, forming supraparticles with specific optical, electronic, and magnetic properties.

Research on self-assembled structures not only focuses on single components but can also be utilized for heterogeneous self-assembly systems, constructing composite materials. This capability allows the introduction of functional molecules or nanoparticles into the self-assemblies with additional and/or synergistic effects, expanding the range of material selections and enhancing their peak performance. For instance, photocatalytic nanoparticles can be combined with light-adsorbing molecules to create composite materials for water pollution treatment, achieving efficient adsorption and degradation of harmful substances.

Overall, research on self-assembled structures is at the forefront and is a hotspot in the fields of materials science and nanotechnology, with significant scientific and practical value. By deeply understanding the principles of intermolecular interactions and rationally designing and controlling molecular structures, it is possible to achieve molecular self-assembly and construct materials with specific functions and properties, targeting new approaches and solutions to major issues such as water pollution.

This chapter aims to systematically summarize and explore the latest research advancements and application technologies of self-assembled structures in the field of water treatment, providing a comprehensive review of the state-of-the-art development. By analyzing the current research status and development trends, the readers will recognize the forefront issues and key challenges, providing insights and guidance for future research directions and priorities. This chapter is suitable for researchers, engineering professionals, and practitioners in the water treatment field. It can also serve as a textbook or reference book for related disciplines in higher education, supporting the development of students' professional knowledge and skills.

9.2 Fundamentals of Self-assembly

9.2.1 Types of Self-assembled Structures

Non-covalent interactions are of the utmost importance for self-assemblies, such as electrostatic forces, hydrogen bonds, and van der Waals forces, and determine the spontaneous organization of molecules into specific structures. This process has extensive applications in materials science and nanotechnology due to its efficiency in constructing complex and precise structures.

9.2.1.1 *Hydrogen Bonding*

The geometric structure of a hydrogen bond is illustrated in Figure 9.1. It involves a proton from a D–H group forming a bond with an atom "A" that has non-bonding electrons. This can be presented as D–H···A, where H stands for a hydrogen atom, and D and A typically refer to electronegative atoms such as nitrogen, oxygen, or sulfur. In this context, D–H is known as the hydrogen bond donor, and A is the hydrogen bond acceptor.

In a hydrogen bond donor, the difference in electronegativity between the D and H atoms creates polarity, resulting in a partial positive charge on the H atom. In a hydrogen bond acceptor, the A atom has lone pairs of electrons in its orbitals. When the hydrogen bond donor and acceptor are close enough, the positive partial charge on the H atom is attracted to the negative partial charge on the A atom, leading to a directional interaction.[1] Table 9.1 lists potential donor and acceptor groups for hydrogen bonds of varying strengths.

The strength of hydrogen bonds is weaker than that of ionic and covalent bonds but stronger than van der Waals forces, and are therefore categorized as intermediate-strength interactions. The typical bond energy of a hydrogen

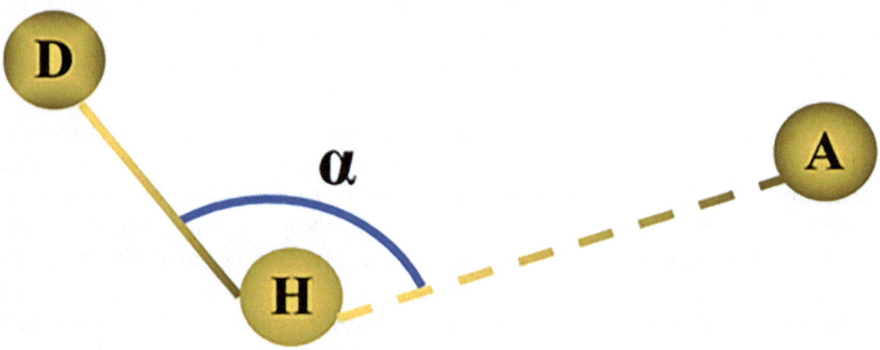

Figure 9.1 Schematic diagram of the hydrogen bond geometry.

Table 9.1 Hydrogen bonds classified by their interaction.[a] [2,3]

	Weak hydrogen bonding	Intermediate hydrogen bonding	Strong hydrogen bonding
Hydrogen bond donor	C–H, S–H, P–H, M–H	O–H, N–H, X–H	F–H
Hydrogen bond acceptor	C=C, X–C, π, S–H, M, X–M, X–H	O=C, O–H, N, S=C, F–H, X	CO_2, O, N, F
Hydrogen bond length	D–H ≪ H···A	D–H < H···A	D–H ≈ H···A
D···A/(Å)	3.2–4.0	2.5–3.2	2.2–2.5
H···A/(Å)	2.2–3.2	1.5–2.2	1.2–1.5
Hydrogen bond angle, α/(°)	90–150	130–180	175–180

[a] Note: X represents halogens with smaller atomic numbers, and M represents transition metals.

bond ranges from 4 to 60 $kJ\,mol^{-1}$.[4] As a short-range non-covalent interaction, hydrogen bonding exhibits both selectivity and directionality, allowing it to regulate the structure of supramolecular materials, molecular assembly processes, chemical reactions, and biological processes. Consequently, hydrogen bonds are of significant research interest in the fields of materials science,[5] chemistry,[6] and biology.[7]

Yoon *et al.*[8] reported a self-assembled structure formed by all-aromatic *p*-mercaptobenzoic acid (Na_4Ag_{44}(*p*-MBA)$_{30}$) nanoparticles. Their research found that the carboxyl groups on the *p*-MBA ligands of these nanoparticles contain both hydrogen bond donors and acceptors, enabling the formation of hydrogen bonds between them. These hydrogen bonds facilitate the aggregation of numerous Na_4Ag_{44}(*p*-MBA)$_{30}$ nanoparticles into multilayer structures. Steiner[1] proposed a synthesis and self-assembly mechanism for naphthalene derivatives. This substance can assemble into stable new supramolecular materials through the synergistic effects of hydrogen bonds and halogen bonds with anions such as SO_4^{2-}, $H_2PO_4^{-}$, and $HP_2O_7^{3-}$.

9.2.1.2 van der Waals Interactions

van der Waals forces exist between all atoms and molecules, and they can be either attractive or repulsive. These forces arise from dipole interactions between adjacent atoms or molecules. There are three types of dipole interactions: permanent dipole–permanent dipole, permanent dipole–induced dipole, and induced dipole–induced dipole interactions.[9] The magnitude of van der Waals forces depends on the composition of the interacting particles, substrates, and media, as well as the distance between them. Additionally, factors such as the surface roughness and geometric shape of the particles and substrates can influence these forces.[10]

By regulating van der Waals interactions within a system, it is possible to achieve self-assembled structures with specific functions. Jiang *et al.*[11] discovered that by utilizing van der Waals interactions between molecules and substrates, the formation of dioctylbenzothienobenzothiophene (C_8-BTBT) molecules into two-dimensional self-assembled organic thin films with varying molecular layers could be precisely controlled. The single-layer films exhibited excellent ambipolar transport capabilities and could serve as interfacial molecular templates for constructing heterojunctions. Bilayer films, used in organic field-effect transistors, demonstrated good electronic properties with maximum mobility of up to 12.8 $cm^2\,V^{-1}\,s^{-1}$. This research provides insights into achieving functional self-assembly by regulating van der Waals interactions between molecules and substrates.

9.2.1.3 π–π Stacking

Since the early 1980s, when the first aromatic π system was reported,[12] π–π stacking interactions have been widely applied in various fields. π–π stacking interactions are mainly achieved through three modes: face-to-face

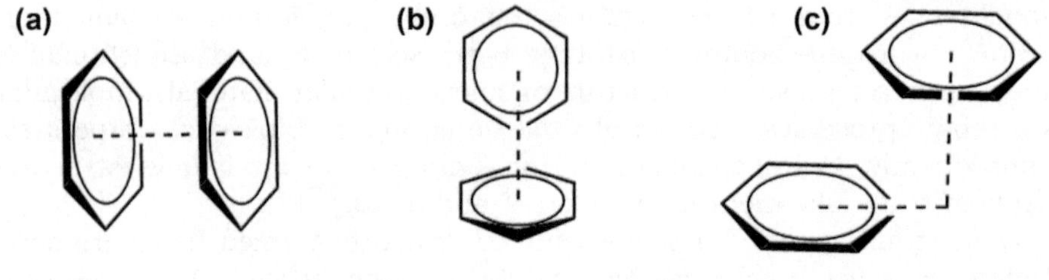

Figure 9.2 π–π stacking interactions are mainly achieved through three modes: (a) face-to-face, (b) edge-to-face, and (c) offset face-to-face stacking. Reproduced from ref. 13 with permission from American Chemical Society, Copyright 2018.

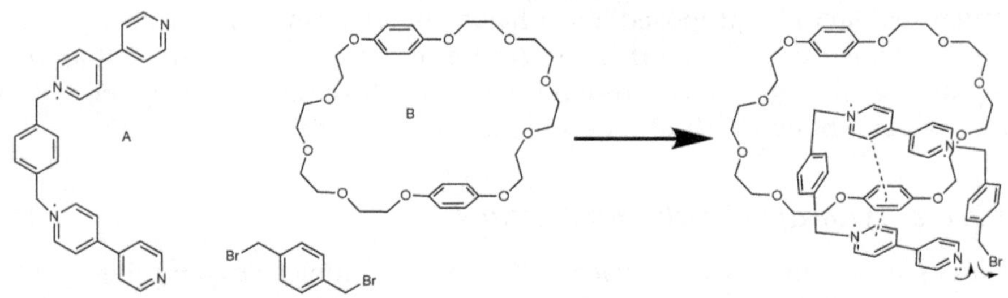

Figure 9.3 Pyridine rings with electron-rich benzene derivatives through π–π stacking interactions. Reproduced from ref. 14 with permission from John Wiley & Sons, Copyright 1989 by VCH Verlagsgesellschaft mbH, Germany.

(Figure 9.2(a)), edge-to-face (Figure 9.2(b)), and offset face-to-face stacking (Figure 9.2(c)). Among these, edge-to-face and offset face-to-face structures are more common. These interactions are similar to non-covalent bonds like hydrogen bonds and van der Waals forces.

An early example of using π–π stacking interactions to drive molecular self-assembly is in the synthesis of rotaxanes. Ashton *et al.*[14] linked electron-poor pyridine rings with electron-rich benzene derivatives through π–π stacking interactions, resulting in a series of designed systems. Figure 9.3 illustrates a schematic of rotaxane synthesis. Additionally, π–π stacking has been employed in the self-assembly of novel nanoscale delivery systems.[15] Wang *et al.*[16] discovered that chitosan, 14-hydroxyrotenone, and poly-dopamine can self-assemble into lollipop-shaped nanostructures through π–π stacking interactions, exhibiting excellent stability. Liu *et al.*[17] reported a new π-conjugated liquid crystal synthesized from oligothiophene (4T) and polyethylene oxide (PEO4), commonly used in flexible electronic devices. π–π stacking interactions determine the molecular ordering of 4T/PEO4 and play a crucial role in controlling the self-assembly of this liquid crystal.

9.2.1.4 Electrostatic Interactions

Due to the positive charge of atomic nuclei and the negative charge of outer-shell electrons, molecules can attract each other, forming electrostatic forces.

The attractive forces between two molecules with opposite charges primarily manifest in three different types: charge–charge interaction, charge–dipole interaction, and dipole–dipole interaction. Electrostatic forces can be calculated using eqn (9.1), where r_{ij} is the distance between two particles carrying charges q_i and q_j, and ε is the dielectric constant of the medium.

$$E_{\mathrm{ele}} = \sum \frac{q_i q_j}{\varepsilon r_{ij}} \tag{9.1}$$

Electrostatic interactions between assembling units represent an important driving force in the process of self-assembly, characterized by their ease of modulation and reversibility. Numerous examples utilize electrostatic self-assembly, including isolated nanostructures,[18] layered surfaces,[19] and macroscopic crystals composed of millimeter-scale polymer particles.[20] Chen *et al.*[21] observed fluorescence enhancement when adding metal ring solutions to heparin, providing evidence for electrostatically driven supramolecular self-assembly. The interaction resulted in the formation of pearl necklace-like structures.

9.2.2 Factors Influencing Self-assembled Structures

The macroscopic properties of self-assembled materials are determined by their microscopic structure. Different external environments and assembly conditions can manipulate the formation of various microscopic structures during self-assembly, thereby further controlling the macroscopic properties. Simultaneously, studying the influence of different external environments and assembly conditions on self-assembly behavior provides profound insights into the mechanisms governing the self-assembly process.

9.2.2.1 *Solvency*

Solvent type is one of the determining factors for self-assembly, and its type and quantity often alter the structure and morphology of assemblies. Gao *et al.*[22] reported on solvent-driven self-assembly to achieve controlled morphology of polymers. By controlling the presence of solvents formaldehyde dimethyl acetal (FDA) and 1,2-dichloroethane (DCE), which alters the solvent polarity and affects hydrophilic/hydrophobic interactions, they customized the self-assembly structure of PEO-*b*-PS (PEO: polyethylene glycol; PS: polystyrene) from aggregated nanoparticles to hollow spheres and mesoporous bulk materials. Similarly, Liu *et al.*[23] demonstrated that adjusting the type and composition of solvents can induce supramolecular self-assembly to achieve different structures.

In addition, solvents can also influence the chirality of supramolecular self-assembly. Xing *et al.*[24] studied a compound containing napthalenediimide and cholesterol units, which, under the influence of solvents with different polarities, formed assemblies of different chirality. Li *et al.*[25] investigated the self-assembly behavior of a derivative of biphenyl dicarboxylic acid, 5-(benzyloxy)-isophthalic acid (BIC-C8), in 2-nonanol solvent. Scanning tunneling microscope images showed that in (*R*)-2-nonanol solvent, the derivative could generate globally homochiral clockwise honeycomb patterns. Conversely, under the influence of (*S*)-2-nonanol solvent, the resulting images formed mirror-symmetric counterclockwise honeycomb grids. These studies demonstrate that the chirality of supramolecular self-assembly is sensitive to solvent conditions.

9.2.2.2 Temperature

Temperature can serve as a significant stimulus for controlling and altering the arrangement of assemblies.[26,27] Differences in temperature can impact the adsorption behavior at interfaces, further influencing the thermodynamic and kinetic processes of the system. This can lead to changes in adsorption energy and weak interactions between molecules within the self-assembled system. At high temperatures, hydrogen bonds within the self-assembled system may break, resulting in structural transformation and disruption of the self-assembled structure. Cheng *et al.*[28] studied the self-assembly of molecules 1 and 2 on graphite surfaces (Figure 9.4), where different temperatures caused variations in their self-assembly configurations. Mukhopadhyay and Datta[29] investigated the self-assembly of three different nucleobases on graphene surfaces at different temperatures. Structural analysis of the nucleobase tetramers showed that high temperatures could disrupt hydrogen bonds between two nucleobase molecules, leading to further disruption of the tetramer structure and affecting the overall arrangement, ultimately resulting in a disordered state.

9.2.2.3 Substrate

The formation of ordered networks of self-assembled monolayers on solid surfaces is a result of the balance between intermolecular interactions and molecule–substrate interactions. Strong molecule–substrate interactions can even disrupt the growth of supramolecular structures. Additionally, the symmetry and periodicity of the substrate also influence the self-assembly process. Research on molecular surface adsorption indicates that substrates not only provide physical support but also control the molecular orientation during growth, thereby affecting adsorption assembly behaviors and driving the formation of specific patterns of assembled units. Liu *et al.*[30]

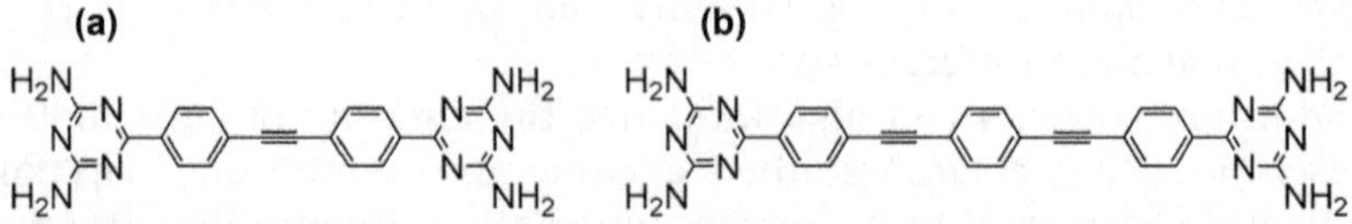

Figure 9.4 Chemical structures of molecules (a) 1 and (b) 2. Reproduced from ref. 28 with permission from American Chemical Society, Copyright 2016.

investigated the impact of different substrates on the self-assembly of star-shaped 2,4,6-tri-(bromophenyl)-(1,3,5)-triazine (BPT) molecules. The study revealed that the substrate influences the primary driving forces of self-assembly. On highly oriented pyrolytic graphite (HOPG) surfaces, the main driving force is Br···H–C bond interactions, whereas on single-layer graphene (SLG) surfaces on copper foil, in addition to the interactions, triple Br–Br–Br bonds and strong substrate–molecule interactions collectively drive the process.

In addition to altering the driving forces of self-assembly to create different patterns, substrates also influence the orderliness of self-assembled systems. Kumar *et al.*[31] studied the effects of three substrates—indium tin oxide (ITO), SiO_x/Si, and gold—on the ordering of self-assembled layers of iron phthalocyanine (FePc) and cobalt phthalocyanine (CoPc) molecules. Their findings indicate that the strength of interaction between molecules and substrates affects molecular stacking, thereby influencing the overall orderliness of the self-assembled system.

9.2.2.4 *Molecular Density*

Compared to other factors like temperature, solvency, and substrate, molecular density is a more cost-effective factor in controlling reaction kinetics. It is well-known that molecular density influences the rate of chemical equilibrium, as well as the size and quality of crystals. In the field of molecular self-assembly, controlling interfacial self-assembly through molecular density has been extensively documented. The initiation of the self-assembly process often requires critical molecular density control, which can also regulate the formation of different configurations. Utilizing differences in molecular density is a common approach to achieving a diverse array of self-assembled structures. Silly *et al.*[32] studied the influence of molecular density on the self-assembly of star-shaped 1,3,5-tri(4-iodophenyl)-benzene molecules at the interface of 1-phenyloctane/graphite. In this self-assembly system, the density of solute molecules significantly affects the halogen bonding between adjacent molecules, thereby causing variations in the internal structure of the self-assembled molecular layers with changes in solute molecular density.

In addition, Cai *et al.*[33] conducted *in situ* electrochemical scanning tunneling microscopy studies on the self-assembly of alkyl-substituted benzo-(1,2-*b*:4,5-*b*′)dithiophene (DDBDT) on Au(111) electrodes. They observed that the solute molecule density of DDBDT is a critical factor influencing its interfacial structure. By varying the surface molecular density, they could adjust the self-assembly arrangement. Similarly, Miao *et al.*[34] investigated the self-assembly of iodine-substituted thiophene derivative (5,10-DITD) at the interface of 1-octanoic acid/graphite. They found that varying the molecular density allows for control over the chiral self-assembly configurations in this system. The structure of the self-assembled system depends on the density of 5,10-DITD molecules.

9.3 Self-assembled Structures for Water Remediation

Self-assembly technology is crucial in water treatment. By forming nanomaterials with specific pore sizes and functional groups, it significantly enhances adsorption, separation, and catalytic efficiency. Both the fabrication and performance of membrane and catalytic materials also benefit from the development of self-assembly constructions, impacting various critical areas, such as desalination, pollutant degradation, and water quality monitoring.

9.3.1 Polymeric Microspheres

Since Rohm and Haas first commercialized K120 core–shell polymer microspheres in 1957 and proposed the concept of "particle design", hollow polymer microspheres have been extensively researched and developed.[35] Throughout decades of developments, researchers can now realize the fabrication of hollow microspheres in a facile manner, utilizing the templates of hard colloidal particles, inorganic nanoparticles, or soft materials, followed by a subsequent stage of core removal. The commonly used templating methods are cumbersome, and are not conducive to large-scale production. Moreover, the template removal process may damage the microsphere shell, compromising the integrity of the microspheric shell without an internally supporting material. To solve these long-standing issues, researchers have explored various novel preparation methods for hollow polymer microsphere fabrications. For instance, block copolymers have been reported to spontaneously form polymer micelles in solution after dissolution, which are then cross-linked or initiated to form hollow polymer microspheres.[36] Another approach involves the use of polymers that are prone to swelling and permeation. Core–shell structured microspheres prepared with such non-swelling polymers are more likely to withstand the etching process, leading to a higher chance of forming hollow structures through polymer swelling.[37]

Hollow polymer microspheres, characterized by high porosity, low density, large surface area, and numerous adsorption sites, are widely applied in the field of water treatment. Gujar *et al.*[38] prepared *N,N,N′,N′*-tetraoctyl diglycolamide/PSF polymer microspheres, using polysulfone (PSF) resin as the matrix material. These microspheres exhibited high-performance adsorption of Am^{3+} ions from wastewater. Wu *et al.*[39] utilized maleic anhydride and itaconic acid to prepare amino-functionalized poly(maleic anhydride-itaconic acid) ($BHP\text{-}NH_3^+$) hollow microspheres *via* emulsion polymerization. They successfully used these microspheres for the selective removal of methyl orange from aqueous solutions, achieving a maximum adsorption capacity of 952 $mg\,g^{-1}$ for methyl orange. The $BHP\text{-}NH_3^+$ hollow microspheres maintained stable adsorption capability over 5 adsorption/desorption cycles (Figure 9.5).

Saleem *et al.*[40] synthesized carboxyl-functionalized mesoporous polymer hollow microspheres for removing heavy metal ions and organic pollutants

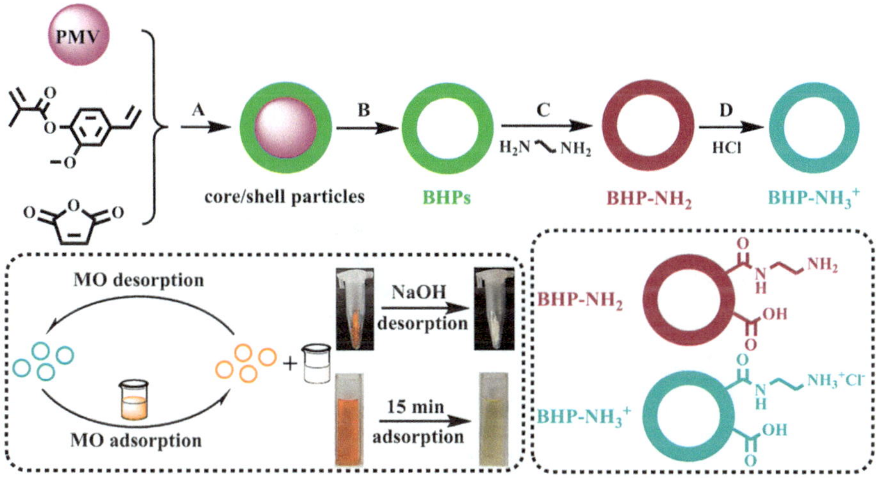

Figure 9.5 Schematic illustration of the adsorption of methyl orange dye on BHP-NH_3^+ microspheres. BHP: poly(maleic anhydride-itaconic acid). Reproduced from ref. 38 with permission from Elsevier, Copyright 2015.

from aqueous solutions, achieving adsorption capacities of 44.5 mg g^{-1} for Cr^{3+}, 48.8 mg g^{-1} for Pb^{2+}, and 50.4 mg g^{-1} for rhodamine B. These hollow microspheres could be reused for up to 5 cycles after simple regeneration. Li *et al.*[41] utilized single-pore HMIP@CD (HMIP: hollow molecularly imprinted polymers; CD: cyclodextrins) polymer microspheres for rapid and selective recognition and removal of Sudan dyes and tetracyclines. Taleb *et al.*[42] prepared microfibrillated cellulose/chitosan (MFC–Chi) microspheres *via* blending for arsenic ion removal from water. Cui *et al.*[43] developed renewable sea urchin-like Fe_3O_4@PDA–Ag (PDA: polydopamine) hollow microspheres as catalysts and adsorbents (Figure 9.6), successfully removing organic dye pollutants from aqueous solutions.

9.3.2 Inorganic Microspheres

Metal oxides and hydroxides, such as ZnO, Fe_3O_4, and MnO_2, have long been investigated to remediate organic waste-containing water due to their low water solubility and minimal environmental impact. The primary objective of employing inorganic microspheres is to avoid secondary pollution from microplastics and heavy metal ions. For this purpose, inorganic microspheres have been the focus of much attention, since they possess large surface areas, excellent chemical stability, and can be functionalized with specific groups. Therefore, they are generally considered as effective adsorbents for removing heavy metal ions and organic dyes. Cheng *et al.*[44] synthesized hierarchical porous $Ni(OH)_2$ and NiO nanosheets *via* simple chemical precipitation methods and studied their adsorption performance toward Congo red aqueous solutions. Wang *et al.*[45] successfully prepared sea

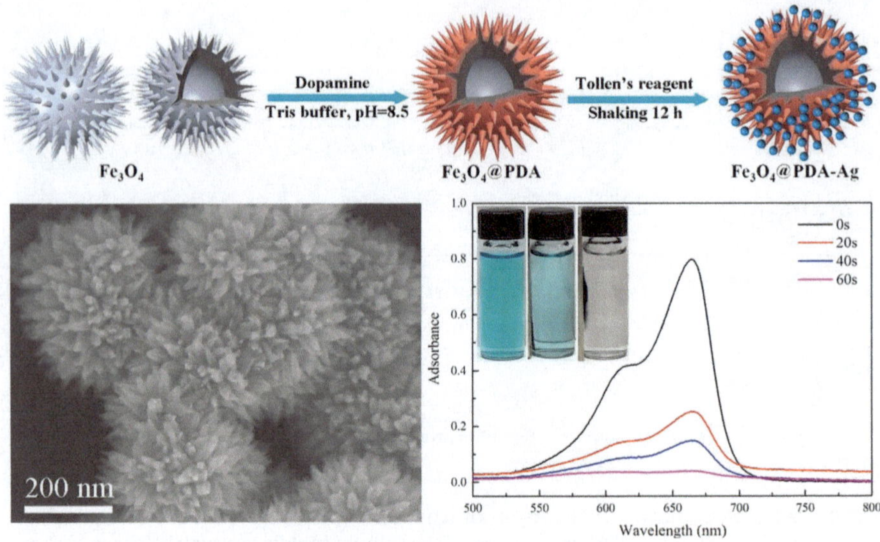

Figure 9.6 Schematic of fabricated urchin-like Fe_3O_4@PDA–Ag hollow microspheres, adsorption UV-vis, and removal volume comparison (PDA: polydopamine). Reproduced from ref. 43 with permission from Elsevier, Copyright 2018.

urchin-like α-FeOOH hollow spheres with nanoneedle structures using self-assembly techniques, demonstrating outstanding capabilities in removing heavy metal ions and organic dyes.

One essential category of inorganic microsphere adsorbents is magnetic nanoparticles. Not only do their large surface areas ensure their adsorption capacity toward contaminants, but also their magneto-responsiveness endows rapid and convenient means for concentrating pollutants and recycling adsorbents, making them widely applicable as novel adsorbents. Yang *et al.*[46] synthesized flower-shaped Fe_3O_4@MnO_2 nanocomposites (Figure 9.7), where the adsorption mechanism primarily involves electrostatic attraction, selectively adsorbing Congo red dye.

9.3.3 Carbon-based Microspheres

Carbon-based materials can be combined with photocatalysts for boosted catalytic performance, as well as improving their stability. Owing to these properties, carbon-based photocatalysts are highly suitable for removing toxic organic pollutants from harsh water environments through photocatalytic degradation and mineralization. In particular, using solar energy for water remediation has the advantages of being safe, non-toxic, energy-efficient, and having a long service life. Not surprisingly, this category of materials has garnered a vast amount of attention from researchers, ranging from a broad aspect of practical application scenarios.

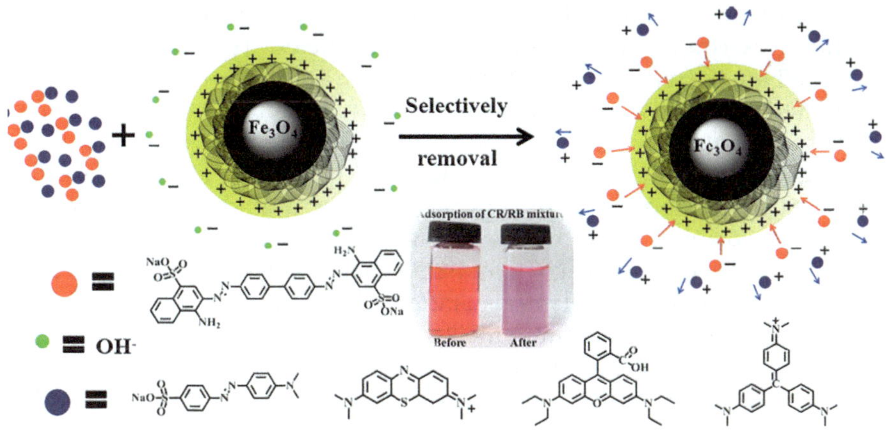

Figure 9.7 Flower-like core–shell MnO_2-coated magnetic composites (Fe_3O_4@MnO_2). Reproduced from ref. 46 with permission from Elsevier, Copyright 2017.

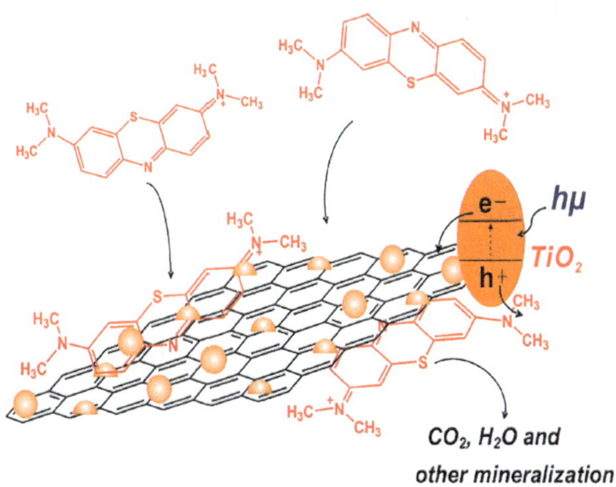

Figure 9.8 Schematic diagram of P25–GR and its mechanism for enhancing methylene blue photodegradation. Reproduced from ref. 47 with permission from the Royal Society of Chemistry.

Two-dimensional carbon-based materials, such as graphene (GR), possess high specific surface area and excellent light transmittance, achieving a transparency of up to 97.7%.[48] As carbon-based modifiers for semiconductor photocatalysts, they offer significant advantages. Zhang *et al.*[47] utilized chemically bonded TiO_2 (P25) and graphene oxide (GO) to prepare P25–graphene (P25–GR) nanocomposite photocatalysts *via* a single-step hydrothermal method, as shown in Figure 9.8. These composites exhibit strong dye adsorption capabilities along with an expanded light absorption range and efficient charge separation properties. Similarly, Sun *et al.* synthesized a GO–CdS composite *via* a two-phase mixing method, demonstrating a higher photocatalytic degradation efficiency for organic dyes compared to pure CdS nanoparticles. This effect should be attributed to the effective inhibition of CdS aggregation when being incorporated with GO, while such a combination also promotes the separation of photogenerated charge carriers. Another example is that Xu *et al.*[49] prepared graphene-hybridized ZnO composite materials, showing excellent photocatalytic performance for organic pollutant degradation in water. The enhanced activity of these composite photocatalysts is mainly due to the electron interactions between GR and ZnO, facilitating rapid electron transfer to GR and promoting efficient separation of photogenerated electron–hole pairs, thus improving quantum efficiency.

Another well-known carbon material is carbon nanotubes (CNTs), which is a unique one-dimensional structure with high specific surface area and excellent light transmittance. CNTs have been widely utilized in the modification of semiconductor photocatalysts. Wang *et al.*[50] employed a modified sol–gel method to prepare multi-walled carbon nanotube (MWNT)/TiO_2 composite photocatalysts. Characterization results indicated that the introduction of MWNTs effectively mitigated TiO_2 particle aggregation, thereby increasing the surface area of the composite photocatalyst. Within the photocatalytic composites, MWNTs act as sensitizers with two major roles: (i) broadening the light absorption range of TiO_2 and (ii) serving as the electron acceptor, both of which originate from the strong interfacial interactions between MWNTs and TiO_2. Collectively, the separation of photogenerated electrons and holes is facilitated, leading to the enhancement of photocatalytic performance (Figure 9.9(a)). Liu *et al.*[51] employed a

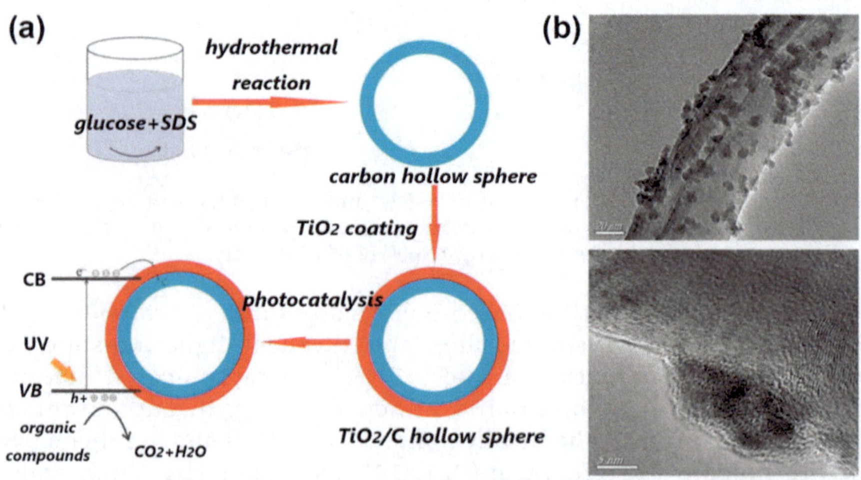

Figure 9.9 (a) Preparation of TiO_2/carbon hollow sphere composite photocatalysts and (b) TEM images of the CdS/CNT composite. CNT: carbon nanotube. Reproduced from ref. 51 with permission from World Scientific Publishing, Copyright 2014.

hydrothermal process to grow CdS nanoparticles *in situ* on functionalized CNTs, resulting in CdS/CNTs composites, as depicted in Figure 9.9(b). Under visible light conditions, the photocatalytic experiment for tetracycline degradation showed approximately 81.2% removal within 60 minutes, and after four cycles, the activity of the composite photocatalyst remained largely unchanged.

9.3.4 Dielectric Microspheres

In recent years, the contact electro-catalysis (CEC) strategy has been successfully proposed as a novel water remediation strategy since its first discovery in 2022.[52] The CEC process utilizes the electron exchange of dielectric microspheres, as driven by the instantaneous contact electrification at liquid–solid or even liquid–liquid interfaces to trigger redox reactions (Figure 9.10).[53,54] Typically, the energy input to activate CEC comes from mechanical agitation, such as ultrasonication. Viable dielectric microspheres can be organic (*e.g.*, FEP and PTFE) and inorganic materials (*e.g.*, SiO_2 and Al_2O_3), among others.[55–57] Unlike traditional catalytic strategies that impose specific requirements on catalysts, CEC opens new avenues to use self-assembled dielectric materials in water treatment.

Applying CEC for water remediation offers several unique advantages.[59–61] Firstly, CEC catalytic materials are unrestricted to traditional catalysis (which is typically expensive), significantly expanding the range of catalyst choices and simplifying catalyst synthesis processes. Natural materials like rocks could potentially serve as CEC catalysts. Secondly, the mechanical stimuli for CEC activation are also generally applicable from various types, such as vibration, sliding, collision, grinding, and ultrasound, which are ubiquitous in the environment but often overlooked. Therefore, CEC enables

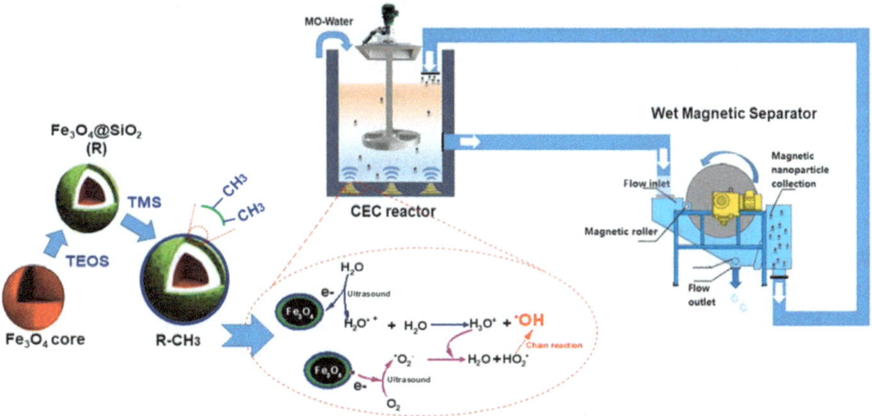

Figure 9.10 Schematic diagram of contact-electro-catalysis degradation of organic wastewater and catalyst recovery. Reproduced from ref. 58 with permission from Elsevier, Copyright 2023.

the degradation of organic pollutants in remote wilderness areas far from electricity, light sources, and synthetic materials.

Self-assembled CEC catalysts have drawn increasing attention because they can be effective both in pollutant remediation and particulate catalyst recovery. Chen *et al.*[58] synthesized a micro–nano catalyst with tunable chemical groups based on magnetic nanoparticles, used for the CEC degradation of methyl orange (MO). As the electronegativity of the chemical group increased (R–F > R–CH_3 > R–NH_2, R = pristine particles), the degradation efficiency of MO also significantly improved. The low surface energy of hydrophobic particles results in a threshold of the exposed total surface area of the catalyst in aqueous solutions (*e.g.*, 20 mg R–F per 50 mL of MO solution), which is a key reason for the nonlinear increase in catalyst mass and CEC efficiency. Using R–CH_3 for CEC, MO degradation efficiency reached 99% within 2 hours. The micro–nano R–CH_3 catalyst could recover almost 100% using a magnetic separator, and after ten CEC cycles, the catalytic efficiency remained above 97%. Although further optimization of system parameters is required, these attractive catalysts are expected to have promising applications in large-scale organic pollutant degradation systems.

9.3.5 Membranes

For water remediation applications, membranes have always been a preferred choice for large-scale and high-throughput applications due to continuous processing and upscaling capabilities. For this, the self-assembly strategy provides a facile tool to modify the original filtering membrane materials with additional functionalities, such as specific adsorption, antibacterial properties, *etc.* Especially with the recent pursuit of cleaner productions, there has been an emerging trend of fabricating sustainable filtration membranes from bio-based materials, which eliminate the possibility of secondary pollution (compared to petroleum-based membranes).[62] Bio-based materials have been investigated extensively for this purpose, including hydrophilic cellulose and hydrophobic lignin compounds. For example, Li *et al.*[63] showcased that the strong cohesive properties of nanocellulose allow it to bind any functional powdery materials into microfibers without losing their original functionalities, which can then be subjected as the substrate of membrane materials. In another approach, Lu *et al.*[64] demonstrated that cellulose-producing bacteria could also interact with nanoparticles and/or hydrophobic membranes, leading to the direct bio-fabrication of thin-film composites and Janus membranes, respectively. This demonstrates that living matter can also be involved in self-assembly procedures, opening a new avenue for studying the interdisciplinary area between microbiology, material, and environmental science.

Guo *et al.*[65] reported the self-assembly of plant-based polyphenol compounds on conventional microporous membranes for ultrafast antibiotic removal (Figure 9.11(a)). Since natural polyphenols contain a high density of dihydroxyphenyl or trihydroxyphenyl groups on their molecular structures,

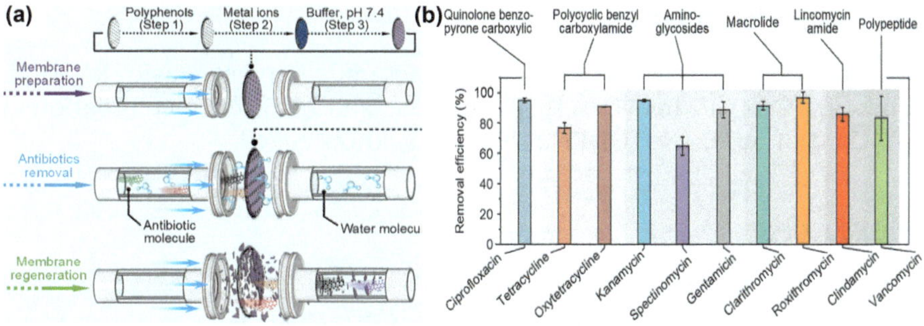

Figure 9.11 (a) Schematic drawing of preparing and regenerating polyphenol-based self-assembled membranes. (b) Adsorption and removal efficiency toward ten common antibiotics. Reproduced from ref. 65 with permission from Elsevier, Copyright 2023.

they can self-assemble into a vast range of membrane substrates through various covalent and non-covalent bindings. It is worth noting that the presence of metal ions (*e.g.*, Fe^{3+}) is essential to incorporate stable complexation formation with polyphenols. Owing to the universal adsorption capability, this self-assembled polyphenolic structure can deplete more than ten types of antibiotics with a removal efficiency ranging from 75–100% (Figure 9.11(b)), whereas the substrate itself only exhibited a removal efficiency of less than 10% under the same condition. Similar strategies have been successfully implemented into microplastics,[66] demonstrating a broad application potential for this technique.

9.4 Conclusion

In conclusion, self-assembled structures play a critical role in the future of water remediation due to their versatility and multi-functional properties. The toolbox of self-assemblies has expanded dramatically in the past decade, enabling rapid advancements in materials engineering and the development of water treatment protocols. As a general rule, intermolecular interactions are key to unleashing the full potential of self-assembled structures, necessitating a deeper fundamental understanding in the future.

In this chapter, we briefly covered the most common types of interactions, including hydrogen bonding, van der Waals forces, π–π stacking, and electrostatic interactions. However, many unconventional colloidal interactions exist in complex multi-component systems, such as cation–π interactions, depletion interactions, and metallic coordination, to name a few. Additionally, the surrounding environment affects the morphology of self-assemblies, which might limit their application scenarios since non-covalent bonds could disassemble under undesired conditions. Nevertheless, tremendous efforts are still needed from both fundamental and practical perspectives to implement self-assembled structures for water remediation applications in real life and on larger scales.

Acknowledgements

The support from the Guangdong Basic and Applied Basic Research Foundation (No. 2024A1515011694), and the Natural Science Foundation of Top Talent of SZTU (No. GDRC202316) is acknowledged.

References

1. T. Steiner, *Angew. Chem., Int. Ed.*, 2002, **41**, 48–76.
2. S. J. Grabowski, *Chem. Rev.*, 2011, **111**, 2597–2625.
3. S. J. Grabowski and W. A. Sokalski, *J. Phys. Org. Chem.*, 2005, **18**, 779–784.
4. K. Wendler, J. Thar, S. Zahn and B. Kirchner, *J. Phys. Chem. A*, 2010, **114**, 9529–9536.
5. R. H. Crabtree, *Science*, 1998, **282**, 2000–2001.
6. C. L. Perrin and J. B. Nielson, *Annu. Rev. Phys. Chem.*, 1997, **48**, 511–544.
7. S. Rej, Y. Ano and N. Chatani, *Chem. Rev.*, 2020, **120**, 1788–1887.
8. B. Yoon, W. Luedtke, R. N. Barnett, J. Gao, A. Desireddy, B. E. Conn, T. Bigioni and U. Landman, *Nat. Mater.*, 2014, **13**, 807–811.
9. R. Forbes, *J. Phys. Colloq.*, 1989, **50**, C8-15–C18-20.
10. I. E. Dzyaloshinskii, E. M. Lifshitz and L. P. Pitaevskii, *Adv. Phys.*, 1961, **10**, 165–209.
11. S. Jiang, J. Qian, Y. Duan, H. Wang, J. Guo, Y. Guo, X. Liu, Q. Wang, Y. Shi and Y. Li, *J. Phys. Chem. Lett.*, 2018, **9**, 6755–6760.
12. H. M. Colquhoun, J. F. Stoddart, D. J. Williams, J. B. Wolstenholme and R. Zarzycki, *Angew. Chem., Int. Ed. Engl.*, 1981, **20**, 1051–1053.
13. T. Chen, M. Li and J. Liu, *Cryst. Growth Des.*, 2018, **18**, 2765–2783.
14. P. R. Ashton, T. T. Goodnow, A. E. Kaifer, M. V. Reddington, A. M. Slawin, N. Spencer, J. F. Stoddart, C. Vicent and D. J. Williams, *Angew. Chem., Int. Ed. Engl.*, 1989, DOI: 10.1002/anie.198912611.
15. W.-R. Zhuang, Y. Wang, P.-F. Cui, L. Xing, J. Lee, D. Kim, H.-L. Jiang and Y.-K. Oh, *J. Controlled Release*, 2019, **294**, 311–326.
16. Y. Wang, Y. Wu, K. Li, S. Shen, Z. Liu and D. Wu, *Adv. Funct. Mater.*, 2019, **29**, 1805582.
17. Z. Liu, B. X. Dong, M. Misra, Y. Sun, J. Strzalka, S. N. Patel, F. A. Escobedo, P. F. Nealey and C. K. Ober, *Adv. Funct. Mater.*, 2019, **29**, 1805220.
18. I. Willerich and F. Gröhn, *J. Am. Chem. Soc.*, 2011, **133**, 20341–20356.
19. M. J. McShane and Y. Lvov, *Dekker encyclopedia of nanoscience and nanotechnology*, 2004, vol. 4, p. 1.
20. B. A. Grzybowski, A. Winkleman, J. A. Wiles, Y. Brumer and G. M. Whitesides, *Nat. Mater.*, 2003, **2**, 241–245.
21. L.-J. Chen, Y.-Y. Ren, N.-W. Wu, B. Sun, J.-Q. Ma, L. Zhang, H. Tan, M. Liu, X. Li and H.-B. Yang, *J. Am. Chem. Soc.*, 2015, **137**, 11725–11735.
22. T. N. Gao, T. Wang, W. Wu, Y. Liu, Q. Huo, Z. A. Qiao and S. Dai, *Adv. Mater.*, 2019, **31**, 1806254.
23. Z. Liu, H. Fan, W. Li, G. Bai, X. Li, N. Zhao, J. Xu, F. Zhou, X. Guo and B. Dai, *Nanoscale*, 2019, **11**, 4751–4758.

24. P. Xing and Y. Zhao, *Acc. Chem. Res.*, 2018, **51**, 2324–2334.
25. S.-Y. Li, T. Chen, J.-Y. Yue, D. Wang and L.-J. Wan, *Chem. Commun.*, 2017, **53**, 11095–11098.
26. Z. Zhang, X. Li, Z. Peng, X. Yan, S. Liu, Y. Hong, Y. Shan, X. Xu, L. Jin and B. Liu, *Nat. Commun.*, 2023, **14**, 4094.
27. R. Chang, L. Zhao, R. Xing, J. Li and X. Yan, *Chem. Soc. Rev.*, 2023, DOI: 10.1039/D2CS00675H.
28. L. Cheng, Y. Li, C.-Y. Zhang, Z.-L. Gong, Q. Fang, Y.-W. Zhong, B. Tu, Q. Zeng and C. Wang, *ACS Appl. Mater. Interfaces*, 2016, **8**, 32004–32010.
29. T. K. Mukhopadhyay and A. Datta, *J. Phys. Chem. C*, 2018, **122**, 28918–28933.
30. C. Liu, L. Yang, Y. Wang, S. Lei and W. Hu, *J. Phys. Chem. C*, 2018, **122**, 12307–12314.
31. A. Kumar, D. Naumenko, L. Cozzarini, L. Barba, A. Cassetta and M. Pedio, *J. Raman Spectrosc.*, 2018, **49**, 1015–1022.
32. F. Silly, *J. Phys. Chem. C*, 2017, **121**, 10413–10418.
33. Z.-F. Cai, H.-J. Yan, D. Wang and L.-J. Wan, *Nanoscale*, 2018, **10**, 3438–3443.
34. X. Miao, J. Li, B. Zha, K. Miao, M. Dong, J. Wu and W. Deng, *Appl. Surf. Sci.*, 2018, **433**, 1075–1082.
35. L. Bai, J. Gu, S. Huan and Z. Li, *RSC Adv.*, 2014, **4**, 27363–27380.
36. R. Grillo, D. Beutel, U. Cataldi, C. Rockstuhl and T. Bürgi, *ACS Appl. Nano Mater.*, 2020, **3**, 6108–6117.
37. Y. Kang, G. Tan, Q. Man, M. Ning, S. Chen, J. Pan and X. Liu, *Mater. Chem. Phys.*, 2021, **266**, 124532.
38. R. Gujar, P. Mohapatra, D. S. Lakshmi and A. Figoli, *J. Chromatogr. A*, 2015, **1422**, 206–212.
39. X. Yong, S. Raza, J. Deng and Y. Wu, *React. Funct. Polym.*, 2018, **132**, 9–18.
40. S. Raza, X. Yong, M. Raza and J. Deng, *Chem. Eng. J.*, 2018, **352**, 20–28.
41. H. Li, L. Zhao, Y. Xu, T. Zhou, H. Liu, N. Huang, J. Ding, Y. Li and L. Ding, *Talanta*, 2018, **185**, 542–549.
42. K. Taleb, J. Markovski, Z. Veličković, J. Rusmirović, M. Rančić, V. Pavlović and A. Marinković, *Arabian J. Chem.*, 2019, **12**, 4675–4693.
43. K. Cui, B. Yan, Y. Xie, H. Qian, X. Wang, Q. Huang, Y. He, S. Jin and H. Zeng, *J. Hazard. Mater.*, 2018, **350**, 66–75.
44. B. Cheng, Y. Le, W. Cai and J. Yu, *J. Hazard. Mater.*, 2011, **185**, 889–897.
45. B. Wang, H. Wu, L. Yu, R. Xu, T. T. Lim and X. W. Lou, *Adv. Mater.*, 2012, **24**, 1111–1116.
46. Q. Yang, H. Song, Y. Li, Z. Pan, M. Dong, F. Chen and Z. Chen, *J. Mol. Liq.*, 2017, **234**, 18–23.
47. N. Zhang, Y. Zhang and Y.-J. Xu, *Nanoscale*, 2012, **4**, 5792–5813.
48. S. Kumar, A. Kumar, A. Bahuguna, V. Sharma and V. Krishnan, *Beilstein J. Nanotechnol.*, 2017, **8**, 1571–1600.
49. T. Xu, L. Zhang, H. Cheng and Y. Zhu, *Appl. Catal., B*, 2011, **101**, 382–387.
50. W. Wang, P. Serp, P. Kalck and J. L. Faria, *J. Mol. Catal. A: Chem.*, 2005, **235**, 194–199.

51. X. Liu, Y. Tang, C. Ma, Y. Yan, P. Lv, G. Yao, Q. Wang and Y. Yan, *Nano*, 2014, **9**, 1450017.
52. Z. Wang, A. Berbille, Y. Feng, S. Li, L. Zhu, W. Tang and Z. L. Wang, *Nat. Commun.*, 2022, **13**, 130.
53. Z. Chen, Y. Lu, R. Manica and Q. Liu, *Nano Energy*, 2021, **89**, 106456.
54. Z. Chen, Y. Lu, R. Manica, J. Lu, D. Shi, J. Li and Q. Liu, *Nanoscale*, 2023, **15**, 8158–8168.
55. Z. Chen, Y. Lu, R. Li, D. Li, B. Xiang, J. Li and Q. Liu, *Nano Energy*, 2023, 108834.
56. X. Zhao, Y. Su, A. Berbille, Z. L. Wang and W. Tang, *Nanoscale*, 2023, **15**, 6243–6251.
57. X. Dong, Z. Wang, A. Berbille, X. Zhao, W. Tang and Z. L. Wang, *Nano Energy*, 2022, **99**, 107346.
58. Z. Chen, Y. Lu, X. Liu, J. Li and Q. Liu, *Nano Energy*, 2023, **108**, 108198.
59. Z. Chen, Y. Lu, R. Hong, Z. Liang, L. Wen, X. Liu and Q. Liu, *Langmuir*, 2024, **40**, 5557–5570.
60. Z. Wang, X. Dong, W. Tang and Z. L. Wang, *Chem. Soc. Rev.*, 2024, DOI: 10.1039/D3CS00736G.
61. Y. Su, A. Berbille, X.-F. Li, J. Zhang, M. PourhosseiniAsl, H. Li, Z. Liu, S. Li, J. Liu and L. Zhu, *Nat. Commun.*, 2024, **15**, 4196.
62. D. Wang, H. Yang, Q. Wang, Y. Lu, J. Yan, W. Cheng, O. J. Rojas and G. Han, *Composites, Part A*, 2023, **164**, 107300.
63. H. Wang, C. Zeng, C. Wang, J. Fu, Y. Li, Y. Yang, Z. Du, G. Tao, Q. Sun and T. Zhai, *Nat. Mater.*, 2024, **23**, 596–603.
64. Y. Lu, Y. Chun, X. Shi, D. Wang, F. Ahmadijokani and O. J. Rojas, *Adv. Mater.*, 2024, 2400311.
65. Y. Wang, Y. He, Q. Wang, X. Wang, B. L. Tardy, J. J. Richardson, O. J. Rojas and J. Guo, *Matter*, 2023, **6**, 260–273.
66. M. Farrokhbin, B. Stojimirovic, M. Galli, M. K. Aminian, Y. Hallez and G. Trefalt, *Phys. Chem. Chem. Phys.*, 2019, **21**, 18866–18876.

CHAPTER 10

Supramolecular Structures Containing Carbon Quantum Dots for Simultaneous Removal and Detection of Contaminants in Water

E. C. SILVA,[a] N. F. DE SOUZA,[b] F. M. MACHADO,[b] A. R. FAJARDO*[c] AND A. J. M. VALENTE*[d]

[a] Laboratório de Estudos de Materiais para Adsorção e Sensoriamento (LEMAS), Federal University of Roraima (UFRR), Campus Paricarana, 69310-000, Boa Vista - RR, Brazil; [b] Technology Development Center, Federal University of Pelotas (UFPel), 1 Gomes Carneiro St., 96010-610, Pelotas-RS, Brazil; [c] Laboratório de Tecnologia e Desenvolvimento de Compósitos e Materiais Poliméricos (LaCoPol), Federal University of Pelotas (UFPel), Campus Capão do Leão, 96010-900, Pelotas-RS, Brazil; [d] CQC-IMS, Department of Chemistry, University of Coimbra, 3004-535, Coimbra, Portugal
*Emails: andre.fajardo@ufpel.edu.br; avalente@ci.uc.pt

10.1 Introduction

Over the last decades, water bodies have been continuously damaged by industrial and domestic dumping, farming pollution, and the deterioration of treatment plants and distribution pipes, exposing millions of people to unsafe water. Additionally, the use of emergent pollutants

Contribution of Colloidal Materials to Air, Water and Soil Environmental Sustainability: From Basic Concepts to Advanced Applications
Edited by Izabel C. Riegel-Vidotti, Lizandra Maria Zimmermann and Heloise R. de Barros

Published by the Royal Society of Chemistry, www.rsc.org

(both in amount and in variety), for example drugs, phytopharmaceuticals and metal ions, has increased, which is also reflected in the quality of water.[1–3] Synthetic substances and non-degradable materials pose serious environmental consequences to water systems and directly or indirectly affect the health of fauna and flora.[1,4] Moreover, the complexity of these contaminants in terms of composition, quantity, stability, and permanence undermines the efficiency of conventional water treatment processes in removing them.[5,6] These same features also make it difficult to identify these contaminants and, therefore, to monitor water quality.[7]

To address this situation and minimize the problems associated with water pollution, new and innovative technologies have been investigated over the years. The most prominent technologies for water treatment can be grouped into methods focused on phase-changing processes, advanced oxidation, and biological processes.[8,9] Among the phase-changing processes, adsorption is the most investigated, likely due to its attractive features, which include simple operation and cost-effectiveness, among others.[10] Adsorption is a surface process in which a solute dispersed in a fluid phase (gas or liquid) migrates to a solid surface. This process can occur driven by physical or chemical forces and its efficiency depends on several parameters, including the chemical nature of the solute (or adsorbate) and the solid (the adsorbent).[11]

A brief review of the literature reveals that a wide range of compounds and materials have been explored and assessed as adsorbents for a myriad of contaminants, including dyes, heavy metals, drugs, hormones, agrochemicals, and other human-made substances.[12–16] In general terms, an ideal adsorbent material should exhibit characteristics such as high surface area, porosity, chemical, thermal and mechanical stabilities, non-toxicity, recycling potential, low cost, and availability.[17,18] Carbonaceous materials partially or fully meet these specifications and are therefore often used in adsorption processes. Although activated carbon is the most widely used carbonaceous material as an adsorbent, due to its high removal efficiencies and maximum adsorption quantities of a vast range of pollutants, this material has two major limitations: its high production cost and low regeneration capacity.[19] Although several strategies have been developed to minimize these drawbacks, including the synthesis of composites,[20,21] other carbon-derived materials have gained attention in recent years for the removal of contaminants from water.[22,23] Among these, CQDs have emerged as a prominent option due to their nanometric size and unique structural and physicochemical characteristics.[24] Additionally, CQDs exhibit attractive photoluminescence (PL) properties, which are influenced by their interactions with other species (ions, molecules, *etc.*) which allows a myriad of applications from sensing to biomedical applications.[25] In short, CQDs have the dual ability to remove and detect certain contaminants from water, making them a promising tool to address the aforementioned challenges.

The straightforward use of CQDs in practical water treatment or monitoring is restricted due to their solubility in water, making their recovery and recycling difficult. Among the strategies to overcome this limitation, those focused on the immobilization of CQDs on support materials,[26] or utilizing CQDs as building blocks, have been developed although they are still at an early stage.[27] In fact, some authors have demonstrated that supramolecular structures based on or containing CQDs exhibit enhanced and stable photophysical properties compared to their free particles.[28] Therefore, the formation of supramolecular structures containing CQDs has been a hot topic in this research field, broadening the range of applicability and efficiency of these remarkable nanomaterials.

This chapter sheds light on the main contributions and advances in the design, development, and application of supramolecular structures containing CQDs for the removal and detection of organic and inorganic contaminants in water. Before discussing these studies in detail, a brief description of the general concepts about CQDs and supramolecular structures is provided in the following sections.

10.2 Overview of Carbon Quantum Dots (CQDs)

Carbon quantum dots (CQDs) are zero-dimensional (0D) carbon materials of small sizes (<10 nm) and are employed particularly in studies where PL is the target property. Typically, CQDs consist of a carbon core with sp^2/sp^3 hybridization (which can exhibit either amorphous or crystalline structures) with oxygen-containing functional groups on its surface.[29,30] Since their discovery in 2004, these nanomaterials have aroused significant interest due to their unique chemical and physical properties compared to conventional fluorescent materials, such as remarkable photostability, tunable PL, and adjustable bandgap.[31] Furthermore, they offer advantages such as low toxicity and excellent biocompatibility, and they can also be synthesized utilizing different methods from cheap and abundant raw materials.[29]

10.2.1 Techniques for the Synthesis of CQDs

The synthesis routes of CQDs play a crucial role in defining their structure and, consequently, their final properties. Although the synthesis of CQDs is relatively straightforward and well-known (especially compared to those for obtaining other carbon-based nanostructures), it still faces some challenges, such as the aggregation of nanoparticles during carbonization and particle size distribution. Since the variation in size, morphology, and type or amount of doping will generate changes in the final properties of these materials,[32] the synthesis method plays a significant role in the entire process to guarantee homogeneity in size and structure.

Various techniques have been employed to synthesize CQDs, which can be categorized into top-down and bottom-up approaches.[33,34] In top-down techniques, CQDs are produced by breaking down or cleaving larger carbon structures into smaller nanoparticles, as is the case when using carbon precursors such as graphite,[35] carbon nanotubes (CNTs),[36] activated carbon[37] or graphene oxide.[38] Some standard top-down techniques used in the synthesis of CQDs in recent years include laser ablation,[39] arc-discharge method,[40] electrochemical oxidation[41] and ultrasonic synthesis.[42] Among the limitations, these methods commonly require numerous steps to decompose carbon materials under harsh conditions, involving strong acids, oxidizing agents and relatively high temperatures. In addition, there is significant heterogeneity in the size distribution and morphology of the CQDs thus obtained.[33]

In a bottom-up way, CQDs are obtained through chemical routes using organic molecular precursors such as citric acid,[43] glucose,[44] and sucrose.[45] The most common bottom-up approaches, used to synthesize CQDs, include microwave,[46,47] hydrothermal[48,49] and pyrolysis.[50] In these methods, the structure and size of CQDs are determined by several factors, for example, the molecular structure of the carbon raw material, the solvent used and the reaction parameters; the latter includes time, temperature, and pressure. These parameters are essential because they significantly influence CQD reactants and the nucleation procedure.[51] Furthermore, bottom-up methods are usually simpler than top-down approaches, allowing for a wider range of raw materials.[30,34] They also enable efficient doping with one or multiple heteroatoms inherited from organic precursors during synthesis[52] and provide a relatively high quantum yield (QY).

In the following sections, various bottom-up and top-down techniques for CQDs will be discussed in greater detail, highlighting their key characteristics.

10.2.2 Top-down Approaches

10.2.2.1 Arc-discharge

The arc-discharge technique is based on the reorganization of carbon atoms, decomposed from bulk carbon precursors in the anode, driven by gas plasma in a sealed reactor. The reactor temperature can reach 4000 K due to the applied current, creating a high-energy plasma. At the cathode, carbon vapor assembles to form CQDs.[53] This technique was used in the first report of the development of CQDs. In 2004, Xu *et al.* synthesized CNTs using the arc-discharge method, and during the separation and purification of these nanomaterials through electrophoresis, they accidentally discovered carbon nanoparticles with fluorescence properties.[54] CQDs produced by this method have good water solubility. However, they typically exhibit a wide particle size distribution due to the formation of carbon particles of varied sizes during the discharge process.[53]

10.2.2.2 Laser Ablation

Laser ablation synthesis is a widely used method for preparing CQDs with a uniform size distribution. This technique uses laser irradiation to break larger carbon source particles, suspended or immersed in a liquid medium, into smaller particles, producing nanoparticles and CQDs.[51] The laser operation may entail continuous wave or pulse modes. Also, laser ablation offers controllable morphology and size but is expensive and involves complex operational steps compared to other synthesis routes.[51] Sun *et al.* developed, for the first time, the laser ablation method for synthesizing CQDs. They started with graphite powder and used laser excitation from a Nd:YAG source (1064 nm, 10 Hz) to generate CQDs within an argon atmosphere at 900 °C and 75 kPa.[35]

10.2.2.3 Electrochemical Oxidation

Electrochemical oxidation is a popular method for producing CQDs, offering benefits such as high purity, low cost, high yield, easy size control, and good reproducibility.[55] In this technique, carbon precursors with larger structures are fragmented into smaller pieces through electrochemical oxidation using powerful oxidizing agents such as H_2O_2, H_2SO_4, and HNO_3.[56] Liu *et al.* were the first to create fluorescent CQDs through chemical oxidation, achieving QYs of 0.8–1.9% and diameters under 2 nm.[57] Recognizing the impact of the microstructure of CQDs on their optical properties, Tan *et al.* produced different CQD microstructures aimed at bioimaging. They achieved this by oxidizing graphitized activated carbon using oxidizing agents such as $HClO_4$ and HNO_3 to produce CQDs with tunable fluorescence.[58]

10.2.3 Bottom-up Approaches

10.2.3.1 Microwave Method

The microwave methods employ microwave (MW) radiation to rapidly heat the carbon source, inducing carbonization and pyrolysis reactions in the raw material to produce CQDs.[33] This technique offers several advantages over previously described methods, including a short reaction time, high yield, low energy consumption, cost-effectiveness, easy scalability, and the absence of organic solvents.[59] The first-time synthesis of QDs by MW-assisted pyrolysis proved to be an economical method, with excellent advantages in terms of material properties, such as bright, stable luminescence and excellent dispersion in water in just a few minutes.[60] In this synthesis, varying amounts of polyethylene glycol (PEG-200) and saccharide are mixed in distilled water to create a homogeneous solution. The solution was then placed into a MW oven, at 500 W for 2 to 10 min, and as the reaction progresses, the solution turns from colorless to yellow and eventually to dark brown, indicating the formation of CQDs.

10.2.3.2 Hydrothermal Method

Hydrothermal synthesis is one of the most used techniques for preparing CQDs. This method is advantageous because it produces particles with a uniform size and high quantum yield in just a single step. Additionally, it is a low-cost, environmentally friendly and easy-to-handle route.[53,61] Normally, a carbon source is dissolved in water or in an organic solvent to form the reaction precursor and then transferred to a Teflon-lined stainless-steel autoclave, where the solution undergoes high-temperature treatment and high pressure.[55,59] In 2010, Zhang *et al.* were pioneers in preparing CQDs using the hydrothermal method. They dissolved ascorbic acid in deionized water, mixed it with ethanol, and then the obtained solution was placed in a Teflon-lined stainless-steel autoclave.[62] The resulting product was extracted and then subjected to dialysis to obtain CQDs. In addition, this technique offers the possibility of easily introducing heteroatoms into CQDs in a controlled manner, which makes it an excellent approach for increasing the quantum yield.[52] More recently, Liang *et al.* prepared silicon-doped carbon quantum dots by the hydrothermal method with a QY of $=0.348$. These results enabled the use of these carbon-based materials for detecting berberine hydrochloride, utilizing its down and upconversion fluorescence.[63]

10.2.3.3 Pyrolysis Method

In the pyrolysis method, organic compounds undergo decomposition at high temperatures in an oxygen-free environment and, in some cases, in the presence of catalysts.[64] Pyrolysis can be conducted in the solid phase, which is generally an effective route to prepare carbon nanoparticles on a large scale without solvents.[65] The effectiveness of this method was demonstrated in 2010 by Pan *et al.*, who synthesized fluorescent CQDs with a high quantum yield (approximately 0.406), prepared by using ethylenediaminetetraacetic acid salts in a one-step pyrolytic route.[66] The method is an economical and scalable technique for preparing CQDs by using several abundant and low-cost raw materials. It is also worth noting that the pyrolysis method enables the encapsulation of CQDs with biocompatible polymers, making them useful for biomedical applications, such as cell imaging and drug delivery.[67]

A comparison of conventional CQD production methods is shown in Table 10.1, detailing the reaction conditions, sizes, QY values, and reported applications in the literature. This comparison highlights the diverse synthesis techniques and their corresponding efficiencies and practical uses.

10.2.4 Optical Properties of CQDs

As previously highlighted, CQDs can play an important role in applications where PL is the target property. In fact, despite the method employed, CQDs consistently exhibit exceptional optical properties. These include a wide spectrum of light emission ranging from ultraviolet (UV) to near-infrared (NIR), adjustable fluorescence, efficient multiphoton up and down conversion, and other characteristics that can be effectively tailored through size, shape, functionalization, surface and edge effects, and doping.[83]

Table 10.1 Production of CQDs from diverse raw materials and techniques for a variety of applications.

Starting material	Technique	Synthesis parameters	Size (nm)	QY (%)	Applications	Ref.
Kiwi peels	Hydrothermal	200 °C for 24 h	5.6	19.0	Detection of Fe^{3+}	68
Citric acid & urea	Hydrothermal	190 °C for 6 h	2.0	35.1	Detection of metronidazole	69
Discarded cigarette butts	Hydrothermal	200 °C for 6 h	3.8	—	Detection of Fe^{3+}	70
Egg whites	Hydrothermal	220 °C for 48 h	2.1	64.0	Detection of metal ions	71
Grass	Hydrothermal	180 °C for 2 h	5.0–10.0[a]	—	Removing organic contaminants	72
Citric acid & sodium sulfide	Hydrothermal	200 °C for 4 h	4.2	21.1	Detection of Hg^{2+}	73
Banana peels	Hydrothermal	200 °C for 24 h	5.0	20.0	Bioimaging	74
Papaya	Pyrolysis	200 °C for 15 min	7.0	23.7	Chromium detection in water	75
Bio-waste lignin	Pyrolysis	300 °C for 2 h	4.0–10.0[a]	8.1	Iron detection & cellular imaging	76
Borassus flabellifer flower	Pyrolysis	300 °C for 2 h	3.0–8.0	14.0	Detection of Fe^{3+}	77
Cotton linter waste	MW-assisted	150 °C for 4 min	10.1	—	Bioimaging	78
Citric acid & urea	MW-assisted	800 W for 5 min	1.9–4.0[a]	13.6	Detection of Bisphenol A	79
Arginine & EDA	MW-assisted	240 °C for 3 min	5.0–13.0[a]	0.15	Detection of Bisphenol A	80
Potato starch	Acid-assisted ultrasound route	400 W for 6 h	3.0–5.0[a]	0.10	Detection of Zn^{2+}	81
Fuel waste	Chemical oxidation	100 °C for 12 h	5.0	3.0	Detection of picric acid and metal ions	82

[a] For CQDs with size range.

Typically, CQDs exhibit intense absorption in the UV range (250–350 nm) and relatively small absorption in the visible spectrum, because of the conjugated π-electrons within the sp^2 hybridized atomic orbital and the chemical groups attached to their edges.[84] A typical absorption spectrum of CQDs is characterized by a peak around 230 nm and a shoulder at *ca.* 300 nm. The former is justified by the π–π* transition of aromatic C=C bonds, while the latter is associated with the n–π* transition of C=O bonds or other attached groups.[85,86] The absorption properties of CQDs can be modified by doping/co-doping with heteroatoms, surface modification or passivation.[87]

Importantly, the maximum absorption wavelength shows a red shift as the particle size increases, indicating an intricate relationship between these two properties.[88]

From the point of view of fundamental research and practical application, PL stands out as a primary characteristic of CQDs.[84,89] This property arises from the absorption of photons by these materials, resulting in the emission of electromagnetic radiation, and, in general, the photoluminescence of CQDs is unique, influenced by both emission wavelength and intensity.[53,90] Nevertheless, the vast array of precursors and production methods used for CQDs results in different sizes, chemical structures, and colors, affecting their optical properties. This complexity indicates that CQDs operate *via* a complex PL mechanism, an area that continues to be actively researched.[91] In recent years, numerous studies have investigated the luminescence mechanisms of CQDs, with prominent theories including surface states,[92,93] quantum confinement effects,[94] and molecular fluorescence,[95,96] with direct impact in sensing, bioimaging, solar devices and energy storage and photocatalysis.[97–100]

Another attractive property of CQDs is upconversion photoluminescence (UCPL), a phenomenon characterized by the emission of high-energy photons due to the absorption of two or more low-energy photons, a non-linear anti-Stokes process.[101] In a groundbreaking 2007 study, Cao *et al.* unveiled that carbon dots exhibit strong visible light emission when excited by two photons, marking the first observation of the up-conversion effect of carbon dots.[102] This discovery paved the way for a multitude of studies exploring the UCPL of CQDs across various fields, as these unique optical properties present a wealth of innovative applications.[103,104]

10.3 Supramolecular Structures – General Concepts

Supramolecular chemistry is, according to J.-M. Lehn (Nobel prize winner in 1987) the chemistry beyond the molecules, which highlights the role of nonbonded interactions in classical organic chemistry.[105] In other words, supramolecular chemistry aims to develop chemical systems, formed by molecules or macromolecules. Overall, these structures result from processes such as inter- and intra-molecular self-assembling, molecular folding, molecular recognition, host–guest complexes, and mechanically interlocked molecular architectures.[106] The formation and spatial organization of these structures depend on the intensity and range of intermolecular forces, including electrostatic interactions, hydrogen bonding, dipole–dipole and ion–dipole interactions, van der Waals forces, π–π interactions, hydrophobic effects, and metal–ligand coordination.[107] In addition to their non-covalent nature, and consequently with interaction energies $<4\ \mathrm{kJ\,mol^{-1}}$ (*e.g.*, hydrophobic effects),[108] supramolecular structures are characterized by reversibility and high thermodynamic stability,[109] with equilibrium constants that, in the case of host–guest structures, can reach values higher than 1×10^6.[110,111]

In terms of dimensions, supramolecular structures commonly remain below the micrometric scale, and in most cases, they are nanometric.

Despite being a cutting-edge research field, the formation of these structures is greatly inspired by nature, which is responsible for notable examples like the DNA structure, enzyme–substrate complexes and protein–protein interactions. By analyzing these examples, it is evident that beyond the interactions, the molecular rearrangement and the formation mechanism are pivotal in defining the stability and functionality of these structures.[112,113] Among supramolecular structures, host–guest complexes involving, in general, an amphiphilic compound (*e.g.*, cyclodextrins or porphyrins – see Figure 10.1), allow control of the properties of guest molecules, including solubility, lability, volatility and sublimation.[114] Considering all these aspects, researchers have engineered intricate and complex structures that have found applications in many areas such as catalysis, electronics, energy and information storage devices, light-harvesting systems, pharmaceutical and biomedical and separation processes.[115]

Focusing on supramolecular structures containing CQDs or made from them, the most common experimental processes to obtain these materials rely on the formation of host–guest complexes[24] or self-assembly processes.[116] In the first case, host molecules with highly functionalized surfaces, often possessing a cavity, are utilized. Among such molecules,

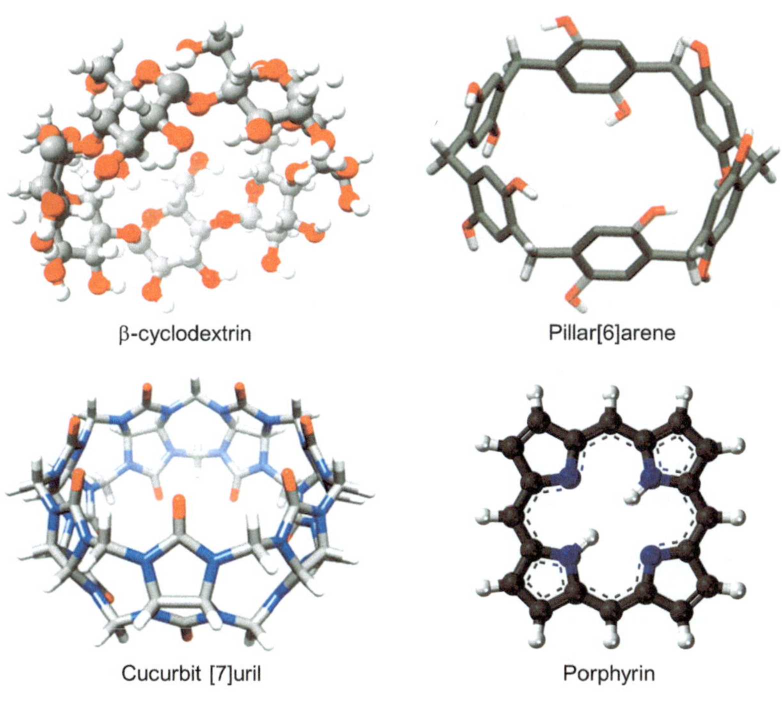

Figure 10.1 General chemical structures of β-cyclodextrin, pillar[6]arene, cucurbit [7]uril and porphyrin.

cyclodextrins (α, β, or γ), cucurbit[*n*]urils, porphyrins and pillar[*n*]arenes are preferred (Figure 10.1).[26,117] Due to their well-known properties and availability, cyclodextrins are one of the most widely used guests for the formation of supramolecular structures. Cyclodextrins are cyclic oligosaccharides consisting of α(1,4)-linked glucopyranose subunits, with a truncated cone (or torus) shape; hydroxyl groups located at the narrower and wider edges of the cone confer a hydrophilic character to this molecule; however, the cavity is lined by glucosidic oxygen bridges, giving the cavity a hydrophobic character, with β-cyclodextrin being one of the most used types. Cucurbit[*n*]urils, pillar[*n*]arenes and porphyrins are families of macrocyclic compounds consisting of *n* glycolurils, *n* hydroquinone and four pyrrole-type units, respectively, bridged by methylene groups.[118–120]

Beyond their characteristic cyclic nature that forms hosting cavities, all the above mentioned compounds can be chemically functionalized, providing an additional advantage in forming stable supramolecular structures with CQDs.[121] This functionalization enhances their ability to interact with CQDs and other molecules, leading to more robust and versatile supramolecular assemblies.[117]

As an alternative to the host–guest approach, CQDs can form self-assembled supramolecular structures in bulk solutions. This process can occur by altering the chemical nature of CQDs using specific precursor compounds and certain experimental conditions, such as concentration. As a result, nanoaggregates can form with different aggregation states due to the inconsistent surface-confined charges present on these structures. This can be exploited to obtain CQD nanoaggregates with different and modulable photophysical properties.[28] Another approach involves functionalizing the surface of CQDs with substituents and macromolecules that can induce self-assembly, typically a spontaneous and entropy-driven process. However, functionalizing CQDs is not a simple task, as noted by some authors,[27] particularly due to the lack of atomic precision in these nanomaterials. Additionally, the rather small dimensions of CQDs hinder asymmetric functionalization, such as with Janus particles.[27]

The grafting of polymers, oligomers, and fatty acids onto CQDs is a recurrent strategy to obtain hairy nanoparticles with amphiphilic properties, allowing their assembly in bulk solutions.[122,123] For instance, Xu *et al.* conducted the functionalization of CQDs with oligoxyethylene groups, resulting in nanoparticles with both non-ionic and ionic amphiphilic natures able to form spontaneously supramolecular structures in water. By controlling certain experimental parameters, the authors demonstrated the ability of these functionalized CQDs to form nanoclusters, nanorods, nanotubes, and vesicles.[27]

Although the use of supramolecular complexes containing CQDs is still in its early days, several studies have been published in previous decades. In what follows, we highlight the use of CQD-based supramolecular structures for different applications. For instance, the development of chiral complexes based on achiral porphyrins and chiral CQDs. The obtained chiral supramolecular composites proved to be able to recognize and discriminate electrochemically the phenylalanine enantiomers.[124] Chirality is a particularly important property in different areas of chemistry and life sciences.

Biomedicine is another area where the application of CQD-containing complexes is highly promising. Due to the relevance of the area, we would like to mention two different studies; in the first one, β-cyclodextrin nanosponges[125] were studied for theranostic application. As will be discussed later, cyclodextrins can interact with CQDs essentially through hydrogen bonds, which leads to the stabilization of the latter with a consequent increase in its PL. The use of nanosponges containing carboxyl-functionalized nitrogen-doped CQDs and loaded with doxorubicin (DOX, a commonly used anticancer drug) has shown potential for application in tumor imaging (due to their strong fluorescence emission at excited wavelengths equal to 365 nm) as well as for the release of DOX, showing a IC_{50} value for HepC2 cells equal to 5.00 $\mu g\, mL^{-1}$.[126] Another example is the synthesis of Se-CQD-based nanoparticles (NPs), by hydrothermal treatment of selenocystine, followed by self-assembly with DSPE-PEG200 and lecithin. The resulting nanoparticles have the advantage of small retention in the liver, but are tremendously effective in ROS scavenging. These two concomitant advantages allow us to improve the post-surgery process in patients with hepatic ischemia-reperfusion injury.[127]

The sensing and catalysis fields are, most probably, the chemistry fields where the development of CQD-containing supramolecular structures is more advanced. Concerning sensing, several approaches have been reported. One of them describes the formation of supramolecular complexes between an anionic porphyrin, the *meso*-tetrakis(4-sulfonatophenyl) porphyrin, and arginine-based carbon nanodots.[128] The supramolecular interaction between these species leads to a decrease in the absorbance of the porphyrin in the presence of CQDs; this phenomenon, electrostatically driven, might be used for sensing other anionic chromophores of biological relevance. The use of these carbon dots in other biologically relevant molecules is suggested, based on the replacement of the porphyrin by biological molecules of interest with optical properties. A different strategy is the encapsulation of CQDs into vesicles, which are further loaded into hydrogel membranes. This not-straightforward process makes it possible to use the fluorescence properties of CQDs, using a solid probe, to, for example, perform semi-quantitative detection of Cu(II) in aqueous solution.[129] Another example involves a different host molecule, pillar[6]arene. This is used to obtain supramolecular adducts with N-doped CQDs for the electrochemical detection of trinitrotoluene.[130] The results were compared with *b*-cyclodextrin-based adducts, showing that the use of pillar[6]arene leads to a decrease in the detection limit of TNT to 0.95 $nmol\, L^{-1}$.

A significant percentage of work related to catalysis is mainly focused on photocatalysis, where the luminescence properties of CQDs are exploited. By way of example, we can describe their use in the catalytic reduction of CO_2; a key point in the reduction of the carbon footprint.[131] Transition metal complexes, such as those containing ruthenium, have been used as catalysts for the reduction of CO_2, taking advantage of their photophysical properties and the ability to modify the surrounding media by changing the ligand. However, these processes exhibit some drawbacks, for example, an exceptionally low F. Thus, authors have used CQDs as an antenna to overcome this issue, *i.e.*, the energy absorbed by the CQD is transferred to Ru increasing the photoluminescence[132] and, consequently, improving the efficiency of the CO_2 reduction.

For a more detailed and in-depth discussion about the particularities of these systems regarding their preparation and characterization, the reader is referred to excellent review papers.[24,107] These reviews provide comprehensive insights into the methodologies and analytical techniques employed in the study of supramolecular structures containing CQDs, highlighting their advancements and potential applications.

Considering these approaches and reported results, the formation of supramolecular structures incorporating CQDs marks a significant advancement in nanomaterials, enhancing their properties and applicability for water treatment and environmental monitoring. The versatility of host–guest complexes and self-assembly processes optimizes CQDs' functionality for practical use. The referenced studies and reviews offer a solid foundation for further exploration, paving the way for applications that could mitigate water pollution and bolster environmental protection efforts. The next sections compile the main studies concerning the use of supramolecular structures containing CQDs in the treatment and monitoring of contaminated water.

10.4 Applications in the Removal and Detection of Water Contaminants

As highlighted in the previous sections, the attractive physical and chemical properties observed for CQDs, and the supramolecular structures formed with them, have significantly boosted their applications across various domains. In the field of environmental management, these materials have been particularly exploited for the removal and detection of a wide range of contaminants from water. The ability of CQDs to form stable, functional supramolecular structures enhances their efficiency and versatility, making them valuable tools in addressing water pollution and improving water quality monitoring.[133,134]

Focusing on this research field, CQDs have been largely studied as adsorbent materials for various water contaminants such as metals,[135] dyes,[136] pharmaceuticals,[137] among others. In this case, the modification of CQDs by functionalizing them with specific functional groups[138] or doping with metal oxides[139] is often carried out to improve their adsorption performance or to impart some feature (photocatalytic ability, for instance). The entrapment or association of CQDs with other materials[140,141] is a common strategy adopted by researchers. Herein, the adsorption performance of these composite/hybrid materials benefits from the synergistic action of CQDs and the supporting material, which commonly exhibits adsorption sites.[142]

More recently, supramolecular structures containing CQDs have emerged as alternative adsorbent materials, with the added benefit of functioning as probes to detect contaminants in water. This dual functionality enhances their applicability in environmental management, offering a promising approach to both remove and monitor pollutants simultaneously. This aspect has gained significant relevance, as many studies demonstrate a direct correlation between exposure to various water contaminants and the development of diseases.[143,144] To assess the local risk of diseases caused by water contaminants, a risk assessment that considers the bioavailability and dosage of these contaminants in water is necessary.

In response to environmental contamination, efficient methods to detect water contaminants are imperative. CQDs offer promising applications in many fields, particularly in fluorescence imaging and the determination of a wide range of contaminants, including complex organic structures. Organic contaminants are particularly challenging to determine and monitor because traditional analytical methods, such as capillary electrophoresis, ion chromatography, mass spectroscopy, electrochemical methods, and liquid chromatography (HPLC), are limited by the high cost of reagents and skilled personnel, and by the potential generation of secondary pollution.[145] Compared to traditional methods, fluorescence detection methods based on CQDs exhibit superior characteristics, including high sensitivity, low toxicity, biocompatibility, robust photostability, easy preparation, low cost, easy surface functionalization, and good dispersibility in water.[146] To provide a perspective on this performance, the following subsections compile recent studies demonstrating the use of supramolecular structures containing CQDs with the dual ability to remove and detect organic and inorganic contaminants in aqueous media.

10.4.1 Organic Contaminants

Among the organic contaminants often detected in water sources, synthetic drugs are particularly problematic due to their chemical characteristics, persistence, bioaccumulation, and the risks posed by their metabolites.[147] The presence of these contaminants in water has become increasingly noticeable due to human habits and health crises, such as the COVID-19 pandemic.[148] To aggravate this scenario, the removal and detection of drugs from water is not a simple task as reported in many studies.[149,150]

Regarding the removal of drug contaminants from wastewater, different strategies have been considered to improve the conventional treatment processes, such as electrocoagulation, ion exchange, and ultrafiltration, which are associated with high-cost techniques for large-scale applications.[151] An alternative solution often mentioned in the literature is the association of treatment methods with adsorption processes mediated by advanced materials. In this sense, supramolecular structures containing CQDs are highly promising. They generally exhibit a high surface area and an abundance of sites for drug adsorption, coupled with the additional advantage of functioning as fluorescent probes for detection. All these aspects were investigated by Yang *et al.* who described the successful design and fabrication of a 3D smart adsorbent based on a supramolecular Zr-MOF with co-doped carbon dots.[152] Through a simple solvothermal method, the functional smart adsorbent was systematically characterized and applied to the selective recognition and removal of tetracycline (TC). The material exhibited excellent recognition performance with a wide sensitivity range (0.08–20.0 $mg\,L^{-1}$) and a low detection limit (0.063 $mg\,L^{-1}$) for TC. As explained, the MOF and CQD supramolecular structure alleviates the diffusion barrier and enhances the mass transfer between the adsorption sites and TC molecules, endowing it with a capacity that can improve the TC adsorption capacity of the parent material to 427.35 $mg\,g^{-1}$. Additionally, the interaction of the TC molecules promoted a charge transfer between the multistructural adsorbent, quenching its fluorescence, thereby allowing drug detection. The data from this research indicate the enormous potential of the smart, bifunctional, and harmless adsorbent for the effective detection and removal of TC residues. Another strategy involves the self-aggregation of melamine–cyanuric acid supermolecules, to form porous carbon nitride, using CQDs as precursors; CQDs play a significant role in the regulation of the microstructure of the material through H-bonds.[153] The obtained supramolecular structures were tested as photocatalysts for the degradation of different drugs, including diclofenac. It has been found that CQD-containing porous carbon nitride exhibits a high performance for diclofenac degradation (for an initial diclofenac concentration equal to 8 ppm, a degradation higher than 90% was obtained after 150 min) and reusability. Meng *et al.* conducted a similar approach where they introduced iron ions into CQD-containing carbon nitride supramolecular structures to allow the occurrence of a photo-Fenton reaction, and were able to degrade 76% of ciprofloxacin in less than 2 min.[154,155]

Another drug with a significant impact on the environment and water is naproxen – an effective anti-inflammatory drug used in the treatment of arthritis and rheumatic diseases. Fang *et al.* addressed the presence of this emerging anthropogenic contaminant in water by doping perylene diimide fibers with carbon quantum dots.[156] The obtained supramolecular fibers allow the photocatalytic degradation of naproxen by forming species such as $^1O_2$, H_2O_2 and $^\bullet OH$. This approach led to a naproxen half-life of just 36 seconds; further application of these doped fibers into cellulose acetate membranes was successfully attempted, by the same authors, increasing the range of applications of these fibers.

Similarly to drugs, the presence of volatile organic compounds (VOCs) poses a substantial risk to humans and the environment.[157] In simple terms, VOCs are a group of organic compounds that contain carbon in their molecular structure, which generally has low molecular weight. These compounds have chemical properties that allow them to easily vaporize into steam or gas form. Once released into the environment, VOCs can persist for extended periods of time, affecting various ecosystems, including water sources.[158] Due to this impact, the development of materials for the removal and detection of these compounds becomes of paramount importance. Silva *et al.* (2024) synthesized and characterized a supramolecular composite comprising CQDs incorporated into poly[(α-cyclodextrin)-*g*-(citric acid)], for the removal and detection of toluene and xylene from aqueous media.[159] The composite exhibited a fluorescence quenching mechanism upon adsorption of both organic compounds. Stern–Volmer constants were utilized to analyze the interaction dynamics, revealing Stern–Volmer extinction constants of 15.4 M^{-1} for toluene and 10.3 M^{-1} for xylene. Consequently, the α-CD/CA/CDot composite demonstrated sensitivity to these volatile organic compounds, with limit of detection (LOD) of 3.7 $mg\,L^{-1}$ for toluene and 4.9 $mg\,L^{-1}$ for xylene. Optimal adsorption conditions were determined, leading to impressive adsorption capacities of 68.9 $mg\,g^{-1}$ for toluene and 48.2 $mg\,g^{-1}$ for xylene. Moreover, the composite achieved removal efficiencies exceeding 70% for both contaminants.

Another class of organic compounds often associated with surface and groundwater contamination is agrochemicals (pesticides, herbicides, *etc.*), which have seen increased use to meet food demands amid population growth. Monitoring and removing these contaminants from water is a high priority due to their deleterious effects on the health of fauna and flora. To address this issue, Yi *et al.* prepared aerogels of CQDs coupled with acetylcholinesterase and choline oxidase through a self-assembling process mediated by copper ions.[160] This composition endowed the aerogels with peroxidase-mimicking activity capable of generating reactive oxygen species (ROS), which promoted rapid Cu^{2+}/Cu^{+} conversion and provided catalytic active sites within a 3D porous structure. This supramolecular architecture demonstrated an impressive ability to detect dichlorvos, an organophosphate insecticide widely used on crops, animals, and in pest strips. The aerogel exhibited an exceptionally low LOD of 7.6 $nmol\,L^{-1}$ within a linear concentration range of 0.02 to 0.30 $\mu mol\,L^{-1}$. Furthermore, the authors demonstrated the efficiency of this aerogel in detecting the insecticide in real samples, including lake water, highlighting the potential of this aerogel in practical uses.

A common aspect in all these systems, independently of the contaminant, is that their detection mechanism is always based on quenching of the PL properties of the CQDs.[97] Typically, after the interaction of the contaminant with the supramolecular structure, *via* an adsorptive process mainly, there is an electron transfer between the adsorbed species or intermediate compounds and the CQDs.[97] This process is favored as electrons transition from excited states in CQDs to the lowest unoccupied molecular orbital of the complex, rather than returning to the ground state, thus quenching fluorescence.[161] This response is readily and conveniently detected by optical or spectroscopic devices, assuring the simplicity of this approach. Indeed, the quenching of the CQDs' fluorescence due to their interaction with these contaminants can be noticed by the naked eye.[159]

10.4.2 Inorganic Contaminants

Waste from various anthropogenic activities often contains high levels of inorganic contaminants, notably toxic metals, which are highly mobile and harmful to living beings, thereby damaging the water supply.[162] Commonly reported metallic species include arsenic, lead, mercury, copper, cadmium, chromium, nickel, silver and magnesium.[163] Additionally, some studies mention the presence of radioactive elements such as uranium, europium, and lanthanum in water sources.[164,165] These contaminants can originate from dyes, fertilizers, pesticides, bleaching agents, certain metal alloys, fossil fuels, and technological waste.

This situation has prompted stricter legislation in some countries regarding the permissible limits of these metals in wastewater, though such regulations are not globally uniform. The detection and quantification of these metals for environmental remediation are becoming more prevalent. Currently, several processes are employed for heavy metal removal, including chemical oxidation, nanofiltration, reverse osmosis, and adsorption.[166] Among these, adsorption is the most viable method as it does not produce byproducts and typically offers high reusability and efficiency under profitable operating conditions. Various compounds and materials, including carbonaceous materials like CQDs, have been assessed for this purpose.[167]

Beyond removing these contaminants, monitoring their presence in water sources is crucial.[168] Traditionally, quantifying metal ions is time-consuming and involves transporting samples to laboratories equipped for high-cost techniques like inductively coupled plasma optical emission spectrometry (ICP-OES) and flame atomic absorption spectrometry (FAAS), which require trained personnel. Consequently, developing methods and probes for real-time detection of metal analytes is essential to reduce costs and labor. The PL properties of CQDs can be used for this task since the interaction between metal species and CQDs can quench their fluorescence.

Indeed, fluorescence quenching is a common strategy to detect and quantify metal species in water because CQDs are extremely sensitive to their surroundings and suppress their fluorescence in the presence of such contaminants. To date, the most common use of CQDs for metal detection involves their use in solid materials,[169,170] though there are few examples in the literature describing the development and application of supramolecular structures with CQDs for this purpose. One such example is the work of Silva *et al.* (2024), which reported the design and use of poly(β-cyclodextrin-*co*-citric acid)/CQD supramolecular composites for the simultaneous adsorption and detection of Ni(II), Cu(II), Cd(II), and Pb(II) ions in an aqueous medium.[26] The composites exhibited high adsorption capacities, reaching 220 mg g^{-1}, and removal efficiencies exceeding 90% for Ni(II) and Cu(II).

The detection process directly depends on the interaction between the contaminant and the composite matrix, indicated by luminescence quenching upon metal ion adsorption, with limit of detection (LOD) values below 6.4 ppm for all investigated metals. These authors also found that CQDs are present in the probe matrix in two different forms as seen by lifetime measurements: CQDs linked by supramolecular interactions to cyclodextrins and CQDs stuck into a polycyclodextrin structure. Using a similar approach, Yuan *et al.* prepared a supramolecular structure based on nitrogen-doped CQDs held on a cucurbit[6]uril structure (host–guest process), which were used as a stable fluorescent probe to detect some metal cations (Fe(III), Ba(II), and Al(III)) and anions (I^- and ClO^-) in an aqueous medium. The detection limits were notably lower, especially for Fe(III) (3.89 $\mu mol\,L^{-1}$) and I^- (6.84 $\mu mol\,L^{-1}$).[171] In a similar work, Xiong *et al.* used tetramethyl-substituted cucurbit[6]uril as a source of carbon and 2,7-dihydroxynaphthalene to form luminescent materials capable of sensing metal ions and anions. The probe proved to be selective for Fe(III), ClO^- and I^-, with detection limits of 0.41 $\mu mol\,L^{-1}$, 0.10 $\mu mol\,L^{-1}$, and 3.71 $\mu mol\,L^{-1}$, respectively.[172]

Zhang *et al.* functionalized CQDs with 3,4-hydroxypyridinone (HOPO) to monitor the presence of uranyl ions in water. During the nuclear fuel cycle, uranium wastewater can be released into the environment, posing threats to humans and other creatures. Uranyl ion (UO_2^{2+}) is the most usual form of uranium with good water solubility. Experimental assays demonstrated that the developed probe exhibited excellent sensitivity (LOD of 6.53 ppb) and selectivity towards uranyl ions with an extremely short response time (30 seconds).[173] The authors attribute this performance to the strong and specific interaction between the HOPO ligand and the uranyl ion, leading to a unique uranyl-induced aggregation of CQDs (*i.e.*, their self-assembling). This phenomenon is highly selective to uranyl ions and is beneficial as it simplifies the separation of the fluorescent probe.

10.5 Concluding Remarks

The use of CQDs in various structures, ranging from micelles to nanofibers, has been explored for numerous applications, including catalysis, sensing, and detection. Although noteworthy progress has been made, much work remains to be done. For instance, combining CQDs with compounds that have amphiphilic characteristics, such as cyclodextrins, enables the synthesis of materials with dual or multifunctional properties. Another future perspective is incorporating CQDs into membranes (*e.g.*, cellulose acetate or polyaniline) or monoliths, such as silica-based aerogels. These approaches could enhance the efficiency and effectiveness of these materials, improving detection and quantification limits and the removal or degradation efficiency of pollutants in water or wastewater treatments.

As demonstrated in this chapter, the attractive properties of CQDs, combined with supramolecular structures, create powerful tools for researchers across various fields, especially those focused on the remediation and monitoring of water bodies. The PL properties of these composite/hybrid materials are particularly effective for detecting and quantifying contaminants of both inorganic and organic nature. Despite this potential, research involving these supramolecular structures in such applications remains limited to fewer than a dozen studies. We anticipate this will change soon with the development of more scalable processes for synthesizing CQDs and creating more sophisticated devices, such as sensors, utilizing these supramolecular structures. Furthermore, CQD supramolecular structures show promise for other applications, such as mediating catalytic and photocatalytic reactions in advanced oxidation processes (AOPs).[153] Unlike adsorption, which only provides the physical removal of contaminants from water, AOPs can effectively degrade organic molecules in water to complete mineralization.[174]

Looking ahead, it is essential that future research involving CQD supramolecular structures closely replicates real-world conditions. Researchers should conduct experiments (adsorption and detection studies) using flow conditions instead of batch processes and real water samples (as opposed to distilled/deionized water) and assess the ability of these structures to provide real-time monitoring of water contaminants. Moreover, the stability of these supramolecular structures must be continuously verified, considering the uncertainties regarding the safety of CQDs toward human health and other living organisms.[175] In addition to practical experimentation, theoretical studies can offer valuable insights into the interaction mechanisms between these supramolecular structures and various contaminants, as well as the detection mechanisms *via* PL quenching. In parallel, machine learning can play a crucial role in optimizing the application of these dual-function devices.

In summary, supramolecular structures containing CQDs represent a promising frontier in the simultaneous removal and detection of contaminants from water bodies. Continued research and development in this area will undoubtedly lead to more efficient, scalable, and practical solutions for water remediation and monitoring.

References

1. R. K. Mishra, S. S. Mentha, Y. Misra and N. Dwivedi, *Water-Energy Nexus*, 2023, **6**, 74.
2. K. Samal, S. Mahapatra and M. A. Hibzur, *Energy Nexus*, 2022, **6**, 100076.
3. J. P. Vareda, A. J. M. Valente and L. Durães, *J. Environ. Manage.*, 2019, **246**, 101.
4. K. Ziani, B. C. Ioniță-Mîndrican, M. Mititelu, M. S. Neacșu, C. Negrei, E. Moroșan, D. Drăgănescu and O. T. Preda, *Nutrients*, 2023, **15**, 617.
5. B. Belete, B. Desye, A. Ambelu and C. Yenew, *Environ. Health Insights*, 2023, **17**, 11786302231195158.
6. A. C. Duarte, S. Rodrigues, A. Afonso, A. Nogueira and P. Coutinho, *Pharmaceuticals*, 2022, **15**, 393.

7. S. Kirschke, T. Avellán, I. Bärlund, J. J. Bogardi, L. Carvalho, D. Chapman, C. W. S. Dickens, K. Irvine, S. B. Lee, T. Mehner and S. Warner, *Environ. Monit. Assess.*, 2020, **192**, 298.
8. H. Hong, Y. Zhao, L. Huang, D. Zhong and D. Shi, *Ecotoxicol. Environ. Saf.*, 2021, **223**, 112605.
9. K. E. O'Shea and D. D. Dionysion, *J. Phys. Chem. Lett.*, 2012, **3**, 2112.
10. N. B. Singh, G. Nagpal, S. Agrawal and Rachna, *Environ. Technol. Innovation*, 2018, **11**, 187.
11. S. Wonorahardio, F. Fajaroh, R. Joharmawan, N. Nazriati and E. Budiasih, *Sci. Rep.*, 2023, **13**, 4213.
12. C. F. De Azevedo, F. M. Machado, N. F. de Souza, L. L. Silveira, E. C. Lima, R. Andreazza and C. P. Bergamnn, *Chem. Eng. J.*, 2023, **454**, 140102.
13. O. Fraiha, N. Hadoudi, N. Zaki, A. Salhi, H. Amhamdi, E. H. Akichouh, F. Mourabit and M. Ahari, *Desalin. Water Treat.*, 2024, **317**, 100114.
14. A. Miron, I. E. Neblea, S. V. Dolana, A. Zaharia, A. M. Gavrila, A. J. M. Valente, T. V. Iordache and A. L. Chiriac, *Proceedings*, 2023, **90**, 6.
15. A. G. B. Pereira, F. H. A. Rodrigues, A. T. Paulino, A. F. Martins and A. R. Fajardo, *J. Cleaner Prod.*, 2021, **284**, 124703.
16. H. S. Rangappa, I. Herath, C. Li and S. Ch, *Environ. Pollut.*, 2024, **343**, 123140.
17. O. O. Ayeleru, H. U. Modekwe, O. R. Onisuru, C. R. Ohoro, C. A. Akinnawo and P. A. Olubambi, *Sustainable Chem. Clim. Action*, 2023, **3**, 100029.
18. S. Satyam and S. Patra, *Heliyon*, 2024, **10**, e29573.
19. K. S. D. Premarathna, A. U. Rajapaksha, B. Sarkar, E. E. Kwonm, A. Bhatnagar, Y. S. Ok and M. Vithanage, *Chem. Eng. J.*, 2019, **372**, 536.
20. P. Ndagijimana, H. Rong, P. Ndokoye, J. P. Mwizerwa, F. Nkinahamira, S. Luo, D. Guo and B. Cui, *J. Cleaner Prod.*, 2023, **417**, 138006.
21. G. Utzeri, L. Verissimo, D. Murtinho, A. A. C. C. Pais, F. X. Perrin, F. Ziarelli, T. V. Iordache, A. Sarbu and A. J. M. Valente, *Molecules*, 2021, **26**, 1426.
22. B. Pan and B. Xing, *Environ. Sci. Technol.*, 2008, **42**, 9005.
23. B. Qiu, Q. Shao, J. Shi, C. Yang and H. Chu, *Sep. Purif. Technol.*, 2022, **300**, 121925.
24. F. Yan, Y. Hou, C. Yi, Y. Wang, M. Xu and J. Xu, *Anal. Chim. Acta*, 2022, **1232**, 340475.
25. X. Gao, C. Du, Z. Zhuang and W. Chen, *J. Mater. Chem. C*, 2016, **4**, 6927.
26. E. C. Silva, C. G. Gomes, J. Pina, R. F. P. Pereira, D. Murtinho, A. R. Fajardo and A. J. M. Valente, *Carbohydr. Polym.*, 2024, **323**, 121464.
27. A. Xu, N. Feng, K. Yin, H. Li and J. Hao, *Nanoscale*, 2022, **15**, 275.
28. C. L. Shen, Q. Lou, K. K. Liu, G. S. Zheng, R. W. Song, J. H. Zang, L. Dong and C. X. Shan, *Carbon*, 2023, **213**, 118217.
29. G. Maduraiveeran and W. Jin, *Mater. Sci. Eng., B*, 2021, **272**, 115341.
30. X. Y. Yuan, Z. C. Yu, J. H. Hu, B. Xiao, T. T. Zhang, K. Li, C. Chen, Z. Tao and X. Xiao, *J. Mater. Chem. C*, 2019, **7**, 6820.

31. H. Zhao, Z. Jiang, C. Wang and Y. Li, *Crystals*, 2023, **13**, 983.
32. M. Li, T. Chen, J. J. Gooding and J. Liu, *ACS Sens.*, 2019, **4**, 1732.
33. P. Kumar, S. Dua, R. Kaur, M. Kumar and G. Bhatt, *RSC Adv.*, 2022, **12**, 4714.
34. U. A. Rani, L. Y. Ng, C. Y. Ng and E. Mahmoudi, *Adv. Colloid Interface Sci.*, 2020, **278**, 102124.
35. Y. P. Sun, B. Zhou, Y. Lin, W. Wang, K. A. S. Fernando, P. Pathak, M. J. Meziani, B. A. Harruff, X. Wang, H. Wang, P. G. Luo, H. Yang, M. E. Kose, B. Chen, L. M. Veca and S. Y. Xie, *J. Am. Chem. Soc.*, 2006, **128**, 7756.
36. J. Zhou, C. Booker, R. Li, X. Zhou, T. K. Sham, X. Sun and Z. Ding, *J. Am. Chem. Soc.*, 2007, **129**, 744.
37. Y. Dong, N. Zhou, X. Lin, J. Lin, Y. Chi and G. Chen, *Chem. Mater.*, 2010, **22**, 5895.
38. Q. Wang, H. Zheng, Y. Long, L. Zhang, M. Gao and W. Bai, *Carbon*, 2011, **49**, 3134.
39. A. Kaczmarek, J. Hoffman, J. Morgiel, T. Mościcki, L. Stobiński, Z. Szymański and A. Małolepszy, *Materials*, 2021, **14**, 729.
40. N. Biazar, R. Poursalehi and H. Delavari, *AIP Conf. Proc.*, 2018, **1920**, 020033.
41. M. Liu, Y. Xu, F. Niu, J. J. Gooding and J. Liu, *Analyst*, 2016, **141**, 2657.
42. S. Y. Park, H. U. Lee, E. S. Park, S. C. Lee, J. W. Lee, S. W. Jeong, C. H. Kim, Y. C. Lee, Y. S. Huh and J. Lee, *ACS Appl. Mater. Interfaces*, 2014, **6**, 3365.
43. S. Chahal, N. Yousefi and N. Tufenkii, *ACS Sustainable Chem. Eng.*, 2020, **8**, 5566.
44. P. Ezati, J. W. Rhim, R. Molaei, R. Priyadarshi, S. Roy, S. Min, Y. H. Kim, S. G. Lee and S. Han, *Sustainable Mater. Technol.*, 2022, **32**, e00397.
45. M. Aslan, H. Eskalen and M. Kavgaci, *Russ. J. Gen. Chem.*, 2023, **93**, 2152.
46. T. V. De Medeiros, J. Manioudakis, F. Noun, J. R. Macairan, F. Victoria and R. Naccache, *J. Mater. Chem. C*, 2019, **7**, 7175.
47. D. Uriarte, E. Vidal, A. Canals, C. E. Domini and M. Garrido, *Talanta*, 2021, **229**, 122298.
48. T. Ogi, K. Aishima, F. A. Permatasari, F. Iskandar, E. Tanabe and K. Okuyamam, *New J. Chem.*, 2016, **40**, 5555.
49. Z. Ye, G. Li, J. Lei, M. Liu, Y. Jin and B. Li, *ACS Appl. Bio Mater.*, 2020, **3**, 7095.
50. M. Rong, Y. Feng, Y. Wang and X. Chen, *Sens. Actuators, B*, 2017, **245**, 868.
51. P. K. Yadav, S. Chandra, V. Kumar, D. Kumar and S. H. Hasan, *Catalysts*, 2023, **13**, 422.
52. X. Kou, S. Jiang, S. J. Park and L. Y. Meng, *Dalton Trans.*, 2020, **49**, 6915.
53. X. Wang, Y. Feng, P. Dong and J. Huang, *Front. Chem.*, 2019, **7**, 671.
54. X. Xu, R. Ray, Y. Gu, H. J. Ploehn, L. Gearheart, K. Raker and W. A. Scrivens, *J. Am. Chem. Soc.*, 2004, **126**, 12736.

55. R. Wang, K. Q. Lu, Z. R. Tang and Y. J. Xu, *J. Mater. Chem. A*, 2017, **5**, 3717.
56. M. Pan, X. Xie, K. Liu, J. Yang, L. Hong and S. Wang, *Nanomaterials*, 2020, **10**, 930.
57. H. Liu, T. Ye and C. Mao, *Angew. Chem., Int. Ed.*, 2007, **46**, 6473.
58. C. Tan, S. Zuo, Y. Zhao and B. Shen, *J. Mater. Res.*, 2019, **34**, 3428.
59. B. D. Mansuriya and Z. Altintas, *Nanomaterials*, 2021, **11**, 2525.
60. H. Zhu, X. Wang, Y. Li, Z. Wang, F. Yang and X. Yang, *Chem. Commun.*, 2009, **34**, 5118.
61. S. Paulo, E. Palomares and E. Martinez-Ferrero, *Nanomaterials*, 2016, **6**, 157.
62. B. Zhang, C. Y. Liu and Y. Liu, *Eur. J. Inorg. Chem.*, 2010, **28**, 4411.
63. M. J. Liang, F. Zhang, Y. L. Zhu, X. Y. Deng, X. P. Chen, Q. J. Zhou and K. J. Tan, *Spectrochim. Acta, Part A*, 2022, **275**, 121139.
64. M. Devi and S. Rawat, *Oxford Open Mater. Sci.*, 2021, **1**, itab014.
65. C. Wang, M. Cong, H. Shi, Z. Yao, E. Liu, X. Hu, P. Guo, W. Xue and J. Fan, *Dyes Pigm.*, 2022, **204**, 110431.
66. D. Pan, J. Zhang, Z. Li, C. Wu, X. Yan and M. Wu, *Chem. Commun.*, 2010, **46**, 3681.
67. C. W. Lai, Y. H. Hsiao, Y. K. Peng and P. T. Chou, *J. Mater. Chem.*, 2012, **22**, 14403.
68. R. Atchudan, T. N. J. I. Edison, S. Perumal, R. Vinodh, A. K. Sundramoorthy, R. S. Babu and Y. R. Lee, *Chemosensors*, 2021, **9**, 166.
69. H. Qi, L. Qiu, X. Zhang, T. Xi, J. Jing, R. Sami, S. F. Alanazi, Z. Alqahtani, M. D. Aljabri and M. M. Rahman, *RSC Adv.*, 2023, **13**, 2663.
70. M. Yang, Y. Yan, F. He, Z. Wang, E. Liu and J. Fan, *Mater. Lett.*, 2024, **357**, 135613.
71. Z. Zhang, W. Sun and P. Wu, *ACS Sustainable Chem. Eng.*, 2015, **3**, 1412.
72. M. Sabet and K. Mahdavi, *Appl. Surf. Sci.*, 2019, **463**, 283.
73. C. Wang, Y. Wang, H. Shi, Y. Yan, E. Liu, X. Hu and J. Fan, *Mater. Chem. Phys.*, 2019, **232**, 145.
74. R. Atchudan, T. N. J. I. Edison, M. Shanmugam, S. Perumal, T. Somanathan and Y. R. Lee, *Phys. E*, 2021, **126**, 114417.
75. D. Pooja, L. Singh, A. Thakur and P. Kumar, *Sens. Actuators, B*, 2019, **283**, 363.
76. Y. Shi, X. Liu, M. Wang, J. Huang, X. Jiang, J. Pang, F. Xu and X. Zhang, *Int. J. Biol. Macromol.*, 2019, **128**, 537.
77. N. Murugan and A. K. Sundramoorthy, *New J. Chem.*, 2018, **42**, 13297.
78. H. Eskalen, S. Urus, S. Cömertpay, A. H. Kurt and Ş. Özgan, *Ind. Crops Prod.*, 2020, **147**, 112209.
79. L. Li, B. Yu, X. Zhang and T. You, *Anal. Chim. Acta*, 2015, **895**, 104.
80. L. Wang, H. X. Cao, C. G. Pan, Y. S. He, H. F. Liu, L. H. Zhou, C. Q. Li and G. X. Liang, *Microchim. Acta*, 2019, **186**, 28.
81. R. Qiang, S. Yang, K. Hou and J. Wang, *New J. Chem.*, 2019, **43**, 10826.

82. S. Venkatesan, A. J. Mariadoss, A. Kathiravan and A. Muthupandian, *Sens. Actuators, B*, 2019, **282**, 972.
83. N. Azam, A. M. Najabat and K. T. Javaid, *Front. Mater.*, 2021, **8**, 700403.
84. T. Yuan, T. Meng, P. He, Y. Shi, Y. Li, X. Li, L. Fan and S. Yang, *J. Mater. Chem. C*, 2019, **7**, 6820.
85. M. J. Molaei, *Sol. Energy*, 2020, **196**, 549.
86. F. Yuan, S. Li, Z. Fan, X. Meng, L. Fan and S. Yang, *Nano Today*, 2016, **11**, 565.
87. S. Anwar, H. Ding, M. Xu, X. Hu, Z. Li, J. Wang, L. Liu, L. Jiang, D. Wang, C. Dong, M. Yan, Q. Wang and H. Bi, *ACS Appl. Bio Mater.*, 2019, **2**, 2317.
88. L. Wang, W. Li, L. Yin, Y. Liu, H. Guo, J. Lai, Y. Han, G. Li, M. Li, J. Zhang, R. Vajtai, P. M. Ajayan and M. Wu, *Sci. Adv.*, 2020, **6**, eabb6772.
89. D. Li, P. Jing, L. Sun, Y. An, X. Shan, X. Lu, D. Zhou, D. Han, D. Shen, Y. Zhai, S. Qu, R. Zbořil and A. L. Rogach, *Adv. Mater.*, 2018, **30**, 1705913.
90. S. Sikiru, T. L. Oladosu, S. Y. Kolawole, L. A. Mubarak, H. Soleimani, L. O. Afolabi and A. O. O. Toyin, *J. Energy Storage*, 2023, **60**, 106556.
91. J. P. Malavika, C. Shobana, S. Sundarraj, M. Ganeshbabu, P Kumar and R. K. Selvan, *Biomater. Adv.*, 2022, **136**, 212756.
92. H. Ding, X. H. Li, X. B. Chen, J. S. Wei, X. B. Li and H. M. Xiong, *J. Appl. Phys.*, 2020, **127**, 231101.
93. L. Bao, C. Liu, Z. L. Zhang and D. W. Pang, *Adv. Mater.*, 2015, **27**, 1663.
94. Z. Liu, H. Zou, N. Wang, T. Yang, Z. Peng, J. Wang, N. Li and C. Huang, *Sci. China: Chem.*, 2018, **61**, 490.
95. M. L. Liu, B. B. Chen, C. M. Li and C. Z. Huang, *Green Chem.*, 2019, **21**, 449.
96. Y. Lou, X. Hao, L. Liao, K. Zhang, S. Chen, Z. Li, J. Ou, A. Qin and Z. Li, *Nano Sel.*, 2021, **2**, 1117.
97. M. C. A. De Oliveira, E. G. Elisângela, I. C. B. Pires, I. C. M. Candido, N. Rakov, H. P. De Oliveira, Y. Xing and G. S. Maciel, *Adv. Fiber Mater.*, 2020, **2**, 302.
98. A. Mehta, A. Mishra and S. Basu, *Plasmonics*, 2018, **13**, 2239.
99. D. S. Su and G. Centi, *J. Energy Chem.*, 2013, **22**, 151.
100. M. Zheng and Z. Xie, *Mater. Today Chem.*, 2019, **13**, 121.
101. A. Vibhute, T. Patil, R. Gambhir and A. P. Tiwari, *Appl. Surf. Sci. Adv.*, 2022, **11**, 100311.
102. L. Cao, X. Wang, M. J. Meziani, F. Lu, H. Wang, P. G. Luo, Y. Lin, B. A. Harruff, L. M. Veca, D. Murray, S. Y. Xie and Y. P. Sun, *J. Am. Chem. Soc.*, 2007, **129**, 11318.
103. D. Hong, X. Deng, J. Liang, J. Li, Y. Tao and K. Tan, *Microchem. J.*, 2019, **151**, 104217.
104. X. Xu, Z. Bao, G. Zhou, H. Zeng and J. Hu, *ACS Appl. Mater. Interfaces*, 2016, **8**, 14118.
105. J. M. Lehn, *Science*, 2002, **295**, 2400.

106. Q. Zhang, Y. X. Deng, H. X. Luo, C. Y. Shi, G. M. Geise, B. L. Feringa, H. Tian and D. H. Qu, *J. Am. Chem. Soc.*, 2019, **141**, 12804.
107. F. Arcudi and L. Đorđević, *Small*, 2023, **19**, 2300906.
108. F. J. M. Hoeben, P. Jonkheijm, E. W. Meijer and A. P. H. J. Schenning, *Chem. Rev.*, 2005, **105**, 1491.
109. D. J. Lundberg, C. M. Brown, E. O. Bobylev, N. J. Oldenhuis, Y. S. Alfarajm, J. Zhao, I. Kevlishvili, H. J. Kulik and J. A. Johnson, *Nat. Commun.*, 2024, **15**, 3951.
110. A. J. M. Valente and O. Söderman, *Adv. Colloid Interface Sci.*, 2014, **205**, 156.
111. Y. You, K. Zhou, B. Guo, Q. Liu, Z. Cao, L. Liu and H. C. Wu, *ACS Sens.*, 2019, **4**, 774.
112. P. J. Cragg, *Supramolecular Chemistry: From Biological Inspiration to Biomedical Applications*, Springer, 2010.
113. B. E. Hirsch, K. P. McDonald, A. H. Flood and S. L. Tait, *J. Chem. Phys.*, 2015, **142**, 101914.
114. J. Szeijtli, *Chem. Rev.*, 1998, **98**, 1743.
115. X. Li, Y. Jin, N. Zhu and L. Y. Jin, *Polymers*, 2023, **15**, 4543.
116. X. Sun, H. Wang, J. Qi, S. Zhou and H. Li, *Dyes Pigm.*, 2021, **190**, 109287.
117. Z. Ran, H. Yang, Z. Li, K. Wang, J. Zhao, X. Ran, G. Du and L. Yang, *ACS Sustainable Chem. Eng.*, 2020, **8**, 10161.
118. H. Cong, X. L. Ni, X. Xiao, Y. Huang, Q. J. Zhu, S. F. Xue, Z. Tao, L. F. Lindov and G. Wei, *Org. Biomol. Chem.*, 2016, **14**, 4335.
119. C. K. P. Monteiro, M. A. F. Faustino and C. Serpa, *Molecules*, 2023, **28**, 7108.
120. K. Yang, Y. Pei, J. Wen and Z. Pei, *Chem. Commun.*, 2016, **52**, 9316.
121. G. T. Williams, C. J. E. Haynes, M. Fares, C. Caltagirone, J. K. Hiscock and P. A. Gale, *Chem. Soc. Rev.*, 2021, **50**, 2737.
122. S. Sarkar, S. Dinda, P. Choudhury and P. Kumar, *Soft Matter*, 2022, **6**, 100076.
123. A. Nicosia, P. Mineo, N. Micali and V. Villari, *Int. J. Mol. Sci.*, 2023, **24**, 15384.
124. X. Niu, S. Yan, R. Zhao, S. Han, K. Cao, H. Li and K. Wang, *Microchim. Acta*, 2023, **190**, 61.
125. G. Utzeri, P. M. C. Matias, D. Murtinho and A. J. M. Valente, *Front. Chem.*, 2022, **10**, 859406.
126. M. Pei, J. Y. Pai, P. Du and P. Liu, *Mol. Pharm.*, 2018, **15**, 4084.
127. B. Bai, S. Qi, K. Yang, X. Yu, R. Jian, T. Zhang, D. Wang, H. Meng, Y. Zhao, Y. Xia, H. Xu, G. Yu and Z. Chen, *Small*, 2023, **19**, 2300217.
128. V. Villari, M. Gaeta, A. D'Urso and N. Micali, *Colloids Surf., A*, 2022, **648**, 129436.
129. X. Sun, G. Li, Y. Yin, Y. Zhang and H. Li, *Soft Matter*, 2018, **14**, 6983.
130. X. Ran, Q. Qu, X. Qian, W. Xie, S. Li, L. Li and L. Yang, *Sens. Actuators, B*, 2018, **257**, 362.
131. A. Bagger, O. Christensen, V. Ivaništšev and J. Rossmeisl, *ACS Catal.*, 2022, **12**, 2561.

132. S. H. Li, M. Y. Qi, Y. Y. Fan, Y. Yang, M. Anpo, Y. M. A. Yamada, Z. R. Tang and Y. J. Xu, *Appl. Catal., B*, 2021, **292**, 120157.
133. N. N. Kadamannil, A. I. Shames, R. Bisht, S. Biswas, N. Shauloff, H. Lee, J. M. Kim and R. Jelinek, *ACS Appl. Mater. Interfaces*, 2024, **6**, 22593.
134. C. S. Tshangana, A. A. Muleia, A. T. Kuvarega, T. J. Malefetse and B. B. Mamba, *J. Water Process Eng.*, 2021, **43**, 102249.
135. T. Li and W. Hu, *Green Energy Environ.*, 2024, **9**, 604.
136. N. U. M. Nizam, M. M. Hanafiah, E. Mahmoudi and A. W. Mohammad, *Sci. Rep.*, 2023, **13**, 12777.
137. H. Ehtesabi, S. Roshani, Z. Bagheri and M. Yaghoubi-Avini, *J. Environ. Chem. Eng.*, 2019, **7**, 103419.
138. H. Wang, L. Ai, Z. Song, M. Nie, J. Xiao, G. Li and S. Lu, *Chem. – Eur. J.*, 2023, **29**, e202302383.
139. Q. Zhou, W. Huang, C. Xu, X. Liu, K. Yang, D. Li, Y. Hou and D. D. Dionysiou, *Chem. Eng. J.*, 2021, **420**, 129582.
140. Z. Song, X. Chen, X. Gong, X. Gao, Q. Dai, T. T. Nguyen and M. Guo, *Opt. Mater.*, 2020, **100**, 109642.
141. T. Teymoorian, N. Hashemi, M. H. Mousazadeh and Z. Entezarian, *SN Appl. Sci.*, 2021, **3**, 305.
142. X. Yan, S. Rahman, M. Rostami, Z. A. Tabasi, F. Khan, A. Alodhayb and Y. Zhang, *ACS Omega*, 2021, **6**, 23504.
143. P. Zhang, M. Yang, J. Lan, Y. Huang, J. Zhang, S. Huang, Y. Yang and J. Ru, *Toxics*, 2023, **11**, 828.
144. S. C. Bondy and A. Campbell, *Int. J. Environ. Res. Public Health*, 2018, **15**, 2.
145. D. Kadadou, L. Tizani, H. Alsafar and S. W. Hasan, *MethodsX*, 2024, **12**, 102582.
146. J. V. Kumar and J. W. Rhim, *J. Environ. Chem. Eng.*, 2024, **12**, 111999.
147. T. Milovac, *J. Med. Humanit.*, 2023, **44**, 245.
148. R. Z. Marques, L. Y. Kochi, M. Guzman-Tordecilla, L. Malage, P. Juneau and M. P. Gomes, *Sci. Total Environ.*, 2023, **892**, 164309.
149. J. Sharma, M. Joshi, A. Bhatnagar, A. K. Chaurasia and S. Nigam, *Environ. Res.*, 2022, **215**, 114219.
150. M. Kotchen, J. Kallaos, K. Wheeler, C. Wong and M. Zahller, *J. Environ. Manage.*, 2009, **90**, 1476.
151. M. Ahmed, M. O. Mavukkandy, A. Giwa, M. Elektorowicz, E. Katsou, O. Khelifi, V. Naddeo and S. W. Hasan, *npj Clean Water*, 2022, **5**, 12.
152. Q. Yang, H. Hong and Y. Luo, *Chem. Eng. J.*, 2020, **392**, 123680.
153. Q. You, Q. Zhang, M. Gu, R. Du, P. Chen, J. Huang, Y. Wang, S. Deng and G. Yu, *Chem. Eng. J.*, 2022, **431**, 133927.
154. D. Meng, Y. L. M. Pang, J. Qiu, C. Fan, Y. Feng and D. Wang, *J. Environ. Chem. Eng.*, 2023, **11**, 110318.
155. P. C. Sharma, A. Jain, S. Jain, R. Pahwa and M. S. Yar, *J. Enzyme Inhib. Med. Chem.*, 2010, **25**, 577.
156. Z. Fang, Y. Liu, P. Chen, D. Li, H. Liu, Z. Xiao, Y. Zheng, Z. Lin, J. Luo, W. Lv and G. Liu, *Chem. Eng. J.*, 2023, **451**, 138571.

157. E. David and V. C. Niculescu, *Int. J. Environ. Res. Public Health*, 2021, **18**, 13147.
158. X. Jin, Y. Wu, M. Santhamoorthy, T. T. Nhi Le, V. T. Le, Y. Yuan and C. Xia, *Chemosphere*, 2022, **308**, 136182.
159. E. C. Silva, J. Pina, R. F. P. Pereira, D. Murtinho, A. J. M. Valente and A. R. Fajardo, *Environ. Pollut.*, 2024, **347**, 123778.
160. G. Yi, Z. Tao, W. Fan, H. Zhou, Q. Zhuang and Y. Wang, *ACS Sustainable Chem. Eng.*, 2024, **12**, 1378.
161. F. Noun, E. A. Jury and R. Naccache, *Sensors*, 2021, **21**, 1391.
162. P. B. Tchounwou, C. G. Yedjou, A. K. Patlolla and D. J. Sutton, *EXS*, 2021, **101**, 133.
163. K. Rehman, F. Fatima, I. Waheed and M. S. H. Akash, *J. Cell. Biochem.*, 2018, **119**, 157.
164. Z. Wang, L. Zhang, K. Zhang, Y. Lu, J. Chen, S. Wang, B. Hu and X. Wang, *Chemosphere*, 2022, **287**, 132313.
165. M. Montaña, A. Camacho, R. Devesa, I. Vallés, R. Céspedes, I. Serrano, S. Blàzquez and V. Barjola, *J. Cleaner Prod.*, 2013, **60**, 77.
166. N. A. A. Qasesm, R. H. Mohammed and D. U. Lawal, *NPJ Clean Water*, 2021, **4**, 36.
167. T. Bhattacharjee, A. Konwar, J. S. Boruah, D. Chowdhury and G. Majumdar, *J. Hazard. Mater. Adv.*, 2023, **10**, 100295.
168. M. Khan, T. Omer, A. Ellahi, Z. Ur Rahman, R. Niaz and S. Ahmad Lone, *Geocarto Int.*, 2023, **38**, 2256313.
169. V. R. Soares, E. C. Silva, C. G. Gomes, M. A. Vieira and A. R. Fajardo, *Chemosphere*, 2024, **350**, 140911.
170. X. Zhang, X. Yi, J. Ouyang, S. Wang, D. Xu, X. Qi, P. Jiang, X. Guo and Y. Wu, *Chem. Eng. J.*, 2024, **479**, 147753.
171. X. Y. Yuan, Z. C. Yu, J. H. Hu, B. Xiao, T. T. Zhang, K. Li, C. Chen, Z. Tao and X. Xiao, *Spectrochim. Acta, Part A*, 2024, **307**, 123632.
172. Y. Xiong, J. H. Hu, R. X. Hou, Z. N. Liu, J. Y. Zhang, X. Y. Yuan, Z. Tao and X. Xiao, *J. Mol. Struct.*, 2024, **1301**, 137301.
173. Z. Zhang, D. Zhang, C. Shi, W. Liu, L. Chen, Y. Miao, J. Diwu, J. Li and and S. Wang, *Environ. Sci.: Nano*, 2019, **6**, 1457.
174. Y. Deng and R. Zhao, *Wastewater Treat. Curr. Pollut.*, 2015, **1**, 167.
175. C. Y. Chung, Y. J. Chen, C. H. Kang, H. Y. Lin, C. C. Huang, P. H. Hsu and H. J. Lin, *Polymers*, 2021, **13**, 1598.

CHAPTER 11

Porous Structures for Water Remediation

DAIANA MARIA FURLAN SANDRINI (0000-0002-1844-5353) AND DENISE FREITAS SIQUEIRA PETRI (0000-0003-4814-8357)*

Institute of Chemistry, University of São Paulo, Av. Prof. Lineu Prestes 748, 05508-000, São Paulo, Brazil, Email: daiana_dmf@hotmail.com
*Email: dfsp@iq.usp.br

11.1 Types of Porous Materials

11.1.1 Metal Organic Frameworks (MOFs)

Metal organic frameworks (MOFs) are characterized by a network structure consisting of metal ions or clusters coordinated to organic ligands to form one-, two-, or three-dimensional structures.[1,2] They are considered the most effective porous materials for treatment and purification of wastewater. Their tunable architecture allows designing materials with physical–chemical properties for specific applications, such as removing heavy metals,[3,4] dyes,[5,6] oils and organic solvents,[7,8] and medicines.[9]

Compared to conventional porous materials, obtaining MOFs is significantly more laborious. Their complex tridimensional porous network involves the self-assembling of organic–inorganic hybrid building units coordinated with di-, tri- or tetradentate organic ligands and a centered metal.[1,10] A wide range of metal ions or clusters, as well as organic ligands can be used in the construction of these materials, thus the preparation of each type of MOF requires unique synthesis conditions, which need to be carefully optimized. Metal ions such as Co^{2+}, Cu^{2+}, Zn^{2+}, Ni^{2+}, Al^{3+}, Mo^{3+},

Contribution of Colloidal Materials to Air, Water and Soil Environmental Sustainability: From Basic Concepts to Advanced Applications
Edited by Izabel C. Riegel-Vidotti, Lizandra Maria Zimmermann and Heloise R. de Barros

Published by the Royal Society of Chemistry, www.rsc.org

Fe^{3+} and Zr^{4+} are among those commonly used in the preparation of MOFs.[10,11] They can offer different coordination geometries with the organic ligands, which can be carboxylate ligands, nitrogen-containing heterocyclic compounds, such as imidazolates, amine and diamine groups and ligands functionalized with hydroxyl, thiol and ester groups.[11,12]

Depending on the structural complexity and the synthesis parameters to be controlled, time-consuming syntheses or those requiring high temperatures may be necessary to obtain MOFs. For instance, Zr-containing MOF crystals (UiO-66) were obtained after 21 h of reaction at 120 °C using dimethylformamide (DMF) as the solvent,[13] whereas Sr based-MOF[8] and Ti based-MOF (MIL-125-NH_2) crystals,[14] took more than 72 h at ~100 °C in DMF for complete crystallization and purification. Toxic solvents such as DMF in addition to dimethylformamide, methanol, ethanol, and acetonitrile are present in several MOF synthesis routes,[15] which represents an environmental concern. Synthesis methods using green solvents such as water, supercritical fluids (SCFs) and ionic liquids (ILs) or solvent-free or solvent-less mechanochemical methods have been proposed as sustainable pathways for the synthesis of MOFs.[15] For example, the composite of $Cu_3(BTC)_2$ (HKUST-1) and graphite oxide with high toluene adsorption capacity and excellent water stability was synthesized by a mechanochemical process using ball milling at room temperature without adding any solvent for 30 min.[16] The preparation of MOF-74 analogs took only 5 min using a minimum amount of liquid exogenous organic base of Hünig (*N*,*N*-diisopropylethylamine) as the solvent, in a ball mill at room temperature.[17] By continuous manufacturing using twin-screw extrusion, Zr based-MOFs (UiO-66, UiO-66-NH_2, MOF-801 and MOF-804) were synthesized in less than 1 hour at room temperature with a production rate equivalent to 1.4 $kg\,h^{-1}$,[18] while the use of a temperature gradient from 30 to 150 °C over 12 h of single screw extrusion produced Zn based-MOFs (ZIF-8) at 4 $kg\,h^{-1}$.[19] In both cases, no solvent or little water was used to promote sufficient lubrication for molecular diffusion.

The intrinsic crystalline coordination network of MOFs is responsible for their highly ordered and uniform structure. Consequently, these materials have high crystallinity and porosity, open metal sites, large pore volume[7,20] and superior specific surface area (SSA) (ranging from 1000 to 8000 $m^2\,g^{-1}$),[11] which is much higher than that of zeolites and carbonaceous materials (<1000 $m^2\,g^{-1}$). These characteristics make MOFs advantageous adsorbents. MOFs such as Zn-based ZIF-8, Co-based ZIF-67, Cr-based MIL series, Fe-based MIL-100, and Zr-based UiO-66 MOFs are among the types of MOFs most employed for the treatment of contaminated water.[20] ZIF-8 MOFs are attractive materials for removing a wide range of pollutants in water due to their high adsorption capacity[2] and their good structural and thermal stability under neutral and alkaline aqueous media.[12,21] ZIF-8 MOF crystals consist of tetrahedral coordinated Zn^{2+} ions and imidazolate linkers forming a reticular structure as illustrated in Figure 11.1(a), analogous to zeolite.[22,23] Figure 11.1(b) shows a ZIF-8 single crystal with a rhombic dodecahedron morphology, 1 mm in size, and a remarkable SAA of

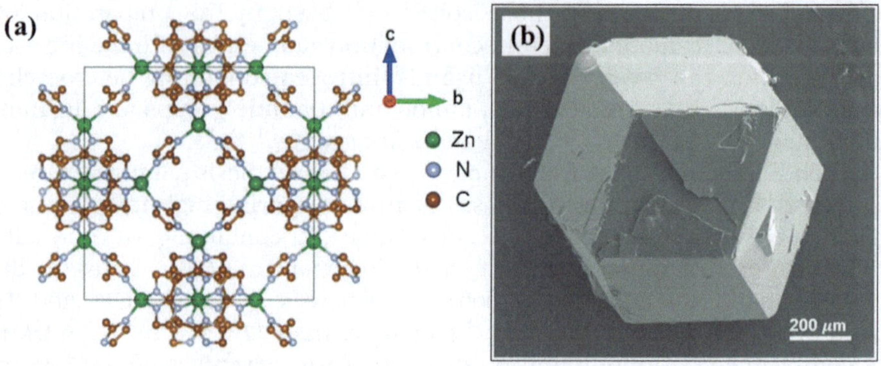

Figure 11.1 (a) Crystal structure of ZIF-8 obtained by X-ray diffraction (XRD), (b) scanning electron microscopy (SEM) images of the millimeter-sized ZIF-8 single crystals with a rhombic dodecahedron shape and a SSA of 1681 $m^2 g^{-1}$. Reproduced from ref. 23, https://doi.org/10.1080/14686996.2023.2292485, under the terms of the CC BY 4.0 license, http://creativecommons.org/licenses/by/4.0/.

1681 $m^2 g^{-1}$.[23] Variations in the sizes, shapes, and pore structure of ZIF crystals have been observed by adjusting synthesis parameters such as the molar ratios of organic ligands to metal ions, temperature, and solvent. Neat ZIF-8 with SSA of 937 $m^2 g^{-1}$ was able to adsorb up to 454.7 $mg g^{-1}$ of Cu^{2+} and 1119.8 $mg g^{-1}$ of Pb^{2+} from wastewater at pH 5.1; an XRD pattern revealed that the architecture of the ZIF-8 crystals remained preserved after the adsorption of the heavy metal, indicating that ZIF-8 were stable adsorbents in water under the study conditions.[3] The exceptional Cu^{2+} and Pb^{2+} ion adsorption capacities demonstrated by ZIF-8 are superior to those reported for most known porous materials; some examples include materials based on functionalized silica,[24–26] natural clays,[27,28] carbon dots,[29] biochar,[30,31] and polymer-based aerogels and foams.[32–34] Additionally, neat ZIF-8 nanoparticles with SSA of 1295 $m^2 g^{-1}$ exhibited a high adsorption capacity for tetracycline (124.6 $mg g^{-1}$) and stability in water in the pH range of 5 to 12.[21]

Since MOFs are commonly synthesized in powder form, the incorporation of these materials into matrices or substrates makes their use as adsorbents for water treatment more practical, and can also contribute to increasing their selectivity and stability in water. Lou *et al.* demonstrated that ZIF-8/chitosan sponges prepared through an *in situ* growth approach showed greater absorption capacity for Malachite green (MG) dye in aqueous solution compared to neat adsorbents (ZIF-8 and chitosan sponges) and excellent selective adsorption towards MG.[35] Zhang and co-workers developed a photocatalytic aerogel by growing ZIF-8 on carbon nitride (C_3N_4) nanosheets (ZIF-8/C_3N_4) and integrating it into an agar matrix using a simple sol–gel process.[36] The resulting ZIF-8/C_3N_4 aerogel demonstrated nearly double the Congo Red adsorption capacity (287.35 $mg g^{-1}$) compared to pristine bulk ZIF-8 (154.87 $mg g^{-1}$). The aerogel network effectively integrated ZIF-8/C_3N_4

into a functional form for water treatment with minimal leaching. Additionally, the adsorbent can be regenerated through photocatalytic degradation of CR under visible light, maintaining stability over five adsorption–degradation cycles.

11.1.2 Zeolites

Zeolites are naturally occurring hydrated aluminosilicates that have been used for decades to remove cations like ammonium,[37] heavy metal ions,[38,39] and radionuclides from water,[40] due to their ability to selectively sorb cations. They exhibit a microporous crystalline structure composed of three-dimensional networks of SiO_4^{4-} and AlO_4^{5-} tetrahedral units interconnected by oxygen atoms,[41,42] as shown in Figure 11.2(a); this crystalline lattice creates cavities in the forms of channels and cages. The isomorphic substitution of Si^{4+} by Al^{3+} in the tetrahedra results in a negative charge in the tetrahedra framework, which is balanced by non-framework monovalent or divalent cations contained in channels and cages, typically Ca^{2+}, Na^{+}

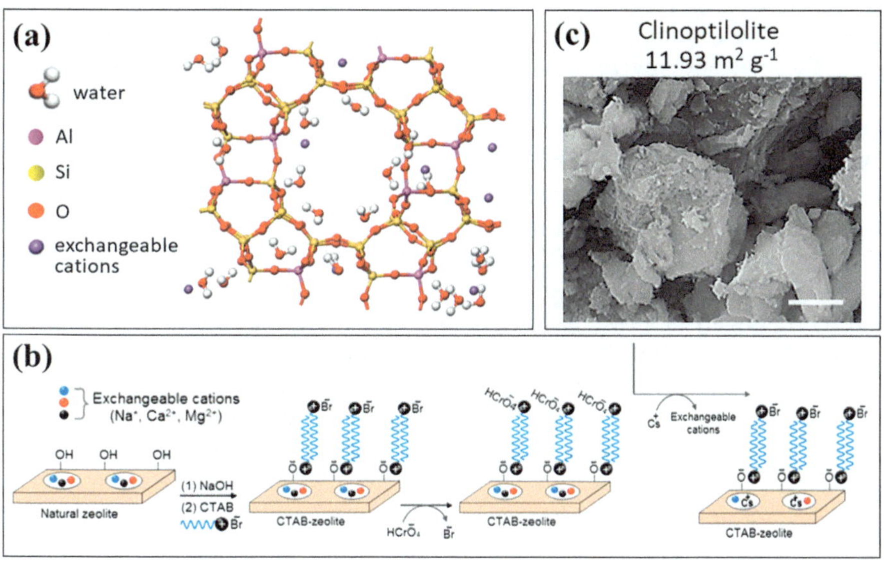

Figure 11.2 (a) Schematic illustration of the chemical structure of a zeolite; (b) schematic showing the adsorption mechanism of Cs^{+} and $HCrO_4^{-}$ species on CTAB-modified natural zeolite; (c) SEM images of the porous structure of the natural zeolite clinoptilolite obtained with scale bars corresponding to 5 μm. (a) Reproduced from ref. 44, https://doi.org/10.3390/pharmaceutics15051352, under the terms of the CC BY 4.0 license, http://creativecommons.org/licenses/by/4.0/. (b) Reproduced from ref. 57, https://doi.org/10.1007/s11356-023-25644-y, under the terms of the CC BY 4.0 license, http://creativecommons.org/licenses/by/4.0/. (c) Reproduced from ref. 58 with permission from Elsevier, Copyright 2019.

and K^+. These ions and water molecules are weakly bound to the framework and can be exchanged by diffusion for a surrounding cation.[42–44] In general, higher aluminum content in the crystal lattice results in zeolites with higher cation exchange capacity (CEC).[45] For instance, clinoptilolite, one of the most studied zeolites in water treatment,[41,42] has a low Al^{3+} content in its tetrahedral sites, resulting in fewer non-framework cations available for ion exchange.[41,44] Through channels and cages, molecules can be selectively captured based on their size and polarity; this characteristic enables zeolites to function as molecular sieves and catalysts.[44] Since natural zeolites exhibit diverse mineralogical compositions, crystalline structure, and pore size, their sorption capacity and selectivity can vary significantly for different types of zeolites. Additionally, zeolites are often found alongside other minerals that act as contaminants, affecting their metal sorption capacity.[46] These characteristics, associated with the non-renewable nature of zeolite deposits, restrict the large-scale use of natural zeolites.

In recent years, natural zeolites and their modified forms have been used to remove anions,[47] organics[48] and as antibacterial agents[49] in the aqueous system. Their use in water separation and purification processes remains promising due to their availability, low cost, high ion-exchange capacity, and molecular sieving properties.[42] Additionally, zeolites show high chemical stability over a wide pH range, thermal and mechanical stability and possess a high surface area per volume, making them one of the most widely used materials for adsorption, capable of achieving complete removal of pollutants in contaminated water.[46]

Modified forms of zeolites are proposed to enhance their removal efficiency through structural changes or by adjusting their adsorption properties using surface modification techniques. Treatment of natural zeolite with acids can remove pore-blocking impurities, dealuminate the structure by hydrolyzing Al–O–Al and Si–O–Al bonds, and eliminate non-framework cations, forming proton-exchanged zeolites,[50] while alkaline treatment may result in dissolution of amorphous materials.[43] In the case of clinoptilolite, Na^+ exchange by H^+ through acid treatment increased the pore volume from 0.13 to 0.25 $mL\,g^{-1}$ and the SSA from 35 to 315 $m^2\,g^{-1}$. This modification facilitated the diffusion of small molecules such as H_2O, O_2, NO, CO_2, and NH_3 by increasing the CEC and opening pores that were partially blocked by non-framework species.[51] Acidic zeolites offer a cost-effective and environmentally friendly solution for degrading toxic organic pollutants in heterogeneous catalytic ozonation processes.[52] It was demonstrated that the catalytic efficiency of zeolites is closely related to the acid site on their surface. Brønsted acid sites, characterized by proton-donating OH groups on the surface of acid-treated natural zeolites, played a crucial role in surface reaction mechanisms that enhanced the removal of methylene blue (MB) from contaminated water. Deprotonated OH groups on the zeolite surface synergistically adsorbed methylene blue and served as catalysts for ozone decomposition reactions into radicals, thus accelerating the oxidation of MB.[53] The NaOH-treated zeolite exhibited a higher adsorption capacity for Mn(II) compared to natural zeolites. The reaction of the zeolite, predominantly composed of clinoptilolite, with NaOH resulted in the formation of hydroxysodalite, increasing the Na content of the zeolite while reducing the Si content due to the cleavage of Si–O–Si bonds. Desilication resulted in the formation of extra-framework aluminum species and enhanced the creation of mesopores, which facilitated the diffusion and uptake of Mn(II) in drinking water.[54] It has been demonstrated that K^+ and Ca^{2+} adhere more strongly to clinoptilolite than Na^+.[55] Since Na^+ is more readily exchanged with other metal cations, the sodium form of zeolite offers improved CEC.

An approach to adjusting the adsorption properties of zeolites involves modifying their surface chemistry with cationic surfactants. The alteration of the net negative structural charge on the zeolite surface can enhance its anion exchange capacity and facilitate the removal of non-polar organic pollutants through partitioning into the hydrophobic surfactant tail group.[41] Surfactant sorption on zeolites is governed by the critical micelle concentration (CMC) of surfactant and CEC of the zeolite, leading to the attachment of monolayers or bilayers of surfactant molecules on the external surface of the zeolite. In bilayers, the newly introduced positive surfaces are balanced by anionic counter ions. This configuration enhances the affinity of the zeolites for sorbing anionic contaminants.[56] For instance, natural zeolite modified with cetyltrimethylammonium bromide (CTAB) showed efficiency to simultaneously adsorb the radionuclides of cesium, $^{134}Cs^+$, and $HCrO_4^-$ species in the pH range of 2.5 to 4.2.[57] In contrast, unmodified natural zeolite was incapable of absorbing $HCrO_4^-$ under the same pH conditions. As illustrated in Figure 11.2(b), $^{134}Cs^+$ is adsorbed *via* ion exchange with exchangeable zeolite cations, while surfactant counter ions (Br^-) are displaced by $HCrO_4^-$ anions. The adsorption capacity of Congo Red, an anionic dye, increased from 16.92 to 200 mg g^{-1} through surface modification of clinoptilolite (Figure 11.2(c)) with hexadecyltrimethylammonium bromide (HDTMA-Br).[58] The incorporation of HDTMA-Br on clinoptilolite reduced its SSA from 11.93 to 7.64 $m^2 g^{-1}$, therefore the enhanced dye adsorption capacity was attributed to the anion exchange between CR and the surfactant. Efficient removal of BTEX (benzene, toluene, ethylbenzene, and xylene isomers) from water was achieved using clinoptilolite-rich tuffs modified with *N*-cetylpyridinium bromide (CPB).[59] Increasing the concentration of CPB deposited on the zeolite surface enhanced the BTEX compounds' adsorption capacity from 47% to 99%. At concentrations exceeding the CMC, a bilayer of surfactant formed on the zeolite surface, facilitating the partitioning of aromatic molecules into the hydrophobic tails of the cationic surfactants, and thereby enhancing their removal from the aqueous medium. In particular, the removal capacity of CPB-modified zeolite for benzene (94.7%) was comparable to that of commercial powdered activated carbon (PAC) (96.6%).

11.1.3 Silica-based Materials

Silica materials predominantly comprise a three-dimensional network of silicon-centered SiO_4 tetrahedra, interconnected by bridging oxygens, resulting in a porous structure with a large SSA. Siloxane groups ($\equiv$Si–O–Si$\equiv$)

are prevalent not only within the bulk but also on the surface of silica, which also features abundant silanol groups (≡Si–OH).[60] These reactive silanol groups play a crucial role in surface modification, making silica an excellent support for functionalization with a wide range of organic and inorganic groups. This characteristic enables the creation of adsorbents with tailored surface physicochemical properties for the removal of specific contaminants. In general, porous inorganic microstructures are widely utilized in water remediation due to their chemical and thermal stability, mechanical robustness,[60] and potential for regeneration.[61] Additionally, silica materials can be synthesized in a variety of shapes and sizes, with varying surface areas and pore sizes; examples of such materials include amorphous silica, mesoporous silica, fumed silica, and silica gels. Due to their high SSA, high porosity, distinctive nano-porous structure and well-defined pore sizes, mesoporous silica supports are among the most commonly used silica materials for adsorption and catalysis in water remediation.[62,63] Although exhibiting remarkable properties for water treatment, mesoporous silica often exhibits strong and sometimes irreversible non-selective adsorption due to the high density of acidic silanol groups on their surface, which can limit their effectiveness.[61] Therefore, modifying the surface with active functional groups is crucial to enhancing adsorption efficiency, improving selectivity, and better controlling surface properties, including hydrophilicity and hydrophobicity.[63,64]

The most used ordered mesoporous materials in water treatment include the Mobil Composition of Matter (MCM) and Santa Barbara Amorphous (SBA) series, particularly MCM-41 and SBA-15. Silica of the MCM-41 type exhibits a hexagonal arrangement of mesopores with diameters of 2 to 10 nm and SSA of 800 to 1200 $m^2\,g^{-1}$. In contrast, the SBA-15 type features uniform hexagonal pores with diameters ranging from 4 to 12 nm, thick framework walls containing micropores, and a SSA of 400 to 900 $m^2\,g^{-1}$.[65] They are synthesized through the condensation of silica precursors, typically sodium silicate or alkoxides such as tetraethyl orthosilicate (TEOS) and tetramethyl orthosilicate (TMOS), around an ordered surfactant used as a template,[66] which directs the formation of a condensed and amorphous polymerized network of siloxanes (Si–O–Si linkages).[60] Reaction time, temperature, ionic strength, the composition of the reaction mixture, and concentrations of silica sources can be adjusted to achieve silica materials with varying morphologies and pore sizes.[61,64] For instance, the synthesis of spherical MCM-41 type silica particles with a combined micro–mesoporous structure and high SSA (1300–1600 $m^2\,g^{-1}$) was achieved by using two silicon sources, TEOS and [3-(methacryloyloxy)propyl]trimethoxysilane (MPTMOS), in the presence of cetyltrimethylammonium bromide (CTAB) and 1,3,5-trimethylbenzene in an alcoholic-aqueous-ammonia medium.[67]

Amino-functionalized mesoporous materials are widely used to improve the removal capacity of inorganic contaminants from water. For instance, MCM-41 modified with *p*-aminobenzoic acid (PABA), PABA-MCM-41, effectively removed up to 98.3% of Cr(VI) at pH 2 from tannery wastewater.

Reusability tests showed that PABA-MCM-41 could eliminate over 90% of Cr(VI) in the first reuse cycle, merely by washing with distilled water, without requiring regeneration.[63] NH_2-MCM-41 was prepared by grafting 3-aminopropyltrimethoxysilane (APTMS) onto the pore walls of MCM-41 mesoporous silica as illustrated in Figure 11.3(a). This material demonstrated 100% selectivity for Cr(VI) in the presence of Ni, Zn, Ag, Co, Cu, Pb, Hg, and Cd ions in an aqueous medium at pH $<$ 3.5, with a maximum adsorption capacity of 94.6 mg g^{-1}. In contrast, unmodified MCM-41 did not adsorb Cr(VI).[68] As depicted in Figure 11.3(b), MCM-41 mesoporous silica, synthesized from an alkaline solution using TEOS and CTAB, displays a uniform particle size of 0.75 μm with SSA of 1070 $m^2 g^{-1}$, a disk-like shape, and a highly ordered hexagonal pore structure (see Figure 11.3(c)). 3-Aminopropyltriethoxysilane ($-NH_2$) and melamine-based dendrimer amines (MDA) were used to functionalize SBA-15 mesoporous silica, resulting in NH_2-SBA-15 and MDA-SBA-15, respectively, for the removal of Pb(II), Cu(II), and Cd(II) from a mixed metal solution.[69] Both functionalized materials showed over 150% and 400% increased adsorption, respectively, compared to unmodified SBA-15. MDA-SBA-15 demonstrated superior removal efficiency due to its multifunctional amine groups, with maximum uptake capacities of 130, 126, and

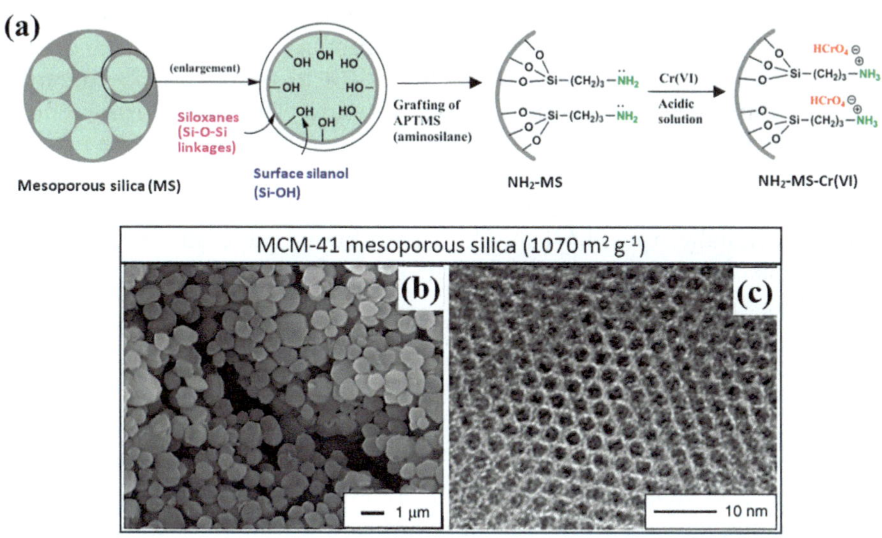

Figure 11.3 (a) Schematic illustrating the modification of mesoporous silica by grafting 3-aminopropyltrimethoxysilane; under acidic conditions, the amine groups become protonated, facilitating the electrostatic attraction of anionic Cr(VI) to the silica support, (b) SEM images showing the microstructure of MCM-41 powder after air calcination, which removes the CTAB template and (c) transmission electron microscopy (TEM) of the MCM-41, revealing the ordered hexagonal arrangement of the pore structure. (a) Reproduced from ref. 60 with permission from American Chemical Society, Copyright 2015. (b) Reproduced from ref. 68 with permission from Elsevier, Copyright 2006.

98 mg g^{-1} for Pb(II), Cu(II), and Cd(II), respectively at pH 4. Additionally, MDA-SBA-15 achieved removal efficiency of 95%, 93%, and 94% for these metals over four regeneration cycles, indicating its effectiveness in wastewater treatment. Amino functionalization of MCM-41 and SBA-15 also showed exceptional removal capacity for anionic dye.[70,71]

Although less frequently explored for organic contaminants, mesoporous silicas also show considerable promise for their removal. Functionalization is mandatory in this context because the silanol groups on the surface of pristine silica adsorbents readily form hydrogen bonds with water, which can impede the adsorption of organic molecules.[64] Bi-functionalized SBA-15 were designed to target specific pollutants such as eosin, 4-nonylphenol (4-NP), and di-*n*-butyl phthalate (DBP).[72] Cooperative adsorbents, diamine/phenyl-SBA-15, diamine/cetyl-SBA-15, and phenyl/cetyl-SBA-15, were synthesized by reacting two organosilanes with SBA-15 simultaneously. The cooperative adsorbents demonstrated higher adsorption capacity for these pollutants at low concentrations compared to monofunctionalized adsorbents. The maximum adsorption capacities for eosin, 4-NP, and DBP were 382.3, 207.3 and 434.2 mg g^{-1}, respectively. On the other hand, simple functionalization of MCM-41 with 3-aminopropyltriethoxysilane achieved 99.25% removal of the antibiotic ciprofloxacin from an aqueous medium.[73] Designing silicate composites also has great promise for pollutant removal. George *et al.* developed a poly(lactic acid) (PLA) composite membrane for separating pentachlorophenol (PCP) and 2,6-dichlorophenol (DCP) from water by incorporating titanium-coated magnetic mesoporous silica (TiO_2-MMS) using a phase inversion method.[74] Their reports were quite interesting because the addition of TiO_2-MMS enhanced the hydrophilicity and porosity of the PLA membranes, improving the permeation flux and retention capacity of PCP and DCP. Additionally, the PLA/TiO_2-MMS membrane achieved a flux recovery ratio of 54% for PCP and 58% for DCP, demonstrating effective antifouling properties.

11.1.4 Biochar

Biochar (BC) is a porous carbonaceous material commonly derived from the pyrolysis of biomass in the absence of oxygen or under oxygen-restricted conditions.[75–77] It is formed alongside biofuel, volatiles, and gases,[78] exhibiting a highly variable chemical composition and yield due to diverse biomass sources.[76,78] In addition to the type of biomass, pyrolysis conditions are fundamental in determining the physicochemical properties of the resulting BC.[76,79] Lignin-rich biomass, such as wood, typically undergoes slower pyrolysis, resulting in a higher yield of BC[80] that is rich in aromatic carbon.[78,81] Higher pyrolysis temperatures promote an increase in the concentration of recalcitrant aromatic structure in BC, but they reduce the yield.[76] More ordered graphitic structures are formed in BC generated above 700 °C.[78] Increasing pyrolysis temperature gradually increases porosity, specific surface area, pH, mineral elements and ash content, aromaticity, and electrical conductivity of the BCs, whereas lower pyrolysis temperatures result in BC with higher volatile matter content and more oxygen-containing functional groups, enhancing cation-exchange capacity.[82,83] Biochars are typically pyrolyzed between 300 °C and 700 °C, with surface areas ranging from tens of $m^2 g^{-1}$ to 800 $m^2 g^{-1}$ under favorable carbonization conditions.[84] Figure 11.4 depicts the microstructure of tea-waste BC obtained at different pyrolysis temperatures. The surface area of the BC increased from 0.204 to 8.105 $m^2 g^{-1}$ as the pyrolysis temperature was raised from 300 °C to 700 °C.[85]

Pyrolysis is the conventional method for producing biochar (BC), but alternative techniques like hydrothermal carbonization (HTC) and gasification are also explored. HTC converts wet biomass at lower temperatures, typically between 180 °C and 260 °C, into hydrochars.[86] During hydrolysis, hydrochars develop numerous acidic groups, including –OH, –COOH, and phenolic groups on their surface, enhancing contaminant adsorption.[87] Biomass gasification is carried out at temperatures above 700 °C in a controlled oxygen atmosphere, producing gases such as CO, H_2 and CO_2, along with smaller quantities of BC rich in alkali and alkaline earth metals.[77,88] This metal-rich BC is useful for the separation of heavy metals by precipitation.[89] At higher reaction temperatures, these BCs may contain polycyclic aromatic hydrocarbons, presenting potential secondary contamination risks.

The interest in utilizing BCs as adsorbents in environmental applications primarily arises from their cost-effective production through biomass valorization, a renewable energy source,[75,77] as well as their thermal and chemical stability. These qualities facilitate secure carbon storage in BC, thereby helping to prevent its release as CO_2 or toxic carbon forms into

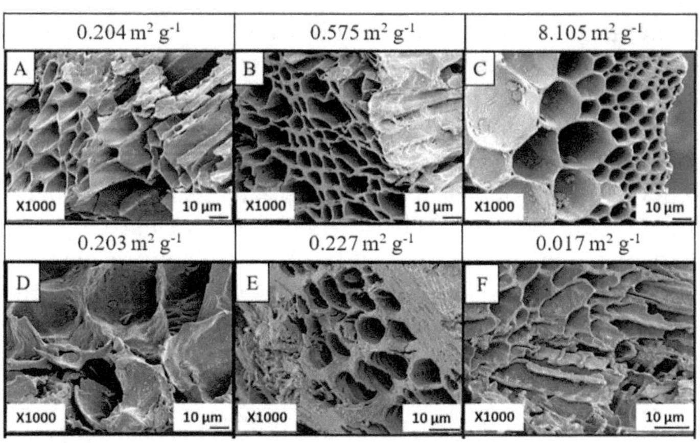

Figure 11.4 SEM images of tea-waste biochar produced at different pyrolysis temperatures, (A) 300 °C, (B) 500 °C and (C) 700 °C and acid modifications of biochar post-pyrolysis at 300 °C using (D) sulfuric acid, (E) nitric acid, and (F) hydrochloric acid. Reproduced from ref. 85 with permission from the Royal Society of Chemistry.

water.[90] Furthermore, BCs are rich in surface functional groups that are derived from biomass such as hydroxyl, carboxyl, acyl, carbonyl, ether, ester, and amino groups, enabling effective immobilization of various pollutants in aqueous media.[89] However, their performance in contaminant removal can be highly variable, depending on their physicochemical properties, the type of biomass used, and additional processes such as biomass pre-treatment and BC post-treatment.[75]

The effectiveness of BCs in removing heavy metals from wastewater largely depends on the pH of the adsorption medium and their surface functional groups.[91] In the pH range of 6.0 to 7.0, ion-exchange mechanisms typically facilitate heavy metal removal, whereas, at pH between 7 and 9, the removal can occur through surface complexation and electrostatic attraction.[92] For instance, Cr(VI) adsorption by pod-based biochar was most effective at pH 6.0 but decreased at pH 11 due to metal hydroxide precipitation.[93] A magnetic biochar composite prepared from pine bark waste demonstrated optimal removal of Pb and Cd from wastewater at pH 2–8,[94] which was reduced below pH 5 due to competition between hydrogen ions and positively charged heavy metals at BC sites. Biochar rich in oxygen-containing functional groups such as –OH, –COOH, and –C–O is particularly more effective for removing metal cations.[89,95] The surplus of negative charges on the surface of BCs boosts its cation exchange capacity, enhancing the electrostatic attraction of BC to positively charged heavy metal ions.[95] Biochar with these characteristics can be obtained by lower temperature pyrolysis or through the HTC technique with the production of hydrochar, as previously mentioned. One approach to enhance metal sorption efficiency involves increasing BC surface functionality through the modification with alkalis (KOH or NaOH), acids or oxidizers (H_2O_2, HNO_3, HCl, and H_2SO_4). This modification can be conducted either *in situ* during carbonization or by adding the chemicals to the pre-formed biochar.[96] Introducing oxidizers and acids adds acid binding sites such as carboxylic and phenolic groups to the BC surface, while alkaline modification enriches the hydroxyl group content.[97] These treatments may either increase or decrease the surface area of the biochar, depending on the feedstock type and processing method. After pyrolysis at 300 °C, acid modification of tea-waste BC revealed that treatments with nitric acid, sulfuric acid, and hydrochloric acid partially degraded the pore structure of the BC. Treatment with nitric and sulfuric acid did not alter the surface area, whereas activation with hydrochloric acid reduced the surface area of the BC from 0.204 to 0.017 $m^2 g^{-1}$ (see Figure 11.4).

H_2O_2-modified peanut-hull biochar exhibited a 20-fold increase in Pb(II) sorption capacity compared to unmodified biochar.[98] HNO_3 activation enhanced the U(VI) sorption ability of cactus biochar,[99] whereas the pre-treatment of date seed biomass with HCl solution resulted in BC with greater sorption capacity for Pb(II), Cu(II) and Ni(II).[97] Post-treatment of BC from waste chicken feathers with KOH solution showed significantly increased sorption capacities of modified BC for Cd(II) and Pb(II).[100] It is worth mentioning that the introduction of more oxygen-containing functional groups on the surface of biochar can be useful in increasing hydrophilicity and consequently its performance in aqueous applications.

High in carbon content, BC shows high capacity in removing organic contaminants from wastewater such as dyes, halogenated hydrocarbons, phenolics, pesticides, aromatics, and antibiotics. Several reports suggest that BCs produced at higher temperatures demonstrate improved efficacy in organic removal, thanks to their larger surface area, pore volume, and heightened aromaticity.[77,92] Post-modification of biochar using ball milling has also been shown to effectively increase its surface area and improve hydrophilicity, resulting in an enhanced adsorption capacity for several volatile organic compounds, including acetone, chloroform, cyclohexane, and toluene,[101] and for methylene blue dye.[102] Organic pollutants are primarily adsorbed through mechanisms such as pore filling, partitioning, and electronic interactions.[89,101] They can also undergo degradation and redox reactions on the BC surface pore due to electron donor–acceptor interactions from its graphitic structure as well as the presence of free radicals generated during the carbonization process.[89] Solutions for degrading organic pollutants in water are environmentally crucial as they avoid the generation of secondary pollutants that are transferred to the solid phase through the adsorbent. Many of these strategies have been achieved with the production of biochar composites; some examples include biochar-supported photocatalytic materials such as BC@$CoFe_2O_4$/Ag_3PO_4 for bisphenol A degradation[103] and BC@TiO_2 for Acid Orange 7 dye degradation.[104]

11.1.5 Activated Carbon

Activated carbon (AC) is one of the oldest and most widely used porous materials in water and wastewater treatment. It is commonly available in powder (PAC) or granulated (GAC) form. AC is predominantly derived from materials such as wood, coal, lignite, peat, coconut shells and fruit pits, which are widely used in its large-scale production.[105] The utilization of non-renewable sources like coal, lignite, and peat for commercial AC production remains an ongoing environmental concern. Unlike biochar, AC production necessarily requires an activation process, which utilizes physical, chemical, or biological agents.[106] This step is essential for enhancing the adsorption properties of ACs by creating well-developed internal porous structures comprising micropores, mesopores, and macropores, resulting in a large SSA (600–2000 $m^2 g^{-1}$),[107,108] generally superior to that of biochar. Before or after activation, biomass undergoes carbonization at temperatures ranging from 800 °C to 1000 °C in the absence of air or with restricted oxygen, aimed at removing non-carbonaceous and volatile components, thereby yielding pure carbon.[107] Therefore, the production process of AC is typically more costly comparative that of biochar. AC derived from compressed wood was chemically activated using H_3PO_4.[109] Following acid impregnation, the wood was pyrolyzed at 900 °C for 30 min to produce the activated carbon, as depicted in Figure 11.5(a). The chemical activation process induced the

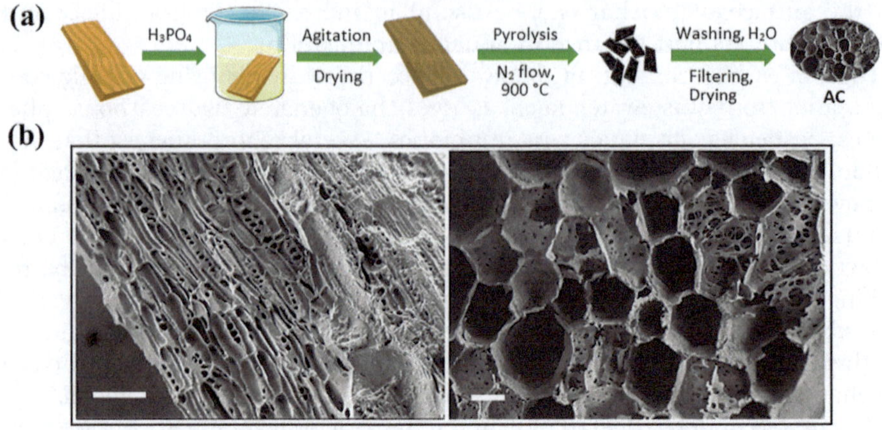

Figure 11.5 (a) Preparation of AC from compressed wood through chemical activation using phosphoric acid (H_3PO_4) as an activating agent; (b) SEM images of the porous structures of resulting AC obtained with scale bars corresponding to 100 μm (left image) and 10 μm (right image). (b) Reproduced from ref. 109, https://eurasianjournals.com/index.php/ej/article/view/451, under the terms of the CC BY 4.0 license, http://creativecommons.org/licenses/by/4.0/.

formation of pores of various sizes and shapes, resulting in a structure with micro, meso, and macroporosity, as illustrated in Figure 11.5(b). The AC adsorbents exhibit the potential to adsorb 311.34 $mg\,g^{-1}$ of crystal violet.

The adsorption capacity of AC depends on its structural characteristics, namely pore volume and surface area, as well as the presence of oxygenated functional groups such as carbonyl, carboxyl, aldehyde, and phenol on surface AC, composed mainly of graphitic layers.[110,111] These functional groups enhance surface reactivity and can originate from the starting biomass or be formed during the activation process.[112] Additionally, these groups can be easily modified to tailor desired properties, thereby enhancing their affinities towards different pollutants present in aqueous solution.

AC solids are effective in removing a wide range of pollutants from water, such as heavy metals,[113] synthetic dyes,[108] small organic compounds like pesticides,[114] microplastics,[115] natural organic matter and algal toxins,[116] and taste and odor compounds,[117] through adsorption and biodegradation mechanisms. Their high adsorption and regeneration capacity are characteristics that make them highly desirable materials for water treatment and purification.[110,118] Although PAC particles typically have higher surface area and adsorption capacity compared to GAC,[119] they are prone to sedimentation and are challenging to regenerate, limiting their application primarily to batch processes, particularly in wastewater treatment facilities. In contrast, GAC particles are well-suited for deployment in columns or beds, facilitating continuous pollutant removal with enhanced process control and easier regeneration,[118] therefore they are the predominant form of AC used in drinking water treatment processes.[120] However, their adsorption capacity is reduced over time due to fouling by natural organic matter and trace organic compounds present in the water, eventually resulting in saturation of the GAC bed and interruption in the treatment process.[118,120,121] Thus, regeneration is usually required to extend the lifespan of the ACs rather than discarding them. While regeneration enhances the sustainability of GAC as an adsorbent by enabling its reuse in multiple adsorption cycles, it can also lead to significant costs associated with degrading or safely disposing of compounds desorbed from the regeneration solution. Furthermore, with an increase in regeneration cycles, the adsorption capacity decreases due to the difficulty in completely removing strongly adsorbed materials from the carbon surface.[121]

Photocatalytic semiconductor-modified AC such as TiO_2 provides an alternative to traditional GAC-based filters by preventing fouling and enabling self-cleaning through UV radiation exposure. For instance, GAC-TiO_2[122] and Bi_2O_3–TiO_2/PAC[123] composites were employed for synergistic removal, degradation, and mineralization of humic acid in water, a major cause of fouling during water filtration. Additionally, in these composites, the target organic molecules are initially adsorbed by the AC and then transferred to the catalytic site where they undergo decomposition. This combination enhances the adsorption capacity by creating additional adsorption sites for pollutants, thus increasing the rate and capacity of degradation. The AC-TiO_2 composite, with a SSA of 1541 $m^2 g^{-1}$, demonstrated superior performance in adsorption and degradation of phenol in seawater under visible light compared to commercial TiO_2 with SSA ranging from 43 to 50 $m^2 g^{-1}$. In contrast, neat AC exhibits catalytic activity only under UV light.[124]

11.1.6 Hydrogels

Hydrogels are three-dimensional networks and porous structures composed of synthetic or natural polymers, whose chains are interconnected by covalent bonds or physical interactions. These structures contain hydrophilic groups or domains that enable them to absorb and retain a substantial amount of water.[125] They can be produced from nearly any hydrophilic polymer, providing a wide range of chemical compositions and bulk physical properties, and molded into a variety of physical forms and sizes, making them suitable for use in different equipment and processes. Additionally, low-cost production and green synthesis routes can be utilized to prepare hydrogels.[126] These characteristics make hydrogel materials advantageous over other adsorbents for removing contaminants in water. One of the most important characteristics of hydrogels for water treatment applications is their swelling capacity and their ability to maintain physical stability over long periods under swelling conditions thanks to its crosslinked network.[126–128] High swelling and wetting not only facilitate the diffusion of pollutants within the hydrogel but also increase the surface area of the polymeric network, exposing more sites for adsorption.[129] Nevertheless, the state of water in the hydrogel might affect its adsorption capacity. For instance, the adsorption capacity of tryptophan modified hydrogels for methylene blue and rhodamine B increased with higher amounts of freezing water.[130]

Hydrogels for water purification are polymeric networks capable of absorbing large amounts of aqueous solution and immobilizing contaminants.[131] However, it is important to emphasize that the volume increase due to swelling can complicate material handling, reduce its mechanical stability, and hinder its regeneration.[132] Some strategies to overcome these challenges include controlling swelling and enhancing the mechanical strength of hydrogels. Swelling can be managed through the synthesis of smart hydrogels[132] or by introducing additional crosslinks in the polymeric structure.[133] Alternatively, using more robust polymers or copolymers[134] and integrating nanostructural reinforcements in the hydrogel matrix[135] are common approaches to increase the mechanical strength of hydrogels. For instance, combining a hydrophilic polymer with lignocellulose particles provides high swelling degree and improved mechanical properties.[136–140]

Beads, films, and membranes are among the most commonly used hydrogel physical forms in water and wastewater treatment.[126,129] They can be synthesized from synthetic polymers including polyacrylic acid (cPAA), polyacrylamide (PAM), poly(vinyl alcohol) (PVA), poly(ethylene glycol) (PEG), poly(*N*-isopropylacrylamide) (PNIPam) and poly(*N*-vinyl-2-pyrrolidone) (PVP) or natural sources such as protein and polysaccharides. Some examples of polysaccharide-based hydrogels investigated for water purification include chitosan, sodium alginate, pectin, starch, chitosan, cellulose, carboxymethyl cellulose (CMC), hydroxypropyl methylcellulose (HPMC) and kappa carrageenan.[141] Figure 11.6 displays the porous structures of adsorbents composed of xanthan gum, chitosan, and HPMC assessed by X-ray microtomography and scanning electron microscopy. They presented SSA of 2.1, 1.3 and 1.6 $m^2 g^{-1}$, and swelling degree values of 6800%, 3300%, and 4800%, respectively.[142] Because of growing environmental concerns, synthetic polymers are increasingly being replaced or combined with natural polymers, especially polysaccharides due to their biodegradability, non-toxicity, cost-effectiveness, availability and renewability.[143] This combination is particularly advantageous, as the functional groups of natural polymers make them effective sorbents, while synthetic polymers can contribute to mechanical strength, chemical and thermal stability or provide additional adsorption sites.[131,143] For example, interpenetrating polymer network hydrogels prepared by mixing CMC solution and crosslinked cPAA demonstrated superior adsorption capacities for Cu(II) and methylene blue compared to their individual components.[144] Hydrogels composed of *N*,*N*,*N*-trimethyl chitosan chloride (N-quaternized chitosan, NQC) combined with cPAA demonstrated a higher removal capacity for Fe(III) and Cd(II) compared to pure chitosan and increased intra- and intermolecular hydrogen bonding interactions among cPAA and NQC offered greater thermal stability for the hydrogels.[145] It was also demonstrated that the addition of PAM to the CMC matrix provided greater

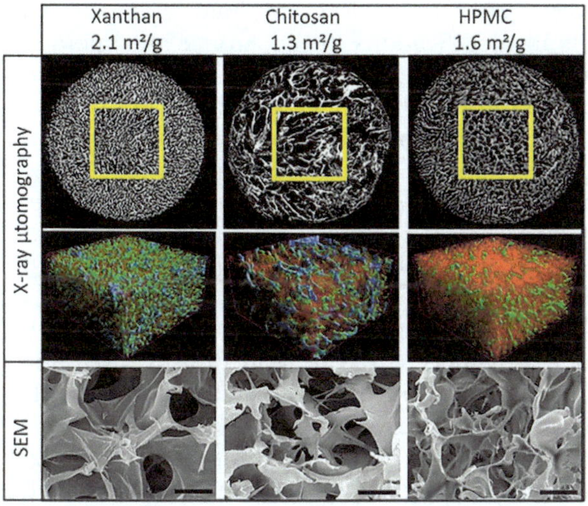

Figure 11.6 X-ray microtomography was used to obtain 2D images (5 mm diameter) and reconstructed 3D images (2 mm × 2 mm × 1 mm) of porous structures composed of xanthan, chitosan, and HPMC. These images allowed for the determination of the surface area of the porous structures. The X-ray attenuation was color-coded as follows: 0 to 10% (void space) in black, 10 to 30% in red, 30 to 70% in green, and 70 to 100% in blue. SEM images of the same porous structures were also obtained, with a scale bar corresponding to 50 μm. Reproduced from ref. 142 with permission from Elsevier, Copyright 2019.

adsorption properties and mechanical robustness. The resulting hydrogel exhibits excellent adsorption performance for Cu(II), Pb(II) and Cd(II).[146]

Bead hydrogels offer distinct advantages for water decontamination. They usually remove large amounts of the pollutant and can be easily recoverable and regenerable from the treatment medium.[147] For example, chitosan beads impregnated with 1-butyl-3-methylimidazolium bromide (BmimBr) removed 86% of methylene blue from contaminated water in just 25 min.[148] Usability testing revealed a loss of just 6% of its adsorption capacity after the 4th reusing cycle. Similarly, beads made from lignosulfonate-grafted acrylic acid with a high capacity for methylene blue adsorption (2013 $mg\,g^{-1}$) were recycled for four cycles and maintained 83% removal efficiency for MB.[149] In another study, chitosan beads impregnated with laccase degraded around 85% of bisphenol A within 150 minutes.[150] Immobilizing the enzyme on the beads was crucial for preserving its catalytic activity; after 28 days of storage at 4 °C, the enzyme retained 90% of its activity, allowing the potential reuse of laccase.

Hydrogels are particularly well-suited for encapsulation and impregnation, acting as an excellent platform for creating novel adsorbents for wastewater treatment. Different fillers, such as metal and metal oxide nanoparticles, inorganic materials, or carbon-based materials (carbon nanotubes, graphene oxide, and biochar), can be incorporated into hydrogels. The incorporation typically results in modifications in the surface area, porosity, stability, specificity, and adsorption capacity of hydrogels.[143] The incorporation of superparamagnetic iron oxide nanoparticles into hydrogels offers the advantage of removing the magnetic hydrogels by simply using a magnet. Furthermore, the magnetite particles can act as additional adsorption sites.[151–153] In an investigation involving montmorillonite (MMT) clay, it was demonstrated that incorporating 2 wt% MMT into poly(acrylic acid-*co*-acrylamide) gel increased the porosity of the composite, and consequently, the methylene blue adsorption capacity increased from 1878 to 1935 mg g^{-1}.[154] Cr(VI) ions and methylene blue adsorbed on porous xanthan gum/TiO_2 (10 wt%) with a SSA of 2.8 $m^2 g^{-1}$ underwent photocatalytic reduction to Cr(III) and photodegradation, respectively, upon UV exposure.[155] Chitosan–silica hydrogels with high porosity (96%) and a SSA of 2.05 $m^2 g^{-1}$ adsorbed 40 mg m^{-2} of Cu(II).[156]

The combination of nanoscale biochar into starch hydrogel was proposed as a strategy for removing Cr(VI) and naproxen drugs from real water matrices. Encapsulating biochar in starch gel increased the surface area significantly, from 48.59 $m^2 g^{-1}$ to 226.94 $m^2 g^{-1}$. The composite exhibited a high swelling capacity (500%) and achieved 50% removal efficiency for Cr(VI) and naproxen even after 40 regeneration cycles.[157] AgNPs embedded in polyacrylamide (PAAm)/starch hydrogels were used to detect trace amounts of phenol in water using surface-enhanced Raman scattering (SERS) spectroscopy. The hydrogel effectively prevented AgNP aggregation, creating numerous hot spots on the hybrid hydrogels and enabling SERS spectra. This system exhibited high reproducibility, stability, and a detection limit for phenol in water of 1×10^{-8} M.[158] Graphene oxide filler in hydrogels usually plays a prominent role in reinforcing the polymer network. Loading 7 wt% tannic acid-modified reduced GO into the chitin matrix not only resulted in excellent removal capacity for Congo red but also greatly improved its mechanical properties. The compressive strength of pure chitin was 22.7 kPa and increased to 72.3 kPa after modified GO loading in the hydrogel.[159]

Table 11.1 summarizes the main advantages and disadvantages of the porous materials used in water remediation discussed in this chapter.

11.2 Methods for Water Treatment

Water treatment methods can be physical, chemical, or biological. The selection of a method or a combination of methods should be based on the specific types and concentrations of contaminants present, the volume of water to be treated, and the desired quality of the treated water. Additionally, it is essential to consider the costs, the durability and regenerability of the materials used in each method, and the scalability of the treatment process. Table 11.2 summarizes the main characteristics of commonly used water remediation methods.

Filtration and sedimentation are the main physical methods for water remediation. The choice of filtration method depends on specific water

Table 11.1 Comparison between the main porous materials applied in water remediation.

Porous material	Advantages	Disadvantages
MOF	• High surface area • Highly adjustable and selective porous structure • High adsorption capacity for gases, metal ions, organic compounds and small molecules • Scale-up capacity	• High synthesis costs • Toxicity potential of organic ligands and solvents used in synthesis • Limited stability under extreme pH and temperature conditions
Zeolites	• High selectivity for cations • High adsorption and ion exchange capacity • Stability over a wide pH range. • High availability	• Complex regeneration depending on the adsorbate; • Non-renewable • High synthesis costs for synthetic zeolites • Natural zeolites are less versatile compared to synthetic ones
Silica based materials	• Excellent thermal and chemical stability • High mechanical resistance • Ease of surface modification • Adjustable porous structure	• Limited selectivity • Adsorption of organic compounds may be limited without functionalization • Non-renewable (*e.g.* mesoporous silica)
Biochar	• High adsorption capacity for organic compounds and heavy metals • Thermal and chemical stability • Carbon sequestration potential • Safe and non-toxic material • High availability • Low synthesis cost	• Variable selective adsorption and effectiveness depending on preparation conditions and feedstock • May require pretreatment to maximize efficiency. • Smaller surface area compared to other porous materials
Activated carbon	• High surface area and adsorption capacity • Effective for a wide range of contaminants • Highly availability • Safe and non-toxic material • Scale-up capacity	• Susceptible to fouling or sedimentation • Limited regeneration • Surface and properties may vary depending on the activation method • Higher synthesis cost than biochar
Hydrogels	• High water retention and high absorption capacity • Biocompatibility • Contaminant encapsulation capacity • Control over the release of agents for specific treatments	• Low mechanical resistance • Potential for chemical degradation • Difficult to regenerate and recycle

Table 11.2 Advantageous and disadvantageous aspects of common water remediation methods.

	Method	Cost	Operation	Efficiency	Energy consumption	Selectivity	Scalability	Versatility	Reg.	Remarks
Physical	Sand filtration	Low	Simple	High only for suspended and large particles	Low	Low	High	Low	No	It requires backwashing, and periodic replacement of sand due to clogging
	Sedimentation	Low	Simple	High only for suspended particles and settleable solids	Low	Low	High	Low	No	Generated sludges require appropriate disposal Sedimentation tank requires periodical cleaning
	Activated carbon filtration	High	Simple	High for organic compounds, chlorine, and certain chemicals, improving taste and odor	Low	High, if specifically modified	High	For drinking water, wastewater, and industrial processes	Thermal regeneration is costly and demands high energy consumption	It requires periodic replacement due to clogging
	Membrane filtration (microfiltration, ultrafiltration, reverse osmose)	High	Suitable for various scales of operation	High for particles, bacteria, viruses, and dissolved substances	High, especially reverse osmosis	High, if specifically modified	High	High, it can be tailored to specific water treatment needs	It produces a concentrated waste stream (brine) that needs appropriate disposal	It provides consistent water quality, meeting stringent regulatory standards Membranes are prone to fouling and scaling, requiring regular cleaning and maintenance

Chemical	Precipitation	Low	Simple	Highly effective in converting dissolved metals into insoluble precipitates	Low	Selectivity requires different solubility product constants	High	A wide range of contaminants, including phosphates, sulfates, and certain organic compounds	No	Precipitates formed may not always settle effectively or may re-dissolve under certain conditions. Sludge may contain toxic metals or other contaminants, necessitating careful handling to prevent environmental pollution
	Coagulation and flocculation	May be high	Simple	Highly effective for colloidal particles	Low	Selectivity requires specific flocculants	High	Surface water, groundwater, and wastewater	No	Flocculants can be costly and require careful handling and dosing May leave residual chemicals in treated water, requiring additional steps for chemical removal or neutralization
	Ion exchange	High	Regular monitoring and maintenance	High towards ions, low towards organic compounds and pathogens	Low compared to reverse osmosis	High	High	High, for ions, heavy metals	It produces waste brine or other chemical waste, which can pose environmental disposal issues	Resins can become fouled or degraded over time by organic matter, iron, manganese, or other substances, reducing their efficiency and requiring more frequent replacement
	Advanced oxidation processes	High	Complex	High	High, unless UV light comes	No	Challenging	High	No	Capable of treating various types of contaminants including

Table 11.2 (*Continued*)

Method	Cost	Operation	Efficiency	Energy consumption	Selectivity	Scalability	Versatility	Reg.	Remarks
				from the sun					pharmaceuticals, pesticides, industrial chemicals, and endocrine-disrupting compounds. AOPs can produce byproducts, some of which may be toxic or require further treatment, such as bromate formation during ozonation
Adsorption	It depends on the type of adsorbent	Easy	High for ions and organic compounds	Low	High, if specifically modified	Fixed-bed columns are scalable	Yes	It produces waste, which can pose environmental disposal issues	Spent adsorbents need to be disposed of properly, which can pose environmental and regulatory challenges
Electrocoagulation	High	Complex	For suspended solids, heavy metals, oils, grease, and certain organic compounds	High		Complex	No	Electrodes can corrode and require regular maintenance or replacement	It can be sensitive to the pH of the water
Electrodialysis	High	Complex	For ionic contaminants, such as salts, nitrates, and heavy metals	High	Low	Complex	No	Yes	Membranes can become fouled by organic matter, scaling, or biological growth, requiring regular cleaning and maintenance.

									It is less effective for removing non-ionic contaminants, such as organic compounds and pathogens, often requiring pre-treatment or additional treatment steps
Biological	Low	Complex	Efficient at removing heavy metals, organic pollutants, pesticides, and solvents.	Low, methane produced can be used as an energy source	Low	High	Yes	No	Sensitive to temperature, pH, and oxygen level. Less effective for removing heavy metals, non-biodegradable organics, and some pharmaceuticals. Sludge as a byproduct, which requires further treatment and disposal

quality requirements, budget constraints, and operational considerations. Slow sand filtration is one of the most efficient and cost-effective methods for providing safe drinking water to people living in remote areas.[160] Overall, sedimentation is a robust and widely used method for removing larger particles and settleable solids from water, but it may not be sufficient for treating water with high turbidity or very fine particles.[161] It is often employed as a primary treatment step in conjunction with other processes like filtration or chemical treatment.

As mentioned in the previous section, AC is widely used for filtration due to its high surface area, pore volume, and ability to remove color, odor and taste from water and wastewater. After saturation, the AC-based filters should be replaced and disposed, increasing cost and environmental impact. However, these drawbacks can be reduced, if active carbon is prepared from agricultural wastes.[162]

Membrane filtration is a widely used method for water treatment that employs semi-permeable membranes to separate from water microorganisms, particulates, and dissolved substances. Microfiltration operates with membranes, which have pore sizes ranging from 0.1 to 10 μm; it removes suspended particles and microorganisms.[163] Ultrafiltration uses membranes that have pore sizes ranging from 0.01 to 0.1 μm, allowing the removal of viruses, bacteria, colloidal silica, and larger organic molecules. It is applied for water and wastewater treatment and protein concentration in food and pharmaceutical industries.[164] Nanofiltration can remove multivalent ions, smaller organic molecules, and some dissolved salts because the membranes have pore sizes ranging from 0.001 to 0.01 μm.[165] Reverse osmosis is used for the desalination of seawater and brackish water, production of ultrapure water for industrial use, and home drinking water purification systems; the pore size is typically smaller than 1 nm.[166]

There are numerous chemical methods for water treatment, each based on different principles. Chemical precipitation remains a valuable and widely used method for removing contaminants from water, particularly metals and certain ions.[167] However, careful consideration of water chemistry, operational practices, and environmental impacts is essential to maximize its effectiveness and minimize drawbacks. Coagulation is the process of adding a coagulant to water to destabilize and aggregate suspended particles, colloids, and dissolved substances; it is thermodynamically irreversible. Aluminum sulfate, ferric chloride, polyaluminum chloride, and organic polymers are commonly used as coagulants.[168] On the other hand, flocculation is the reversible destabilization of particles. Generally, the flocs are formed by the addition of a polymer chain (charged or uncharged), which binds to different particles simultaneously, forming "bridges" among them; this mechanism is called "bridging flocculation". After coagulation, a flocculant may be added to help form larger flocs, which settle at the bottom of the tank.[169] Ion exchange is an effective and versatile method for water treatment, particularly for specific ion removal and water softening. Nevertheless, it involves operational complexities, chemical handling, and potential environmental impacts that need to be carefully managed.[170] Advanced oxidation processes (AOPs) are a set of chemical treatment procedures designed to remove organic and some inorganic materials from water and wastewater by oxidation through reactions with hydroxyl radicals ($^{\bullet}OH$). These processes are highly effective in degrading pollutants that are otherwise difficult to remove. AOPs can break down contaminants to carbon dioxide, water, and inorganic ions, leading to complete mineralization rather than just transformation into less harmful substances.[171] They offer a powerful and versatile approach to water treatment, capable of degrading a wide range of pollutants and providing disinfection. However, they also involve high operational costs, complexity, and potential byproduct formation, necessitating careful consideration and management to ensure their effective and safe use. Adsorption is a widely used process in water treatment where contaminants are removed from water by adhering to the surface of a solid material, typically activated carbon, zeolites, biomass, or synthetic resins. Adsorption is an effective and versatile method for water treatment, offering high efficiency, selectivity, and low energy consumption.[172] However, it also has limitations related to adsorbent saturation and disposal, cost, and the potential for desorption. Appropriate management and optimization are necessary to maximize the benefits of adsorption in water treatment applications. In the presence of complex contaminant mixtures, the adsorption process may become less efficient due to competitive adsorption among different contaminants. Electrocoagulation involves applying an electric current to water, causing the dissolution of metal electrodes (usually iron or aluminum), which then form coagulants that help remove contaminants. Electrocoagulation is versatile and chemical-free, but it requires careful management of energy consumption and electrode maintenance.[173] Electrodialysis uses an electric potential to move ions through selective membranes, separating them from water.[174] It is efficient for ion removal and desalination but involves high initial costs and membrane maintenance.

Biological methods for water treatment involve using living organisms to degrade or transform contaminants. Compared to chemical or physical methods, bioremediation is effective, environmentally friendly, and can be more cost-efficient, but it can be slower. Aerobic and anaerobic bioreactors use aerobic microorganisms and oxygen, and anaerobic microorganisms in the absence of oxygen, respectively. In aerobic treatment, the wastewater flows into tanks or ponds containing microorganisms that break down the contaminants in the presence of oxygen. If the tanks or ponds are shallow the aeration process is easier. In contrast, the anaerobic process should take place in a closed tank to avoid oxygen. Anaerobic treatment can generate methane as a byproduct, which can be used as fuel in other processes.[175] Phytoremediation involves the use of plants to (i) absorb contaminants through their roots and store them in their stems, leaves, or roots; (ii) metabolize contaminants within their tissues; (iii) absorb through their roots, concentrate, and precipitate heavy metals from contaminated water; and (iv) stabilize contaminants in the soil or water, avoiding their migration.[176] Biosorption refers to the utilization of bacteria, fungi, algae, or plant biomass to adsorb and concentrate heavy metals or other pollutants from aqueous solutions.[177] Algae and plants can remove excess nutrients, such as nitrogen and phosphorus, from agricultural runoff and wastewater, helping to prevent eutrophication.

11.3 Challenges and Considerations

Real-world scenarios involving the removal of contaminants are essential for ensuring safe drinking water, cleaning industrial and municipal wastewater, and remediating polluted lakes, rivers, and groundwater. The choice of materials and methods for treating wastewater is often challenging due to factors such as the large volume of wastewater, cost, feasibility, stability, scalability, and environmental impact.

Using porous materials for wastewater treatment offers numerous advantages. Their large surface area and porosity enhance contact with contaminants, increasing treatment efficiency. Additionally, porous materials can be tailored to target specific contaminants and many can be regenerated and reused, reducing operational costs. Porous materials derived from biomass, such as biochar, are sustainable and environmentally friendly. On the other hand, high-performance materials like MOFs can be expensive to produce. Dealing with the fate of the spent adsorbents and membranes without causing secondary pollution is challenging.[178]

The circular economy aims to create a closed-loop system where resource input, waste, emissions, and energy leakage are minimized by slowing, closing, and narrowing material and energy loops. Products, materials, and resources are kept in use for as long as possible, extracting maximum value from them while in use, then recovering, regenerating, or transforming them at the end of their service life.[179] This approach promotes environmental sustainability, economic resilience, and long-term resource security. Thus, developing applications for spent adsorbents and exhausted membranes, instead of disposing of them, is important to contribute to the circular economy and to avoid secondary pollution.

Regenerating contaminated membranes and adsorbents by desorption through rinsing with strong acids occurs due to the replacement of cationic metal ions by H^+ ions. However, this process produces a leachate that is more concentrated than the original wastewater, which, despite its reduced volume, still requires appropriate treatment or disposal. At the end of their lifecycle (adsorption/desorption), spent membranes and adsorbents should be disposed of properly through incineration or landfilling.[180]

Thermal regeneration is generally applied for contaminated activated carbon.[180,181] Upon heating under an inert atmosphere, the adsorbed molecules undergo desorption and decomposition.[182] In the presence of oxygen and/or in the wet state the desorption and oxidation of the contaminants may take place. However, the toxicity of the byproducts must be evaluated.[183]

Biological regeneration consists of applying living microorganisms to extract contaminants from the adsorbent's pores or surfaces. This method is cost-effective, suitable for biodegradable pollutants, and can result in complete regeneration of the adsorbent.[184]

In the electrochemical regeneration the contaminants desorbed from the surface of the spent adsorbent undergo oxidation into non-toxic products and/or mineralization.[185,186] Electro-Fenton regeneration, a redox process involving Fe^{2+} and H_2O_2, produces hydroxyl radicals ($^{\bullet}OH$) that react with the organic contaminants, yielding CO_2 and H_2O.[187,188]

To mitigate potential risks associated with the disposal of sludge, adsorbents, and membranes contaminated with organic molecules and toxic metals, incorporating these materials into cementitious pastes could be a viable strategy. This approach can effectively immobilize contaminants, reducing their environmental impact and providing a sustainable disposal method.[136,189,190] The immobilization of toxic metals, such as lead and mercury, in cementitious paste offers a potential method for isolating them from the environment. However, to be effective, these heavy metals should not leach upon contact with water or during cycles of rain and sun exposure. Furthermore, using such cementitious paste in civil construction is not advisable, as future demolition could potentially expose the immobilized contaminants to the environment.

Spent adsorbents can be reused as a nutrient source. Soil fertilizers contain essential macronutrients, such as calcium (Ca), magnesium (Mg), phosphorus (P), nitrogen (N), potassium (K), and sulfur (S), as well as micronutrients, including boron (B), chlorine (Cl), copper (Cu), iron (Fe), manganese (Mn), molybdenum (Mo), nickel (Ni), and zinc (Zn).[191] Kinetic studies on the release of these elements from spent adsorbents into the soil can help evaluate their potential as a nutrient source for agriculture. For instance, it is recommended that at 25 °C, there should be up to 15% release of these nutrients within 24 h and up to 75% release within 28 days.[192]

Transforming heavy metal-contaminated adsorbents into catalysts is an elegant method for recycling materials for new applications.[193] For instance, copper-loaded adsorbents have proven to be efficient catalysts for the reduction of 4-nitrophenol to 4-aminophenol.[146,194] Lead-loaded[195] and mercury-loaded[196] adsorbents demonstrated efficient catalytic activity for drug degradation. Ni-loaded biochar was modified and applied as supercapacitors, demonstrating that metal loaded spent adsorbents can be reused in energy producing processes.[197] This approach not only provides an effective means of recycling waste materials, but also contributes to the development of sustainable energy technologies. Figure 11.7 depicts possible strategies to regenerate spent adsorbents and membranes and some strategies to transform them into products for other applications.

Life Cycle Assessment (LCA) is important for evaluating the environmental impacts of adsorbents and membranes throughout their entire lifecycle.[198] This assessment includes all stages from raw material extraction, manufacturing, distribution, use, and disposal or recycling. LCA provides a

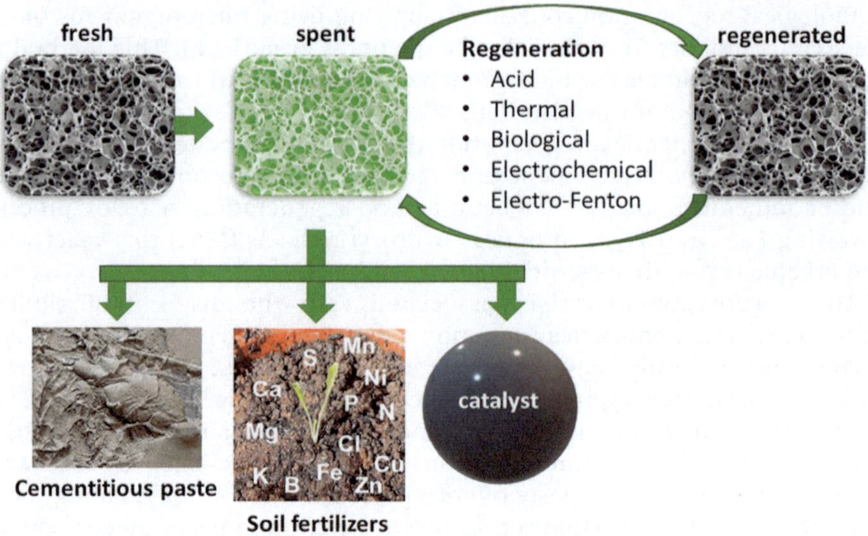

Figure 11.7 Strategies to regenerate spent adsorbents and methods to repurpose them into products for other applications.

comprehensive view of the environmental aspects and potential impacts associated with the adsorbents and membranes, and their related processes, enabling better decision-making towards sustainability.[199]

Acknowledgements

D. M. F. S. is grateful for a CAPES fellowship. Finance Code 001. D. F. S. P. acknowledges for the financial support from Fundação de Apoio a Pesquisa do Estado de São Paulo (Grant 2018/13492-2) and Conselho Nacional de Desenvolvimento Científico e Tecnológico (CNPq Grant 304017/2021).

References

1. G. Férey, *Chem. Soc. Rev.*, 2008, **37**, 191.
2. R. C. G. Frem, G. Arroyos, J. B. D. Flor, R. C. Alves, G. N. Lucena, C. M. da Silva and M. F. Coura, *Quim. Nova*, 2018, **41**, 1178.
3. Y. Huang, X. Zeng, L. Guo, J. Lan, L. Zhang and D. Cao, *Sep. Purif. Technol.*, 2018, **194**, 462.
4. C. Lei, J. Gao, W. Ren, Y. Xie, S. Y. H. Abdalkarim, S. Wang, Q. Ni and J. Yao, *Carbohydr. Polym.*, 2019, **205**, 35.
5. T. Wang, L. Han, X. Li, T. Chen and S. Wang, *Front. Chem.*, 2022, **10**, 962383.
6. Q. Zhang, D. Wang, R. Yu, L. Luo, W. Li, J. Cheng and Y. Zhang, *Green Process Synth.*, 2023, **12**, 20228123.
7. J.-Q. Chen, Z. Sharifzadeh, F. Bigdeli, S. Gholizadeh, Z. Li, M.-L. Hu and A. Morsali, *J. Environ. Chem. Eng.*, 2023, **11**, 109469.

8. A. Raj, R. M. Rego, K. V. Ajeya, H.-Y. Jung, T. Altalhi, G. M. Neelgund, M. Kigga and M. D. Kurkuri, *Chem. Eng. J.*, 2023, **453**, 139757.
9. S. Kashif, S. Akram, M. Murtaza, A. Amjad, S. S. A. Shah and A. Waseem, *Diamond Relat. Mater.*, 2023, **136**, 110023.
10. M. K. Batrakova, O. V. Solovtsova, A. A. Fomkin, A. Y. Tsivadze, A. V. Shkolin, A. A. Shiryaev and V. V. Vysotskii, *Prot. Met. Phys. Chem. Surf.*, 2017, **53**, 961–966.
11. F. Yang, M. Du, K. Yin, Z. Qiu, J. Zhao, C. Liu, G. Zhang, Y. Gao and H. Pang, *Small*, 2022, **18**, 2105715.
12. T. Rasheed, *Chemosphere*, 2023, **313**, 137607.
13. A. Jrad, P. Damacet, Z. Yaghi, M. Ahmad and M. Hmadeh, *ACS Appl. Nano Mater.*, 2022, **5**, 10795.
14. S. Rojas, J. García-González, P. Salcedo-Abraira, I. Rincón, J. Castells-Gil, N. M. Padial, C. Marti-Gastaldo and P. Horcajada, *Sci. Rep.*, 2022, **12**, 14513.
15. A. Al Obeidli, H. Ben Salah, M. Al Murisi and R. Sabouni, *Int. J. Hydrogen Energy*, 2022, **47**, 2561.
16. Y. Li, J. Miao, X. Sun, J. Xiao, Y. Li, H. Wang, Q. Xia and Z. Li, *Chem. Eng. J.*, 2016, **298**, 191.
17. Z. Wang, Z. Li, M. Ng and P. J. Milner, *Dalton Trans.*, 2020, **49**, 16238.
18. B. Karadeniz, A. J. Howarth, T. Stolar, T. Islamoglu, I. Dejanović, M. Tireli, M. C. Wasson, S.-Y. Moon, O. K. Farha, T. Friščić and K. Užarević, *ACS Sustainable Chem. Eng.*, 2018, **6**, 15841.
19. D. Crawford, J. Casaban, R. Haydon, N. Giri, T. McNally and S. L. James, *Chem. Sci.*, 2015, **6**, 1645.
20. R. M. Rego, G. Kuriya, M. D. Kurkuri and M. Kigga, *J. Hazard. Mater.*, 2021, **403**, 123605.
21. C. Wu, Z. Xiong, C. Li and J. Zhang, *RSC Adv.*, 2015, **5**, 82127.
22. Y. Liu, H. Cheng, M. Cheng, Z. Liu, D. Huang, G. Zhang, B. Shao, Q. Liang, S. Luo, T. Wu and S. Xiao, *Chem. Eng. J.*, 2021, **417**, 127914.
23. A. Alowasheeir, N. L. Torad, T. Asahi, S. M. Alshehri, T. Ahamad, Y. Bando, M. Eguchi, Y. Yamauchi, Y. Terasawa and M. Han, *Sci. Technol. Adv. Mater.*, 2024, **25**, 2292485.
24. H.-T. Fan, X.-T. Sun, Z.-G. Zhang and W.-X. Li, *J. Chem. Eng. Data*, 2014, **59**, 2106.
25. N. Ahmad, H. Sereshti, M. Mousazadeh, H. Rashidi Nodeh, M. A. Kamboh and S. Mohamad, *Mater. Chem. Phys.*, 2019, **226**, 73.
26. Md. R. Awual, Md. M. Hasan, J. Iqbal, A. Islam, Md. A. Islam, A. M. Asiri and M. M. Rahman, *Microchem. J.*, 2020, **154**, 104585.
27. T. S. Anirudhan and P. S. Suchithra, *Chem. Eng. J.*, 2010, **156**, 146.
28. M. Ahrouch, J. M. Gatica, K. Draoui, D. Bellido and H. Vidal, *J. Hazard. Mater.*, 2019, **365**, 519.
29. A. M. Bakry, F. S. Awad, J. A. Bobb and M. S. El-Shall, *ACS Omega*, 2020, **5**, 33090.
30. W. Liu, K. Li, X. Hu, X. Hu, R. Zhang and Q. Li, *Front. Environ. Chem.*, 2021, **2**, 693509.

31. J. Peng, Z. Zhang, Z. Wang, F. Zhou, J. Yu, R. Chi and C. Xiao, *J. Taiwan Inst. Chem. Eng.*, 2024, **156**, 105363.
32. L. Pan, Z. Wang, Q. Yang and R. Huang, *Nanomaterials*, 2018, **8**, 957.
33. R. Wang, Y. Li, X. Shuai, J. Chen, R. Liang and C. Liu, *Foods*, 2021, **10**, 3127.
34. Y. Liu, Y. Meng, X. Qiu, F. Zhou, H. Wang, S. Zhou and C. Yan, *Ceram. Int.*, 2023, **49**, 7030–7039.
35. J. Lou, Q. Fu, L. Yu, H. Yuan, J. Zhao, X. Wei, T. Wang and C. Mo, *Carbon Lett.*, 2023, **33**, 2029.
36. W. Zhang, S. Shi, W. Zhu, L. Huang, C. Yang, S. Li, X. Liu, R. Wang, N. Hu, Y. Suo, Z. Li and J. Wang, *ACS Sustainable Chem. Eng.*, 2017, **5**, 9347.
37. A. Hedström, *J. Environ. Eng.*, 2001, **127**, 673.
38. M. J. Zamzow, B. R. Eichbaum, K. R. Sandgren and D. E. Shanks, *Sep. Sci. Technol.*, 1990, **25**, 1555.
39. U. Wingenfelder, C. Hansen, G. Furrer and R. Schulin, *Environ. Sci. Technol.*, 2005, **39**, 4606.
40. A. Godelitsas, P. Misaelides, A. Filippidis, D. Charistos and I. Anousis, *J. Radioanal. Nuc. Chem. Artic.*, 1996, **208**, 393.
41. P. J. Reeve and H. J. Fallowfield, *J. Environ. Manage.*, 2018, **205**, 253.
42. Ž. Z. Tasić, G. D. Bogdanović and M. M. Antonijević, *J. Min. Metall., Sect. A*, 2019, **55**, 67.
43. S. Wang and Y. Peng, *Chem. Eng. J.*, 2010, **156**, 11.
44. I. Souza, F. García-Villén, C. Viseras and S. Pergher, *Pharmaceutics*, 2023, **15**, 1352.
45. F. A. Mumpton, *Proc. Natl. Acad. Sci. U. S. A.*, 1999, **96**, 3463.
46. L. F. de Magalhães, G. R. da Silva and A. E. C. Peres, *Adsorpt. Sci. Technol.*, 2022, 4544104.
47. W. Lu, C. Zhang, P. Su, X. Wang, W. Shen, B. Quan, Z. Shen and L. Song, *Environ. Sci.: Water Res. Technol.*, 2022, **8**, 2170.
48. M. S. Hosseini Hashemi, F. Eslami and R. Karimzadeh, *J. Environ. Manage.*, 2019, **233**, 785.
49. A. Nikolov, L. Dobreva, S. Danova, J. Miteva-Staleva, E. Krumova, V. Rashev and N. Vilhelmova-Ilieva, *Acta Microbiol. Bulg.*, 2023, **39**, 147.
50. B. Paul, J. J. Dynes and W. Chang, *Desalination*, 2017, **419**, 141.
51. H. Kurama, A. Zimmer and W. Reschetilowski, *Chem. Eng. Technol.*, 2002, **25**, 301.
52. X. Li, L. Fu, F. Chen, S. Zhao, J. Zhu and C. Yin, *Catalysts*, 2023, **13**, 342.
53. H. Valdés, R. F. Tardón and C. A. Zaror, *Chem. Eng. J.*, 2012, **211–212**, 388.
54. A. Ates and G. Akgül, *Powder Technol.*, 2016, **287**, 285.
55. R. T. Pabalan and F. P. Bertetti, *Rev. Mineral. Geochem.*, 2001, **45**, 453.
56. M. Jiménez-Castañeda and D. Medina, *Water*, 2017, **9**, 235.
57. M. A. Hamoud, S. F. Abo-Zahra, M. A. Attia, H. H. Someda and M. R. Mahmoud, *Environ. Sci. Pollut. Res.*, 2023, **30**, 53140.

58. R. Nodehi, H. Shayesteh and A. R. Kelishami, *Microchem. J.*, 2020, **153**, 104281.
59. A. Torabian, H. Kazemian, L. Seifi, G. N. Bidhendi, A. A. Azimi and S. K. Ghadiri, *Clean: Soil, Air, Water*, 2010, **38**, 77.
60. M. K. Dinker and P. S. Kulkarni, *J. Chem. Eng. Data*, 2015, **60**, 2521.
61. S. Zeb, N. Ali, Z. Ali, M. Bilal, B. Adalat, S. Hussain, S. Gul, F. Ali, R. Ahmad, S. khan and H. M. N. Iqbal, *J. Water Process Eng.*, 2020, **38**, 101675.
62. A. L.-T. Pham, D. L. Sedlak and F. M. Doyle, *Appl. Catal., B*, 2012, **126**, 258.
63. J. A. S. Costa, R. A. de Jesus, D. O. Santos, J. B. Neris, R. T. Figueiredo and C. M. Paranhos, *J. Environ. Chem. Eng.*, 2021, **9**, 105259.
64. P. N. E. Diagboya and E. D. Dikio, *Microporous Mesoporous Mater.*, 2018, **266**, 252.
65. V. Chaudhary and S. Sharma, *J. Porous Mater.*, 2016, **24**, 741.
66. S. Kumar, M. M. Malik and R. Purohit, *Mater. Today: Proc.*, 2017, **4**, 350.
67. D. A. Kurdyukov, D. A. Eurov, D. A. Kirilenko, J. A. Kukushkina, V. V. Sokolov, M. A. Yagovkina and V. G. Golubev, *Microporous Mesoporous Mater.*, 2016, **223**, 225.
68. K. F. Lam, K. L. Yeung and G. Mckay, *Microporous Mesoporous Mater.*, 2007, **100**, 191.
69. A. Shahbazi, H. Younesi and A. Badiei, *Chem. Eng. J.*, 2011, **168**, 505.
70. M. Mirzaie, A. Rashidi, H.-A. Tayebi and M. E. Yazdanshenas, *J. Chem. Eng. Data*, 2017, **62**, 1365.
71. H. Zeidan, M. Can and M. E. Marti, *Res. Chem. Intermed.*, 2022, **49**, 221.
72. Y. Zhang, Z.-A. Qiao, Y. Li, Y. Liu and Q. Huo, *J. Mater. Chem.*, 2011, **21**, 17283.
73. G. Abu Rumman, T. J. Al-Musawi, M. Sillanpaa and D. Balarak, *Environ. Nanotechnol., Monit. Manage.*, 2021, **16**, 100536.
74. J. George, S. K. Jha, D. Chakrabarty, A. Chakraborty and V. K. Vaidyanathan, *J. Polym. Environ.*, 2024, **32**, 2325.
75. W. Xiang, X. Zhang, J. Chen, W. Zou, F. He, X. Hu, D. C. W. Tsang, Y. S. Ok and B. Gao, *Chemosphere*, 2020, **252**, 126539.
76. M. F. Aller, *Crit. Rev. Environ. Sci. Technol.*, 2016, **46**, 1183.
77. M. Kamali, L. Appels, E. E. Kwon, T. M. Aminabhavi and R. Dewil, *Chem. Eng. J.*, 2021, **420**, 129946.
78. S. Nanda, A. K. Dalai, F. Berruti and J. A. Kozinski, *Waste Biomass Valorization*, 2016, **7**, 201.
79. L. R. Ortiz, E. Torres, D. Zalazar, H. Zhang, R. Rodriguez and G. Mazza, *Renew. Energy*, 2020, **155**, 837.
80. D. Lv, M. Xu, X. Liu, Z. Zhan, Z. Li and H. Yao, *Fuel Process. Technol.*, 2010, **91**, 903.
81. S. K. Das, G. K. Ghosh, R. K. Avasthe and K. Sinha, *J. Environ. Manage.*, 2021, **278**, 111501.

82. A. Tomczyk, Z. Sokołowska and P. Boguta, *Rev. Environ. Sci. Bio/Technol.*, 2020, **19**, 191.
83. A. A. Albalasmeh, M. Z. Quzaih, M. A. Gharaibeh, M. Rusan, O. E. Mohawesh, S. R. Rababah, A. Alqudah, A. G. Alghamdi and A. Naserin, *Sci. Rep.*, 2024, **14**, 7012.
84. K. Weber and P. Quicker, *Fuel*, 2018, **217**, 240.
85. C. Peiris, O. Nayanathara, C. M. Navarathna, Y. Jayawardhana, S. Nawalage, G. Burk, A. G. Karunanayake, S. B. Madduri, M. Vithanage, M. N. Kaumal, T. E. Mlsna, E. B. Hassan, S. Abeysundara, F. Ferez and S. R. Gunatilake, *RSC Adv.*, 2019, **9**, 17612.
86. M. Z. Yameen, S. R. Naqvi, D. Juchelková and M. N. A. Khan, *Biochar*, 2024, **6**, 25.
87. N. Saha, A. Saba and M. T. Reza, *J. Anal. Appl. Pyrolysis*, 2019, **137**, 138.
88. H. S. Kambo and A. Dutta, *Renewable Sustainable Energy Rev.*, 2015, **45**, 359.
89. H. Yu, W. Zou, J. Chen, H. Chen, Z. Yu, J. Huang, H. Tang, X. Wei and B. Gao, *J. Environ. Manage.*, 2019, **232**, 8.
90. S. K. Karan, D. Woolf, E. S. Azzi, C. Sundberg and S. A. Wood, *GCB BioEnergy*, 2023, **15**, 1424.
91. B. K. Biswal and R. Balasubramanian, *J. Environ. Chem. Eng.*, 2023, **11**, 110986.
92. P. R. Yaashikaa, P. S. Kumar, S. Varjani and A. Saravanan, *Biotechnol. Rep.*, 2020, **28**, e00570.
93. A. Singh, N. Srivastava, M. Shah, A. Hashem, E. F. A. Allah and D. B. Pal, *Biomass Convers. Biorefin.*, 2021, **13**, 13369.
94. D. H. K. Reddy and S.-M. Lee, *Colloids Surf., A*, 2014, **454**, 96.
95. B. Qiu, X. Tao, H. Wang, W. Li, X. Ding and H. Chu, *J. Anal. Appl. Pyrolysis*, 2021, **155**, 105081.
96. S. Anto, M. P. Sudhakar, T. Shan Ahamed, M. S. Samuel, T. Mathimani, K. Brindhadevi and A. Pugazhendhi, *Fuel*, 2021, **285**, 119205.
97. Z. Mahdi, A. El Hanandeh and Q. J. Yu, *J. Environ. Chem. Eng.*, 2019, **7**, 103379.
98. Y. Xue, B. Gao, Y. Yao, M. Inyang, M. Zhang, A. R. Zimmerman and K. S. Ro, *Chem. Eng. J.*, 2012, **200–202**, 673.
99. L. Hadjittofi and I. Pashalidis, *J. Radioanal. Nucl. Chem.*, 2014, **304**, 897.
100. H. Chen, X. Yang, Y. Liu, X. Lin, J. Wang, Z. Zhang, N. Li, Y. Li and Y. Zhang, *Waste Manage.*, 2021, **130**, 82.
101. F. Sadegh, N. Sadegh, W. Wongniramaikul, R. Apiratikul and A. Choodum, *Process Saf. Environ. Prot.*, 2024, **182**, 559.
102. H. Lyu, B. Gao, F. He, A. R. Zimmerman, C. Ding, J. Tang and J. C. Crittenden, *Chem. Eng. J.*, 2018, **335**, 110.
103. Y. Zhai, Y. Dai, J. Guo, L. Zhou, M. Chen, H. Yang and L. Peng, *J. Colloid Interface Sci.*, 2020, **560**, 111.
104. S. Silvestri, M. G. Gonçalves, P. A. da Silva Veiga, T. T. da, S. Matos, P. Peralta-Zamora and A. S. Mangrich, *J. Environ. Chem. Eng.*, 2019, **7**, 102879.

105. H. Marsh and F. Rodríguez-Reinoso, in *Activated Carbon*, Elsevier, 2006, pp. 13–86.
106. J. Rivera-Utrilla, M. Sánchez-Polo, V. Gómez-Serrano, P. M. Álvarez, M. C. M. Alvim-Ferraz and J. M. Dias, *J. Hazard. Mater.*, 2011, **187**, 1–23.
107. P. Le Cloirec, C. Brasquet and E. Subrenat, *Energy Fuels*, 1997, **11**, 331.
108. Y. A. B. Neolaka, A. A. P. Riwu, U. O. Aigbe, K. E. Ukhurebor, R. B. Onyancha, H. Darmokoesoemo and H. S. Kusuma, *Results Chem.*, 2023, **5**, 100711.
109. Synthesis of Microporous Activated Carbon (MAC) Using Compressed Wood as Waste Biomass Via Pyrolysis for Adsorption of Crystal Violet, https://www.researchgate.net/publication/332739750_Synthesis_of_Microporous_Activated_Carbon_MAC_Using_Compressed_Wood_as_Waste_Biomass_Via_Pyrolysis_for_Adsorption_of_Crystal_Violet (accessed July 21, 2024).
110. C. Jiang, S. Cui, Q. Han, P. Li, Q. Zhang, J. Song and M. Li, *IOP Conf. Ser.: Earth Environ. Sci.*, 2019, **237**, 022049.
111. A. Larasati, G. D. Fowler and N. J. D. Graham, *J. Environ. Chem. Eng.*, 2021, **9**, 105555.
112. A. Bhatnagar, W. Hogland, M. Marques and M. Sillanpää, *Chem. Eng. J.*, 2013, **219**, 499.
113. P. Sherugar, M. Padaki, N. S. Naik, S. D. George and D. H. K. Murthy, *Chemosphere*, 2022, **287**, 132085.
114. B. Zieliński, P. Miądlicki and J. Przepiórski, *Sci. Rep.*, 2022, **12**, 20869.
115. P. Mulindwa, J. S. Kasule, F. Nantaba, J. Wasswa and A. J. Expósito, *Int. J. Sustainable Eng.*, 2024, **17**, 13.
116. G. Newcombe, Removal of natural organic material and algal metabolites using activated carbon, in *Interface Science in Drinking Water Treatment*, ed. G. Newcombe and D. Dixon, Elsevier, 2006, pp. 133–153.
117. F. Wang, X. Li, T. Liu, X. Li, Y. Cui, L. Xu, S. Huo, B. Zou, J. Qian, A. Ma and G. Zhuang, *Catalysts*, 2023, **13**, 1356.
118. J. Jjagwe, P. W. Olupot, E. Menya and H. M. Kalibbala, *J. Bioresour. Bioprod.*, 2021, **6**, 292.
119. S. Deng, Y. Nie, Z. Du, Q. Huang, P. Meng, B. Wang, J. Huang and G. Yu, *J. Hazard. Mater.*, 2015, **282**, 150.
120. J. Yuan, E. Passeport and R. Hofmann, *Water Res.*, 2022, **210**, 118026.
121. S. M. Korotta-Gamage and A. Sathasivan, *Chemosphere*, 2017, **167**, 120.
122. C. Orha, R. Pode, F. Manea, C. Lazau and C. Bandas, *Process Saf. Environ. Prot.*, 2017, **108**, 26.
123. Z. Zhou, J. Yan, X. Li, J. Ren, H. Liang and Y. Liu, *J. Membr. Sci.*, 2024, **693**, 122353.
124. Y. J. O. Asencios, V. S. Lourenço and W. A. Carvalho, *Catal. Today*, 2022, **388–389**, 247.
125. S. K. H. Gulrez, S. Al-Assaf and O. Glyn, Hydrogels: Methods of Preparation, Characterisation and Applications, in *Progress in Molecular and Environmental Bioengineering – From Analysis and Modeling to Technology Applications*, ed. A. Carpi, InTech, London, 2011.

126. Md. Ahmaruzzaman, P. Roy, A. Bonilla-Petriciolet, M. Badawi, S. V. Ganachari, N. P. Shetti and T. M. Aminabhavi, *Chemosphere*, 2023, **331**, 138743.
127. A. C. Alavarse, E. C. G. Frachini, R. L. C. G. da Silva, V. H. Lima, A. Shavandi and D. F. S. Petri, *Int. J. Biol. Macromol.*, 2022, **202**, 558.
128. J. F. Resende, I. M. R. Paulino, R. Bergamasco, M. F. Vieira and A. M. S. Vieira, *Braz. J. Chem. Eng.*, 2022, **40**, 23.
129. V. Van Tran, D. Park and Y.-C. Lee, *Environ. Sci. Pollut. Res.*, 2018, **25**, 24569.
130. P. V. O. Toledo, O. D. Bernardinelli, E. Sabadini and D. F. S. Petri, *Carbohyd. Polym.*, 2020, **248**, 116765.
131. L. Weerasundara, B. Gabriele, A. Figoli, Y.-S. Ok and J. Bundschuh, *Crit. Rev. Environ. Sci. Technol.*, 2020, **51**, 1970.
132. D. Li, W. Zhan, W. Zuo, L. Li, J. Zhang, G. Cai and Y. Tian, *Chem. Eng. J.*, 2022, **450**, 138417.
133. L. Wang, Y. Long, H. Ding, J. Geng and B. Bai, *Chem. Eng. J.*, 2017, **317**, 952.
134. Md. T. I. Mredha, S. K. Pathak, V. T. Tran, J. Cui and I. Jeon, *Chem. Eng. J.*, 2019, **362**, 325.
135. L. Yan, X. Yang, H. Zeng, Y. Zhao, Y. Li, X. He, J. Ma and L. Shao, *J. Membr. Sci.*, 2023, **668**, 121243.
136. V. H. A. Braga, D. P. Fuentes, P. V. Oliveira and D. F. S. Petri, *J. Environ. Chem. Eng.*, 2024, **12**, 113306.
137. M. P. Callisaya, D. P. Fuentes, V. H. A. Braga, C. M. Finzi-Quintão, P. V. Oliveira and D. F. S. Petri, *Environ. Res.*, 2024, **252**, 118970.
138. I. P. Meneses, S. D. Novaes, R. S. Dezotti, P. V. Oliveira and D. F. S. Petri, *J. Hazard. Mater.*, 2022, **421**, 126804.
139. L. M. Furtado, D. P. Fuentes, R. A. Ando, P. V. Oliveira and D. F. Siqueira Petri, *J. Cleaner Prod.*, 2022, **380**, 134969.
140. S. D. Novaes, P. V. Oliveira and D. F. S. Petri, *Environ. Sci. Pollut. Res.*, 2022, **29**, 63936.
141. P. V. O. Toledo and D. F. S. Petri, in *Sustainable Textiles: Production, Processing, Manufacturing, Chemistry*, ed. S. S. Muthu and A. Khadir, Springer Nature Singapore, Singapore, 2022, ch. 8, pp. 185–238.
142. P. V. O. Toledo and D. F. S. Petri, *Int. J. Biol. Macromol.*, 2019, **123**, 1180.
143. A. G. B. Pereira, F. H. A. Rodrigues, A. T. Paulino, A. F. Martins and A. R. Fajardo, *J. Cleaner Prod.*, 2021, **284**, 124703.
144. P. V. O. Toledo, D. P. C. Limeira, N. C. Siqueira and D. F. S. Petri, *Cellulose*, 2019, **26**, 597.
145. R. R. Mohamed, M. H. A. Elella and M. W. Sabaa, *Int. J. Biol. Macromol.*, 2017, **98**, 302.
146. C. B. Godiya, X. Cheng, D. Li, Z. Chen and X. Lu, *J. Hazard. Mater.*, 2019, **364**, 28.
147. S. Kondaveeti, D. F. S. Petri and H. E. Jeong, *Int. J. Biol. Macromol.*, 2022, **204**, 345.

148. H. Karimi-Maleh, S. Ranjbari, B. Tanhaei, A. Ayati, Y. Orooji, M. Alizadeh, F. Karimi, S. Salmanpour, J. Rouhi, M. Sillanpää and F. Sen, *Environ. Res.*, 2021, **195**, 110809.
149. C. Yu, F. Wang, C. Zhang, S. Fu and L. A. Lucia, *React. Funct. Polym.*, 2016, **106**, 137.
150. M. Bilal, Z. Jing, Y. Zhao and H. M. N. Iqbal, *Biocatal. Agric. Biotechnol.*, 2019, **19**, 101174.
151. E. Frachini and D. Petri, *J. Braz. Chem. Soc.*, 2019, **30**, 2010.
152. S. D. Novaes, P. V. Oliveira and D. F. S. Petri, *ChemPlusChem*, 2017, **82**, 904.
153. A. T. Paulino, A. G. B. Pereira, A. R. Fajardo, K. Erickson, M. J. Kipper, E. C. Muniz, L. A. Belfiore and E. B. Tambourgi, *Carbohyd. Polym.*, 2012, **90**, 1216.
154. C. Wei, Z. Xu, F. Han, W. Xu, J. Gu, M. Ou and X. Xu, *Colloid Polym. Sci.*, 2018, **296**, 653.
155. P. V. O. Toledo, L. R. Marques and D. F. S. Petri, *Int. J. Polym. Sci.*, 2019, **2019**, 1–13.
156. J. P. Vareda, P. M. C. Matias, J. A. Paixão, D. Murtinho, A. J. M. Valente and L. Durães, *Gels*, 2024, **10**, 192.
157. A. K. Mohamed and M. E. Mahmoud, *Bioresour. Technol.*, 2020, **308**, 123263.
158. M. A. L. Dutra, N. do N. Marques, R. da S. Fernandes, M. de S. M. de Souza Filho and R. de C. Balaban, *Ecotoxicol. Environ. Saf.*, 2020, **200**, 110771.
159. C. Liu, H. Liu, K. Tang, K. Zhang, Z. Zou and X. Gao, *J. Polym. Environ.*, 2020, **28**, 984.
160. K. Abdiyev, S. Azat, E. Kuldeyev, D. Ybyraiymkul, S. Kabdrakhmanova, R. Berndtsson, B. Khalkhabai, A. Kabdrakhmanova and S. Sultakhan, *Water*, 2023, **15**, 2007.
161. K. Y. Prasad, *Handbook of Water and Used Water Purification*, Springer International Publishing, Cham, 2019.
162. M. J. Rajan and C. I. Anish, Role of Activated Carbon in Water Treatment, in *Water Quality – New Perspectives*, ed. S. Dincer, IntechOpen, 2022.
163. T. A. S. Committee on Periodical Publications of the Membrane Process Committee, Microfiltration and ultrafiltration Membranes for Drinking Water, *J. AWWA*, 2008, **100**, 84–97.
164. R. Shoshaa, M. Y. Ashfaq and M. A. Al-Ghouti, *Environ. Technol. Innovation*, 2023, **31**, 103162.
165. Z. Samavati, A. Samavati, P. S. Goh, A. Fauzi Ismail and M. Sohaimi Abdullah, *Chem. Eng. Res. Des.*, 2023, **189**, 530.
166. K. P. Lee, T. C. Arnot and D. Mattia, *J. Membr. Sci.*, 2011, **370**, 1–22.
167. L. K. Wang, D. A. Vaccari, Y. Li and N. K. Shammas, Chemical Precipitation, in *Physicochemical Treatment Processes*, ed. L. K. Wang, Y.-T. Hung and N. K. Shammas, Humana Press, Totowa, NJ, 2005, pp. 141–197.

168. R. M. El-taweel, N. Mohamed, K. A. Alrefaey, S. Husien, A. B. Abdel-Aziz, A. I. Salim, N. G. Mostafa, L. A. Said, I. S. Fahim and A. G. Radwan, *Curr. Res. Green Sustainable Chem.*, 2023, **6**, 100358.
169. C. S. Lee, J. Robinson and M. F. Chong, *Process Saf. Environ. Prot.*, 2014, **92**, 489.
170. A. Ali, M. Sadia, M. Azeem, M. Z. Ahmad, M. Umar and Z. Ul Abbas, *Futuristic Biotechnol.*, 2023, **3**, 12.
171. D. Ma, H. Yi, C. Lai, X. Liu, X. Huo, Z. An, L. Li, Y. Fu, B. Li, M. Zhang, L. Qin, S. Liu and L. Yang, *Chemosphere*, 2021, **275**, 130104.
172. R. Rashid, I. Shafiq, P. Akhter, M. J. Iqbal and M. Hussain, *Environ. Sci. Pollut. Res.*, 2021, **28**, 9050.
173. S. Boinpally, A. Kolla, J. Kainthola, R. Kodali and J. Vemuri, *Water Cycle*, 2023, **4**, 26.
174. S. Al-Amshawee, M. Y. B. M. Yunus, A. A. M. Azoddein, D. G. Hassell, I. H. Dakhil and H. A. Hasan, *Chem. Eng. J.*, 2020, **380**, 122231.
175. M. Jain, S. S. Kumar and L. Goswami, Aerobic and anaerobic bioreactor systems for wastewater treatment, in *Techno-economics and Life Cycle Assessment of Bioreactors*, ed. P. Mishra, L. Singh and P. Ghosh, Elsevier, 2022, pp. 13–22.
176. S. Barathi, J. Lee, R. Venkatesan and A. A. Vetcher, *Plants*, 2023, **12**, 3816.
177. E. Torres, *Processes*, 2020, **8**, 1584.
178. S. Satyam and S. Patra, *Heliyon*, 2024, **10**, e29573.
179. Circular economy: definition, importance and benefits, https://www.europarl.europa.eu/topics/en/article/20151201STO05603/circular-economy-definition-importance-and-benefits (accessed July 15, 2024).
180. A. V. Baskar, N. Bolan, S. A. Hoang, P. Sooriyakumar, M. Kumar, L. Singh, T. Jasemizad, L. P. Padhye, G. Singh, A. Vinu, B. Sarkar, M. B. Kirkham, J. Rinklebe, S. Wang, H. Wang, R. Balasubramanian and K. H. M. Siddique, *Sci. Total Environ.*, 2022, **822**, 153555.
181. J. M. Hua, K. M. Wei and Q. Zheng, *Hydrometallurgy*, 2012, **117–118**, 86.
182. T. Van Truong and D.-J. Kim, *Environ. Res.*, 2021, **196**, 110877.
183. W. Zhou, X. Meng, J. Gao, H. Zhao, G. Zhao and J. Ma, *Chem. Eng. J. Adv.*, 2021, **5**, 100083.
184. M. El Gamal, H. A. Mousa, M. H. El-Naas, R. Zacharia and S. Judd, *Sep. Purif. Technol.*, 2018, **197**, 345.
185. B. Ferrández-Gómez, R. Ruiz-Rosas, S. Beaumont, D. Cazorla-Amorós and E. Morallón, *Chemosphere*, 2021, **264**, 128399.
186. D. H. S. Santos, J. P. T. S. Santos, J. L. S. Duarte, L. M. T. M. Oliveira, J. Tonholo, L. Meili and C. L. P. S. Zanta, *Process Saf. Environ. Prot.*, 2022, **159**, 1150.
187. Z. Liu, B. Ren, H. Ding, H. He, H. Deng, C. Zhao, P. Wang and D. D. Dionysiou, *Water Res.*, 2020, **171**, 115456.
188. X. Meng, C. Zhang, J. Zhuang, G. Zheng and L. Zhou, *Chemosphere*, 2020, **244**, 125523.
189. J. B. Rial and M. L. Ferreira, *Sci. Total Environ.*, 2022, **823**, 153370.

190. M. A. B. S. Nunes, D. P. Fuentes, J. A. F. S. Mesquita, R. C. O. Romano, R. G. Pileggi, P. V. Oliveira and D. F. S. Petri, *J. Hazard. Mater. Adv.*, 2023, **12**, 100366.
191. S. Kumar, S. Kumar and T. Mohapatra, *Front. Plant Sci.*, 2021, **12** 665583.
192. D. Harikishore Kumar Reddy, K. Vijayaraghavan, J. A. Kim and Y.-S. Yun, *Adv. Colloid Interface Sci.*, 2017, **242**, 35.
193. T. Velempini, M. Ahamed and K. Pillay, *Results Chem.*, 2023, **5**, 100901.
194. I. C. Arunachellan, V. S. Sypu, N. H. Kera, K. Pillay and A. Maity, *J. Environ. Chem. Eng.*, 2021, **9**, 105242.
195. E. C. Umejuru, E. Prabakaran and K. Pillay, *ACS Omega*, 2021, **6**, 11155.
196. N. Ballav, R. Das, S. Giri, A. M. Muliwa, K. Pillay and A. Maity, *Chem. Eng. J.*, 2018, **345**, 621.
197. Y. Wang, Y. Zhang, L. Pei, D. Ying, X. Xu, L. Zhao, J. Jia and X. Cao, *Sci. Rep.*, 2017, **7**, 41523.
198. S. Wang, W. Lu, S. Esakkimuthu, H. Chen, J. Yang, M. Mu and X. Gong, *J. Cleaner Prod.*, 2023, **427**, 139269.
199. C. Sabando-Fraile, M. Corral-Bobadilla, R. Lostado-Lorza and S. Cabredo-Pinillos, *J. Cleaner Prod.*, 2024, **450**, 141938.

CHAPTER 12

Nanocomposites for Magnetic Assisted Water Treatment Adsorption Technologies

ANA L. DANIEL-DA-SILVA, SOFIA F. SOARES, CLÁUDIA B. LOPES AND TITO TRINDADE*

Department of Chemistry and CICECO-Aveiro Institute of Materials, University of Aveiro, 3810-193 Aveiro, Portugal
*Email: tito@ua.pt

12.1 Introduction

Adsorption technologies, among other procedures, have allowed the treatment of water supplied from various sources, which resulted in safe supplies for most of the world population. However, this is still not the reality for a significant part of the world's population. In fact, according to the United Nations it is estimated that 26% of the world's population (2 billion people) does not have access to safe drinking water and 46% (3.6 billion people) lack access to basic sanitation.[1] The availability of water varies greatly from region to region and is strongly conditioned by natural circumstances such as geological and climatic factors. This scenario is worsened in certain regions of the globe, due to the occurrence of natural disasters or following armed conflicts. Furthermore, lacking access to safe drinking-water negatively impacts all aspects of societies, primarily because it is the cause of several health problems, with the implementation of guidelines for drinking-water quality regarded as an international priority.[2] Although water consumption has increased globally and the real scenarios vary considerably in the globe,

Contribution of Colloidal Materials to Air, Water and Soil Environmental Sustainability: From Basic Concepts to Advanced Applications
Edited by Izabel C. Riegel-Vidotti, Lizandra Maria Zimmermann and Heloise R. de Barros

Published by the Royal Society of Chemistry, www.rsc.org

it is forecasted that the global consumption of drinking water will increase in the future due to population growth and changes in consumption habits, considering an increasingly industrialized society, with intensive access to goods and products. On the other hand, these factors are also the origin of new types of contaminants present in water sources, for which conventional water treatment technologies are not efficient. These pollutants are called contaminants of emerging concern (CEC), which include, among others, chemical species containing heavy metals, pesticides, and traces of pharmaceutical products, which even in trace amounts represent a risk to human health and the ecosystem.[3–10] Furthermore, the uptake of certain water contaminants has been placed as a priority due to environmental protection guidelines in waste management but also due to economic criteria, such as the recovery of critical elements that include lanthanides and precious metals.[11–15] Hence, there is an imperative calling on scientists and engineers, whom with their specialized knowledge and predictive capacity based on solid scientific knowledge, contribute to the development of technologies that make it possible to face such societal challenges.

Although the field of water treatment keeps evolving, with several promising technologies emerging as front runners for innovative and efficient approaches (*e.g.* advanced oxidation processes, membrane filtration with nanomaterials, electrochemical technologies, or even microbial fuel cells that combine wastewater treatment with energy generation), adsorption technologies continue to be in high demand and present several challenges.[16] The appeal of this technology over others is due to its high efficiency and selectivity in removing different types of contaminants, even at low concentrations, and its simple design, cost effectiveness and applicability to different water sources, making it a very attractive option. Furthermore, unlike other methods, adsorption does not generate by-products, and it offers the ability to use different types of adsorbents, including natural materials or environmentally friendly nanomaterials, which enhances its environmentally friendly appeal. The principle of adsorption is based on the interplay between surface forces, intermolecular interactions, and the specific properties of the adsorbent and adsorbate molecules, leading to the preferential accumulation of certain molecules at the adsorbent interface, resulting in their effective separation from the solution. With emergent nanotechnologies, the integration of nanomaterials with their unique and tailorable properties has opened new avenues for further advancements in adsorption technologies.[17–19] Hence, colloidal nanomaterials have been extensively investigated by exploiting size and shape effects, along with methods of surface functionalization envisaging efficient and selective sorbents for innovative water treatments, among many other applications.[20,21]

Nanostructured materials showing high adsorption capacity are an obvious option for developing new technologies in water treatment. Indeed, these materials are characterized by having nanometric dimensions and high specific surface areas. The huge surface area per volume of adsorbent material leads to a high number of surface sites with potential for the adsorption of chemical species dissolved in water, particularly in the capture of target contaminants. Nanostructured adsorbent materials can be applied as colloids dispersed in the contaminated aqueous medium or in the form of immobilized materials, such as porous membranes.[19] Although immobilized materials may have a lower surface area compared to similar materials in colloidal form, they have the great advantage of being more easily removed and regenerated for later reuse. On the other hand, colloidal nanosorbents combine high specific surface area and more extensive surface chemical functionalization, an important aspect when taking into account chemical selectivity for certain target contaminants. The development of colloidal magnetic nanosorbents combines both worlds: high surface area per volume of adsorbent and easy removal by means of magnetic assisted separation.

Numerous classes of colloidal nanomaterials exhibit potential for applications as sorbents with magnetic properties and also to develop associated technologies in water treatment.[22,23] This can be illustrated by considering that well-known colloidal systems, such as aqueous magnetite nanoparticles, not only standalone as nanosorbent materials but can be coupled with multiple structures, resulting in a plethora of composite nanostructures tailored for specific purposes.[17,24,25] These composite structures exhibit great diversity in terms of structural dimensionality, chemical composition, and morphological characteristics. In this context, it is essential to note that this chapter does not aim to provide an exhaustive list of magnetic nanosorbents reported in the literature, but rather to explore recent advancements in the chemistry underlying the development of such materials. Consequently, this chapter reviews advances in magnetic bionanocomposites as new colloidal sorbents. Nevertheless, common principles and concepts to many other sorbents are recognized in these composite nanomaterials, particularly when considering their application in nanotechnologies for water treatment. The subsequent section reviews fundamental aspects that take into consideration the nature of the colloidal nanosorbents that are then explored in the following sections.

12.2 Structure, Magnetism, and Surface Chemistry in Magnetic Nanosorbents

12.2.1 Fundamental Aspects of Magnetic Nanoparticles

Adsorption is a surface phenomenon, in which the existence of active sites at the sorbent's surfaces allows interaction with certain chemical species that are intended to be captured and removed from the contaminated aqueous medium. Therefore, adsorbent nanomaterials (termed here as nanosorbents) have several advantages over other types of materials for treating contaminated water, such as bulk powders or membranes. Among these advantages, obviously included is the high surface area per volume of solid, which offers a huge solid/liquid interface area that provides active sites for adsorption. Furthermore, a large surface area contributes to extensive chemical functionalization of nanosorbents, for example aimed at selective adsorption processes of contaminating species, and also enhances the colloidal stability of sorbents in the dispersing medium. These characteristics make nanomaterials attractive for adsorption techniques, considering not only technological criteria, for example based on adsorption capacity, but also in terms of costs, since the volume of solid adsorbent material to be used is substantially reduced. Furthermore, the energy costs associated with the dispersion of colloidal nanosorbents in the aqueous medium are also lower, compared to the dispersion of powders with larger particle sizes or aggregates. However, the use of nanomaterials in the context of adsorption for water treatment also poses practical challenges, such as their efficient separation from the aqueous medium. Magnetic nanosorbents thus emerge as an advantageous alternative, as they can be integrated into adsorption technologies assisted by magnetic separation, allowing their removal by applying an external magnetic field. Furthermore, the development of safe adsorption technologies using nanomaterials requires the consideration of nanotoxicity issues related to their residence and fate in the treated water.[26–29]

It seems appropriate at this point to make a reference to the magnetic properties of materials, considering that magnetism is a functionality required for the adsorbents described in this chapter. This topic is briefly reviewed here for its pertinence, but several bibliographic sources are available for the reader wishing a deeper approach on magnetism in nanoscale materials.[30–32]

The magnetic field ($\boldsymbol{B}$) inside a material subjected to a field ($\boldsymbol{H}$) is given by

$$\boldsymbol{B} = \boldsymbol{H} + 4\pi\boldsymbol{M} \tag{12.1}$$

The magnetization $\boldsymbol{M}$ of a material is the total magnetic moment over its entire volume, which is associated with the orbital and spin electron momentum. Eqn (12.1) is a field equation† that gives the magnetic induction ($\boldsymbol{B}$) in a material as a combined effect of $\boldsymbol{H}$ and $\boldsymbol{M}$. It can be rearranged to provide the magnetic permeability of the material as

$$\mu = \frac{\boldsymbol{B}}{\boldsymbol{H}} = 1 + 4\pi\chi_V \tag{12.2}$$

Eqn (12.2) indicates that materials with large values of permeability (*e.g.* iron) subjected to an applied field $\boldsymbol{H}$ experience inside, a large magnetic induction $\boldsymbol{B}$ due to the enhancement factor μ. The effectiveness of magnetization per volume unit of material due to an applied field is translated in the value of χ_V, which is the volume susceptibility

$$\chi_V = \frac{\boldsymbol{M}}{\boldsymbol{H}} \tag{12.3}$$

† In cgs (emu) units.

The magnetic susceptibility can be expressed either per gram of material by dividing by its density, giving values for the commonly used mass magnetic susceptibility,

$$\chi = \frac{\chi_V}{\rho} \tag{12.4}$$

or as the molar magnetic susceptibility dividing χ by the molecular weight

$$\chi_M = \frac{\chi}{M_W} \tag{12.5}$$

Table 12.1 summarizes in simplified terms the characteristics of materials depending on their magnetic behavior. There are other kinds of magnetism not shown in Table 12.1 but they are not relevant to consider in this context, with the exception of superparamagnetism which will be briefly explained below. In fact, the most relevant materials in magnetism-assisted adsorption applications, typically at ambient temperature, are those materials called ferromagnets and ferrimagnets, as well as nanomaterials that exhibit superparamagnetism. In the case of ferromagnetic materials (*e.g.* Fe), the value of χ is high and positive at room temperature (RT); the magnetic moments of the atoms in the magnetic domains are aligned in parallel resulting in a net magnetic moment. There is a short-range exchange energy that is minimized in this situation and at RT the thermal energy is not sufficient to disrupt the alignment of the magnetic moments. The so-called ferrimagnetic materials (*e.g.* Fe_3O_4) have magnetic domains with both parallel and anti-parallel magnetic moments but the net balance is a non-zero magnetic moment. Just like ferromagnetic materials, they have high and positive values of magnetic susceptibility.

In the case of nanocrystalline materials, we also must consider the phenomenon of superparamagnetism, which occurs in particles with dimensions of a critical size, below which the particle itself constitutes a single-magnetic domain, and above the system's blocking temperature. The size range in which this type of behavior occurs is in the order of nanometers, which is why superparamagnetism is associated with nanoscale materials, such as colloidal iron oxide nanoparticles (SPIONs: superparamagnetic iron oxides). Iron oxide nanoparticles have been employed in different applications, including as sorbents, due to their magnetic, structural and surface properties.[20,33] The magnetic susceptibility of superparamagnetic nanomaterials is lower than that of ferromagnetic/ferrimagnetic materials. However, these systems can be advantageous in some situations, as they do not exhibit magnetic remanence, hence they do not keep magnetic memory after removal of the external magnetic field. This magnetic behavior limits aggregation occurring in magnetic particles, with its concomitant specific surface area decrease, which can be a determinant aspect for certain adsorption technologies. Finally, there are also materials with magnetism not relevant for assisting the separation in adsorption processes, yet it is instructive to

Table 12.1 Classification of materials according to common types of magnetic behavior.

Magnetism	Atomic origin	Magnetic susceptibility (χ)	Effect of applied magnetic field ↑		Examples of macrocrystalline solids
			$H = 0$	$H \neq 0$	
Diamagnetism	Atoms with paired electrons resulting in no magnetic moment.	Low and negative; very weak magnetic response; No temperature dependence.		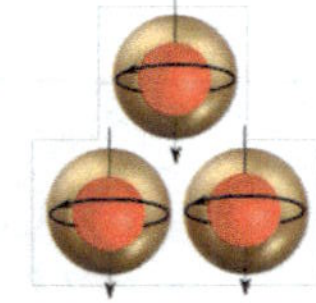	Gold Quartz Graphite
Paramagnetism	Atoms having unpaired electrons with randomly oriented magnetic moments (magnetic disorder).	Low and positive; Decreases by increasing T.	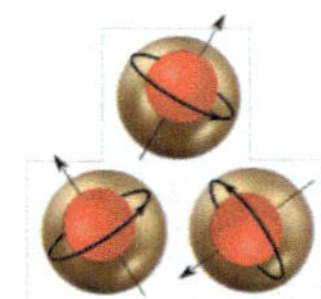	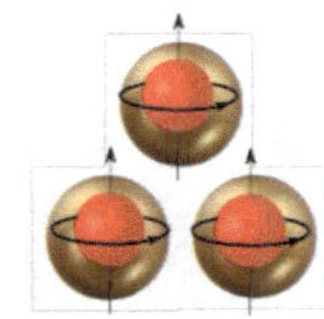	Platinum Aluminum Tin
Ferromagnetism	Magnetic moments in the same domain are aligned parallel leading to a net high magnetic moment (magnetic ordered state).	High and positive; temperature and field-dependent; Become paramagnetic above T_c.	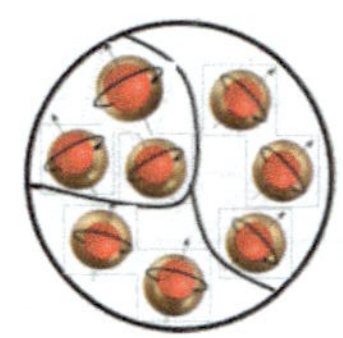	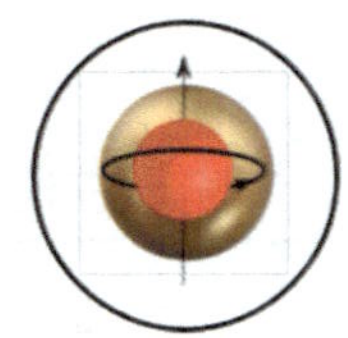	Iron Cobalt Gadolinium

Table 12.1 (*Continued*)

Magnetism	Atomic origin	Magnetic susceptibility (χ)	Effect of applied magnetic field ↑ $H = 0$	$H \neq 0$	Examples of macrocrystalline solids
Ferrimagnetism	Magnetic moments in the same domain are aligned parallel and anti-parallel but result in a high magnetic moment.	High and positive; temperature and field-dependent; Become paramagnetic above T_c.	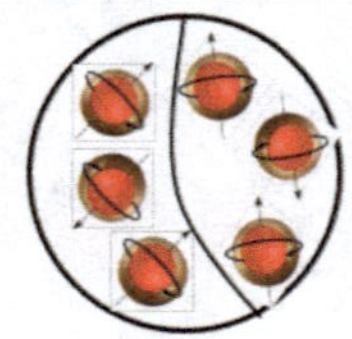	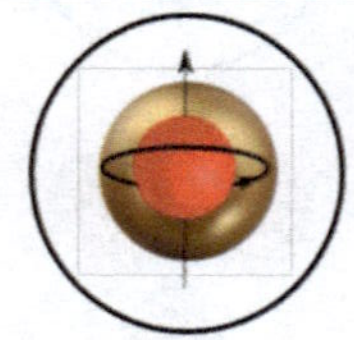	Magnetite Maghemite Greigite (Fe_3S_4)
Antiferromagnetism	Magnetic moments are aligned parallel and anti-parallel resulting in null magnetic moment.	Low and positive; temperature dependent; Become paramagnetic above T_N.	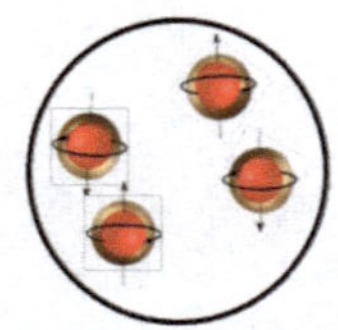	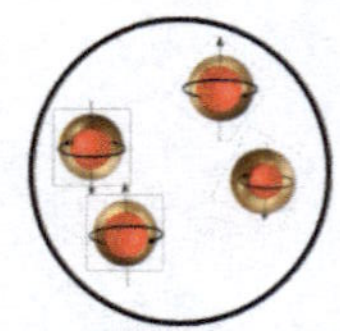	Chromium Wüstite (FeO) Manganese sulfide

know their behavior (Table 12.1). Thus, diamagnetic materials are characterized by low and negative magnetic susceptibility values. Amorphous silica is an example of a diamagnetic system, such as certain natural biopolymers (*e.g.* carrageenan), both used in the functionalization of magnetic adsorbents discussed later. At a given temperature, when the arrangement of atoms in a material leads to anti-parallel magnetic moments in the absence of an external field, the total magnetic moment is null and the system is called an antiferromagnet, characterized by low values of magnetic susceptibility. The common iron oxide hematite (Fe_2O_3) is an antiferromagnet (or weakly ferromagnetic) until 675 °C (Néel temperature, T_N), above which it becomes paramagnetic. A paramagnetic material is characterized by positive but very small magnetic susceptibility. Magnetite, which is ferrimagnetic at RT, becomes paramagnetic above 585 °C (Curie temperature, T_C), as the magnetic moments of the atoms become disordered and non-interacting.

12.2.2 Structure and Morphological Characteristics

Materials applied as magnetic adsorbents present different configurations in terms of their nanostructure, morphology (shape and size) and surface composition. The simplest type of this nanostructure corresponds to a single-phase material, whose nanoparticles combine relevant magnetic properties and high adsorption capacity. The paradigmatic example of this type of system is colloidal magnetite (Fe_3O_4), which has been widely investigated in adsorption processes assisted by magnetic separation, namely as bare particles with non-modified surfaces.[24,34] Magnetite has an inverse cubic spinel structure, based on a cubic closed packed array of oxide anions, with eight formula units (Fe_3O_4) per unit cell. The cationic sites in this structure are occupied differently as compared to a normal spinel. Hence the tetrahedral sites and half of the octahedral sites are occupied by ferric ions, while the other half of octahedral sites are occupied by ferrous ions. The sites occupancy originates two magnetic sublattices with antiparallel spins in relation to each other, with all ferrous ions contributing to the magnetic moment while cancellation occurs for all ferric ions. Hence, the ferrimagnetism in magnetite arises from the magnetic moments in the ferrous ions that occupy octahedral sites. Maghemite (γ-Fe_2O_3) is also an important iron oxide for integrating ferrimagnetic nanosorbents, however its saturation magnetization is lower (74–80 emu g^{-1}) than that of bulk magnetite (92 emu g^{-1}). Maghemite has a spinel structure similar to that of magnetite, but with all metal ions in the trivalent state and cation vacancies in the octahedral sites. Hence, it is an oxidized form of magnetite, with the charge neutrality of the unit cell being achieved by the presence of these cation vacancies.[33]

Despite their simplicity and usefulness, the encapsulation of these magnetic nanoparticles with a second material, for example amorphous silica or a polymer, allows the creation of core/shell nanostructures that are more advantageous in several scenarios. Indeed, this type of composite nanostructure is among the most investigated nanosorbents in this context, as it gives nanosorbents with improved chemical robustness and the possibility of diverse surface chemical functionalization. As a matter of example, Figure 12.1 shows TEM images of magnetite nanoparticles and derived core/shell nanostructures that have been explored in our research group as magnetic nanosorbents.[35,36] This subject will be discussed in more detail in the following section, but for the moment we note that despite having the same type of magnetic core (magnetite), these systems behave very differently in terms of sorption for chemical species dissolved in water. The magnetic behavior in the core/shell structures is determined essentially by the magnetic core, though the diamagnetic coating causes a slight decrease in the magnetization as compared to the bare magnetite particles. However, other configurations can be explored by judicious control of the

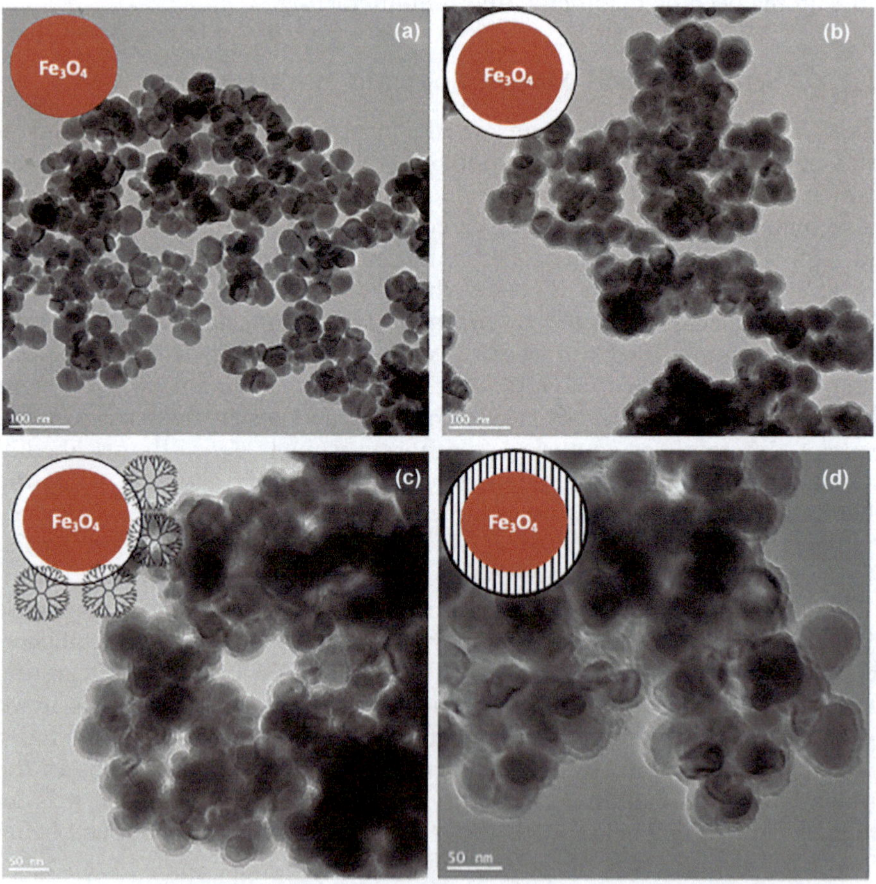

Figure 12.1 TEM images of magnetic nanostructures investigated in our laboratories as colloidal sorbents: magnetite (a); silica coated magnetite (b); silica coated magnetite with surfaces functionalized with dendrimer molecules (c); magnetite coated with hybrid shells of silica and κ-carrageenan (d).

morphological characteristics of nanostructured magnetic adsorbents. For example, the development of Janus particles for magnetic-assisted nanotechnologies has been reported, although their application in the environmental field still offers great potential that has yet to be explored.[37] The contaminant species to be removed, normally present as complex mixtures, have affinities for the adsorbents that depend on different operational parameters and, naturally, on the characteristics of the adsorbents. In this case, in addition to surface chemistry of each type of sorbent, nanoformulations comprising blends of such sorbents can also be explored.

Other iron oxides showing ferromagnetic behavior have been explored although to considerably less extent, such as the class of spinel-type ferrites (MFe_2O_4, M a divalent metal) to which, by the way, magnetite (inverse spinel) also belongs but that is often considered as a special case. These materials are well known for their magnetic properties, which make them useful for magnetic recording supports and magnetic sensors. Depending on the magnetic behavior, a ferrite can be classified as a hard ferrite or a soft ferrite. The former is characterized by high coercivity and high magnetic remanence, hence are quite useful as permanent magnets, while soft ferrites show low coercivity and low magnetic remanence, making them useful for applications requiring fast changes in the magnetic flux. In comparison to magnetite, considerably less literature has been devoted to such spinel-type ferrite nanosorbents. Some practical criteria have made these materials less attractive for magnetic assisted adsorption technologies, in comparison to magnetite, such as issues associated with the presence of toxic transition metals in some of the ferrites or even the higher cost associated with their production. On the other hand, they also offer advantages, such as high magnetization values in certain ferrites (*e.g.* $CoFe_2O_4$) and better chemical resistance to oxidation, namely when dispersed in acidic aqueous medium. Hence, several studies have reported the usefulness of diverse types of spinel ferrite nanosorbents and their nanocomposites for the uptake of water contaminants.[38,39]

12.2.3 Surface Chemical Modification of Magnetic Iron Containing Nanoparticles

For certain chemical species present in aqueous solutions, the magnetic nanoparticles of the above-mentioned compounds can be used as nanosorbents without any surface treatment. However, composite systems that use this type of magnetic nanomaterial together with other phase(s) are of great relevance in adsorption technologies. Since the other phase is diamagnetic, as often happens, a decrease in the magnetization of the material is expected, when compared to the pure magnetic nanosorbent. This aspect must be considered in the development of magnetic nanosorbents, noting that it is possible to adjust their chemical composition so that the application of the nanocomposites is not compromised. Nevertheless, surface chemical functionalization of magnetic nanoparticles is an important step in the development of magnetic nanosorbents. Firstly, it promotes the colloidal stability in an aqueous medium, thus ensuring that the magnetic properties are not significantly altered due to agglomeration processes resulting from van der Waals interactions or following the application of an external magnetic gradient, as happens with ferromagnetic nanosorbents. In addition, the functionalization of the particles' surfaces provides several chemical strategies for conferring to the nanosorbents selectivity for target species to be removed from the contaminated medium.

The specific surface area and magnetic behavior are regarded as determining properties in the development of efficient magnetic nanosorbents. However, other factors should be considered, such as their eco-toxicity and raw-materials cost. Overall, by considering the different aspects involved, it is not surprising that iron-based nanomaterials are the systems of choice for magnetically assisted adsorption technologies, specifically, iron and its oxides, magnetite and maghemite. Although not the focus of this chapter, it should be noted that iron-based nanomaterials can also be employed in water treatment procedures, but which do not involve adsorption technologies assisted by magnetic separation. Two illustrative examples are as follows, the use of the non-magnetic iron oxide ferrihydrite in Pb^{2+} adsorption polymeric membranes[40] and the application of zero valent iron nanoparticles as reducing agents for industrial wastewater containing heavy metal ions (*e.g.* Cu^{2+}) and arsenic species.[41] This section provides examples of several applications of iron-containing magnetic nanoparticles, highlighting the importance of surface functionalization for achieving stable and selective nanosorbents. Furthermore, it introduces a general context for applying chemical methods to design magnetic nanosorbents, from the bare colloidal particles to polymer based nanosystems; the latter will be explored in detail in the subsequent chapters. For these purposes, it is instructive to illustrate surface modifications in zero-valent iron (ZVI) nanoparticles, which as bare colloids are easily prone to oxidation in ambient conditions, but once their surfaces have been modified offer great potential in several applications.[42]

Zero-valent iron nanoparticles are praised for their ferromagnetic properties and high magnetization saturation values but require surface modifications to avoid their fast oxidation in aqueous solutions. Furthermore, they have a high tendency for aggregation but that can be limited by surface modification strategies. The composition of the coating can be adjusted based on practical scenarios, influencing operational sorption parameters like salinity, pH, and chemical speciation, such as organic matter and other dissolved contaminants. Various coatings, derived from both inorganic and organic sources, have been extensively investigated, and their properties depend on the method used for surface modification.[43] Examples of core/shell nanostructures for adsorption include ZVI coated with mesoporous hydrated silica applied to a multi-component system containing Pb(II), Cd(II), and As(V).[44] Although iron lixiviation has been limited due to the silica coating, this work does not provide information concerning the potential use of the materials in magnetic assisted procedures. Also, carbon coated ZVI nanoparticles have been obtained by diverse methods, such as the carbonization of Fe(III)-crosslinked sodium alginate, to achieve magnetic nanosorbents for the removal of Pb(II) and its surface reduction into elemental Pb.[45] Core/shell nanostructures containing ZVI have also been applied to wastewater envisaging the recovery of precious metals. For example, Ling *et al.* have reported the application of iron nanoparticles with their surfaces passivated with iron oxides/hydroxides for recovery of Au and Ag, among other metals.[46] Noteworthy, the surface chemistry of such nanostructures plays a determinant role not only in promoting the adsorption of metal species but also in mediating redox processes that result in the clustering of the reduced metals at the particle surfaces.

Polymers encapsulating ZVI are examples of surfactant coatings often explored to produce core/shell ferromagnetic nanosorbents with improved colloidal stability.[47,48] Among the organic coatings, there has been interest in using polymers of natural sources or passivation layers that derive from the green synthesis of ZVI, for example using plant extracts.[49–51] To limit particle aggregation, ZVI nanoparticles have been supported on several substrates or coupled to other inorganic phases, which might also contribute to decrease the oxidation rate of the metal.[52] Hence, extensive research has been carried out in the past years for developing multifunctional nanocomposites, that not only assist magnetic separation but can also participate in chemical reactions envisaging the contaminants elimination.[53] In fact, when applying ZVI nanoparticles for water decontamination, additional mechanisms come into play namely in the removal of heavy metal ions present in aqueous systems. These mechanisms encompass participation in redox processes, which result in the alteration of the oxidation state of the contaminant species, and eventually in their co-precipitation.[54] Oxidized arsenium species, As(III)/As(V), have been reduced to As(0) in anoxic conditions using ZVI nanoparticles.[55] The surface reduction process contributes to the sorption efficiency and promotes the encapsulation of the As species in between the iron core and iron(III) oxide outer shell, thus limiting the secondary release of such toxic species. Overall, composite nanostructures containing ZVI offer great promise for magnetic assisted adsorption technologies, based on their adsorption performance and reuse ability, with additional chemical stability of the magnetic phase against oxidation.[56]

Besides ZVI nanoparticles, there are other iron containing nanoparticles of great interest for magnetic adsorption technologies. In an earlier report of magnetic nanosorbents for heavy metal ions, ferrimagnetic iron sulfide nanoparticles were produced *via* metabolic processes occurring in sulfate-reducing bacteria present in iron containing solutions.[57] Experimental conditions should be judiciously controlled to achieve strongly magnetic iron sulfides, in the form of ferrimagnetic phases $Fe_{1+x}S$ (mackinawite) and Fe_3S_4 (greigite). The crystal structure of greigite is basically the same as magnetite, in which sulfide anions replace the oxides; it is the common iron sulfide with ferrimagnetic order present in fine-grained sediments of natural anaerobic environments.[58] However, as noted above, magnetite and maghemite are iron oxides with suitable properties as sorbents in magnetic assisted water treatment technologies. Particle size distribution and surface chemistry can be controlled to provide systems that can be either ferrimagnetic or superparamagnetic systems for magnetic assisted adsorption procedures. Hence, Trindade *et al.* have demonstrated that ferrimagnetic Fe_3O_4 nanoparticles can be surface modified with silica/shells enriched in dithiocarbamate moieties, that confer to the colloidal particles high efficiency as sorbents for the removal of aqueous Hg(II) in different operational conditions.[59–61] On the other hand, superparamagnetic Fe_3O_4 nanoparticles functionalized with dimercaptosuccinic acid were also investigated as sorbents for ionic species of metals such as Hg, Ag, Pb, Cd, and Tl. The superparamagnetic nanosorbents could be separated from the treated solutions within a minute, using a 1.2 T magnet. While the surface thiol ligands effectively bind to the aforementioned soft metal species, the exposed bare iron oxide surface sites coordinate to As species.[62] The removal of heavy metal cations can also be achieved through outer layers of metal–organic frameworks (MOF), for example in the adsorption of Pb^{2+} using ZIF-8.[63] The removal of anions can be successful by functionalizing the surfaces using metalloporphyrins, given their selectivity for anions, such as the binding to fluoride ions by the metal center of the porphyrin.[64]

Superparamagnetic iron oxides have been investigated for various nanotechnological applications.[20] Biomedicine is one of the key areas where these nanomaterials find important uses, such as contrast agents in magnetic resonance imaging (MRI) or in phototherapies against cancer.[65] However, these colloidal systems have also been explored as nanosorbents for removing different pollutants from water, including magnetic separation processes under low magnetic fields. Currently, several protocols are available for the synthesis of SPIONs with different particle sizes and particle shapes, which in some of the cases also include chemical surface treatments.[66,67] These methodologies allow the adsorbent to be chemically tailored for various water treatment scenarios. Because superparamagnetic nanoparticles only exhibit magnetization when an external magnetic gradient is applied, these nanosystems can be collected from the treated medium, regenerated if required, and then reused in subsequent sorption processes. This is clearly an advantage found in SPIONs applied to adsorption technologies. However, very small magnetite/maghemite nanoparticles present decreased chemical stability due to their susceptibility to air oxidation. Therefore, surface coating is crucial in developing SPIONs as nanosorbents. This step not only prevents oxidation but also enhances selectivity for targeting water contaminants. The following examples illustrate the potential of SPIONs in adsorption technologies applied to water purification.

Hexavalent chromium species are very toxic and can be found in aqueous effluents from metal-plating and tanning industries. Thus, implementing magnetic-assisted adsorption technologies for Cr(VI) can be a promising strategy for such industries, to comply with environmental requirements. For example, the sorption performance of superparamagnetic maghemite was favorably compared with conventional sorbents like activated carbon and clays in the presence of several competing ionic species.[68] The removal of radioactive contaminants from water and subsequent treatment of waste is another area where magnetic nanosorbents are particularly attractive. SPIONs are especially useful because such colloidal systems can operate with minimal stirring mechanisms, limiting particle aggregation and ultimately leading to relatively low volumes of solid radioactive material. Calì *et al.* investigated oleic acid-coated superparamagnetic Fe_3O_4 obtained by a hot-injection method (*ca.* 12 nm), as efficient and selective nanosorbents for capturing U(VI) aqueous species (1690 mg U per g Fe_3O_4) in the presence of alkaline-earth metal ions.[69] The initially hydrophobic Fe_3O_4 nanoparticles were subjected to a ligand-exchange reaction, replacing oleic acid with phosphate ligands. The anionic layered phosphate shells coating the SPIONs account for the high adsorption capacity and selectivity for U(VI) species. Other nanoarchitectures of superparamagnetic iron oxides have been explored for the removal of uranium radionuclides from water.[70,71] For instance, nanoengineered SPIONs (11 nm) have been applied for uranium separation and sensing, demonstrating superior sorption performance compared to commercial magnetite particles (10–50 nm).[70] This study highlights the role of nanosurface chemistry in the overall adsorption process of uranium species onto the nanoengineered SPIONs. These were made water-compatible by the presence of oleic acid bilayers on the particle surfaces, which can bind up to 50 wt% uranium. Furthermore, this organic coating mediates a redox process that reduces U(VI) to U(IV), which is less soluble and helps limit the desorption mechanisms.

This section presented some fundamental aspects on the use of magnetic nanosorbents and it demonstrated that there are several advantages in exploring, in particular, composite nanostructures as compared to the application of the bare magnetic nanosorbents. Most of the illustrative cases reviewed here so far involve the surface modification and chemical functionalization of magnetic cores, typically with silica shells and/or functional capping molecules. Other possibilities emerge in developing composite nanostructures that comprise polymeric coatings around the magnetic cores (*e.g.* Fe_3O_4), which include the versatility of chemical functionalization, aiming at selective adsorption of target contaminants, and the viability of regeneration and reuse processes of the sorbents following their magnetic separation. These more practical aspects and other properties will be addressed in the following sections using examples selected from the literature and of our own research.

12.3 Magnetic Nanocomposites Based on Polymers of Natural Origin

12.3.1 Key Aspects in the Design of Magnetic Bionanocomposites for Adsorption

Biopolymers, derived from living organisms, are sustainable materials characterized by low toxicity, abundant availability, and structural functional groups with an inherent affinity toward diverse pollutants. Given these attributes, there has been a growing interest in exploring the applications of biopolymers and their composites in water treatment, particularly as adsorbents for water decontamination.[72,73] However, biopolymers typically exhibit water solubility and necessitate crosslinking or immobilization onto insoluble supports for effective use in water treatment applications. Moreover, a significant challenge associated with the adsorption-treatment process lies in the recovery of spent sorbents, which ideally should be facile and cost-effective.[74] Integration of biopolymers with magnetic components results in composites that can be easily retrieved using magnetic separation, thereby facilitating their reuse in multiple cycles and rendering them highly attractive for water treatment *via* adsorption.[75,76]

As shown in the previous section, the magnetic component comprises typically iron oxide nanoparticles, that can be incorporated *via in situ* synthesis methods conducted in the presence of biopolymers under mild experimental conditions compatible with biopolymers, such as those utilized in co-precipitation.[77] An alternative approach offering greater control over the composite properties involves the surface modification of pre-prepared magnetic nanoparticles or their dispersion within the biopolymer matrix to yield the desired bionanocomposite. A detailed description of the methods for magnetic bionanocomposite production falls beyond the scope of this work and can be found elsewhere.[76]

In the design of bionanocomposite sorbents, it is essential to ensure strong bonding of the biopolymer to the magnetic counterpart to withstand long-term wetting, pH changes, and reuse cycles. The establishment of weak bonding between these two components, such as physical adsorption of the biopolymer around magnetic nanoparticles, is insufficient to prevent the leaching of the biopolymer, resulting in limited reuse.[78] The literature abounds with studies on magnetic bionanocomposites with weak bonding between the two components, wherein the aspect of sorbent reusability is often overlooked and unexplored. To prevent biopolymer and nanoparticle leaching, several strategies exist, including the use of chemical crosslinkers such as glutaraldehyde, epichlorohydrin, and genipin.[79–81] Alternatively, the biopolymer can be covalently linked onto the surface of the nanoparticles. Having this strategy in mind, our research group has devised a method for preparing magnetic biosorbents wherein the biopolymer is grafted onto a siliceous network that acts as a surface coating of Fe_3O_4 nanoparticles during the encapsulation process. This yields an organic–inorganic hybrid shell enriched in the biopolymer component. In the first stage, the biopolymer is modified with a silane coupling agent. Subsequently, a mixture of the modified biopolymer and tetraethylorthosilicate undergoes hydrolysis and condensation to form the hybrid shell around the magnetic nanoparticles.[82] This method creates more robust coatings with a higher degree of functionalization in comparison to conventional techniques such as non-covalent coating and surface-grafting approaches. These features are essential for the production of reusable adsorbents with improved adsorption capacities. A series of magnetic biosorbents have been prepared using this approach, employing polysaccharides with distinct ionic character and variable chemical functionalities.[35,83–85] These biosorbents have demonstrated efficacy in the uptake of target CEC, such as pharmaceuticals and pesticides. The choice of biopolymer facilitates the fabrication of magnetic nanosorbents featuring a spectrum of chemical functionalities and tunable surface charges.[19]

12.3.2 Water Remediation Using Magnetic Biosorbents

This section describes the latest advances in magnetic nanocomposites derived from natural polymers for the removal of pollutants from water using adsorption technologies, namely in the adsorption of organic and inorganic pollutants. The discussion includes pertinent details such as the pollutant targeted, the maximum adsorption capacity, and the underlying adsorption mechanisms. Applications in the removal of micro- and nanoplastics are also discussed.

12.3.2.1 Organic Pollutants Removal

In recent years, there has been significant attention towards magnetic chitosan-based nanocomposites, to fulfill the demanding requirements of water remediation applications, including high treatment efficiency and rapid magnetic separation.[86,87] These nanocomposites have demonstrated great potential in removing several organic pollutants from water, as elucidated in Table 12.2. As already mentioned, iron oxides such as Fe_3O_4 and γ-Fe_2O_3 are commonly used as magnetic phases, with tetracyclines emerging as the predominant class of pharmaceutical contaminants targeted, showing adsorption capacities ranging from 84.0 $mg\,g^{-1}$ to 834.0 $mg\,g^{-1}$ (Table 12.2). Functionalized magnetic chitosan-based adsorbent (Fe_3O_4@CAA) with a core–shell structure has been proposed for the clean-up of tetracycline from water.[88] These materials were prepared *via* ultrasound-initiated radical grafting copolymerization and cross-link reaction, wherein the chitosan was modified with acrylamide monomer and 2-acrylamido-2-methyl-1-propanesulfonic acid monomer to enhance the creation of active adsorption sites for the target pollutant tetracycline. The adsorption efficiency was determined to be 325.0 $mg\,g^{-1}$. The notable adsorption capacity of the adsorbent was attributed to the presence of amino and hydroxyl groups. The authors elucidated that the adsorption mechanism primarily involves electrostatic interactions and hydrogen bonding (Figure 12.2).

A mesoporous magnetic nanocomposite (CTM@Fe_3O_4) was fabricated by polycondensation of chitosan with thiobarbituric acid using malondialdehyde as a linker, in the presence of Fe_3O_4 nanoparticles. The bionanocomposite was effective in adsorbing tetracycline with a notable adsorption capacity of 215.3 $mg\,g^{-1}$.[89] The adsorption process was governed

Table 12.2 Magnetic bionanocomposites for the removal of various organic pollutants in water (based on the literature of the past five years).

Magnetic adsorbent	Target pollutant	Maximum adsorption capacity ($mg\,g^{-1}$)	Adsorption mechanism	Ref.
Magnetic chitosan-based adsorbent (Fe_3O_4@CAA)	Tetracycline	325.0	Electrostatic interaction and hydrogen bonding	88
Chitosan modified with thiobarbituric acid and malondialdehyde magnetic nanoparticles (CTM@Fe_3O_4)	Tetracycline	215.3	Electrostatic interaction, π–π stacking, hydrogen bonding and van der Waals forces	89
Ethanediamine-modified magnetic (Fe_3O_4) chitosan	Tetracycline	84.0	Electrostatic interaction, pore filling, π–π interactions and hydrogen bonding	99
Magnetic chitosan/microalgae biocomposites	Tetracycline, ciprofloxacin, and amoxicillin	834.0, 394.9, and 150.8	Electrostatic interaction	90
$ZnFe_2O_4$/chitosan magnetic particles	Diclofenac	188.0	Electrostatic interaction	100
Quaternary chitosan-based magnetic nanosorbents	Diclofenac, naproxen, ketoprofen	188.5, 438.1, and 221.5	Electrostatic interaction, hydrogen bonding, and hydrophobic effect	101
Poly(sodium 4-styrenesulfonate-*co*-maleic acid) sodium salt-functionalized-chitosan magnetic composite	Norfloxacin and sulfadiazine	412.6 and 214.0	Electrostatic interaction, π–π stacking, hydrogen bonding, and hydrophobic effect	102
Trimethyl chitosan/siloxane-hybrid coated Fe_3O_4 nanoparticles	Sulfamethoxazole	42.0	Electrostatic interaction	103
Magnetic hydrochar grafted-chitosan	Malachite green	420.0	Electrostatic interaction, hydrogen bonding, and π–π interactions	91
Magnetic carboxyl-functionalized chitosan composite	Methyl orange	1541.5	Electrostatic interaction and hydrogen bonding	92

Table 12.2 (*Continued*)

Magnetic adsorbent	Target pollutant	Maximum adsorption capacity (mg g^{-1})	Adsorption mechanism	Ref.
Magnetic (Fe_3O_4) chitosan composite	Methyl orange	638.6	Electrostatic interaction and π–π stacking	93
Magnetic Fe_3O_4 embedded chitosan-crosslinked-polyacrylamide composites	Sunset yellow	359.7	Electrostatic interaction and ion exchange	94
Magnetic γ-Fe_2O_3/Al^{3+}@chitosan-derived biochar	Congo red	636.9	Electrostatic interaction and flocculation/coagulation	95
Magnetic chitosan-benzil/ZnO/Fe_3O_4 nanocomposite	Remazol Brilliant Blue R	620.5	Electrostatic interaction and π–π interactions	96
Chitosan-functionalized $ZnFe_2O_4$	Dichlorvos	119.8	Electrostatic interaction	97
Quaternary chitosan magnetite nanosorbents	Glyphosate	3.04	Electrostatic interaction and hydrophobic effect	83
Magnetic chitosan/activated carbon@UiO-66 bio-nanocomposite	Imidacloprid	25.2	Electrostatic interaction, hydrogen bonding, π–π and n–π interactions	98
Carrageenan-grafted-poly(acrylamide) magnetic nanocomposite modified with graphene oxide	Ciprofloxacin	1146.4	Electrostatic interaction, π–π interactions and hydrogen bonding	104
κ-, ι-, and λ-carrageenan magnetic hybrid nanoparticles	Ciprofloxacin	878.0, 969.0 and 865.0	Electrostatic interaction, hydrogen bonding and ion exchange	85
Magnetic nanosorbents with siliceous hybrid shells of alginic acid and κ-, and λ-carrageenan	Ciprofloxacin	464.0, 423.0 and 1350.0	Electrostatic interaction	35
Sulfonic acid-amide functionalized magnetic sodium alginate	Ciprofloxacin and moxifloxacin	365.6 and 149.5	Electrostatic interaction and hydrogen bonding	105

Table 12.2 (*Continued*)

Magnetic adsorbent	Target pollutant	Maximum adsorption capacity ($mg g^{-1}$)	Adsorption mechanism	Ref.
Magnetic ferroferric oxide-polyvinyl alcohol-alginate coated UiO-67 nanospheres	Methylene blue	1371.8	Hydrogen bonding, π–π bonds, electrostatic interaction, and mesoporous pore flow	106
Sodium alginate-modified alkali-activated eggshell/ Fe_3O_4 nanoparticles	Crystal violet and methylene blue	332.5 and 304.4	Electrostatic interaction	107
Magnetic Fe_3O_4 nanoparticles loaded guava leaf powder impregnated into calcium alginate hydrogel beads	Methylene blue	136.7	Pore diffusion, hydrogen bonding, and electrostatic interaction	108
Starch-composite magnetic material modified with polyethyleneimine	Diclofenac, methyl orange, and amaranth	620.5, 470.8 and 193.7	Electrostatic interaction	109
Starch-containing magnetic montmorillonite	Tetracycline	135.2	van der Waals forces, electrostatic interaction, complexation, hydrogen bonding and π–π bonds	110
Magnetic starch-based adsorbent	Crystal violet	49.1	Electrostatic interaction and hydrogen bonding	111
Magnetic (Fe_3O_4) cellulose nanocrystal	Doxycycline	8.5–12.5	Electrostatic interaction, hydrogen bonding and π–π stacking.	112
Magnetic (Fe_3O_4) porous cellulose molecularly imprinted polymer	Tetracycline, demeclocycline, and minocycline	14.5, 10.1 and 11.0	Hydrogen bonding and electrostatic interaction	113
Chitosan-gelatin/ cetyltrimethyl-ammonium bromide magnetic polymer composite	Alizarin yellow R	344.6	Electrostatic interaction, hydrogen bonding, and π–π bonds	114

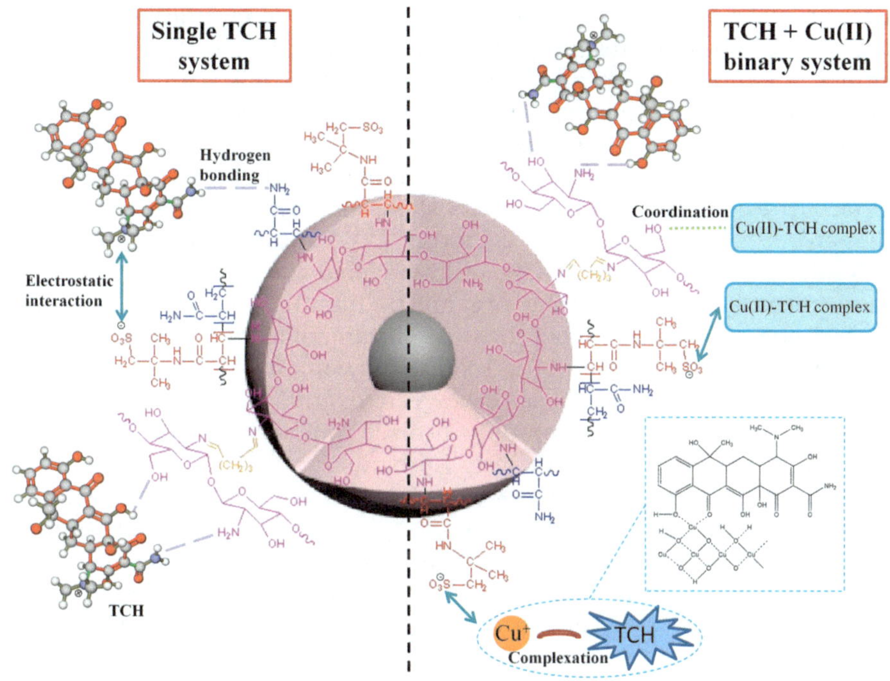

Figure 12.2 Adsorption mechanism of tetracycline (TCH) by magnetic chitosan-based adsorbent (Fe_3O_4@CAA).[88] Reproduced from ref. 88 with permission from Elsevier, Copyright 2023.

by electrostatic attraction, π–π stacking, hydrogen bonding and van der Waals forces.

Recent advances include the development of eco-friendly biosorbents comprising magnetite, chitosan and two different microalgae strains, *Chlorella vulgaris* (MCC) and *Arthrospira platensis* (MCA), exhibiting an excellent adsorption performance towards three antibiotics.[90] Magnetite formation was achieved *via in situ* co-precipitation, in the presence of a mixture of chitosan and the microalgae, crosslinked with sodium tripolyphosphate. The experimental maximum adsorption capacities were 834.0, 394.9, and 150.8 mg g^{-1} for tetracycline, ciprofloxacin and amoxicillin, respectively, using MCA-based bionanocomposites. Electrostatic interactions were identified as the primary interaction governing the adsorption mechanism.

Furthermore, chitosan-based magnetic biosorbents have demonstrated efficacy in the removal of various organic pollutants, including dyes (*e.g.* malachite green, methyl orange, sunset yellow, Congo red, and Remazol brilliant blue R)[91–96] and pesticides (*e.g.* dichlorvos, glyphosate, and imidacloprid)[83,97,98] *via* adsorption. Among the reported systems, we highlight a magnetic carboxyl-functionalized chitosan composite that was synthesized by grafting diethylenetriamine pentaacetate acid, a polycarboxylic amino acid, onto the surface of a composite comprising Fe_3O_4 microspheres and crosslinked chitosan, resulting in efficient removal of methyl orange dye from wastewater.[92] With an impressive maximum adsorption capacity of 1541.5 mg g^{-1} for methyl orange, the adsorption mechanism was attributed to electrostatic interactions and hydrogen bonding.

Besides chitosan, several polymers of natural origin have been employed to fabricate magnetic bionanocomposites, including carrageenan,[35,85,104] alginate,[35,105–108] starch,[109–111] cellulose,[112,113] and gelatin.[114] As detailed in Table 12.2, several studies have investigated the performance of these nanocomposites in adsorbing organic pollutants from water. Seaweed-derived polysaccharides, noted for their sustainability and biocompatibility, have garnered attention and are increasingly employed for organic pollutant adsorption. Among these, carrageenan and alginate, derived from seaweed, are the most commonly utilized polysaccharides in the production of eco-friendly magnetic biosorbents.[35,85,104–108] According to recent findings, magnetic carrageenan-based adsorbents have exhibited considerable efficacy in the removal of ciprofloxacin from water, showing high adsorption capacities ranging from 423 mg g^{-1} to 1350 mg g^{-1}.[35,85,104] This performance is attributed to the presence of anionic ester sulfate groups in carrageenan, enabling interactions with cationic pollutants, thereby predominantly governing the adsorption mechanism through electrostatic interactions. Alginate, primarily extracted from brown seaweeds, is characterized by its abundant carboxylate groups, enabling interactions with cationic pollutants such as pharmaceutical compounds (ciprofloxacin and moxifloxacin)[35,105] and dyes (methylene blue and crystal violet).[106–108] The results revealed noteworthy maximum adsorption capacities from 136.7 mg g^{-1} to 1371.8 mg g^{-1} using alginate-based magnetic biosorbents.

Starch and cellulose, which are neutral polysaccharides, have been employed in the production of magnetic biosorbents for the remediation of pharmaceuticals (diclofenac, tetracycline, doxycycline, demeclocycline, and minocycline)[109,110,112,113] as well as certain dyes (methyl orange, amaranth, and crystal violet).[109,111] Their adsorption efficiency ranged from 8.5 to 620.5 mg g^{-1}. Their adsorption capacity was attributed to the abundance of hydroxyl groups capable of establishing hydrogen bonds with the pollutants (Figure 12.3).

Gelatin, a mixture of peptides and proteins derived from the controlled hydrolysis of collagen, can enhance the mechanical properties of the adsorbent and contribute to increasing its adsorption capacity.[114] For example, magnetic chitosan/gelatin composites were produced by reacting chitosan with gelatin using glutaraldehyde as crosslinker, in the presence of carboxyl modified Fe_3O_4 nanoparticles, for the removal of alizarin yellow R dye from water.[114] The maximum adsorption capacity achieved was 344.0 mg g^{-1} and the adsorption mechanism involved electrostatic interactions, hydrogen bonding and π–π stacking between the dye and the composite material.

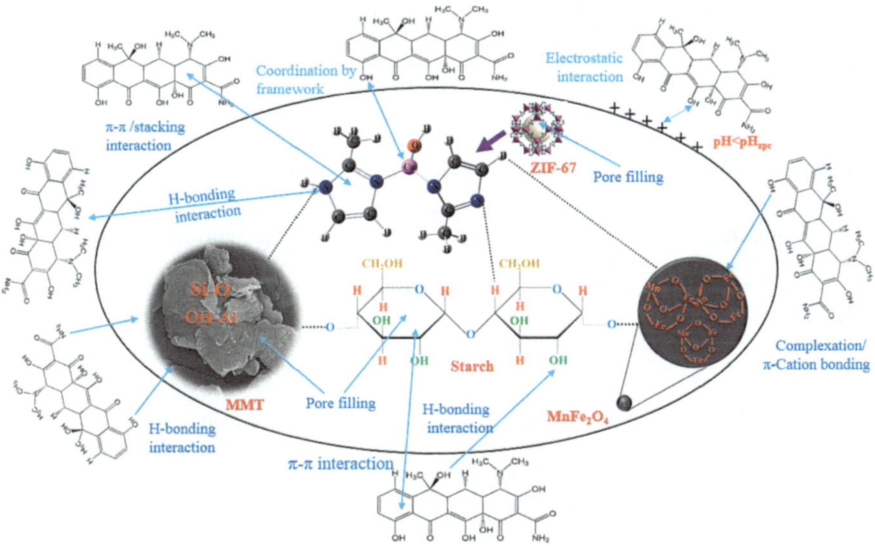

Figure 12.3 Schematic mechanism of tetracycline adsorption by starch-containing magnetic montmorillonite nanocomposite adsorbent.[110] Reproduced from ref. 110 with permission from Elsevier, Copyright 2024.

12.3.2.2 Inorganic Pollutants Removal

Heavy metals have been extensively utilized in various human activities, leading to a growing concern regarding their presence in wastewater.[115] Their propensity to accumulate in the environment and within living organisms, even at trace levels, poses significant risks to human health.[116] To address this challenge, several types of sorbents have been explored, including the application of magnetic nanoparticles as sorbents, as noted in the previous section. Also, a wide variety of magnetic bionanocomposites have been proposed for the effective removal of heavy metal species from water as detailed in Table 12.3.

Chitosan-based magnetic nanocomposites have been particularly explored for the removal of various heavy metal ions from water, including Ni(II), Co(II), Cr(IV), Cu(II), Cd(II), Pb(II), Zn(II), and Hg(II).[117–121,129] For example, a chitosan–magnetite nanocomposite synthesized *via* the coprecipitation method exhibited high adsorption capacities for nickel and cobalt ions from aqueous solutions.[117] Under optimized conditions, the nanocomposite demonstrated adsorption capacities of 30.0 $mg\,g^{-1}$ for Ni^{2+} and 53.1 $mg\,g^{-1}$ for Co^{2+}. The adsorption was ascribed to interactions between chitosan amino groups and metal cations. In another study, zirconium-doped carbon shell-protected magnetic Fe_3O_4 composite was synthesized by a hydrothermal method with chitosan and applied in the removal of Cr(VI) from water.[118] The material achieved a notable maximum adsorption capacity of 132.7 $mg\,g^{-1}$. The surface functional groups (–OH and –NH_2) underwent protonation in an acidic environment, thereby facilitating electrostatic

Table 12.3 Recently reported magnetic bionanocomposites for the removal of heavy metal ions.

Magnetic adsorbent	Target pollutant	Maximum adsorption capacity ($mg g^{-1}$)	Adsorption mechanism	Ref.
Fe_3O_4/chitosan nanocomposite	Ni(II) and Co(II)	30.0 and 53.1	Electrostatic interaction and ion exchange	117
Magnetic Zr-doped chitosan carbon-shell protected magnetic composites	Cr(IV)	132.7	Electrostatic interaction and ligand exchange	118
Magnetic alginate coated chitosan core@shell beads	Cu(II), Cd(II), Co(II), Ni(II), Pb(II), Zn(II), and Hg(II)	12 to 20	Metal ions chelation, complexation, and electrostatic interaction	119
Fe_3O_4–chitosan derived from *Prosopis farcta* biomass extract	Pb(II)	30.8	Electrostatic interaction	120
Fe_3O_4–graphene oxide–chitosan	Zn(II)	386.9	Electron donation	121
Graphene oxide@Fe_3O_4-decorated iota-carrageenan composite	Pb(II)	454.6	Electrostatic interaction, precipitation mechanism, and Lewis acid–base interaction	122
Magnetic Fe_3O_4 sulfonated nanoparticle caged alginate microbeads	Cd(II) and Pb(II)	37.0 and 85.0	Ion exchange and complexation	123
Magnetic biosorbent derived from sodium alginate and activated graphite schist	Mn(II)	333.1	Electrostatic interaction and ion complexation	124
Starch-composite magnetic material modified with polyethyleneimine	Cr(IV)	164.6	Electrostatic interaction, ion exchange, hydrogen bonding, and van der Waals forces	109
Polymer-grafted magnetic starch adsorption material St/Fe_3O_4-*g*-p(GMA-r-HEMA)	Cu(II)	94.1	Ion exchange	125
Starch-coated $CuFe_2O_4$-modified magnetic biochar	Pb(II) and Cd(II)	184.2 and 201.4	Complexation and electrostatic interaction	126

Table 12.3 (*Continued*)

Magnetic adsorbent	Target pollutant	Maximum adsorption capacity ($mg\,g^{-1}$)	Adsorption mechanism	Ref.
Magnetic adsorbent based on κ-carrageenan and N-doped carbon dots	Pb(II), Cu(II), Hg(II), and Cd(II)	114.6, 107.6, 99.2, and 94.2	Electrons exchange	127
Magnetic cellulose/ chitosan nanocomposites	Zr(IV) and Hf(IV)	180.0 and 103.0	Ion exchange	128

attraction and binding of Cr(VI) anionic species. Additionally, the authors observed that Cr(VI) could also be adsorbed through ligand exchange processes.

Numerous studies have documented the synthesis of alginate-based magnetic composites and their utilization in wastewater remediation.[119,123,124,129] For instance, core@shell magnetic beads comprising an alginate–bentonite core decorated with Fe_3O_4 nanoparticles and coated with a chitosan shell layer, were produced employing a non-thermal co-precipitation approach for *in situ* Fe_3O_4 production on the alginate–bentonite surface.[119] This material exhibited a high efficacy in magnetically removing a spectrum of organic and toxic inorganic pollutants from simulated wastewater, including Cu(II), Cd(II), Co(II), Ni(II), Pb(II), Zn(II), and Hg(II). The observed sequence of metal ions removal efficiency was Pb(II) > Cd(II) > Cu(II) > Zn(II) > Ni(II) > Hg(II) > Co(II) with adsorption capacities ranging from 12 to 20 $mg\,g^{-1}$. The magnetic alginate/chitosan beads comprised a variety of functional groups from both the biopolymers and clay permanent negative charges, which facilitated chelation, complexation, and electrostatic interactions with heavy metal ions (Figure 12.4).

Furthermore, modified starch-based magnetic biosorbents have been developed for similar applications.[109,125,126] Recently, a starch-coated $CuFe_2O_4$-modified magnetic biochar composite was developed. This involved modifying oxidized sawdust biochar with Fe/Cu loading for $CuFe_2O_4$ formation, followed by starch-coating/amination, aiming to enhance its adsorption performance for the removal of two metal ions, Pb(II) and Cd(II), from aqueous solutions.[126] The maximum adsorption capacities were 184.2 $mg\,g^{-1}$ for Pb(II) and 201.4 $mg\,g^{-1}$ for Cd(II). The results suggest that complexation and electrostatic attraction were the predominant mechanisms involved. Additionally, ion exchange coupled with physical adsorption also occurred during the adsorption process.

The use of other polymers of natural origin, such as κ-carrageenan and cellulose, for the adsorption of heavy metals, has received comparatively less attention.[127,128] In a recent study, magnetic cellulose/chitosan nanocomposites

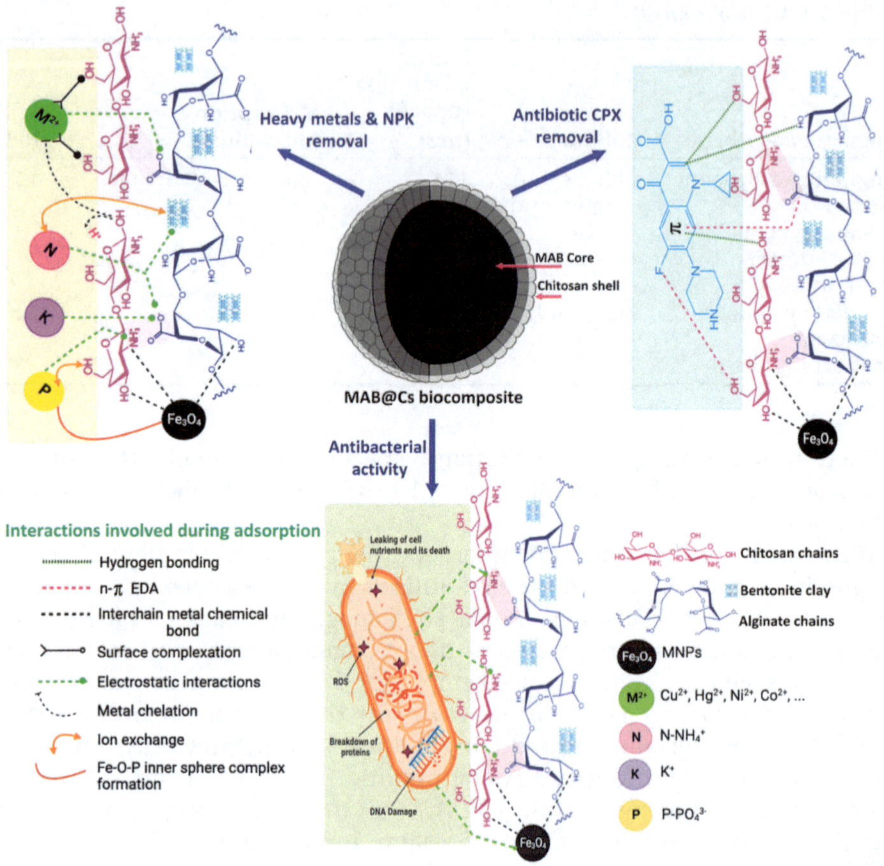

Figure 12.4 Proposed adsorption mechanism of organic and inorganic pollutants by magnetic alginate coated chitosan core@shell beads and their antibacterial activity.[119] CPX: ciprofloxacin; MAB: magnetic alginate–bentonite; Cs: chitosan; EDA: electron donor–acceptor, MNPs: magnetic nanoparticles. Reproduced from ref. 119 with permission from Elsevier, Copyright 2024.

functionalized with norvaline and 8-hydroxyquinoline were investigated for the uptake of zirconium and hafnium ions from both synthetic and waste effluents.[128] The maximum adsorption capacities of Zr(IV) and Hf(IV) ions on the nanocomposites were assessed to be 180.0 and 103.0 $mg\,g^{-1}$, respectively. The uptake mechanism was attributed to chelation and ion exchange, providing evidence for the potential occurrence of chemisorption interactions. The polysaccharide κ-carrageenan was employed to produce magnetic sorbents comprising N-doped carbon dots and Fe_3O_4 magnetic nanoparticles, to remove several heavy metal ions from water.[127] The maximum adsorption capacities for Pb(II), Cu(II), Hg(II), and Cd(II) were determined to be 114.6, 107.6, 99.2, and 94.2 $mg\,g^{-1}$, respectively. Chemisorption was identified as the rate-controlling step in the adsorption process.

12.3.2.3 Microplastics and Nanoplastics Removal

The presence of microplastics and nanoplastics in the environment has emerged as a pressing concern, prompting significant attention among the scientific community and the public in general. Water pollution resulting from these minuscule plastic particles originates from various sources, including personal care products and synthetic textiles, as well as the degradation of plastic items. Microplastics and nanoplastics, characterized by their extended half-lives, propensity for fragmentation, and capacity to adsorb heavy metals and organic contaminants, pose significant threats to aquatic and terrestrial ecosystems. These particles contribute to the degradation of environmental quality and entail serious ecological risks for organisms and human health. Given their adverse effects on ecosystems and public health, it has become imperative to devise effective and sustainable methodologies for removing microplastics and nanoplastics from water and wastewater.[130–132]

Recent studies have demonstrated the potential of biopolymers in the removal of micro- and nanoplastics. For instance, the addition of chitosan to the traditional inorganic flocculant poly aluminum chloride enhanced the flocculation removal of polyethylene terephthalate microplastics from water.[133] The observed effect was attributed to an enhanced charge neutralization and adsorption, resulting from the introduction of chitosan. Porous alginate sponges demonstrated efficacy in the removal of polystyrene microplastics ($\leq$5 μm in size), even in the presence of other contaminating ions adsorbed onto the microplastics, such as Cu^{2+}.[134] In another study, graphene oxide (GO) membranes were modified on both the front and back surfaces using distinct biopolymers, namely chitosan and alginate. This process yielded positive and negative surface charge-modified GO heterogeneous composite membranes, facilitating the simultaneous removal of several contaminants, including nanoplastics.[135] It was found that the composite membrane exhibited a notable removal efficiency for polystyrene nanoplastics of very small particle size (50 nm) and could retain them under the influence of gravity.

Despite the demonstrated potential, the exploration of biopolymers in the form of magnetic bionanocomposites for the removal of micro- and nanoplastics has been limited, with only a few studies reported in the literature. Recently, aerogels composed of chitosan and polydopamine, incorporating Fe_3O_4 nanoparticles as the magnetic component, have been synthesized, resulting in a porous recyclable sorbent with magnetic properties that facilitate easy collection *via* an external magnetic field.[136] The magnetic aerogels were highly efficient in removing polyethylene terephthalate, polyethylene and polystyrene microplastics from water, achieving removal percentages in the range of 92–97%. This remarkable adsorption capacity was ascribed to a combination of factors, including high porosity, electrostatic interactions between the positively charged chitosan amino groups and the negatively charged surface of microplastics, and physical adhesion facilitated by polydopamine. Very recently, alkylated nanoscale zerovalent iron grown onto naturally derived cellulose nanofibers, has been shown to rapidly enrich and trap micro- and nanoplastics from aqueous environments.[137] The observed high removal efficiency, reaching approximately 98% within a brief contact time of 1 minute, for particles of polystyrene (100 nm), polyvinyl chloride (2 μm), and poly(methyl methacrylate) (10 μm) in artificial seawater, was attributed to the synergistic interplay of attractive electrostatic and hydrophobic interactions between the fabricated magnetic composite and the plastic particles.

12.4 Resource Recovery and Valorization of Wastewater Using Magnetic Nanosorbents

12.4.1 Recovery of Precious Metals and Technological Critical Elements

In recent years, magnetic nanosorbents have received considerable attention for their application in resource recovery from wastewater and industrial effluents *via* magnetic solid-phase extraction. The escalating growth of industries and technological advancements has led to an increased demand for precious metals (*e.g.*, Au, Ag, and Pt) and technology-critical elements such as lanthanides, among others. The increased demand for these elements anticipated a crisis due to their relative scarcity in the earth's crust, increased costs, and geopolitical constraints. At the same time, the disposal of electronic and electrical waste containing these critical elements has increased significantly. As a result, the recovery of these elements from secondary sources such as electronic waste, industrial wastewater, and mine water has been regarded as an alternative with a positive impact on economics and environment.[138,139]

Magnetic adsorbents engineered with a high affinity for such elements can be effectively used to capture them. Once loaded, the magnetic adsorbents are then subjected to an elution process of the adsorbed element using a suitable solvent or stripping agent to facilitate further purification. This step is fundamental to the successful practical implementation of the methodology for separating and recovering these elements and to the reusability of the adsorbent material. Desorption offers several advantages, including the ability to obtain pre-concentrated solutions of recovered elements, the reusability of the magnetic sorbent, thereby reducing the process costs, and a reduction in sludge. Several eluent solutions with different concentrations have been used to desorb the target elements from loaded sorbents and to regenerate adsorption sites. Some examples include inorganic acids (HCl, HNO_3, or H_2SO_4), inorganic bases (*e.g.*, NaOH), salts (*e.g.*, NaCl), acids, organic compounds (*e.g.*, CH_3COOH, thiourea), and chelating agents (*e.g.*, EDTA).[140] Despite the importance of desorption, this process has received relatively little attention.[141]

Although the recovery of precious metal ions and critical elements is currently a hot topic, the use of magnetic nanocomposites for this purpose remains uncommon. For instance, Ilankoon *et al.*[142] used polyethyleneimine-coated iron oxide magnetic nanoparticles to extract gold and copper from a gold ore thiosulfate leachate. Separation of the metals was achieved using different eluent solutions.[142] Nearly 100% elution of copper was achieved using 0.02 mol L^{-1} EDTA, while the selective elution of gold was achieved using 0.01 mol L^{-1} NaOH solution. Magnetic silica composite with the –SH functional group (Fe_3O_4@SiO_2–SH), prepared by a one-pot synthesis method, was used for Au(I) adsorption.[143] The attachment of –SH to the surface of silica-coated Fe_3O_4 promotes the chelating interaction between –SH and Au(I), which enhances the adsorption of the metal on the magnetic composite. The performance of this adsorbent was pH dependent and showed a saturated adsorption capacity of 15.1 mg g^{-1}.[143] The Fe_3O_4@SiO_2–SH adsorbent demonstrated the ability to recover over 90% of the Au present in both simulated and real thiosulfate-based gold leaching solutions. As an eluent, a $Na_2S_2O_3$ solution (3 mol L^{-1}) was found to be highly effective in desorbing Au (95.5%) without compromising the structural integrity of the magnetic composite. A magnetic Fe_3O_4/MgAl-layered double hydroxide (LDH) nanocomposite was synthesized by conventional co-precipitation in the dispersed Fe_3O_4 suspension with the aim of recovering gold and iridium from the soils around the mining area. The maximum adsorption capacities of this magnetic composite for gold and iridium were 124 mg g^{-1} and 115 mg g^{-1}, respectively.[144] The recovery of the precious elements from the composite reached more than 80% using a H_2SO_4 solution 2.0 mol L^{-1}, and the composite showed efficiency up to ten consecutive adsorption–desorption cycles. An amine-functionalized magnetic nanocomposite was synthesized by a silica coating on the $MnFe_2O_4$ core for the recovery of Pt(IV) from aqueous solutions. This nanocomposite with spherical morphology and particle diameter of about 50 nm exhibited a high sorption capacity of 380 mg g^{-1} for Pt(IV).[145] In contrast to bare $MnFe_2O_4$, the composite exhibits a much lower saturation magnetization value, 0.357 emu g^{-1} *versus* 41 to 66 emu g^{-1}. Nevertheless, the composite was easily recovered from solution. For the recyclability of the composite, the authors used an acidified thiourea solution (thiourea 0.05 mol L^{-1} + HCl 0.01 mol L^{-1}) due to its affinity for noble metal ions and did not observe any significant loss of efficiency during three sorption–desorption cycles.[145] In addition, this composite $MnFe_2O_4$@SiO_2–NH_2 also showed good stability with no relevant loss of Mn or Fe from the spinel structure of the $MnFe_2O_4$ nanoparticles.

Nowadays, not only precious metals are targeted for recovery. With the huge demand for critical elements and the enrichment of low concentration lanthanide ions in waste waters, their recovery has become a crucial research area to optimize the use of these resources, and some studies are appearing in this field. In a recent study, Javadian *et al.*[146] synthesized a polyamide with methine-thiophene and pyrimidine linkages (poly(pyrimidine-thiophene-amide) (PPTA)) and then prepared a magnetic PPTA (M-PPTA) using a hydrothermal technique with different ions (Fe^{3+}, Zn^{2+}, and Ni^{2+}) in acid medium in the presence of PPTA. This magnetic material, with a magnetic saturation of 50.5 emu g^{-1}, showed spherical morphology with particle sizes below 100 nm, and it was used to absorb dysprosium, terbium, and neodymium (50 mg L^{-1}), achieving percentages of adsorption ranging between 95.6 and 98.4% for Nd and Dy, respectively. The adsorption efficiency decreased to values around 85–89% after four adsorption–desorption cycles, using a 0.2 mol L^{-1} HNO_3 solution to desorb the target elements. The authors attribute this result to the significant chemical interactions between Nd, Dy, and Tb and the functional groups of M-PPTA, resulting in incomplete ion desorption.[146] A magnetic composite named g-C_3N_4/Fe_3O_4@ZIF-8 was synthesized by first using g-C_3N_4 as the substrate and Fe_3O_4 as the magnetic component to prepare a magnetic graphitic carbon nitride material, which was then incorporated *in situ* with a zeolitic imidazolate framework (ZIF-8) in aqueous phase and at RT.[147] This composite showed high selectivity and large adsorption capacity for La(III) (177 mg g^{-1}), and the main adsorption mechanisms are ion exchange and surface complexation. In addition, this composite, which has a saturation magnetization of only 8.1 emu g^{-1}, retained 81% of its initial adsorption efficiency after five cycles of reuse.[147]

Research on magnetic biosorbents for the recovery of metal ions and critical elements remains scarce. For instance, magnetic chitosan microspheres, with an average size of 2 μm, prepared *via* the suspension cross-linking technique with Fe_3O_4 nanoparticles as the magnetic component, were effectively used as reusable adsorbents for the selective separation of Ag nanoparticles (with diameters ranging from 10 to 100 nm) from environmental waters. The Ag nanoparticles could be recovered by desorption using thiourea, which forms a stable complex with silver.[148] Magnetic composites of ion-imprinted chitosan were used for selective adsorption and recovery of Ni(II) from solutions. After Ni(II) adsorption, almost complete elution and recovery of the adsorbed Ni(II) ions was achieved using a 0.5% EDTA–Na solution. In addition, the sorption capacity of the recycled ion-imprinted magnetic composites showed only a slight decrease of about 10% after 15 cycles of adsorption and desorption.[149] A novel calcium alginate/carboxymethyl chitosan/$Ni_{0.2}Zn_{0.2}Fe_{2.6}O_4$ magnetic bionanocomposite was synthesized for the adsorption of some critical elements (Nd, Tb, and Dy) from single solutions.[150] The magnetic bionanocomposite showed a very good adsorption capacity and after four cycles of reuse, the adsorption values were still high, about 95% for all the elements.[150] The ions were easily desorbed from the biocomposite using 0.2 mol L^{-1} HNO_3 solution. This bionanocomposite with a particle size below 100 nm has a saturation magnetization lower than other magnetic composites found in the literature (14.1 emu g^{-1}), but it could still be easily separated from the solution by applying an external magnetic field. In other work, iron oxide-loaded calcium alginate beads were synthesized to adsorb lanthanum from aqueous solutions.[151] In addition to the large adsorption capacity for La(III) (123.5 mg g^{-1}), the magnetic calcium alginate beads showed a magnetic saturation of 63.2 emu g^{-1}, confirming their easy separation from solution by an external magnet.[151]

Furthermore, the magnetic component itself, without additional surface modification or combination to form nanocomposites, has been explored for the recovery of various metals. For example, magnetite particles have demonstrated selectivity towards Au, Pt, and Pd from chloride solutions.[152,153] More recently, magnetic nickeliferous pyrrhotite from mine tailings has shown remarkable efficiency in the recovery of gold from chloride solutions.[154] Manganese ferrite nanoparticles have been studied for the removal of several rare earth elements, including Y, La, Ce, Pr, Nd, Eu, Gd, Tb, and Dy.[155] The experimental parameters such as adsorbent mass, ionic strength, and pH have a strong influence on the sorption efficiency, especially pH, since the main adsorption mechanism identified was electrostatic interactions. Under the best conditions, pH range 6–8 and low ionic strength, these nanoparticles show very high adsorption efficiencies (>90%). However, these studies focus primarily on the adsorption process and do not address desorption for the recovery of metal species.

12.4.2 Recovery and Reuse of Nutrients

Anthropogenic activities frequently result in the excessive inputs of nutrients (*e.g.* phosphate, nitrate) to aquatic ecosystems, exacerbating eutrophication and impairing water quality. Conversely, the global supply of nutrients such as phosphate is decreasing. Considering both the need to decrease eutrophication and the need for sustainable nutrient management, the removal and recovery of these nutrients from wastewater is of great importance. Nanocomposites based on biopolymers such as alginate, cellulose, and chitosan have shown remarkable efficacy in nutrient recovery through adsorption–desorption processes, particularly for the recovery of phosphate (PO_4^{3-}) and nitrate (NO_3^-).[156] Some examples include magnetic composite chitosan beads containing fillers such as kaolin and biochar.[157,158] In these studies, magnetite formation occurred in the presence of the filler, followed by dispersion of the magnetic filler particles in chitosan and subsequent crosslinking. In addition, the surface of the biochar-based composite was further modified with amine groups using ethylenediamine and triethylenetetramine. After adsorption from swine effluent, the phosphate-loaded biochar-based magnetic composites were evaluated as a fertilizer in pot experiments, with results comparable to those obtained with conventional phosphate fertilizers. This suggests that a potential solution for spent phosphate adsorbents is their use as alternative fertilizers in agriculture. In another study, amine surface-functionalized Fe_3O_4/alginate composite beads have demonstrated effective uptake of phosphate and nitrate. This performance was attributed to the interaction of these species with Fe^{3+} sites and protonated amine groups *via* surface complexation and electrostatic attraction.[159] The desorption process for sorbent regeneration was performed using a KOH solution, but the amount of phosphate and nitrate recovered was not quantified.

Other magnetic composites without the incorporation of biopolymers have also been investigated for nutrient recovery. One example is a nanostructure consisting of a mesoporous ZrO_2 shell and a magnetite core.[160] The synthesis of this nanocomposite involves several steps, starting with the synthesis of Fe_3O_4 nanoparticles by the co-precipitation of Fe^{2+} and Fe^{3+} salts in an ammoniacal medium under an inert atmosphere, and their further encapsulation in SiO_2. The surface coating protects the magnetic core and allows the uniform coating of mesoporous ZrO_2.[160] The mesoporous ZrO_2 shell in this nanocomposite provides excellent adsorption capability for phosphate ions while the Fe_3O_4 core allows an easy and fast magnetic separation. The Fe_3O_4@mZrO_2 magnetic composite showed a maximum phosphate adsorption capacity of 39 mg g^{-1}, which was dependent on both the initial phosphate concentration and the initial pH of the solution. Moreover, the regeneration of this adsorbent was effective using a mixture of KOH and KCl (about 90% of phosphate desorption), and at the end of four consecutive cycles, the Fe_3O_4@mZrO_2 composite retained more than 80% of its adsorption capacity and about 80% of phosphate recovery after the desorption step.

Taking advantage of the high affinity of lanthanum(III) (a moderately hard trivalent Lewis acid) for phosphate (hard base) and the high affinity of EDTA for Fe_3O_4 nanoparticles, Yang *et al.*[161] reported the synthesis of a La–EDTA–Fe_3O_4 composite, prepared by a chemical co-precipitation method for the adsorption of phosphate from water. First, the EDTA-coated Fe_3O_4 magnetic nanoparticles were prepared by the co-precipitation of Fe^{2+} and Fe^{3+} salts in a medium containing ammonia and EDTA, and then the resulting nanoparticles were brought into contact with a La(III) solution under stirring for a few hours. In this study, the La–EDTA–Fe_3O_4 composite was able to remove 99% of the phosphate in the water at pH values between 6 and 7, performing better than the bare EDTA–Fe_3O_4 composite. In addition, the maximum adsorption capacity found for an initial phosphate concentration of 5 mg L^{-1} was 4.19 mg g^{-1}.[161] The desorption process for sorbent regeneration and reusability was performed using a NaOH solution as the eluent. After the second cycle, the adsorption efficiency decreased by 2% and after three consecutive adsorption–desorption cycles, the removal percentage decreased by about 26%, reaching 68% of phosphate removal. The low regeneration capacity of La–EDTA–Fe_3O_4 is attributed to the strong affinity between La and phosphate.[161] Another La-based sorbent for phosphate removal was reported by Baile *et al.*[162] The author prepared magnetic $La(OH)_3/Fe_3O_4$ nanocomposites with different La to Fe mass ratios by a precipitation and hydrothermal method.[162] The 4 : 1 ($La(OH)_3/Fe_3O_4$) nanocomposite shows high sorption capacity (83.5 mg g^{-1}) and good reusability (70%) after 5 consecutive cycles with 1 mol L^{-1} NaOH solution. As mentioned in the previous section, in addition to being incorporated into composites, the bare magnetic particles have also been used as sorbents for the removal and recovery of phosphate from water. Some examples include magnetite, hematite, and mixtures of goethite plus maghemite.[163–165] In general, the bare nanoparticles have lower efficiencies than the magnetic composites described above, and among them, magnetite seems to perform better. According to the study of Tu *et al.*,[164] Fe_3O_4 nanoparticles prepared by the ferrite process have a maximum adsorption capacity of 3.65 mg g^{-1} for phosphorus at pH 2.77, a value higher than those reported for hematite and a mixture of goethite and maghemite.[163,165] The regeneration of the sorbent can be performed using NH_4OH and NaOH solutions as eluents, while the phosphorus from the enriched solution can be recovered as calcium phosphate by precipitation with $CaCl_2$. During three adsorption–desorption cycles, the desorption efficiency from the magnetic nanoparticles slightly decreases from 99.1% P in the first cycle to 92.8% in the third cycle, and this decrease is attributed to the occupation of the adsorption sites by the undesorbed P or the collection loss of the synthesized adsorbent.[164] However, these magnetite nanoparticles showed a high potential for reuse in successive cycles.

12.5 From Laboratory to Pilot-scale Studies

Despite the increasing interest in the use of magnetic nanocomposites for pollutant adsorption in water, limited attention has been given to their application in real-world wastewater treatment scenarios. Moreover, the majority of existing studies remain confined to laboratory settings. While laboratory-scale studies offer valuable insights into optimal adsorption parameters and mechanisms, pilot-scale studies are crucial for evaluating the performance of the nanocomposites under realistic environmental and operational conditions. The transition of laboratory-scale investigations to pilot-scale studies of new sorbents poses always significant challenges. In the case of magnetic nanosorbents, the up-scale procedures also raise fundamental questions regarding their applicability on an industrial scale. Furthermore, as a new nanotechnology, it has several implications in terms of required infrastructures. The required technology should integrate specifically magnetic separation procedures, effectively addressing challenges associated with continuous operation, particle size suitability, and ease of cleaning.[166]

Most of the adsorption studies performed at the laboratory-scale have been conducted in discontinuous ("batch") mode, with limited literature available on the potential utilization of continuous adsorption for pollutant removal using magnetic nanocomposites.[23,167] Batch trials are essential to unveil fundamental insights into the adsorbent's characteristics and factors influencing the adsorption process, including initial adsorbate concentration, adsorbent dosages, particle size, temperature, and solution pH. On the other hand, continuous tests aim at assessing the adsorbent's capacity for ongoing contaminant removal. While researchers often prefer batch mode for its minimal adsorbent material requirement and time efficiency, continuous systems are more suitable for industrial-scale design and scaling purposes due to their simple configuration and ease of scaling.[168,169]

Table 12.4 summarizes key physicochemical properties, target pollutants, and operating conditions of pilot-scale magnetic adsorbent

Table 12.4 Pilot-scale adsorption studies using magnetic adsorbents for the removal of water pollutants: physico-chemical properties of the adsorbent, target pollutant, and operating conditions.

Magnetic adsorbents	Physico-chemical properties	Target pollutant	Operating conditions	Ref.
Commercial bare Fe_3O_4 nanoparticles	Fe_3O_4 diameter = 15 nm;	Chromium (tannery wastewater)	Pilot scale experiment with a reactor volume = 150 L (batch); magnetic separator at a flow rate = 18 L h^{-1} (continuous)	170
Commercial modified Fe_3O_4 nanoparticles	Fe_3O_4 diameter = 15 nm; modified surface with amine groups (–NH_2)	Chromium (tannery wastewater)	Pilot scale experiment; treatment capacity = 16 m^3 per day	176
Magnetic composite nanoparticles (γ-Fe_2O_3/polymer)	γ-Fe_2O_3 diameter = 10 nm; γ-Fe_2O_3 nanoparticles imbedded into the acrylamide hydrogel with a final diameter = 10 μm	Chromium	Pilot scale experiment; prototype with a reactor volume = 5 L and a 5 L magnetic separation unit	178
Sn-based magnetic nanocomposite	Fe_3O_4 diameter = 40 nm	Chromium	Pilot scale experiment with a reactor volume = 2 L and flow rate = 30–140 mL min^{-1} (continuous); rotary magnetic separator = 18 L h^{-1}	172
ZnFeZr coupled with Fe_3O_4/SiO_2 nanoparticles	Composition: ZnO; Fe_3O_4/SiO_2; oxy/hydroxides of Zr	Phosphate (secondary wastewater effluent)	Pilot scale experiment with a reactor volume = 100 L (batch); prototype with a magnetic separation unit; separation efficiency for 20 cycles of treatment	180
Iron oxide nanoparticle-loaded chitosan composites	Diameter of the final composites = 1.2 mm; magnetic phases: γ-Fe_2O_3 and α-FeOOH	Phosphate	Pilot scale experiment; fixed-bed column experiments (without magnetic separation) (height = 100 cm, inner diameter = 45 cm, flow rate = 7.05 L min^{-1}); reactor volume = 80 L	179

Table 12.4 (*Continued*)

Magnetic adsorbents	Physico-chemical properties	Target pollutant	Operating conditions	Ref.
Fe_3O_4/La(OH)$_3$ composite	Composition: commercial Fe_3O_4 nanoparticles with a diameter of 1000 nm; $LaCl_3$; NaOH	Phosphate	Pilot scale experiment with flow rate = 2 m^3 per day	173
Commercial bare Fe_3O_4 nanoparticles	Fe_3O_4 diameter = 15–20 nm; specific surface area = 82 $m^2\,g^{-1}$	Arsenic	Pilot scale experiment with flow rate = 12 L, 1 L min^{-1}; reactor volume = 1.11 L)	171
Commercial $MnFe_2O_4$ nanoparticles	$MnFe_2O_4$ diameter = 150 nm; specific surface area = 106.1 $m^2\,g^{-1}$	Arsenic antimony	Pilot scale experiment with flow rate = 1, 2 and 3 L min^{-1}; superconducting magnetic separation; magnetic separation chamber = 0.126 L	175
Magnetic clay (bentonite)	Bentonite modification with Fe_3O_4 nanoparticles; bentonite diameter = 6 μm; Fe_3O_4 nanoparticles prepared by co-precipitation	Direct black 22 (dye)	Pilot scale experiment with flow rate = 0.08 L min^{-1}; magnetic drum separator	177

applications. In terms of composition, magnetic adsorbents primarily consisted of magnetite[170–173] or other magnetic iron oxides such as maghemite[174] or ferrites.[175] These materials may undergo functionalization[176] or remain unmodified,[170] and can be combined with other components such as clays[177] or polymers,[178,179] to form composite materials. Notably, the dimensions of the final adsorbent in these systems typically fall within the micrometer[178] or millimeter[179] range, positioning them beyond the nanoscale. Target pollutants primarily include ionic heavy metals or metalloids, such as chromium from tannery wastewater,[170,172,176,178] as well as arsenic and antimony found in some minerals and industrial wastewater, posing risks to groundwater and surface water.[171,175] Additionally some studies have focused on the removal of phosphates present in wastewater effluents.[173,179,180] However, concerning the removal of organic micropollutants, only one study addressing the removal of the dye Direct Black 22 from textile wastewater was identified.[177]

12.6 Conclusions and Outlook

This chapter provides an up-date review on the development of magnetic nanosorbents for water treatment technologies. Among the several classes of colloidal magnetic nanoparticles, the most common systems in adsorption applications exhibit ferromagnetic/ferrimagnetic or superparamagnetic behavior. In this context, iron containing materials (Fe, Fe_3O_4 and γ-Fe_2O_3) have been intensely investigated, which also form the main topic of this chapter in terms of selected inorganic materials. The former systems show high saturation magnetization values and high magnetophoretic mobility, which makes them especially useful in adsorption technologies assisted by magnetic separation. However, unlike superparamagnetic nanoparticles, they are characterized by a non-null remanent magnetization, which induces their easy aggregation under an external magnetic field and afterwards its removal. In certain scenarios this aggregation process can be favorable because it facilitates the nanosorbents' separation from the dispersing medium, but the decrease in the available surface area for sorption can be critical, requiring assessment. Mechanical stirring can circumvent to some extent this drawback and facilitate their redispersion in the aqueous medium. Superparamagnetic nanoparticles, in turn, do not present this limitation, as they are characterized by zero magnetic coercivity and do not have magnetic memory once the magnetic gradient is removed. Thus, in the absence of an external magnetic field they are more easily redispersed, which is clearly advantageous in terms of nanosorbent regeneration and reuse. On the other hand, superparamagnetic nanosorbents require longer removal times for separation from the medium under the effect of an external magnetic field, as compared to ferromagnetic or ferrimagnetic nanomaterials, which might be critical for the implementation of this type of nanotechnology in certain water treatment units or in resources recovery.

A considerable part of this work was devoted to the chemistry of surfaces of the magnetic particles employed as the cores of the nanosorbents. This was approached here as a crucial aspect in developing nanosorbents, namely for the development of composite nanostructures containing polymeric coatings that comply with several criteria, that include technical performance, costs, sustainability and eco-toxicological assessment. Hence, the development of effective and sustainable magnetic sorbents is not a trivial task. Resorting to surface modifications of magnetic nanoparticles using biopolymers or derived hybrid shells are examples of very promising strategies. Nevertheless, achieving high adsorption capacity and selectivity toward target pollutants demands high specific surface area and adequate surface functionalization of the colloids, keeping the magnetic properties required for their separation from aqueous systems. Furthermore, to comply with sustainability criteria, magnetic sorbents should enable multiple decontamination cycles at high efficiencies, with negligible environmental impact. As such, fundamental research on magnetic nanosorbents should be kept as an important aspect for the development of magnetic assisted adsorption technologies. For example, there has been recent interest in magnetic nanosorbents that aim for multifunctionality.[181,182] Typically, these multifunctional sorbents contain nanophases that give them possibilities beyond magnetic separation under an external magnetic gradient. Thus, multifunctional nanosystems were developed that allow, for example, optical monitoring based on the plasmonic properties conferred by plasmonic nanoparticles.[183,184]

Finally, to bridge the gap between laboratory-scale studies and real-world applications, the continuous adsorption mode and pilot-scale implementation of magnetic nanocomposites need to be further explored. There is a clear need for nanoengineering projects that tackle specifically up-scale procedures envisaging the application of magnetic nanosorbents, that allows us to unlock the full potential of magnetic adsorbents in water treatment and resource recovery in a logic of circular economy. As noted in the previous section of this chapter, the implementation of magnetic separation at pilot- or full-scale requires specialized equipment. In addition to magnetic separators, advanced systems such as magnetic extractors and fluidized beds have emerged as noteworthy innovations, but more studies are required on the integration of different equipment in real scenarios, that take into consideration environmental impacts.

Acknowledgements

This work was developed within the scope of the project CICECO-Aveiro Institute of Materials, UIDB/50011/2020 (DOI: 10.54499/UIDB/50011/2020), UIDP/50011/2020 (DOI: 10.54499/UIDP/50011/2020) & LA/P/0006/2020 (DOI: 10.54499/LA/P/0006/2020), financed by national funds through the FCT/MCTES (PIDDAC). S. F. S. thanks FCT for her research contract (2023.06221.CEECIND). C. B. L. thanks FCT for her researcher contract 2021.03739.CEECIND/CP1659/CT0025 (DOI: 10.54499/2021.03739. CEECIND/CP1659/CT0025, PTCRIS, 2022–2024a). A. L. D.-d-S. acknowledges FCT for funding (CEECIND/03075/2018/CP1559/CT0020; DOI: 10.54499/CEECIND/03075/2018/CP1559/CT0020). We thank Dr João Amaral (U. Aveiro) for the helpful discussion on the magnetism of materials.

References

1. United Nations, *The United Nations World Water Development Report 2023: Partnerships and Cooperation for Water*, UNESCO, Paris, 2023.
2. *Guidelines for drinking-water quality: fourth edition incorporating the first and second addenda*, World Health Organization, Geneva, 2022.
3. United Nations Environment Programme, Pharmaceuticals and personal care products in the environment: An Emerging Issue, Global Programme of Action for the Protection of the Marine Environment from Land-Based Activities, 2017

4. T. Deblonde, C. Cossu-Leguille and P. Hartemann, *Int. J. Hyg. Environ. Health*, 2011, **214**, 442–448.
5. P. Verlicchi, M. Al Aukidy and E. Zambello, *Sci. Total Environ.*, 2012, **429**, 123–155.
6. J.-J. Jiang, C.-L. Lee and M.-D. Fang, *Mar. Pollut. Bull.*, 2014, **85**, 391–399.
7. *Contaminants of Emerging Concern in the Environment: Ecological and Human Health Considerations*, ed. R. U. Halden, American Chemical Society, Washington, DC, 2010, vol. 1048.
8. N. R. Maddela, B. Ramakrishnan, D. Kakarla, K. Venkateswarlu and M. Megharaj, *RSC Adv.*, 2022, **12**, 12396–12415.
9. M. H. Hassan, R. Khan and S. Andreescu, *Electrochem. Sci. Adv.*, 2022, **2**, 1–27.
10. Marine Pollution – Monitoring, Management and Mitigation, in *Springer Textbooks in Earth Sciences, Geography and Environment*, ed. A. Reichelt-Brushett, 2023.
11. United Nations Environment Programme, *Global Waste Management Outlook 2024: Beyond an age of waste – Turning rubbish into a resource*, Nairobi, 2024.
12. T. Dutta, K.-H. Kim, M. Uchimiya, E. E. Kwon, B.-H. Jeon, A. Deep and S.-T. Yun, *Environ. Res.*, 2016, **150**, 182–190.
13. J. C. K. Lee and Z. Wen, *Nat Sustainability*, 2018, **1**, 598–605.
14. C. E. D. Cardoso, J. C. Almeida, C. B. Lopes, T. Trindade, C. Vale and E. Pereira, *Nanomaterials*, 2019, **9**, 814.
15. O. A. Kazi, W. Chen, J. G. Eatman, F. Gao, Y. Liu, Y. Wang, Z. Xia and S. B. Darling, *Adv. Mater.*, 2023, **35**, 2300913.
16. G. L. Dotto and G. McKay, *J. Environ. Chem. Eng.*, 2020, **8**, 103988.
17. M. Khajeh, S. Laurent and K. Dastafkan, *Chem. Rev.*, 2013, **113**, 7728–7768.
18. R. Das, C. D. Vecitis, A. Schulze, B. Cao, A. F. Ismail, X. Lu, J. Chen and S. Ramakrishna, *Chem. Soc. Rev.*, 2017, **46**, 6946–7020.
19. A. C. Estrada, A. L. Daniel-da-Silva, C. Leal, C. Monteiro, C. B. Lopes, H. I. S. Nogueira, I. Lopes, M. J. Martins, N. C. T. Martins, N. P. F. Gonçalves, S. Fateixa and T. Trindade, *Front. Chem.*, 2022, **10**, 1011186.
20. A. L. Daniel-da-Silva and T. Trindade, in *Surface Chemistry of Colloidal Nanocrystals*, The Royal Society of Chemistry, 2021, pp. 209–257.
21. Q. Liu, Q. Zhou and G. Jiang, *TrAC, Trends Anal. Chem.*, 2014, **58**, 10–22.
22. M. M. Khin, A. S. Nair, V. J. Babu, R. Murugan and S. Ramakrishna, *Energy Environ. Sci.*, 2012, **5**, 8075.
23. S. Muthu Prabhu, N. R. Rane, X. Li, S. V. Otari, S. D. Girawale, A. R. Palake, K. M. Kodam, Y.-K. Park, Y.-H. Ha, K. Kumar Yadav, M. Ali Khan and B.-H. Jeon, *Chem. Eng. J.*, 2023, **468**, 143474.
24. C. T. Yavuz, J. T. Mayo, W. W. Yu, A. Prakash, J. C. Falkner, S. Yean, L. Cong, H. J. Shipley, A. Kan, M. Tomson, D. Natelson and V. L. Colvin, *Science*, 2006, **314**, 964–967.

25. F. L. Sousa, A. L. Daniel-da-Silva, N. J. O. Silva and T. Trindade, in *Eco-friendly Polymer Nanocomposites*, 2015, pp. 279–310.
26. M. Bundschuh, J. Filser, S. Lüderwald, M. S. McKee, G. Metreveli, G. E. Schaumann, R. Schulz and S. Wagner, *Environ. Sci. Eur.*, 2018, **30**, 6.
27. N. Wilson, *Bioscience*, 2018, **68**, 241–246.
28. A. Kazemi, N. Bahramifar, A. Heydari and S. I. Olsen, *J. Cleaner Prod.*, 2018, **174**, 527–537.
29. J. C. Almeida, C. E. D. Cardoso, D. S. Tavares, T. Trindade, C. Vale, R. Freitas and E. Pereira, *Environ. Sci. Pollut. Res.*, 2024, **31**, 28789–28802.
30. K. M. Krishnan, *Fundamentals and Applications of Magnetic Materials*, Oxford University Press, 2016.
31. M. Reis, *Fundamentals of Magnetism*, Academic Press, Elsevier Inc., Oxford, UK, 2013.
32. Z. Ma, J. Mohapatra, K. Wei, J. P. Liu and S. Sun, *Chem. Rev.*, 2023, **123**, 3904–3943.
33. T. Radu, C. Iacovita, D. Benea and R. Turcu, *Appl. Surf. Sci.*, 2017, **405**, 337–343.
34. R. S. Carvalho, A. L. Daniel-da-Silva and T. Trindade, *Part. Part. Syst. Charact.*, 2016, **33**, 150–157.
35. S. F. Soares, M. J. Rocha, M. Ferro, C. O. Amorim, J. S. Amaral, T. Trindade and A. L. Daniel-da-Silva, *Int. J. Biol. Macromol.*, 2019, **139**, 827–841.
36. T. Fernandes, H. I. S. Nogueira, C. O. Amorim, J. S. Amaral, A. L. Daniel-da-Silva and T. Trindade, *Chem. – Eur. J.*, 2022, **28**, e202202382.
37. P. N. A. L. Pottail, A. Chithambharan and S. C. Sharma, *Chem. Phys. Impact*, 2023, **7**, 100250.
38. D. H. K. Reddy and Y.-S. Yun, *Coord. Chem. Rev.*, 2016, **315**, 90–111.
39. D. S. Tavares, C. B. Lopes, J. C. Almeida, C. Vale, E. Pereira and T. Trindade, *Environ. Sci. Pollut. Res. Int.*, 2020, **27**, 22523–22534.
40. J. He, D. Xiong, P. Zhou, X. Xiao, F. Ni, S. Deng, F. Shen, D. Tian, L. Long and L. Luo, *Chem. Eng. J.*, 2020, **393**, 124696.
41. S. Li, W. Wang, F. Liang and W. Zhang, *J. Hazard. Mater.*, 2017, **322**, 163–171.
42. K. Li, J. Li, F. Qin, H. Dong, W. Wang, H. Luo, D. Qin, C. Zhang and H. Tan, *J. Cleaner Prod.*, 2023, **415**, 137812.
43. M. Stefaniuk, P. Oleszczuk and Y. S. Ok, *Chem. Eng. J.*, 2016, **287**, 618–632.
44. Q. Ma, W. Teng, Y. Sun, Y. Chen, Y. Xue, X. Chen, C. Zhang, H. Zhang, J. Fan, Y. Qiu and R. Fu, *Sci. Total Environ.*, 2022, **827**, 154329.
45. Y. Shi, C. Yu, M. Liu, Q. Lin, M. Lei, D. Wang, M. Yang, Y. Yang, J. Ma and Z. Jia, *RSC Adv.*, 2021, **11**, 36826–36835.
46. L. Ling, X. Huang and W. Zhang, *Adv. Mater.*, 2018, **30**, 1705703.
47. Y.-P. Sun, X.-Q. Li, W.-X. Zhang and H. P. Wang, *Colloids Surf., A*, 2007, **308**, 60–66.

48. J. Soukupova, R. Zboril, I. Medrik, J. Filip, K. Safarova, R. Ledl, M. Mashlan, J. Nosek and M. Cernik, *Chem. Eng. J.*, 2015, **262**, 813–822.
49. R. Hao, D. Li, J. Zhang and T. Jiao, *Nanomaterials*, 2021, **11**, 650.
50. B. Desalegn, M. Megharaj, Z. Chen and R. Naidu, *Heliyon*, 2019, **5**, e01750.
51. A. M. Abdelfatah, M. Fawzy, A. S. Eltaweil and M. E. El-Khouly, *ACS Omega*, 2021, **6**, 25397–25411.
52. H. Jabeen, V. Chandra, S. Jung, J. W. Lee, K. S. Kim and S. B. in Kim, *Nanoscale*, 2011, **3**, 3583.
53. D. Tesnim, B. A. Hedi and J. Simal-Gandara, *ACS Omega*, 2023, **8**, 34364–34376.
54. L. Di, X. Chen, J. Lu, Y. Zhou and Y. Zhou, *J. Water Process Eng.*, 2023, **53**, 103913.
55. J. Tuček, R. Prucek, J. Kolařík, G. Zoppellaro, M. Petr, J. Filip, V. K. Sharma and R. Zbořil, *ACS Sustainable Chem. Eng.*, 2017, **5**, 3027–3038.
56. M. Bhaumik, H. J. Choi, R. I. McCrindle and A. Maity, *J. Colloid Interface Sci.*, 2014, **425**, 75–82.
57. J. H. P. Watson and D. C. Ellwood, *Miner. Eng.*, 1994, **7**, 1017–1028.
58. S. Mann, N. H. C. Sparks, R. B. Frankel, D. A. Bazylinski and H. W. Jannasch, *Nature*, 1990, **343**, 258–261.
59. P. I. Girginova, A. L. Daniel-da-Silva, C. B. Lopes, P. Figueira, M. Otero, V. S. Amaral, E. Pereira and T. Trindade, *J. Colloid Interface Sci.*, 2010, **345**, 234–240.
60. P. Figueira, C. B. Lopes, A. L. Daniel-da-Silva, E. Pereira, A. C. Duarte and T. Trindade, *Water Res.*, 2011, **45**, 5773–5784.
61. D. S. Tavares, A. L. Daniel-da-Silva, C. B. Lopes, N. J. O. Silva, V. S. Amaral, J. Rocha, E. Pereira and T. Trindade, *J. Mater. Chem. A*, 2013, **1**, 8134.
62. W. Yantasee, C. L. Warner, T. Sangvanich, R. S. Addleman, T. G. Carter, R. J. Wiacek, G. E. Fryxell, C. Timchalk and M. G. Warner, *Environ. Sci. Technol.*, 2007, **41**, 5114–5119.
63. N. Rafie, M. Khodadadi, M. Zamani, A. Zarepour and A. Zarrabi, *Environ. Res.*, 2023, **226**, 115694.
64. T. Poursaberi, M. Hassanisadi, K. Torkestani and M. Zare, *Chem. Eng. J.*, 2012, **117–125**, 189–190.
65. Y. Xiao and J. Du, *J. Mater. Chem. B*, 2020, **8**, 354–367.
66. T. Hyeon, S. S. Lee, J. Park, Y. Chung and H. Bin Na, *J. Am. Chem. Soc.*, 2001, **123**, 12798–12801.
67. D. Amara, J. Grinblat and S. Margel, *J. Mater. Chem.*, 2012, **22**, 2188–2195.
68. J. Hu, G. Chen and I. M. C. Lo, *Water Res.*, 2005, **39**, 4528–4536.
69. E. Calì, J. Qi, O. Preedy, S. Chen, D. Boldrin, W. R. Branford, L. Vandeperre and M. P. Ryan, *J. Mater. Chem. A*, 2018, **6**, 3063–3073.
70. W. Li, J. T. Mayo, D. N. Benoit, L. D. Troyer, Z. A. Lewicka, B. J. Lafferty, J. G. Catalano, S. S. Lee, V. L. Colvin and J. D. Fortner, *J. Mater. Chem. A*, 2016, **4**, 15022–15029.

71. P. Zhu, S. Luo, H. Li and J. Pan, *Sep. Purif. Technol.*, 2023, **317**, 123715.
72. P. R. Yaashikaa, P. Senthil Kumar and S. Karishma, *Environ. Res.*, 2022, **212**, 113114.
73. A. Subash, M. Naebe, X. Wang and B. Kandasubramanian, *J. Hazard. Mater.*, 2023, **443**, 130168.
74. A. V. Baskar, N. Bolan, S. A. Hoang, P. Sooriyakumar, M. Kumar, L. Singh, T. Jasemizad, L. P. Padhye, G. Singh, A. Vinu, B. Sarkar, M. B. Kirkham, J. Rinklebe, S. Wang, H. Wang, R. Balasubramanian and K. H. M. Siddique, *Sci. Total Environ.*, 2022, **822**, 153555.
75. M. I. A. Abdel Maksoud, A. M. Elgarahy, C. Farrell, A. H. Al-Muhtaseb, D. W. Rooney and A. I. Osman, *Coord. Chem. Rev.*, 2020, **403**, 213096.
76. S. F. Soares, T. Fernandes, T. Trindade and A. L. Daniel-da-Silva, *Environ. Chem. Lett.*, 2020, **18**, 151–164.
77. A. L. Daniel-da-Silva, T. Trindade, B. J. Goodfellow, B. F. O. Costa, R. N. Correia and A. M. Gil, *Biomacromolecules*, 2007, **8**, 2350–2357.
78. A. M. Salgueiro, A. L. Daniel-da-Silva, A. V. Girão, P. C. Pinheiro and T. Trindade, *Chem. Eng. J.*, 2013, **229**, 276–284.
79. W. Yang, Y. Liu, Y. Zhu, W. Jiang, F. Shi, J. Hu, S. Jiang and S. Jian, *Chem. Eng. J.*, 2023, **476**, 146907.
80. A. A. Kadam and D. S. Lee, *Bioresour. Technol.*, 2015, **193**, 563–567.
81. Y. Liu, R. Liu, M. Li, F. Yu and C. He, *Carbohydr. Polym.*, 2019, **220**, 141–148.
82. S. F. Soares, T. R. Simões, M. António, T. Trindade and A. L. Daniel-da-Silva, *Chem. Eng. J.*, 2016, **302**, 560–569.
83. S. F. Soares, C. O. Amorim, J. S. Amaral, T. Trindade and A. L. Daniel-da-Silva, *J. Environ. Chem. Eng.*, 2021, **9**, 105189.
84. S. F. Soares, M. I. Rodrigues, T. Trindade and A. L. Daniel-da-Silva, *Colloids Surf., A*, 2017, **532**, 305–313.
85. S. F. Soares, J. Nogueira, T. Trindade and A. L. Daniel-da-Silva, *J. Nanostruct. Chem.*, 2023, **13**, 283–302.
86. M. Saiyad, N. Shah, M. Joshipura, A. Dwivedi and S. Pillai, *Mater. Today: Proc.*, 2023, **7**, 4891–4907.
87. X. Zang, R. Jiang, H.-Y. Zhu, Q. Wang, Y.-Q. Fu, D.-X. Zhao, J.-B. Li and H. Liu, *Sep. Purif. Technol.*, 2024, **330**, 125521.
88. X. Zheng, C. Pan, S. Zheng and Y. Guo, *Sep. Purif. Technol.*, 2024, **335**, 126212.
89. T. Ahamad, M. Naushad, T. Al-Shahrani, N. Al-hokbany and S. M. Alshehri, *Int. J. Biol. Macromol.*, 2020, **147**, 258–267.
90. S. Mirizadeh, C. Solisio, A. Converti and A. A. Casazza, *Sep. Purif. Technol.*, 2024, **329**, 125115.
91. J. S. Algethami, M. A. M. Alhamami, A. A. Alqadami, S. Melhi and A. F. Seliem, *Int. J. Biol. Macromol.*, 2024, **254**, 127767.
92. Q. Wang, X. Jia, M. Jin, R. Guo, B. Niu, H. Yan and H. Wang, *Int. J. Biol. Macromol.*, 2023, **253**, 126631.
93. B. Chen, F. Long, S. Chen, Y. Cao and X. Pan, *Chem. Eng. J.*, 2020, **385**, 123926.

94. R. Jiang, T.-T. Shen, H.-Y. Zhu, Y.-Q. Fu, S.-T. Jiang, J.-B. Li and J.-L. Wang, *Int. J. Biol. Macromol.*, 2023, **227**, 1234–1244.
95. R. Jiang, H.-Y. Zhu, Y.-Q. Fu, X. Li, S.-T. Jiang and J.-B. Li, *Environ. Sci. Pollut. Res.*, 2023, **30**, 44985–44998.
96. A. Reghioua, D. Barkat, A. H. Jawad, A. S. Abdulhameed, A. A. Al-Kahtani and Z. A. ALOthman, *J. Environ. Chem. Eng.*, 2021, **9**, 105166.
97. A. I. Adeogun, O. A. Osideko, M. A. Idowu, O. A. Akinloye and E. A. Ofudje, *J. Dispersion Sci. Technol.*, 2023, **44**, 51–60.
98. H. Motaghi, P. Arabkhani, M. Parvinnia and A. Asfaram, *Sep. Purif. Technol.*, 2022, **284**, 120258.
99. Y. Liu, X. Zhang and L. Zhao, *Water Cycle*, 2023, **4**, 179–191.
100. J. M. N. dos Santos, C. R. Pereira, E. L. Foletto and G. L. Dotto, *Int. J. Biol. Macromol.*, 2019, **131**, 301–308.
101. S. F. Soares, T. Trindade and A. L. Daniel-da-Silva, *Coatings*, 2021, **11**, 964.
102. M. Wang and X. You, *Chem. Eng. J.*, 2023, **454**, 140417.
103. S. F. Soares, T. Fernandes, T. Trindade and A. L. Daniel-da-Silva, *Molecules*, 2019, **24**, 1958.
104. F. Hassanzadeh-Afruzi, M. Forouzandeh-Malati, F. Ganjali, M. Mehdi Salehi, A. Maleki and E. Nazarzadeh Zare, *Alexandria Eng. J.*, 2023, **82**, 503–517.
105. J. Jiang, X. Jiang, Y. Zou, J. Zhai, W. Ding, H. Li and H. Zheng, *J. Cleaner Prod.*, 2023, **391**, 136122.
106. Z. Jing, Y. Li, Q. Du, X. Pi, Y. Wang, S. Zhao and Y. Jin, *Int. J. Biol. Macromol.*, 2023, **253**, 126967.
107. A. S. Elshimy, M. Mobarak, J. S. Ajarem, S. N. Maodaa, A. Bonilla-Petriciolet, Z. Li, M. A. Korany, D. S. Ammar, D. G. Awad, S. A. Elberbash and M. K. Seliem, *Int. J. Biol. Macromol.*, 2024, **256**, 128528.
108. V. S. Munagapati, H.-Y. Wen, A. R. K. Gollakota, J.-C. Wen, K.-Y. A. Lin, C.-M. Shu, V. Yarramuthi, P. K. Basivi, G. M. Reddy and G. V. Zyryanov, *Int. J. Biol. Macromol.*, 2023, **246**, 125675.
109. J. Tan, L. Kong, J. Fang, X. Liang and Y. Yin, *Environ. Adv.*, 2024, 100489.
110. J. Sakhaei Niroumand, S. J. Peighambardoust and R. Mohammadi, *Int. J. Biol. Macromol.*, 2024, **259**, 129263.
111. Q. Tan, X. Jia, R. Dai, H. Chang, M. W. Woo and H. Chen, *Sep. Purif. Technol.*, 2023, **320**, 124157.
112. A. I. A. Soliman, J. A. Díaz Baca and P. Fatehi, *Carbohydr. Polym.*, 2023, **308**, 120619.
113. D. Li, Z. Wen, J. Lin, J. Zeng, Z. Li, F. Ke, D. Gao and D. Wang, *Mater. Today Chem.*, 2023, **33**, 101742.
114. F. Lu, G. Ding, X. Ma, B. Huang and L. You, *J. Mol. Liq.*, 2023, **391**, 123303.
115. N. A. A. Qasem, R. H. Mohammed and D. U. Lawal, *npj Clean Water*, 2021, **4**, 36.
116. A. U. Rehman, S. Nazir, R. Irshad, K. Tahir, K. ur Rehman, R. U. Islam and Z. Wahab, *J. Mol. Liq.*, 2021, **321**, 114455.

117. M. J. Sharifi, A. Nouralishahi and A. Hallajisani, *Int. J. Biol. Macromol.*, 2023, **248**, 125984.
118. J. Li, J. Su, Q. Yang and Z. Yang, *Mater. Chem. Phys.*, 2023, **306**, 128057.
119. S. Chkirida, N. El Mernissi, N. Zari, A. el kacem Qaiss and R. Bouhfid, *Int. J. Biol. Macromol.*, 2024, **260**, 129389.
120. N. Karami, A. Mohammadpour, M. R. Samaei, A. M. Amani, M. Dehghani, R. S. Varma and J. N. Sahu, *Int. J. Biol. Macromol.*, 2024, **254**, 127663.
121. C. Bulin, *Int. J. Biol. Macromol.*, 2023, **241**, 124588.
122. A. M. Omer, M. El-Sayed, E. M. Abd El-Monaem, G. M. El-Subruiti and A. S. Eltaweil, *Int. J. Biol. Macromol.*, 2023, **253**, 127437.
123. A. Sami Abboud, A. Ghaffarinejad and A. Mollahosseini, *Environ. Nanotechnol., Monit. Manage.*, 2023, **20**, 100874.
124. M. A. Ali, M. Mobarak, A. M. Salah, A. Yehia, E. C. Lima, A. Q. Seliem, A. S. Elshimy, M. Al-Dossari, N. S. A. EL-Gawaad, H. I. Bendary and M. K. Seliem, *Int. J. Biol. Macromol.*, 2024, **261**, 129692.
125. C. Hou, W. Yan, H. Ma, Y. Bai and Z. Yuan, *Colloids Surf., A*, 2024, **685**, 133255.
126. Q. Chen, Y. Zhang, H. Xia, R. Liu and H. Wang, *Int. J. Biol. Macromol.*, 2024, **258**, 128973.
127. Z. Rahmani, M. Ghaemy and A. Olad, *Hydrometallurgy*, 2022, **213**, 105915.
128. A. E. Mubark, *J. Water Process Eng.*, 2024, **58**, 104759.
129. H. Zeng, S. Sun, K. Xu, W. Zhao, R. Hao, J. Zhang and D. Li, *React. Funct. Polym.*, 2022, **177**, 105328.
130. A. Verma, G. Sharma, A. Kumar, P. Dhiman, G. T. Mola, A. Shan and C. Si, *Chemosphere*, 2024, **352**, 141365.
131. A. T. Adeleye, M. M. Bahar, M. Megharaj and M. M. Rahman, *J. Water Process Eng.*, 2023, **53**, 103777.
132. G. Chellasamy, R. M. Kiriyanthan, T. Maharajan, A. Radha and K. Yun, *Environ. Res.*, 2022, **208**, 112724.
133. L. Huang, W. He, Y. Zhang, X. Wang, K. Wu, Z. Yang and J. Zhang, *J. Water Process Eng.*, 2023, **54**, 103929.
134. R. Ma, Y. Feng, J. Yu, X. Zhao, Y. Du and X. Zhang, *Environ. Sci. Pollut. Res.*, 2023, **30**, 104135–104147.
135. Z. Li, W. Xie, Z. Zhang, S. Wei, J. Chen and Z. Li, *Int. J. Biol. Macromol.*, 2023, **245**, 125524.
136. B. Zheng, B. Li, H. Wan, X. Lin and Y. Cai, *J. Hazard. Mater.*, 2022, **431**, 128611.
137. Y. Zhang, Z. Dong, Z. Deng, S. Shi, C. Tang and X. Hu, *Environ. Sci.: Nano*, 2023, **10**, 2566–2577.
138. L. Molina-Calderón, C. Basualto-Flores, V. Paredes-García and D. Venegas-Yazigi, *Sep. Purif. Technol.*, 2022, **299**, 121708.
139. A. Abidli, Y. Huang, Z. Ben Rejeb, A. Zaoui and C. B. Park, *Chemosphere*, 2022, **292**, 133102.
140. L. Molina-Calderón, C. Basualto-Flores, V. Paredes-García and D. Venegas-Yazigi, *Sep. Purif. Technol.*, 2022, **299**, 121708.

141. A. Chatterjee and J. Abraham, *Biotechnol. Lett.*, 2019, **41**, 319–333.
142. N. D. Ilankoon, C. Aldrich, E. A. Oraby and J. J. Eksteen, *Miner. Process. Extr. Metall. Rev.*, 2020, **41**, 311–322.
143. J. Li, S. Chen, Y. Chen, L. Zhao, X. Li, H. Wang, X. Hu and F. Zi, *J. Alloys Compd.*, 2023, **964**, 171251.
144. N. R. Biata, S. Jakavula, G. P. Mashile, A. Nqombolo, R. M. Moutloali and P. N. Nomngongo, *Hydrometallurgy*, 2020, **197**, 105447.
145. D. H. K. Reddy, W. Wei, L. Shuo, M.-H. Song and Y.-S. Yun, *ACS Appl. Mater. Interfaces*, 2017, **9**, 18650–18659.
146. H. Javadian, M. Taghavi, M. Ruiz, I. Tyagi, M. Farsadrooh and A. M. Sastre, *J. Rare Earths*, 2023, **41**, 1796–1804.
147. J. Wang, M. Shi, Y. Qin, Y. Huang, X. Li and X. Cao, *Appl. Surf. Sci.*, 2023, **608**, 155141.
148. T. Tolessa, X.-X. Zhou, M. Amde and J.-F. Liu, *Talanta*, 2017, **169**, 91–97.
149. Y. Chen, X. Ma and J. Peng, *Carbohydr. Polym.*, 2021, **271**, 118435.
150. H. Javadian, M. Ruiz, T. A. Saleh and A. M. Sastre, *J. Mol. Liq.*, 2020, **306**, 112760.
151. D. Wu, J. Zhao, L. Zhang, Q. Wu and Y. Yang, *Hydrometallurgy*, 2010, **101**, 76–83.
152. R. D. Alorro, N. Hiroyoshi, H. Kijitani, M. Ito and M. Tsunekawa, *Miner. Process. Extr. Metall. Rev.*, 2010, **31**, 201–213.
153. P. Homchuen, R. D. Alorro, N. Hiroyoshi, R. Sato, H. Kijitani and M. Ito, *Miner. Process. Extr. Metall. Rev.*, 2016, **37**, 246–254.
154. A. R. M. Calderon, R. D. Alorro, B. Tadesse, K. Yoo and C. B. Tabelin, *Resour., Conserv. Recycl.*, 2020, **161**, 104971.
155. J. Pinto, D. Branco, L. Carvalho, B. Henriques, R. Freitas, T. Trindade, D. Tavares and E. Pereira, *Environ. Technol. Innovation*, 2023, **32**, 103432.
156. A. M. Ahmed, M. L. Mekonnen and K. N. Mekonnen, *Carbohydr. Polym. Technol. Appl.*, 2023, **6**, 100386.
157. P. Karthikeyan and S. Meenakshi, *Int. J. Biol. Macromol.*, 2021, **168**, 750–759.
158. Y. Zhang, G. Shi, W. Wu, A. Ali, H. Wang, Q. Wang, Z. Xu, W. Qi, R. Li and Z. Zhang, *Colloids Surf., A*, 2022, **634**, 127980.
159. I. A. Kumar and N. Viswanathan, *J. Taiwan Inst. Chem. Eng.*, 2019, **102**, 283–296.
160. A. Sarkar, S. K. Biswas and P. Pramanik, *J. Mater. Chem.*, 2010, **20**, 4417.
161. J. Yang, Q. Zeng, L. Peng, M. Lei, H. Song, B. Tie and J. Gu, *J. Environ. Sci.*, 2013, **25**, 413–418.
162. B. Wu, L. Fang, J. D. Fortner, X. Guan and I. M. C. Lo, *Water Res.*, 2017, **126**, 179–188.
163. C. Colombo, V. Barrón and J. Torrent, *Geochim. Cosmochim. Acta*, 1994, **58**, 1261–1269.
164. Y.-J. Tu, C.-F. You, C.-K. Chang and M.-H. Chen, *J. Taiwan Inst. Chem. Eng.*, 2015, **46**, 148–154.
165. C. Liu and P. M. Huang, *Can. J. Soil Sci.*, 2000, **80**, 445–454.
166. M. Franzreb, *Curr. Opin. Colloid Interface Sci.*, 2020, **46**, 65–76.

167. A. Mudhoo and M. Sillanpää, *Environ. Chem. Lett.*, 2021, **19**, 4393–4413.
168. A. B. Dichiara, S. J. Weinstein and R. E. Rogers, *Ind. Eng. Chem. Res.*, 2015, **54**, 8579–8586.
169. H. Patel, *Appl. Water Sci.*, 2019, **9**, 45.
170. T. Li, T. Yang, Z. Yu, G. Xu, Q. Han, G. Luo, J. Du, Y. Guan and C. Guo, *Chem. Eng. Process.*, 2021, **169**, 108611.
171. C. D. Powell, A. J. Atkinson, Y. Ma, M. Marcos-Hernandez, D. Villagran, P. Westerhoff and M. S. Wong, *J. Nanopart. Res.*, 2020, **22**, 48.
172. N. Maniotis, K. Kalaitzidou, E. Asimoulas and K. Simeonidis, *J. Water Process Eng.*, 2023, **53**, 103825.
173. S. Ahmed, Y. Zhang, B. Wu, Z. Zheng, C.-F. Leung, T. Choy, Y. Kwok and I. M. C. Lo, *Sci. Total Environ.*, 2021, **791**, 148281.
174. S. C. N. Tang, D. Y. S. Yan and I. M. C. Lo, *Ind. Eng. Chem. Res.*, 2014, **53**, 15718–15724.
175. Z. Qi, T. P. Joshi, R. Liu, Y. Li, H. Liu and J. Qu, *J. Hazard. Mater.*, 2018, **343**, 36–48.
176. T. Li, Y. Guan, C. Guo, T. Yang, Z. Yu and G. Xu, *Process Saf. Environ. Prot.*, 2021, **149**, 575–580.
177. T. Salinas, I. Durruty, L. Arciniegas, G. Pasquevich, M. Lanfranconi, I. Orsi, V. Alvarez and S. Bonanni, *J. Environ. Manage.*, 2018, **218**, 562–568.
178. S. C. N. Tang, D. Y. S. Yan and I. M. C. Lo, *Ind. Eng. Chem. Res.*, 2014, **53**, 15718–15724.
179. J.-H. Kim, S.-B. Kim, S.-H. Lee and J.-W. Choi, *Environ. Technol.*, 2018, **39**, 770–779.
180. A. Drenkova-Tuhtan, M. Schneider, M. Franzreb, C. Meyer, C. Gellermann, G. Sextl, K. Mandel and H. Steinmetz, *Water Res.*, 2017, **109**, 77–87.
181. P. Westerhoff, P. Alvarez, Q. Li, J. Gardea-Torresdey and J. Zimmerman, *Environ. Sci.: Nano*, 2016, **3**, 1241–1253.
182. X. Zhang, K. Zhang, Y. Shi, H. Xiang, W. Yang and F. Zhao, *Sci. Total Environ.*, 2024, **920**, 170951.
183. P. C. Pinheiro, S. Fateixa, A. L. Daniel-da-Silva and T. Trindade, *Sci. Rep.*, 2019, **9**, 19647.
184. S. F. Soares, N. M. A. S. Silva, J. Brenheiro, S. Fateixa, A. L. Daniel-da-Silva and T. Trindade, *Sens. Diagn.*, 2024, **3**, 1177–1189.

Section III: Soil

CHAPTER 13

Colloidal Materials and Their Crucial Role in Soil Sustainability: An Integrated Approach

GABRIELA PATRICIA UNIGARRO VILLARREAL,*[a]
ESTEFÂNIA VANGELIE RAMOS CAMPOS[a,b] AND
LEONARDO FERNANDES FRACETO[a]

[a] Institute of Science and Technology, São Paulo State University (UNESP), Av. Três de Março 511, Sorocaba, SP 18087-180, Brazil; [b] B. Nano Soluções Tecnológicas Ltda, Avenida Itavuvu, 11.777, Sorocaba, 18078-005, Brazil
*Email: gabriela.u.villarreal@unesp.br

13.1 Introduction

Although the world's population has grown significantly over time, this growth has slowed in recent years. It is also undergoing urbanization due to increased internal migration from rural to urban areas. However, there is uncertainty regarding future population growth owing to uncertainties in future projections of population growth and change, considering the wide range of factors that can influence fertility rates, mortality patterns, and migration flows over time.[1] According to the latest revision of the World Population Prospects conducted in 2022, the global population is projected to reach 8.5 billion in 2030, 9.7 billion in 2050, and 10.4 billion in 2100,[2] with direct consequences for food production due to increased demand,

Contribution of Colloidal Materials to Air, Water and Soil Environmental Sustainability: From Basic Concepts to Advanced Applications
Edited by Izabel C. Riegel-Vidotti, Lizandra Maria Zimmermann and Heloise R. de Barros

Published by the Royal Society of Chemistry, www.rsc.org

requiring greater quantities of natural resources and exerting greater pressure on the soil.

Hunger is a common problem due to population growth and the increased need for food. It is estimated that one in eight people, or approximately 12.5% of the global population, experiences chronic undernourishment, with the majority residing in developing countries.[3] It is projected that by 2050, 77 million people will be at risk of food insecurity.[4] Although many people may not be "hungry" to the point of experiencing physical discomfort, they may still be in a situation of food insecurity (at a moderate/low level) as even when having access to food that could meet their needs, they are without certainty that it will last in the medium and long term. They may also be compelled to reduce the quality and/or quantity of the food they consume to survive, potentially leading to various forms of malnutrition and serious consequences for the health and well-being of these individuals.[5] Therefore, the need to sustainably meet the growing and changing demand for food while ensuring that no one goes hungry has become one of the world's greatest challenges.[6] Considering the above and the latest projections, it is anticipated that an increase in agricultural output between 25% and 70% will be necessary to meet the food demand by 2050.[7]

Although several measures have been implemented to mitigate hunger globally, food insecurity and malnutrition remain significant.[8] The current situation and trends in global food security are presented in the "State of Food Security and Nutrition in the World" report. The most recent report (from 2023) highlights growing concerns regarding food insecurity in developing countries due to factors such as climate change and conflicts.[9]

Climate change and extreme weather events will negatively affect food security and increase hunger.[10] This may be due to direct effects on agricultural production through adverse impacts on key natural resources used for this purpose, such as soil and water, reducing access to food (either through availability or price increases),[11] thereby reducing the worldwide production of key crops by approximately 5%.[12] This issue becomes a fundamental axis in the debate on sustainable development and the Sustainable Development Goals (SDGs), specifically SDG 13 "Climate Action," which aims to "Take urgent action to combat climate change and its impact," and SDG 2 "Zero Hunger," which aims to "End hunger, achieve food security, improve nutrition, and promote sustainable agriculture, ensuring equal access to healthy and sufficient food for all".[13]

In this context, agriculture plays a fundamental role, as it is both a victim of and a contributor to climate change.[14] It simultaneously contributes to greenhouse gas emissions, with a share of around 24% of total emissions, but is also affected by severe meteorological conditions such as droughts, floods, and heatwaves, which can reduce crop productivity and threaten food security.[15] Furthermore, climate change can alter the distribution patterns and proliferation of pests, leading to more frequent and severe outbreaks of crop diseases and prompting farmers to resort to the use of large quantities of pesticides, resulting in contaminated soil, water, and air, which are significantly toxic to humans and wildlife.[16] Crop losses due to pests account for between 20% and 40% of the total potential production, including an estimated 13% damage due to plant pathogens, 13% to weeds, and 14% to insect pests.[17]

After outlining the challenges faced by agriculture in this context, it is essential to consider the crucial role of soil. Soil provides an incalculable number of vital resources to meet the growing demands of the current and future population, with the primary resource being the food supply.[18] However, soils can also retain diverse pollutants, including pesticides, heavy metals, or polycyclic aromatic hydrocarbons (PAHs), thus acting as pollution absorbers. Pesticides are the most common xenobiotic contaminants, resulting in significant accretion in the topsoil, accumulation, and persistence in the environment.[19]

According to the Food and Agriculture Organization of the United Nations (FAO), millions of tons of pesticides are used worldwide, including 29% insecticides, 47% herbicides, 17% fungicides, and 5% other pesticides.[20] Globally, the countries that consume the largest quantities of pesticides are Bolivia, Germany, India, Colombia, Chile, Turkey, and Ecuador.[21] Due to their high toxicity and persistence, these substances affect organisms such as microbes, algae, invertebrates, plants, and humans, accumulating in them and causing serious problems.[22] These substances also contribute to pollution, contamination, and degradation of the soil, representing a significant risk to ecological security, as 64% of cultivated land worldwide is exposed to various extremely harmful active ingredients, with approximately 31% of them facing a very high risk.[23]

A significant portion of the sprayed pesticides bind to soil particles through sorption and adsorption *via* cationic–anionic interactions, playing a crucial role in the overall process. However, the soil type also plays a key role in pesticide adsorption, with clay soils adsorbing a higher concentration due to the availability of organic matter compared to loamy, clayey, and sandy soils.[24] Similarly, the biomagnification and bioaccumulation of pesticides pose a significant risk to soil biota, further increasing the ecotoxicity of these substances in the soil.[25] Soil biota, including microorganisms and fauna, play a vital role in maintaining soil functions, such as preserving soil quality and biodiversity, and regulating nutrient cycling. When this biota is altered, it can disrupt nitrogen, carbon, and phosphorus cycles, reduce nutrient availability for plants, and contribute to soil impoverishment, diminishing fertility, stability, and the ability to sustain plant and animal life.[26]

Considering everything that has already been established, it is imperative to develop innovations in the field of agrochemicals to minimize the disadvantages of conventional pesticides and their harmful impact, in which colloidal materials appear. Nanotechnology offers a boost for sustainable agricultural development, presenting innovative strategies to mitigate the harmful effects of pesticides through their specific nanoscale properties.[27] Colloidal materials are micrometer or nanometer particles dispersed in a continuous medium, such as liquids, gases, or solids.[28] Dispersion systems are classified according to the dispersion medium used: when it is benzene, it is called benzenol; when it is alcohol, it is also called alcohol; when it is a gas or liquid, it is called aquasol; when the dispersion medium is solid, they are gels; and when it is air, they are aerosols, although there are also hybrid materials such as hydrosols.[29] These colloidal particles exhibit unique and useful characteristics because of their size and large surface-to-volume ratios.[30]

Thus, the development and use of colloidal materials offers a wide and varied range of opportunities to improve the delivery and performance of agrochemicals in agriculture, leveraging innovations to advance sustainable and efficient agricultural practices, maximize food production, and achieve the SDGs.

13.2 Nanotechnology and Colloidal Suspensions: An Integrated Approach

Sustainability and soil well-being are fundamental aspects of food security and environmental health; however, they are threatened by various factors including soil degradation, nutrient loss, declining fertility, and contamination. With the ongoing quest to conserve and enhance the natural environment, nanotechnology has become a crucial way to provide innovative solutions to address these challenges.[31]

Nanotechnology and colloidal suspensions may not appear to have a close relationship, yet they converge in a field where different materials have the capacity to be manipulated at the nanometric scale. This allows not only understanding, but also precise and specific control, modification, and improvement of their characteristics.[32] In the quest for environmental remediation, sustainable agriculture, and soil sustainability, the properties of colloidal suspensions have endless possibilities and mark a new research horizon, opening new opportunities for innovative and more environmentally friendly solutions in various fields ranging from advanced monitoring and analysis methods to the design and implementation of intelligent soil management systems. Nanoparticles are a specific type of colloid, in which the dispersion phase is composed of nanoparticles, which due to their exceptionally small size can behave as dynamic systems due to their well-known colloidal interactions with other particles and among themselves, called "surface forces," and exhibit unique properties related to their size, such as the ability to absorb and transport substances; however, the medium in which they are suspended also has a great influence.[33]

Nanoscience and technology are considered the primary fields of the 21st century, serving as facilitative processes and offering significant solutions to meet the demands of sustainable development, food security, and soil sustainability as well as various other issues in the agricultural sector.[34] However, one can discuss the transformation of the agricultural sector through precision agriculture, which adopts an approach based on the use of information technology in commercial farming management, which aims to maximize agricultural production while minimizing the use of herbicides, pesticides, and fertilizers. This provides numerous potential benefits in terms of environmental protection, sustainability, food security, and productivity.[35] Precision agriculture integrates nanotechnology through an intelligent distribution system based on nanosensors and nanoparticles, assisting farmers in reducing the usage of nutrients, water, and agrochemicals and enhancing productivity through closely monitored supervision.[36]

Nanotechnology offers several strategies for the efficient delivery of active ingredients in agriculture through various nanosystems, such as nanoparticles, allowing for sustained and targeted release into the soil or to plants.[37] This technique enhances the efficacy of active ingredients, protects them from environmental degradation, and enables their gradual release, thereby reducing the need for frequent applications. The nanoparticles used in active ingredient delivery must be synthesized from organic or inorganic compounds and can be classified into four categories based on their chemical composition: carbon-based nanoparticles (single walled carbon nanotubes, multi-walled carbon nanotubes, fullerenes, and graphene), polymeric compounds, metal oxides (ceramic, a metal oxide is a crystalline solid that contains metal cations and oxide anions), and metallic nanoparticles (Au, Ag, Pt, Zn, and Ni).[38] There are three synthesis methods: biological, chemical, and physical. Biological methods include biosynthesis using microorganisms (fungi, algae, cyanobacteria, and bacteria), enzymes, agricultural residues and plant extracts. Chemical methods include sol–gel, microemulsion, co-precipitation, microwave, and hydrothermal synthesis. Physical methods include spark discharge, ultrasound, irradiation, electrochemistry, laser ablation, ion sputtering, evaporation, and condensation.[39]

However, the use of these nanomaterials has both advantages and disadvantages. There are several uncertainties about how they are applied, absorbed, and internalized in plant cells, as well as regarding the penetration mechanisms in target organisms and the associated risks.[40] Because of their size and properties, they can generate undesirable ecotoxicological effects.[41] One of the safest options used for active delivery is polymer-based nanoparticles, especially those of natural origin (such as polysaccharides), as they have the significant advantage of being degradable by soil microorganisms, resulting in products with low ecotoxicological potential.[42]

13.3 Advanced Applications in the Interaction of Nanotechnology and Colloidal Suspensions in Soil

Advanced applications in the interaction of nanotechnology and colloidal suspensions in soil are part of an extensive and promising research field, advancing towards significant implications in environmentally sustainable and precision agriculture through the use of nanomaterials to enhance quality and functionality.[43] Thus transitioning from an experimental approach to a practical phase, which contributes significantly to the agricultural sector by enabling soil diagnosis and monitoring, stabilization, and enhancement; the distribution of pesticides, fertilizers, and nutrients; and the remediation of contaminated soils.[44] Consequently, a large number of multidisciplinary studies related to nanomaterials and nanotechnology have been published in the last century, covering a wide range of scientific research, providing potential strategies within the field of environmental sustainability of soil, and focusing on its protection, conservation, and remediation.

Figure 13.1 provides an overview of the many applications of agro-nanotechnology in various facets of agriculture, primarily in the integration of nanomaterials and nanoengineering techniques in agricultural production, from crop protection to the modulation of metabolic processes to increase productivity and quality. This graphic offers a promising view of how agro-nanotechnology will open new frontiers in agriculture, driving sustainability and productivity in a constantly changing world.

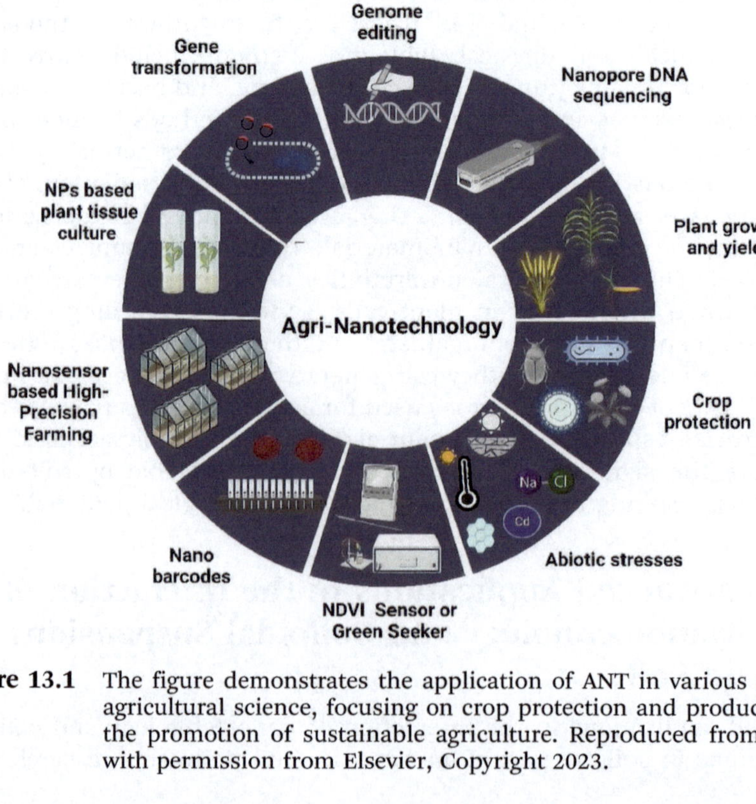

Figure 13.1 The figure demonstrates the application of ANT in various areas of agricultural science, focusing on crop protection and production for the promotion of sustainable agriculture. Reproduced from ref. 38 with permission from Elsevier, Copyright 2023.

13.3.1 Soil Diagnosis and Monitoring

Within this field, significant advancements have been made in the development of nanosensors and detection technology. This has enabled real-time monitoring of key soil parameters, such as humidity, acidity, and nutrient concentration, to improve the dynamic balance of the soil's physical, chemical, and biological properties, along with precise nutrient management, for the development of precision agriculture, supporting sustainability through the appropriate use of resources, and enhancing crop productivity.[45] For example, nanosensor systems based on carbon nanodots and quantum dots have been developed to detect heavy metals, multimodal nanosensors for the detection and elimination of mercury, and nanobiosensors for the detection and monitoring of enzymes.[46]

Figure 13.2 shows the application and use of nanosensors for monitoring certain elements present in the soil, such as pesticides, heavy metals, plant pathogens, and toxins.

Functionalized colloidal materials with specific probes can be used to develop highly sensitive and selective monitoring systems, providing valuable information for decision-making in agricultural and environmental management.

Table 13.1 summarizes some studies on soil diagnosis and monitoring, highlighting the role and importance of colloidal materials in this area of study.

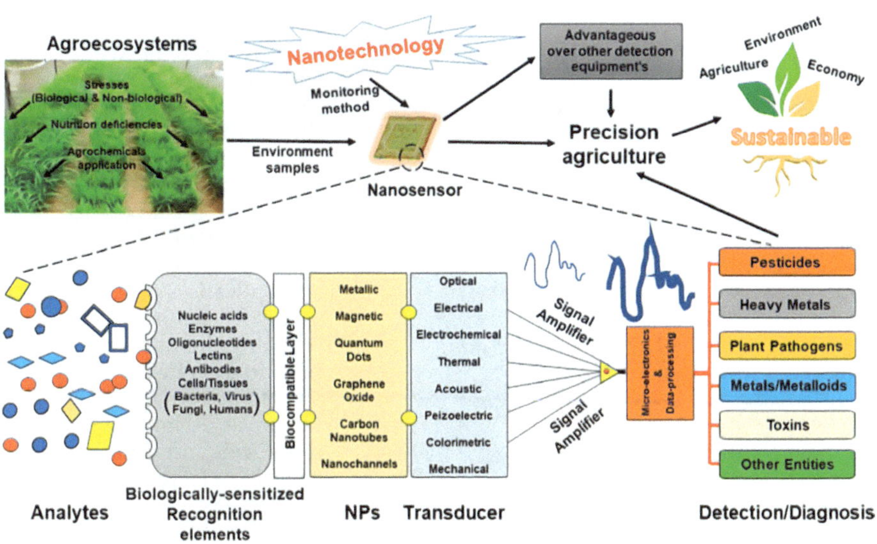

Figure 13.2 Simplified representation illustrating the component of nanosensors to monitoring agroecosystems. Reproduced from ref. 47, https://doi.org/10.1186/s11671-021-03593-0, under the terms of the CC BY 4.0 license, http://creativecommons.org/licenses/by/4.0/.

Table 13.1 Studies with some colloidal materials applied in soil diagnosis and monitoring.

Type of colloidal material	Monitored parameters	Key results	Ref.
Electrochemical nanosensors (three types): nanocomposites of carbon nanotube and hematite nanostructures; addition of nitrate reductase enzyme; treatment with bovine serum albumin.	Nitrate	The performance of the nanosensor with bovine serum albumin was the best in terms of electrochemical performance with high sensitivity and accuracy, low detection limit, good specificity and reproducibility, and better shelf life.	48
Hybrid electropolymerizable thiophene and electroreceptive ferrocene nanosensors based on azide-functionalized carbon nanotubes	Parathion and chlorantraniliprole pesticides	The study demonstrated that the nanosensor system is an efficient and precise option for determining pesticides in soil and food. It saves time, is cost-effective, and practical. There was good agreement between the increased and recovered concentrations of parathion and chlorantraniliprole, indicating the system's ability to detect and quantify these pesticides with precision.	49
Colorimetric sensor based on gold nanoparticles functionalized with trithiocyanuric acid	Cadmium	It features a low detection limit and rapid monitoring of cadmium contamination. This sensor has great potential for quantifying the hazardous Cd^{2+} in soil.	50
Copper oxide nanoparticles	Rhizobacteria (*Bacillius megaterium*)	The application of nanoparticles helped in reducing the production of indole acetic acid as a stress marker. Nanoparticles can adhere to the cell and thus can generate a toxic effect in a species. Even if the concentration did not manage to inhibit or kill the organisms completely, it had an undesirable effect on microorganisms.	51

13.3.2 Soil Stabilization and Improvement

Colloidal materials can be used to stabilize soil structure and prevent erosion and nutrient loss. Nanotechnology enables the design of nanoparticles with specific properties that enhance soil cohesion and improve the ability to retain water and nutrients, thereby promoting environments conducive to plant growth.

In Table 13.2, some examples of studies for soil stabilization and improvement are highlighted, emphasizing the role and importance of colloidal materials in this area of study.

13.3.3 Improved Delivery of Herbicides, Pesticides, and Fertilizers

Current farming systems rely heavily on synthetic fertilizers to provide necessary nutrients; however, their continued and extended use poses a risk to the environment, soil quality, and the nutritional balance of the microorganisms in the root zone of plants.[56] Nanotechnology enables the design of sustained nutrient release systems, utilizing colloidal materials as vehicles for the precise delivery of fertilizers to the soil, increasing nutrient use efficiency and reducing leaching and groundwater pollution.

Nanofertilizers are crucial in precision and sustainable agriculture as they optimize crop development, quality, and yield while increasing nutrient uptake efficiency, photosynthesis, and dry matter production, thereby protecting plants from biotic and abiotic stress, and reducing waste, application frequency, and fertilizer costs (see Figure 13.3).[57] Additionally, the mixture of biofertilizers and nanoparticles strengthens plant growth and their ability to withstand stress, becoming a more sustainable agricultural tool from economic and environmental standpoints, being more versatile and durable, with promising potential to transform traditional agricultural techniques into smart agriculture.[56]

These nanodelivery systems have been successfully employed as efficient carriers of various active ingredients and can improve the effectiveness and targeted release of agrochemicals and fertilizers, potentially having a positive impact on pest and disease control in agriculture, as well as increasing nutrients while simultaneously reducing the negative environmental effects.[59] The use of nanopesticides and nanoherbicides has numerous advantages such as improved absorption and transport by plants, greater efficacy, reduced residues, delayed resistance, lower dosage, and protection against natural enemies and beneficial insects.[60]

Nanopesticides have been demonstrated to be a more beneficial option than conventional pesticides. The application of conventional pesticides causes a loss of soil biodiversity, a decrease in pollinator populations, and negative effects on non-target organisms. In contrast, nanopesticides are superior as they provide high adsorption, better tissue permeability, lower

Table 13.2 Studies with some colloidal materials applied in the stabilization and improvement of soil.

Type of colloidal material	Monitored parameters	Key results	Ref.
Iron and magnesium oxide nanoparticles	Water retention in soil and the tensile strength of aggregates	The levels of nanoparticles showed a considerable increase in water retention relative to the control, which is likely attributed to the increased adsorption surfaces in the treated soils. Permanent wilting point conditions, field capacity, and the water content in the nanoparticle treatment were reduced compared to other treatments due to the soil's higher susceptibility and the effect on the fine soil pores.	52
Biocompatible silica nanoparticles	Atterberg limits, the maximum dry density, ideal water content, and shear strength of soil	Reduction of the plasticity index. There was a significant increase in shear strength, cohesion, maximum dry unit weight, and angle of internal friction. The California Bearing Ratio (CBR) strength of the soil cured with nanosilica exhibits a substantial increase compared to the control. There were maximum increases in the strength parameters of the soil modified with the optimal amount of nanosilica (1.5%).	53
Carbon nanotubes based on montmorillonite and amorphous silica dioxide nanoparticles	Elastic modulus and equivalent damping ratio of the soil	The increase in hydrostatic pressure incrementally increases the soil's stiffness, increases the resistance to deformation, and reduces the capacity to absorb impacts. As the amount of hydrostatic pressure increases, silica nanoparticles have a positive impact on increasing the soil's Young's modulus. The influence of hydrostatic pressure on increasing the elastic modulus is more significant in the case of using silica nanoparticles than in the other cases researched.	54

Table 13.2 (*Continued*)

Type of colloidal material	Monitored parameters	Key results	Ref.
Carbon-based nanoparticles	Soil specific gravity, pH	An increase in soil pH and dry density was observed, and the soil's plastic and liquid limits were improved. Carbon nanofibers and carbon nanotubes have good potential for stabilizing degraded soils.	55

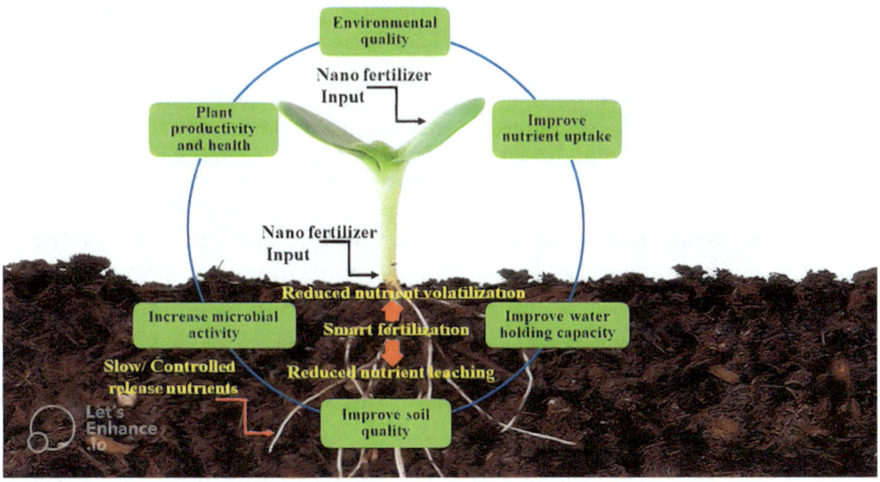

Figure 13.3 The cycle of nanofertilizers. Reproduced from ref. 58 with permission from Elsevier, Copyright 2022.

volatilization, and sustained release.[61] The beneficial effects of nano-technology-based pesticide formulations are presented in Figure 13.4.

In Table 13.3, some examples of studies aimed at improving the delivery of pesticides, herbicides, and fertilizers are highlighted, emphasizing the role and importance of colloidal materials in this area of research.

13.3.4 Remediation of Contaminated Soils

Nanotechnology offers tools for remediating contaminated soils by synthesizing nanomaterials capable of adsorbing and degrading organic and inorganic contaminants. Colloidal materials are particularly well suited for this purpose because of their large surface area and ability to react with pollutants. The combination of nanotechnology and colloidal materials enables the efficient removal of soil contaminants, restoration of soil quality, and reduction of environmental risks.[31]

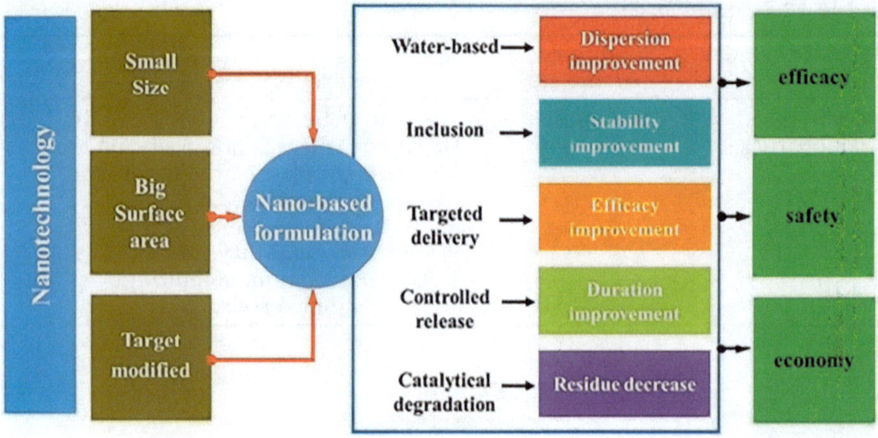

Figure 13.4 Beneficial improvements of nano-based pesticide formulations. Reproduced from ref. 62, https://www.irjse.in/2019/vol_7/Issue_6/article_1.php, under the terms of the CC BY 4.0 license, http://creativecommons.org/licenses/by/4.0/.

These nanoscale materials play a fundamental role in defining the trajectory, mobility, reactivity, and toxicity of contaminants in soil, providing significant tools for decontamination, capable of altering the physicochemical characteristics of the soil, abiotic stress, and the morphological, physiological, cellular, and molecular responses of plants, whose effectiveness depends not only on their properties but also on their interactions with the soil. However, their application is still at an experimental stage on a laboratory scale, although their potential is very promising.[74]

In Table 13.4, examples of studies on the remediation of contaminated soils are highlighted, emphasizing the role and importance of colloidal materials in this area of study.

13.4 Conclusions and Future Perspectives

In a world where population growth continues, hunger persists, and climate change is an ongoing challenge, agriculture is facing unprecedented challenges. It is becoming increasingly important to increase food production sustainably and ensure food security. However, this need is hindered by various interrelated problems such as the overuse of pesticides to combat pests, pollution, soil degradation, and increasing environmental toxicity.

In this context, new hope is emerging in the form of colloidal systems, an advanced technology that promises to transform modern agriculture. These systems offer a promising alternative to the traditional methods of applying agrochemicals and address the fundamental challenges faced by the agricultural sector.

Among these colloidal systems, nanotechnology offers various applications beyond simply providing the active ingredients. They are used to

Table 13.3 Studies with some colloidal materials applied to improve the delivery of pesticides, herbicides, and fertilizers.

Type of colloidal material	Type of pesticide or fertilizer	Key results	Ref.
Metal–organic framework PCN-777 nanoparticles with large mesopores	Avermectin	The nanopesticides prevented the UV photolysis of arteriovenous malformation due to the organic ligands PCN-777 and the sodium lignosulfonate coating. The nanopesticides proved to be highly effective as insecticides and for the control of *Spodoptera litura* in cabbage fields compared to the traditional pesticide formulation. In addition, the nanopesticides did not affect plant growth and showed high biological safety in relation to non-target organisms and human cells.	63
Chitosan nanoparticles	*Mentha spicata* essential oil	The nanopesticides exhibited a slower release rate and proved to be more effective than pure essential oils. Cell viability assays suggest that nanopesticides are safe and suitable for use in bioinsecticidal applications, as over 90% of cells remained alive after 24 hours of treatment at the lowest concentration.	64
Zein nanoparticles	Curcumin and carvacrol	The mortality rates of *Tetranychus urticae* are $77.1 \pm 11.5\%$, and emulsified ingredients show a lower repellency ($51.1 \pm 5.4\%$) in comparison with nanoencapsulated ingredients ($61.1 \pm 8.8\%$). Nanoencapsulated compounds demonstrate potential biological activity against soybean caterpillars, pests, and mites without producing phytotoxicity. Cytotoxicity tests report approximately 60% viability for fibroblast cell lines *versus* 20% for keratinocyte cell lines.	65
Chitosan nanoparticles	Botanical pesticide PONNEEM	Treatment with nanoparticles encapsulating the botanical pesticide exhibited 88.5% antifeedant activity and 90.2% larvicidal activity against *Helicoverpa armigera*, and the weight of its pupae was considerably lower. The formulation of this chitosan-based nanopesticide is very promising in the management of *H. armigera*.	66
Mesoporous silica nanoparticles chelated with copper ions	Azoxystobin	The application of these nanoparticles showed a higher antifungal efficacy against *Pyricularia oryzae* compared to the technical azoxystrobin applied with the same dose of active ingredient.	67

Table 13.3 (*Continued*)

Type of colloidal material	Type of pesticide or fertilizer	Key results	Ref.
		Dynamic contact angle measurement evidenced that a polydopamine coating can enhance the adhesion ability and deposition efficiency of growing plant leaves.	
Chitosan/sodium tripolyphosphate nanoparticles	Paraquat	The herbicide encapsulated in the nanoparticles was found to have lower toxicity compared to the pure compound, as evidenced by genotoxicity and cytotoxicity tests. After encapsulation, paraquat maintained its efficiency, according to the results of herbicide activity measurements. The research concluded that the inclusion of paraquat in nanoparticles can be an effective tool to reduce the negative impacts on human health and the environment. Therefore, the formulation is considered to have potential for use in agriculture.	68
Alginate nanoparticles	Dicamba	The alginate nanoparticles allowed for continuous and extended release of dicamba over a ten-day period, showing the feasibility of this preparation method for use in applications outside of the laboratory.	69
Poly(epsilon-caprolactone) nanoparticles	Altrazine	Nanoparticle formulations have proven effective in controlling target species. Soil column experiments revealed that the use of nanoparticles reduced atrazine mobility in soil. The chromosome aberration study on *Allium cepa* revealed that nanoparticles can decrease the genotoxicity of the herbicide.	70
Zinc sulphate nanoparticles	Potassium	Nanoparticles could help increase corn biomass, lower abiotic stress, and increase corn nutrition under cadmium- and water-limited stresses. Nanoparticles mediated Cd immobilization in soil and plant roots, showing great potential for soil improvement in the presence of cadmium and water stress.	71
Nanoparticles of chitosan sensitive to pH	Nitrogen	The studied nanoparticles constitute a promising platform for water absorption and pH for the controlled release and absorption of nitrogen-based fertilizers	72

Table 13.3 (*Continued*)

Type of colloidal material	Type of pesticide or fertilizer	Key results	Ref.
		for soil fertilization. It improves the efficiency of fertilizer use and reduces the environmental impact due to nitrogen loss. In a confined equilibrium state, the N fertilizer remained trapped within the nanoparticles. The nanoparticles demonstrate controlled release properties that conformed to swelling patterns influenced by environmental pH.	
Nano-zeolite, raw coir pith, and composted coir pith	Potassium	The encapsulated fertilizer slowly released potassium over a longer period compared to the control fertilizer. The data strongly indicate that nano-scale zeolite will be effective in increasing potassium utilization efficiency, especially when combined with composted coconut husk. The slow release of K may result in a probable reduction of environmental contamination.	73

Table 13.4 Examples of studies on the remediation of contaminated soils.

Type of colloidal material	Target contaminants	Key results	Ref.
Zerovalent iron nanoparticles	Arsenic and mercury	The use of nanoparticles succeeded in binding arsenic in the soil and increased mercury uptake by the plant, despite reducing oxidative stress. An increase in TPN content (total potassium, phosphorus, and nitrogen) was also observed in the roots of plants in the treated soil. The application of organic fertilizer on the nanoparticles mobilized arsenic, which improved bioaccumulation factors and mitigated arsenic accumulation. This nanoscale material shows potential for phytomanagement of As/Hg polluted soils and soil stabilization.	75

Table 13.4 (*Continued*)

Type of colloidal material	Target contaminants	Key results	Ref.
Hydroxyapatite and biochar nanoparticles	Arsenic, copper, selenium, and zinc	Only the soil treated with nanoparticles showed a higher pH than the control soil and revealed an increase in labile concentrations of carbon, total nitrogen, and available phosphorus. The soil treatment resulted in increased concentrations of exchangeable non-acid cations and reduced sodium levels, which contributed to improved soil fertility, decreased the risk of sodicity, and increased ryegrass biomass. In addition, an increase in enzyme activity was observed, especially in dehydrogenase and glucosidase. Soil management also had a positive effect on the levels of exchangeable non-acid cations, decreasing sodium levels, improving soil fertility, and increasing ryegrass biomass. In addition, enzymatic activity was enhanced, particularly dehydrogenase and glucosidase. The treatment led to a significant reduction in the number of metals in the biomass, mitigating contamination risks.	76
Magnetic lignin-based nanoparticles	Phosphate and iron	The recovered phosphate and chelated iron from phosphate and chelated iron, present in magnetic adsorbent nanoparticles, can be reused as a multi-element fertilizer. The cyclic batch adsorption tests showed that the nanoparticles have high recyclability and reusability, decreasing costs and preventing further soil contamination.	77

Table 13.4 (*Continued*)

Type of colloidal material	Target contaminants	Key results	Ref.
Zinc/zero valent iron nanoparticles	Diesel and gas oil	The stabilization of nanoparticles increased foamability, foam stability, and overall diesel removal efficiency from the foams. By applying the treatment, a maximum elimination efficacy of 83.8% of the present diesel and 73% of the gas oil was achieved. There is considerable potential for future ecological applications in remediating and stabilizing soils contaminated with foam using nanoparticles.	78
Cerovalent iron nanoparticles stabilized with carboxymethylcellulose	Nitrobenzene	Stabilized nanoparticles degraded the nitrobenzene absorbed by the soil quickly and rapidly converted nitrobenzene into nitrosobenzene. However, the subsequent conversion of intermediates was slower. This material has the potential to facilitate *in situ* remediation of soils and groundwater contaminated with nitrobenzene.	79
Nanocomposite catalyst comprised of polymer-stabilized Fe_2O_3–Zn	Organophosphate pesticide (profenofós)	Increasing the catalyst dosage, initial concentration of profenofos, and reaction temperature leads to a gradual increase in the amount of profenofos adsorbed on the nanocomposite catalyst. The removal efficiency of the used pesticide was maintained for up to four cycles using the nanocatalyst.	80

remediate contaminated soils and improve soil quality and long-term productivity. They are also used for soil stabilization and enhancement, promoting healthier and more degradation-resistant structures. Furthermore, these systems enable precise soil diagnosis and monitoring, thereby providing farmers with vital information for land management and conservation, which are key areas for future research.

Nanotechnology within soil sustainability brings significant benefits, although it is important to consider challenges such as environmental safety and environmental persistence. Addressing these issues through comprehensive risk-and-benefit assessments and promoting greater collaboration among scientists, farmers, and policymakers is essential. In the future, it will be crucial to tackle these challenges holistically by recognizing the interconnectedness between food security, environmental health, and agricultural sustainability.

Integrating emerging technologies, such as nanomaterial-based colloidal systems, can fundamentally transform agricultural processes into more resilient, equitable, and sustainable models. However, it is crucial to address the ethical, regulatory, and safety issues associated with the use of nanotechnology in agriculture to ensure that these advancements are used responsibly.

An important focus for future research is the exploration and optimization of new colloidal nanomaterials and synthesis techniques that allow for greater efficiency in soil restoration and conservation. These materials are more flexible and durable, allowing them to cope with a wide range of soil challenges, such as erosion, salinization, and compaction.

Another fundamental element is the incorporation of colloidal nanomaterials into precision agriculture strategies and innovative technologies, such as digital agriculture and artificial intelligence. It is also important to consider the combination of geospatial information, predictive modeling, and application systems for colloidal materials, because this may have the potential to transform soil management and monitoring, thereby enabling more efficient, sustainable, and customized agriculture.

The development of security systems from conception to design is crucial to address the complex and interconnected challenges faced by humanity in the 21st century, as highlighted in this section. As we strive to increase food production to ensure global food security, it is essential to consider the effectiveness and efficiency of agricultural methods, and the environmental and health impacts associated with these practices. Applying colloidal systems to agriculture is a promising approach that offers safer and more sustainable alternatives to traditional agrochemicals. By designing safe agricultural systems as a core priority, the risks of contamination, soil degradation, and negative effects on human health and the environment can be reduced, thus ensuring more sustainable and climate-resilient food production.

In summary, nanotechnology as a colloidal technology offers an innovative and promising method, but requires a collaborative and multidisciplinary approach to achieve its full potential, thus facing current and medium/long-term challenges related to soil and cultivating a more prosperous and equitable future for future generations.

Acknowledgements

This work was financed in part by the São Paulo State Research Foundation (FAPESP, grants, CEPID-CBioClima – #2021/10639-5 and #2017/21004-5) and

INCT Nanotechnology for Sustainable Agriculture, the National Council for Scientific and Technological Development (MCTI-CNPq – INCTNanoAgro #405924/2022-4); Coordenação de Aperfeiçoamento de Pessoal de Nível Superior – Brasil (MEC-CAPES INCTNanoAgro #88887.953443/2024-00).

References

1. D. Gu, K. Andreev and M. E. Dupre, *China CDC Wkly*, 2021, **3**, 604–613.
2. Naciones Unidas, World Population Prospects 2022, https://population.un.org/wpp/.
3. FAO, The State of Food Security and Nutrition in the World, https://www.fao.org/publications/home/fao-flagship-publications/the-state-of-food-security-and-nutrition-in-the-world/en (accessed 6 May 2024).
4. C. Janssens, P. Havlík, T. Krisztin, J. Baker, S. Frank, T. Hasegawa, D. Leclère, S. Ohrel, S. Ragnauth, E. Schmid, H. Valin, N. Van Lipzig and M. Maertens, *Nat. Clim. Change*, 2020, **10**, 829–835.
5. FAO, Hunger and food insecurity, http://www.fao.org/hunger/en/.
6. H. C. J. Godfray, J. R. Beddington, I. R. Crute, L. Haddad, D. Lawrence, J. F. Muir, J. Pretty, S. Robinson, S. M. Thomas and C. Toulmin, *Science*, 2010, **327**, 812–818.
7. M. C. Hunter, R. G. Smith, M. E. Schipanski, L. W. Atwood and D. A. Mortensen, *BioScience*, 2017, **67**, 386–391.
8. FAO, Hunger and food insecurity, https://www.fao.org/hunger/en.
9. FAO, Urbanization, agrifood systems transformation and healthy diets across the rural–urban continuum, https://www.fao.org/publications/home/fao-flagship-publications/the-state-of-food-security-and-nutrition-in-the-world/en.
10. Z. Fang and C. Cao, *IEEE J. Sel. Top. Appl. Earth Obs. Remote Sens.*, 2014, **7**, 3157–3166.
11. P. A. Williams, O. Crespo, M. Abu and N. P. Simpson, *Environ. Res. Lett.*, 2018, **13**, 103004.
12. C. Zhao, B. Liu, S. Piao, X. Wang, D. B. Lobell, Y. Huang, M. Huang, Y. Yao, S. Bassu, P. Ciais, J.-L. Durand, J. Elliott, F. Ewert, I. A. Janssens, T. Li, E. Lin, Q. Liu, P. Martre, C. Müller, S. Peng, J. Peñuelas, A. C. Ruane, D. Wallach, T. Wang, D. Wu, Z. Liu, Y. Zhu, Z. Zhu and S. Asseng, *Proc. Natl. Acad. Sci. U. S. A.*, 2017, **114**, 9326–9331.
13. H. El Bilali, I. Bassolé, L. Dambo and S. Berjan, *Agric. For.*, 2020, **66**, 197–210.
14. J. Streimikis and T. Baležentis, *Sustainable Dev.*, 2020, **28**, 1702–1712.
15. S. Lenka, N. K. Lenka, V. Sejian and M. Mohanty, in *Climate Change Impact on Livestock: Adaptation and Mitigation*, ed. V. Sejian, J. Gaughan, L. Baumgard and C. Prasad, Springer India, New Delhi, 2015, pp. 37–48.
16. K. E. Ukhurebor, U. O. Aigbe, A. S. Olayinka, W. Nwankwo and J. O. Emegha, *AU EJournal Interdiscip. Res.*, 2020, **5**, 44–49.

17. L. Tyler Gula, Researchers Helping Protect Crops From Pests|NIFA, https://www.nifa.usda.gov/about-nifa/blogs/researchers-helping-protect-crops-pests.
18. C. Malheiro, M. Prodana, D. N. Cardoso, R. G. Morgado and S. Loureiro, *Environ. Pollut.*, 2024, 123989.
19. E. Wołejko, A. Jabłońska-Trypuć, U. Wydro, A. Butarewicz and B. Łozowicka, *Appl. Soil Ecol.*, 2020, **147**, 103356.
20. FAO, *Pesticides use, pesticides trade and pesticides indicators*, FAO, 2022.
21. Y. S. Chaudhari, P. Kumar, S. Soni, A. Gacem, V. Kumar, S. Singh, V. K. Yadav, V. Dawane, S. Piplode, B.-H. Jeon, H. A. Ibrahium, R. A. Hakami, M. T. Alotaibi, M. H. Abdellattif, M. M. S. Cabral-Pinto, P. Yadav and K. K. Yadav, *Toxicol. Appl. Pharmacol.*, 2023, **466**, 116449.
22. B. Durdu, F. Gurbuz, H. Koçyiğit and M. Gurbuz, *Environ. Monit. Assess.*, 2023, **195**, 1002.
23. F. H. M. Tang, M. Lenzen, A. McBratney and F. Maggi, *Nat. Geosci.*, 2021, **14**, 206–210.
24. N. Bošković, K. Brandstätter-Scherr, P. Sedláček, Z. Bílková, L. Bielská and J. Hofman, *Chemosphere*, 2020, **261**, 127637.
25. Y. Kianpoor Kalkhajeh, B. Huang, W. Hu, C. Ma, H. Gao, M. L. Thompson and H. C. Bruun Hansen, *Agric., Ecosyst. Environ.*, 2021, **307**, 107230.
26. G. Bhandari, K. Atreya, J. Vašíčková, X. Yang and V. Geissen, *Sci. Total Environ.*, 2021, **788**, 147921.
27. J. He, J. Li, Y. Gao, X. He and G. Hao, *J. Hazard. Mater.*, 2023, **456**, 131599.
28. Z. Li, Q. Fan and Y. Yin, *Chem. Rev.*, 2022, **122**, 4976–5067.
29. A. Tresa Sunny, P. Velayudhan and S. Thomas, in *Colloidal Metal Oxide Nanoparticles*, Elsevier, 2020, pp. 3–13.
30. I. S. Aranson, *C. R. Phys.*, 2013, **14**, 518–527.
31. A. Wahab, M. Muhammad, S. Ullah, G. Abdi, G. M. Shah, W. Zaman and A. Ayaz, *Sci. Total Environ.*, 2024, **926**, 171862.
32. G. Lövestam, H. Rauscher, G. Roebben, B. S. Klüttgen, N. Gibson, J.-P. Putaud and H. Stamm, *Considerations on a definition of nanomaterial for regulatory purposes*, Publications Office, LU, 2010.
33. N. M. Kovalchuk, D. Johnson, V. Sobolev, N. Hilal and V. Starov, *Adv. Colloid Interface Sci.*, 2019, **272**, 102020.
34. Y. Huang and L. Wang, *Environ. Earth Sci.*, 2016, **75**, 497.
35. M. Zain, H. Ma, S. Ur Rahman, Md. Nuruzzaman, S. Chaudhary, I. Azeem, F. Mehmood, A. Duan and C. Sun, *Plant Physiol. Biochem.*, 2024, **206**, 108244.
36. R. Prasad, V. Kumar and K. S. Prasad, *Afr. J. Biotechnol.*, 2014, **13**, 705–713.
37. R. Bhattacharjee, L. Kumar, N. Mukerjee, U. Anand, A. Dhasmana, S. Preetam, S. Bhaumik, S. Sihi, S. Pal, T. Khare, S. Chattopadhyay, S. A. El-Zahaby, A. Alexiou, E. P. Koshy, V. Kumar, S. Malik, A. Dey and J. Proćków, *Biomed. Pharmacother.*, 2022, **155**, 113658.

38. A. Singh, V. D. Rajput, A. Varshney, K. Ghazaryan and T. Minkina, *Plant Stress*, 2023, **10**, 100253.
39. H. Naaz, K. Rawat, P. Saffeullah and S. Umar, *Environ. Chem. Lett.*, 2023, **21**, 539–559.
40. R. P. Singh, R. Handa and G. Manchanda, *J. Controlled Release*, 2021, **329**, 1234–1248.
41. R. Grillo, L. F. Fraceto, M. J. B. Amorim, J. J. Scott-Fordsmand, R. Schoonjans and Q. Chaudhry, *J. Hazard. Mater.*, 2021, **404**, 124148.
42. M. Vemula and A. V. B. Reddy, *Nanotechnol. Environ. Eng.*, 2023, **8**, 805–814.
43. A. Usman, U. A. Gbanguba, A. Ismaila and A. Umar, *J. Underutilized Legum.*, 2020, **2**, 1–6.
44. M. Usman, M. Farooq, A. Wakeel, A. Nawaz, S. A. Cheema, H. ur Rehman, I. Ashraf and M. Sanaullah, *Sci. Total Environ.*, 2020, **721**, 137778.
45. S. Beegum and S. Das, in *Agricultural Nanobiotechnology*, ed. S. Ghosh, S. Thongmee and A. Kumar, Woodhead Publishing, 2022, pp. 465–478.
46. N. Mandal, S. Adhikary and R. Rakshit, in *Soil Analysis: Recent Trends and Applications*, ed. A. Rakshit, S. Ghosh, S. Chakraborty, V. Philip and A. Datta, Springer, Singapore, 2020, pp. 285–304.
47. P. Sharma, V. Pandey, M. M. M. Sharma, A. Patra, B. Singh, S. Mehta and A. Husen, *Nanoscale Res. Lett.*, 2021, **16**, 136.
48. M. Kundu, P. Krishnan, K. A. Chobhe, K. M. Manjaiah, R. P. Pant and G. Chawla, *J. Soil Sci. Plant Nutr.*, 2022, **22**, 2777–2792.
49. S. O. Tümay, A. Şenocak, E. Sarı, V. Şanko, M. Durmuş and E. Demirbas, *Sens. Actuators, B*, 2021, **345**, 130344.
50. T. Liu, X. Huang, C. Wang, J. Liu and X. Mao, *Talanta*, 2024, **274**, 126055.
51. G. D. Tripathi, Z. Javed and K. Dashora, *Emerging Contam.*, 2024, **10**, 100280.
52. E. Daraei, H. Bayat and P. Zamani, *Pedosphere*, 2024, **34**, 1136–1145.
53. F. Buazar, *Sci. Rep.*, 2019, **9**, 15147.
54. L.-C. Zhao and L. Xu, *Constr. Build. Mater.*, 2024, **415**, 134203.
55. A. AlObaid, K. Ur Rehman, S. Andleeb, K. O. Erinle and A. Mahmood, *J. King Saud Univ., Sci.*, 2022, **34**, 101716.
56. B. Sharma, S. Tiwari, K. C. Kumawat and M. Cardinale, *Sci. Total Environ.*, 2023, **860**, 160476.
57. A. Qureshi, D. K. Singh and S. Dwivedi, *Int. J. Curr. Microbiol. Appl. Sci.*, 2018, **7**, 3325–3335.
58. A. M. Jakhar, I. Aziz, A. R. Kaleri, M. Hasnain, G. Haider, J. Ma and Z. Abideen, *NanoImpact*, 2022, **27**, 100411.
59. L. Djaya, Hersanti, N. Istifadah, S. Hartati and I. M. Joni, *Biocatal. Agric. Biotechnol.*, 2019, **19**, 101153.
60. J. Yin, X. Su, S. Yan and J. Shen, *Nanomaterials*, 2023, **13**, 1255.
61. M. Kannan, N. Bojan, J. Swaminathan, G. Zicarelli, D. Hemalatha, Y. Zhang, M. Ramesh and C. Faggio, *Int. J. Environ. Sci. Technol.*, 2023, **20**, 10507–10532.
62. R. Urkude, *Int. Res. J. Sci. Eng.*, 2019, **7**(6), 151–156.

63. J. Liu, D. Xu, G. Xu, X. Li, J. Dong, X. Luan and X. Du, *Chem. Eng. J.*, 2023, **475**, 146312.
64. A. Choudhary, R. K. Salar and R. Thakur, *Biocatal. Agric. Biotechnol.*, 2024, **55**, 102989.
65. P. L. de, F. Proença, E. V. R. Campos, T. G. Costa, R. de Lima, A. C. Preisler, H. C. de Oliveira, C. M. da Rocha, D. J. de Andrade, K. C. Goncalves, R. A. Polanczyk and L. F. Fraceto, *Plant Nano Biol.*, 2024, **8**, 100067.
66. M. Gabriel Paulraj, S. Ignacimuthu, M. R. Gandhi, A. Shajahan, P. Ganesan, S. M. Packiam and N. A. Al-Dhabi, *Int. J. Biol. Macromol.*, 2017, **104**, 1813–1819.
67. C. Xu, Y. Shan, M. Bilal, B. Xu, L. Cao and Q. Huang, *Chem. Eng. J.*, 2020, **395**, 125093.
68. R. Grillo, A. E. S. Pereira, C. S. Nishisaka, R. de Lima, K. Oehlke, R. Greiner and L. F. Fraceto, *J. Hazard. Mater.*, 2014, **278**, 163–171.
69. F. Artusio, D. Casà, M. Granetto, T. Tosco and R. Pisano, *Processes*, 2021, **9**, 1641.
70. A. E. S. Pereira, R. Grillo, N. F. S. Mello, A. H. Rosa and L. F. Fraceto, *J. Hazard. Mater.*, 2014, **268**, 207–215.
71. M. Umair, S. H. Zafar, M. Cheema, R. Minhas, A. M. Saeed, M. Saqib and M. Aslam, *Sci. Total Environ.*, 2023, **896**, 165147.
72. T. Piroonpan, E. Huajaikaew, N. Kurantowicz, P. Potiyaraj and W. Pasanphan, *Eur. Polym. J.*, 2024, **203**, 112670.
73. Bhavishya, K. S. Subramanian, M. Gopal, R. Bhat, S. S. Nair, M. Radhakrishnan and M. K. Rajesh, *Biocatal. Agric. Biotechnol.*, 2024, **57**, 103054.
74. Z. Ulhassan, A. R. Khan, Y. Hamid, W. Azhar, S. Hussain, M. S. Sheteiwy, A. Salam, K. R. Hakeem and W. Zhou, in *Metals Metalloids Soil Plant Water Systems*, ed. T. Aftab and K. Hakeem, Academic Press, 2022, pp. 287–308.
75. D. Baragaño, R. Forján, N. Álvarez, J. R. Gallego and A. González, *J. Hazard. Mater.*, 2022, **433**, 128748.
76. R. Forján, M. Arias-Estévez, J. L. R. Gallego, E. Santos and D. Arenas-Lago, *Sci. Total Environ.*, 2024, 172451.
77. T. Li, S. Lü, Z. Wang, M. Huang, J. Yan and M. Liu, *Sci. Total Environ.*, 2021, **765**, 142745.
78. I. Priyadarshini, A. Chowdhury, A. Rao, B. Roy and P. Chattopadhyay, *J. Environ. Manage.*, 2023, **325**, 116596.
79. Z. Cai, J. Fu, P. Du, X. Zhao, X. Hao, W. Liu and D. Zhao, *Chem. Eng. J.*, 2018, **332**, 227–236.
80. S. Ullah, H. Nasir, T. Akhtar, S. Mahboob, S. A. B. Bukhari and K. H. Thebo, *Appl. Surf. Sci.*, 2024, **662**, 160027.

CHAPTER 14

Colloid-mediated Transport of Contaminants Through Soils

ALANA DEDUCK CICILINSKI,*[a] VANDER FREITAS MELO[b] AND PATRICIO PERALTA-ZAMORA[a]

[a] Chemistry Department, Federal University of Paraná, 81531-980, Curitiba-PR, Brazil; [b] Department of Soils and Agricultural Engineering, Federal University of Paraná, 33505-658, Curitiba-PR, Brazil
*Email: alanadeduck8@gmail.com

14.1 Introduction

Since the Stockholm Conference in 1972, the world has been discussing the foundations of sustainable development, seeking solutions that allow economic growth, without sacrificing the environment or neglecting human needs, particularly those of the poorest populations. Unfortunately, after 50 years of agendas, protocols, and agreements, the health of the planet has declined substantially, causing well-documented deleterious effects in practically all environmental compartments.

Due to this reality, efforts are made to implement environmental monitoring and remediation programs, activities that prove to be complex, due to the complex dynamics of contaminants in the environment. This complexity is particularly evident in contaminated soils, in which the dynamics of the contaminant depends not only on its physical–chemical properties or climatic peculiarities, but also on the characteristics of the soil.

Depending on its physicochemical properties, a contaminant deposited in the soil can volatilize, be degraded by biotic and abiotic processes, be retained in the organic and inorganic components of the soil or leach through the soil column toward groundwater. In general, all these processes

Contribution of Colloidal Materials to Air, Water and Soil Environmental Sustainability: From Basic Concepts to Advanced Applications
Edited by Izabel C. Riegel-Vidotti, Lizandra Maria Zimmermann and Heloise R. de Barros

Published by the Royal Society of Chemistry, www.rsc.org

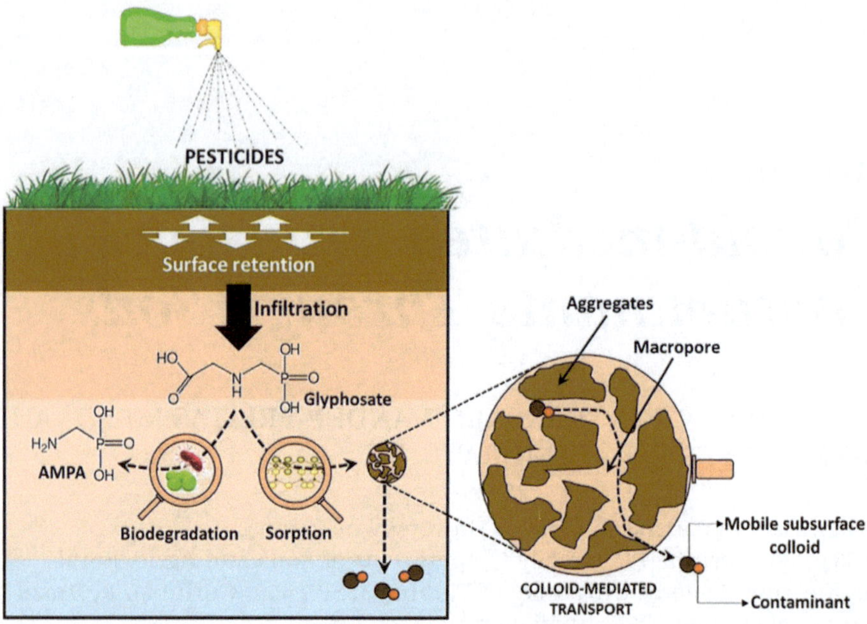

Figure 14.1 Representation of glyphosate dynamics in soil emphasizing colloid-mediated transport.

influence the dynamics of the contaminant in the soil, affecting its half-life, mobility, and transport. However, it is important to emphasize that, in addition to the processes that can be intuitively recognized, there are others that, in a less obvious way, also influence the dynamics of soil contaminants. In this context, colloid-mediated transport shows relevance, particularly in fractured or macoporous soils (see the illustration in Figure 14.1).

This chapter aims to elucidate the main aspects associated with colloid-mediated transport, emphasizing its relevance to the dynamics and propagation of pollutants in the soil. The discussion is conducted using glyphosate as a model, not only due to its environmental relevance, but mainly due to its physical–chemical characteristics that enable this type of transport.

14.2 Glyphosate: Characteristics and Environmental Implications

In the last 40 years, Brazilian grain production has practically increased six-fold, reaching approximately 236 million tons.[1] At the same time, the cultivated area only doubled in size, which implies a notable increase in productivity. In fact, agricultural productivity was facilitated by the improvement of techniques, the introduction of resistant species and, concomitantly, the greater consumption of pesticides. According to data provided by the Brazilian Institute of the Environment and Renewable Natural

Resources,[2] glyphosate (GLY) was the best-selling active ingredient in Brazil in 2018, totaling 195 056.2 tons, basically for the management of weeds in soybean, corn, and rice crops.

Glyphosate [*N*-(phosphonomethyl)glycine] is a broad-spectrum herbicide used in agriculture and for controlling aquatic weeds. It is classified as a non-selective, systemic, and post-emergent herbicide, which acts as a potent inhibitor of the activity of 5-enolpyruvylshikimate-3-phosphate synthase (EPSPS), an enzyme that participates in the synthesis of aromatic amino acids and, consequently, hormones and other critical plant metabolites, including flavonoids, lignins, and other phenolic compounds.[3]

In the environment, the primary pathway for glyphosate degradation is associated with soil bacteria, in particular Pseudomonas, which results in the production of its main metabolite, aminomethylphosphonic acid (AMPA), and glyoxylic acid. An alternative and less common degradation pathway was identified in the presence of specific species of soil bacteria, which metabolize glyphosate to inorganic phosphate and sarcosine and subsequently convert sarcosine to glycine.[4]

Since glyphosate is applied directly to the soil, its transport is strongly influenced by soil properties, including biodiversity and type of vegetation cover, biomass and microbial activity, content and mineral composition of the clay fraction, content, and quality of organic matter, in addition to other characteristics, such as temperature, humidity and aeration. On the other hand, transport also depends on the physicochemical characteristics of GLY, particularly its acidic character. Glyphosate is a zwitterion that has 4 ionizable groups, which makes the predominant ionic form largely dependent on the pH of the soil solution. In Brazilian soil conditions, GLY can have 2 or 3 ionized groups, which influences its interactions with soil components and, consequently, its mobility.

Based on the values of the octanol–water partition coefficient ($\log K_{OW}$: -4.59 to -0.94)[5] it is inferred that GLY is a hydrophilic herbicide, which provides high solubility in water (S_w: 10 000–15 700 $mg\,L^{-1}$ at 25 °C).[5] The high solubility of GLY should favor its mobility in the soil. However, studies show a high tendency for glyphosate adsorption in the soil, which implies reduced mobility.[6–10]

According to Glass (1987), glyphosate adsorption is related to ion exchange processes involving the clay fraction,[7] while Piccolo, Celano, and Pietramellara (1992) suggest that glyphosate retention is due to a ligand exchange mechanism that involves the metal–humic acid fraction (Fe–HA).[11] Although the sorption mechanism depends on soil properties, the low mobility of glyphosate is very well documented, which implies limited vertical leaching through the soil column and, consequently, low availability to plants or other soil organisms.[12]

In view of the characteristics outlined above, an important issue to be assessed is the possible risk of pollution of surface and underground waters by glyphosate. The movement of GLY and its metabolite from the application site to groundwater and surface waters can follow several pathways depending on the field configuration (soil class, type, and quality of colloids, shape and size of soil aggregates, proportion and continuity of macropores, *etc.*), agricultural practices (application, direct planting), and meteorological conditions (intensity and magnitude of rainfall).

Surface runoff is the movement of pesticides on the surface of the soil, together with soil particles, after intense rains in steeper reliefs (smooth and strong undulating). In general, pesticides that have greater persistence and greater sorption to soil particles have greater surface runoff potential. According to the method proposed by Goss (1992),[13] the molecules that show high transport potential associated with sediments are those that have long persistence ($t_{1/2} \geq 40$ days) and high sorption in the soil ($K_{OC} \geq 1000$ L kg^{-1}), or high sorption ($K_{OC} \geq 500$ L kg^{-1}) associated with low solubility in water ($S_w \leq 0.5$ mg L^{-1}). Thus, glyphosate falls into the first situation, by presenting long persistence ($t_{1/2}$ 20–100 days)[5] and high sorption to the soil (K_{OC} 3–4000 L kg^{-1}),[5] which justifies preliminarily stating that this herbicide has a significant tendency to run off superficially in the soil.

In 1989, Gustafsson proposed an index to classify pesticides according to their leaching potential, the GUS index (eqn (14.1)), considering persistence ($t_{1/2}$) and soil sorption (K_{OC}).[14]

$$\mathrm{GUS} = \log t_{1/2}(4 - \log K_{OC}) \tag{14.1}$$

where GUS is a dimensionless empirical number, $t_{1/2}$ is the persistence (days), and K_{OC} is the soil sorption (L kg^{-1}).

Using the extremes of the reported ranges of $t_{1/2}$ and K_{OC}, GLY presents a GUS index of 5.7 for the lowest values, which classifies it as leachable, and a GUS of −0.8 for the upper limits of the ranges, classifying the herbicide as non-leachable. It is important to highlight that such discrepancies are common when evaluating parameters associated with soil, primarily due to the significant variability in its properties. In any case, it is estimated that the risk of groundwater pollution by glyphosate and AMPA is low, especially in deep and unstructured soils, with a low presence of macropores. Additionally, it is estimated that the occurrence of glyphosate in drainage water does not necessarily indicate leaching into groundwater, because deeper layers of the soil can absorb and even degrade the herbicide before it reaches groundwater.[15]

Nevertheless, many reports indicate the presence of GLY in groundwater, which implies that its transport may be enhanced under certain conditions, including the presence of macro pores, cracks, and colloidal particles.[16–18]

14.3 Colloid-mediated Transport

14.3.1 Nature of Soil Colloids

Colloids are operationally defined as particles with a diameter of less than 10 μm.[19] In soil, this colloidal fraction is formed from the decomposition (or weathering) of precursor materials, such as organic residues and primary minerals.

Organic colloids involve a non-humic fraction, consisting of carbohydrates, amino acids, proteins and lipids, and a humic fraction, resulting from the transformation (humification) of organic matter from plant and animal remains. On the other hand, inorganic colloids correspond mainly to secondary minerals, such as aluminosilicates (kaolinite) and oxides and hydroxides of iron (hematite and goethite) and aluminum (gibbsite). In soils, secondary colloidal minerals in the clay fraction have an upper size limit of 2 μm, while organic humic compounds are even smaller. It is important to highlight that this distinction is a significant simplification, because colloidal material is normally a mixture of both organic and inorganic materials.[19] This complex organo-mineral association can occur in the form of microaggregates and macroaggregates (occluded colloids) or as water-dispersible colloids (free colloids), which are potentially mobile.

In general, the colloidal fraction has a high and heterogeneous surface area, involving a variety of functionalities and charges. Thus, this fraction presents high chemical reactivity, favoring interactions with polluting substrates, which has relevant implications for its transport and destination. Oxides and hydroxides of Fe (hematite and goethite) and Al (gibbsite) are important constituents of the clay fraction of weathered tropical soils.[20] These minerals have high pH values at zero charge point (pH_{ZCP} between 7.5 and 8.5), which confers excess positive charges for soil pH values below these limits.[21] Under these conditions there is a higher concentration of biprotonated ferrol and aluminol groups ($-FeOH_2^{+0.5}$ and $-AlOH_2^{+0.5}$),[22] which favors the adsorption of glyphosate, which under these conditions can present two or three negative charges (see Figure 14.2).

The adsorption of glyphosate on minerals from the clay fraction occurs through two mechanisms: (i) external sphere complexation[23] and (ii) internal sphere complexation.[24] In the first case, the negative functional groups of glyphosate are hydrated and adsorption occurs by electrostatic attraction in the Stern layer or in the double diffuse layer, surrounding the mineral colloid. In the second case, the negative functional groups of glyphosate in the Stern layer lose their water of hydration, which favors the exchange of ligands involving the biprotonated group $-OH_2^{+0.5}$, which causes the herbicide's oxygen to bind directly to the Fe or Al from the surface of the oxides. This bond has a high covalent character, being more stable under environmental conditions.

Studies carried out by Barja and Afonso (2005)[25] show that AMPA and glyphosate form complexes on the goethite surface, the extent of which depends on the pH and concentration of the ligand in the solution. Two predominant inner sphere complexes were identified, one monodentate at low pH values and the other bidentate at high pH values, both involving the phosphonate group of the molecules under study.[25] Regarding the interaction between GLY and iron oxides and hydroxides, it is worth noting that phosphate and glyphosate compete for sorption sites on the goethite surface, with clear preferential adsorption of phosphate.[26] This is a relevant

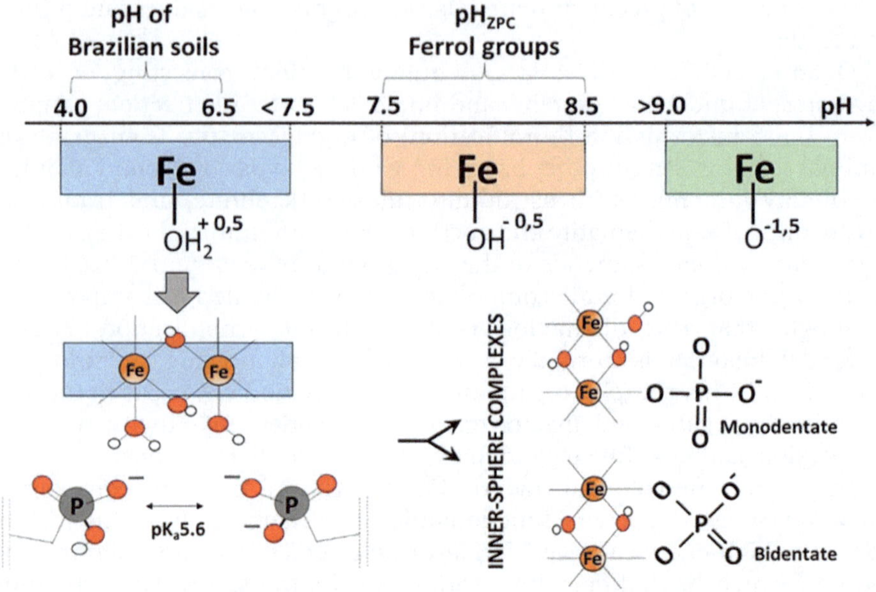

Figure 14.2 Illustration of the ferrol group protonation, possible interactions with glyphosate and formation of inner-sphere complexes.

antecedent, due to phosphate being able to desorb GLY,[27] contributing to its mobility in the soil.

Regarding organic colloids, it is important to mention that there is extensive documentation about their interaction with glyphosate, involving different types of associations with different degrees of stability. However, it is estimated that the organic fraction has a secondary role in the sorption of GLY in soils.[28]

When the glyphosate molecule encounters soil aggregates, all the adsorption reactions mentioned above can occur. As a result, herbicide migration in the soil profile is very limited. On the other hand, glyphosate can also be adsorbed in the dispersion of organic and mineral colloidal particles present in soil macropores, which can increase its leaching in the soil profile towards the water table.

14.3.2 Transport of Colloids Through Soil

Normally, the mobility of pollutants is evaluated according to their distribution between two phases. A dissolved phase, which passes through a filter medium with controlled pore size, and a retained phase, considered as particulate material. However, the dissolved phase includes a wide spectrum of colloidal particles (see Figure 14.3), which, having adsorption sites for organic molecules, can influence the mobility, reactivity, and bioavailability of pollutants.[29] Under specific conditions, the colloidal phase can be mobile, potentially transporting pollutants as

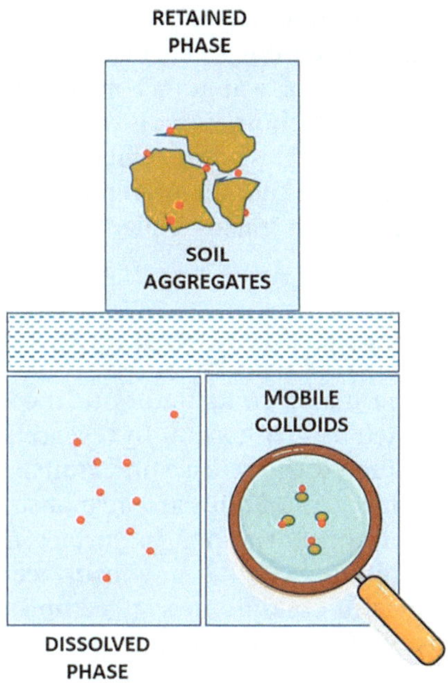

Figure 14.3 Representation of colloidal phases that may be inappropriately considered soluble fractions.

efficiently as the aqueous phase (colloid-mediated transport). Although this process significantly contributes to the transport associated with surface runoff, greater emphasis will be placed on the leaching process, which transports pollutants from the vadose zone to groundwater.

Given its great heterogeneity, it is quite difficult to predict the movement of solids, liquids, and gases through the soil. However, it is understood that there are transport pathways associated with the presence of interconnected macropores and preferential flows associated with the presence of fractures and cracks. In this category, flows along fissures, flows through channels created by fauna, finger-like flows caused by water infiltration between two soil layers and lateral flow along inclined layers are recognized.[30] Although it is estimated that around 90% of water infiltrated into the soil is mobilized through macropores, preferential flows operate actively within the soil profile, which is relevant in the transport of pollutants towards groundwater.[31,32] Preferential flows have often been described as true shortcuts for infiltrated water, allowing it to bypass a large part of the soil volume and come into contact with groundwater.

In recent years, great emphasis has been given to the study of interactions between dissolved contaminants and mobile colloids, understanding that these adsorptive interactions during infiltration events and other transient flow systems facilitate the transport of pollutants in porous media. Thus, field-scale studies have confirmed the rapid migration of radionuclides, heavy metals, pharmaceutical products and pesticides adsorbed by colloids, increasing the flow of contaminants into the groundwater.[33] Additionally, many studies have been carried out to evaluate the transport of pesticides through the preferential flow of water. According to results compiled by Zhang *et al.* (2018),[34] this phenomenon is responsible for most of the pesticide transport in soil profiles, even when strongly adsorbed by soil particles. It should also be noted that the preferential flow of water is vital for recharging groundwater, which raises warnings about the transport of pollutants.[31,34]

14.3.3 Glyphosate Transport by Soil Colloids

Colloids are known to be major contributors to the migration of strongly sorbed contaminants, such as glyphosate. In research conducted in China, Zhang *et al.* (2016)[31] evaluated the release and transport of natural colloids *via* surface and subsurface runoff in rain response events. The authors observed that the colloid concentrations in surface flow were consistently 1 to 2 orders of magnitude higher than in subsurface flows. They also concluded that the dynamics of colloids in surface and subsurface flows was governed by the intensity of precipitation and its temporal distribution and that colloidal particles smaller than 10 μm dominated the surface flow. According to the authors, these results indicate the need to avoid the application of highly sorbent agrochemicals, such as glyphosate, immediately before rain, as their transport in association with colloids can occur quickly over long distances, both by surface and subsurface runoff.[31]

In agreement, previous studies have demonstrated that preferential flow and colloidal transport represent important mechanisms in the transfer of this herbicide to water bodies. The contribution of colloid-facilitated transport (particles with a diameter smaller than 0.24 μm) was evaluated by de Jonge *et al.* (2000).[35] The authors found that the total transport of glyphosate was in the range of <1 to 27% for sandy clay soil and <1 to 52% for sandy soil, depending on the soil treatment. They also concluded that the risk of glyphosate leaching from soil is limited to flow conditions in the macropores that occur shortly after its application.[35]

Results obtained by Gjettermann *et al.* (2009)[36] show that structural conditions can be very important for the mode of glyphosate transport, with most of the leached glyphosate (on average 68%) being linked to particles larger than 20 nm.[36] Kjaer *et al.* (2011)[37] observed that transport facilitated by particles smaller than 0.24 μm was responsible for 13–16% of glyphosate leaching, and this was only transported vertically in soils with macropores. The colloidal transport of glyphosate and AMPA in clayey soils structured with macropores was also demonstrated by Borgaard (2011)[38] in several lysimetric studies and in field trials; the author concluded that leaching is increased by heavy rains shortly after glyphosate application.

According to Daouk *et al.* (2013),[39] the transfer of glyphosate from fields to the adjacent surface occurs mainly by surface runoff, in a dissolved state or bound to small colloids, while Lupi *et al.* (2015)[40] emphasize that vertical transport of glyphosate may also be associated with colloidal particles. According to antecedents compiled by Vereecken (2005),[41] the adsorption of glyphosate in soil preferentially involves the mineral phase, with a clear emphasis on iron oxides. In this way, the association with colloidal particles favors the transport of glyphosate through the soil, mainly through preferential flows in conditions of intense rain. This transport mobilizes glyphosate to the deepest layers of the soil and, more concerningly, to groundwater.[41]

14.3.4 Presence of Glyphosate in Groundwater

Contrary to many studies that describe the strong sorption of glyphosate in the soil and, consequently, its low leaching capacity, there are several studies that report the presence of glyphosate (and AMPA) in groundwater. Sanchis *et al.* (2012),[42] for example, evaluated the presence of glyphosate in 169 groundwater samples (69 sampling sites) in the Catalonia region (Spain), finding the herbicide in 41% of the samples, with an average concentration of 200 $ng\,L^{-1}$, with maximum values of 2.5 $\mu g\,L^{-1}$. Average concentrations of around 500 $ng\,L^{-1}$ were also reported by Mortl *et al.* (2013),[43] in work involving the analysis of groundwater samples in Hungary.

A broad study regarding the environmental occurrence of glyphosate and AMPA in the United States was conducted by Battaglin *et al.* (2014).[44] Samples from different matrices were collected in 38 states, between 2001 and 2010, including 1171 groundwater samples. Of these, 68 tested positive for the presence of glyphosate (maximum value detected 2.03 $\mu g\,L^{-1}$ and average results <0.02 $\mu g\,L^{-1}$) and 168 tested positive for AMPA (maximum value 4.88 $\mu g\,L^{-1}$ and average results <0.02 $\mu g\,L^{-1}$). According to the authors, these and other results confirm that glyphosate and AMPA are mobile compounds, which occur widely in the environment and have agricultural and urban sources. They also found that it was not common for glyphosate to be detected without AMPA, while AMPA was detected without glyphosate in 17.9% of the total samples from all matrices analyzed.[44]

Groundwater samples from 5 locations in Canada were analyzed by Van Stempvoort *et al.* (2014),[45] with glyphosate and AMPA detected in 4 of the locations and in just over 10% of the total of 281 samples. According to the authors, the results suggest that riverine groundwater containing glyphosate and AMPA may be common in some areas along urban streams and therefore may contribute to the stream's pesticide load. Osten and Dzul-Caamal (2017)[46] conducted studies regarding the presence of glyphosate and AMPA in groundwater in intensive agricultural locations (Mexico), totaling 29 samples. Glyphosate was detected in all agricultural communities studied and the detection of glyphosate in groundwater samples showed variations that depended on the sampling locations. The highest concentration obtained was 1.41 $\mu g\,L^{-1}$, while the lowest level was 0.44 $\mu g\,L^{-1}$.

Recently, glyphosate has been detected in groundwater samples from agricultural areas in China, at maximum concentrations of the order of 2 $\mu g\,L^{-1}$.[18]

The presence of glyphosate has also been reported in groundwater from wine-growing regions of northeastern Italy, in many cases exceeding the maximum permitted limits.[16] The presence of glyphosate and AMPA was detected in groundwater in Sweden, due to the use of this herbicide to combat the proliferation of weeds on railways. In some cases, the GLY concentration exceeds the maximum permitted limit in the European Union (0.1 $\mu g\,L^{-1}$).[47]

In Brazil, the presence of glyphosate was detected in all groundwater samples analyzed (52) in the region of the Rio Preto Hydrographic Basin (Alto Rio Jardim Hydrographic Unit – Federal District), with maximum levels of around 1.6 $\mu g\,L^{-1}$.[48]

Finally, it is important to highlight that the rapid flow of water through fissures can be very pronounced in some types of soil, which favors infiltration and contamination of groundwater. Thus, according to observations by Hartmann *et al.* (2018), the transport of pollutants through fast flows increases the risk of contamination of groundwater reserves, making predictions that neglect this type of transport underestimated.[17]

14.4 Final Considerations

Based on the findings discussed, glyphosate interacts strongly with the inorganic constituents of soil, particularly iron oxides and hydroxides (such as hematite and goethite) and aluminum oxides (gibbsite). These colloids can be dispersed in the soil and transported by preferential flows (macropores), reaching, in certain soil types and under specific conditions, groundwater.

Although comprehensive reviews of the literature provide strong evidence of the impact of glyphosate on fish and mammals, there is a need for more in-depth studies to establish some type of correlation between the presence of glyphosate and effects on human health. Nonetheless, it is important to emphasize that the colloidal fraction acts as a potential reservoir for pollutants in aquatic environments, which can compromise the quality of the water offered to the population.

References

1. EMBRAPA, *Brazilian Agricultural Research Company. Trajectory of Brazilian agriculture, 2018*, 2018, https://www.embrapa.br/visao/trajetoria-da-agricultura-brasileira (accessed in March 2024).
2. IBAMA, *Brazilian Institute of the Environment and Renewable Natural Resources. Pesticide marketing reports*, 2022, https://www.gov.br/ibama/pt-br/assuntos/quimicos-e-biologicos/agrotoxicos/relatorios-de-comercializacao-de-agrotoxicos (accessed in March 2024).
3. S. O. Duke and S. B. Powles, *Pest Manage. Sci.*, 2008, **64**, 319.
4. O. K. Borgaard and A. L. Gimsing, *Pest Manage. Sci.*, 2008, **64**, 441.
5. D. Mackay, W. Y. Shiu, K. C. Ma and S. C. Lee, *Handbook of Physical-chemical Properties and Environmental Fate for Organic Chemicals*, CRC Press, Boca Raton, 2nd edn, 2006.

6. P. Sprankle, W. F. Meggitt and D. Penner, *Weed Sci.*, 1975, **23**, 229.
7. R. L. Glass, *J. Agric. Food Chem.*, 1987, **35**, 97.
8. S. Autio, K. Siimes, P. Laitinen, S. Rämö, S. Oinonen and L. Eronen, *Chemosphere*, 2004, **55**, 215.
9. L. Carretta, A. Cardinali, A. Onofri, R. Masin and G. Zanin, *Int. J. Environ. Res.*, 2021, **15**, 1037.
10. E. de Gerónimo and V. C. Aparicio, *Eur. J. Soil Sci.*, 2022, **73**, e13188.
11. A. Piccolo, G. Celano and G. Pietramellara, *Sci. Total Environ.*, 1992, **123/124**, 77.
12. D. N. Roy, S. K. Konar, S. Banerjee, D. A. Charles, D. G. Thompsom and R. Prasad, *J. Agric. Food Chem.*, 1989, **37**, 437.
13. W. D. Goss, *Weed Technol.*, 1992, **6**, 701.
14. D. I. Gustafsson, *Environ. Toxicol. Chem.*, 1989, **8**, 339.
15. R. H. Coupe, S. Kalkhoff, P. D. Capel and C. Gregoire, *Pest Manage. Sci.*, 2012, **68**, 16.
16. N. Suciu, E. Russo, M. Calliera, G. P. Luciani, M. Trevisan and E. Capri, *Sci. Total Environ.*, 2023, **866**, 161171.
17. A. Hartmann, S. Jasechko, T. Gleeson, Y. Wada, B. Andreo, J. A. Barberá, H. Brielmann, L. Bouchaou, J. Charlier, W. G. Darling, M. Filippini, J. Garvelmann, N. Goldscheider, M. Kralik, H. Kunstmann, B. Ladouche, J. Lange, G. Lucianetti, J. F. Martín, M. Mudarra, D. Sánchez, C. Stumpp, E. Zagana and T. Wagener, *Proc. Natl. Acad. Sci. U. S. A.*, 2021, **118**, 1.
18. Y. Geng, L. Jiang, D. Zhang, B. Liu, J. Zhang, H. Cheng, L. Wang, Y. Peng, Y. Wang, Y. Zhao, Y. Xu and X. Liu, *Sci. Total Environ.*, 2021, **769**, 144396.
19. D. J. Lapworth, D. C. Gooddy, I. Harrison, A. Kim and C. Vane, *Colloidal Phase Transport of Pesticides: A Review with Special Reference to Major UK Aquifers*, British Geological Survey, Nottingham, 2005.
20. R. M. Cornell and U. Schwertmann, *The Iron Oxides: Structure, Properties, Reactions, Occurrences and Uses*, Wiley-VCH, Weinheim, 2nd edn, 2003.
21. M. P. Gaigeot, M. Sprik and M. J. Sulpizi, *J. Phys.: Condens. Matter*, 2012, **24**, 124106.
22. C. G. Albuquerque, F. Gavelaki, V. F. Melo, A. C. V. Motta, A. J. G. Zarbin and C. M. Ferreira, *Rev. Bras. Cienc. Solo*, 2022, **46**, DOI: 10.36783/18069657rbcs20210146.
23. L. Stolze, D. Zhang, H. Guo and M. Rolle, *Geochim. Cosmochim. Acta*, 2019, **248**, 274.
24. J. D. Kubicki, K. W. Paul, L. Kabalan, Q. Zhu, M. K. Mrozik, M. Aryanpour and D. R. Strongin, *Langmuir*, 2012, **28**, 14573.
25. B. C. Barja and M. S. Afonso, *Environ. Sci. Technol.*, 2005, **39**, 585.
26. A. L. Gimsing and O. Borggaard, *Clays Clay Miner.*, 2001, **49**, 270.
27. C. V. Waiman, M. J. Avena, A. E. Regazzoni and G. P. Zanini, *J. Colloid Interface Sci.*, 2013, **394**, 485.
28. L. H. da Cruz, H. de Santana, C. T. B. V. Zaia and D. A. M. Zaia, *Braz. Arch. Biol. Technol.*, 2007, **50**, DOI: 10.1590/S1516-89132007000300004.
29. D. Cheng, X. Liu, S. Zhao, B. Cui, J. Bai and Z. Li, *Sci. Total Environ.*, 2017, **578**, 649.

30. S. E. Allaire, S. Roulier and A. J. Cessna, *J. Hydrol.*, 2009, **378**, 179.
31. W. Zhang, X. Y. Tang, Q. S. Xian, N. Weisbrod, J. E. Yang and H. L. Wang, *J. Hydrol.*, 2016, **542**, 101.
32. S. M. Franklin, A. N. Kravchenko, R. Vargas, B. Vasilas, J. J. Fuhrmann and Y. Jin, *Soil Biol. Biochem.*, 2021, **161**, 108398.
33. D. Deb and S. Chakma, *Int. J. Environ. Sci. Technol.*, 2023, **20**, 6955.
34. Y. Zhang, Z. Zhang, Z. Ma, J. Chen, J. Akbar, S. Zhang, C. Che, M. Zhang and A. Cerdà, *Can. J. Soil Sci.*, 2018, **98**, 604.
35. H. de Jonge, L. W. Jonge and O. H. Jacobsen, *Pest Manage. Sci.*, 2000, **56**, 909.
36. B. Gjettermann, C. T. Petersen, C. B. Koch, N. H. Spliid, C. Gron, D. L. Baun and M. Styczen, *J. Environ. Qual.*, 2009, **38**, 2382–2393.
37. J. Kjaer, V. Ernstsen, O. H. Jacobsen, N. Hansen, L. W. De Jonge and P. Olsen, *Chemosphere*, 2011, **84**, 471.
38. O. K. Borgaard, *Trends Soil Plant Sci. J.*, 2011, **2**, 16.
39. S. Daouk, L. F. Alencastro and H. R. Pfeifer, *J. Environ. Sci. Health, Part B*, 2013, **48**, 725.
40. L. Lupi, K. S. B. Miglioranza, V. C. Aparicio, D. Marino, F. Bedmar and D. A. Wunderlin, *Sci. Total Environ.*, 2015, **536**, 687.
41. H. Vereecken, *Pest Manage. Sci.*, 2005, **61**, 1139.
42. J. Sanchis, L. Kantiani, M. Llorca, F. Rubio, A. Ginebreda, J. Fraile, T. Garrido and M. Farré, *Anal. Bioanal. Chem.*, 2012, **402**, 2335.
43. M. Mortl, G. Németh, J. Juracsek, B. Darvas, L. Kamp, F. Rubio and A. Székács, *Microchem. J.*, 2013, **107**, 143.
44. W. A. Battaglin, M. T. Meyer, K. M. Kuivila and J. E. Dietze, *J. Am. Water Resour. Assoc.*, 2014, **50**, 275.
45. D. R. van Stempvoort, J. W. Roy, S. J. Bronw and G. Bickerton, *Chemosphere*, 2014, **95**, 455.
46. J. R. Osten and R. Dzul-caamal, *Int. J. Environ. Res. Public Health*, 2017, **14**, 595.
47. H. Cederlund, *Sci. Total Environ.*, 2022, **811**, 152361.
48. N. L. Pires, E. P. de Araújo, E. C. Oliveira-Filho and E. D. Caldas, *Sci. Total Environ.*, 2023, **875**, 162499.

CHAPTER 15

Contribution of Colloids to Soil and Sustainability

ALANA DEDUCK CICILINSKI* AND
PATRICIO PERALTA-ZAMORA

Federal University of Paraná, Chemistry Department, Curitiba-PR, Brazil
*Email: alanadeduck8@gmail.com

15.1 Introduction

Soil is a highly complex matrix due to its physical and chemical characteristics. It is composed of a heterogeneous mixture of air, water, inorganic solids (such as clay minerals), organic solids (humic and fulvic acids, humin), and microorganisms of plant and animal origin.[1] It exhibits important physical properties, such as the size distribution and arrangement of particles in aggregates, which constitute the texture and structure of the soil, respectively. These particles may possess a colloidal nature, comprising mainly the clay fraction of the soil—that is, the fraction with a particle size smaller than 2 μm—and the fine silt fraction, 2–10 μm.[2]

Due to the heterogeneous nature of soil constituents and their colloidal aggregates, there are numerous potential binding sites for a wide range of materials with diverse chemical characteristics. Various types of interactions can occur with other components, such as substances from anthropogenic and agrochemical activities used to enhance agricultural productivity. Given that the contaminants reaching this compartment are also, for the most part, highly complex, it is necessary to study their behaviour over the long term and under varying conditions.

For this reason, there is a significant demand for environmental remediation techniques that can be used to treat contaminated soils. However, this challenge

Contribution of Colloidal Materials to Air, Water and Soil Environmental Sustainability: From Basic Concepts to Advanced Applications
Edited by Izabel C. Riegel-Vidotti, Lizandra Maria Zimmermann and Heloise R. de Barros

Published by the Royal Society of Chemistry, www.rsc.org

is particularly complex, not only due to the complexity of the environmental matrices involved but also due to the diverse ways in which interactions between contaminants and contaminated matrices occur.

15.2 Soil Contamination

Environmental problems related to soil are linked to its degradation, primarily due to human activity. Degradation can be defined as changes or disturbances that are deleterious or undesirable.[3]

Among the various types of degradation, contamination is particularly significant, as it involves the introduction of chemical compounds that alter the natural state of the soil. The most common and emerging contaminants are hydrocarbons, pharmaceuticals, and pesticides. These contaminants negatively affect human health and ecosystems, causing significant environmental impacts.

As far as these contaminants go, hydrocarbons are compounds whose molecules contain only carbon and hydrogen atoms. These compounds can be found in the soil, as a result of both natural combustion sources (*e.g.*, wildfires) and polluting sources linked to anthropogenic activities. They come from fuels, petroleum derivatives, waxes, paints, *etc.*, and soil contamination occurs mainly due to accidents during the transport of these substances and leaks in underground storage tanks at gas stations.

Most of these contaminants can be carcinogenic and mutagenic and, depending on their fate and transport, can pose a serious threat to human health and environmental quality, particularly aromatic solvents (basically represented by benzene, toluene, ethylbenzene, and xylenes), polycyclic aromatic hydrocarbons (PAHs), and total petroleum hydrocarbons (TPH).[4]

Drugs, in turn, are chemotherapeutic agents used in the prevention and treatment of human diseases, or to increase productivity in the agricultural industry.[5] There are several types of drugs, which can be classified according to their chemical structure and functional groups, mechanism of action, metabolic pathways, and routes of administration. The main types include antibiotics, hormones, anti-inflammatories, antidepressants, and blood pressure regulators.[6]

Despite controlled use, drugs are found in various environments, especially in the soil. Deposition occurs primarily through two routes: (a) elimination of metabolites and (b) inappropriate disposal.[7] The first route occurs after the ingestion, metabolization, and excretion of these molecules in the sewage system. The drugs reach the treatment stations, where they are subjected to effluent cleaning protocols. However, the efficiency of these processes varies according to the chemical properties of the drugs and often tends to be inadequate.[8] As a consequence, the residual content is high, and the water remains contaminated with pharmaceuticals. After this, the drugs enter the soil environment primarily due to rain activity (transporting effluents between locations) and through the use of contaminated water in the irrigation of plantations.[9] Additionally, sludge produced in the treatment of effluents containing pharmaceuticals can be landfilled, further contributing to soil contamination.

The second scenario occurs due to the disposal of unused or expired drugs directly into the sewage system or in inappropriate disposal locations. As a consequence, pharmaceuticals end up in landfills, where they contaminate the soil. Similarly, in the agricultural industry, the improper disposal of excrement and fertilizers also leads to indirect contamination of soil by residual pharmaceuticals.[10]

The presence of pharmaceuticals in the soil leads to adverse effects on the organisms that inhabit it and, indirectly, on human health. Psychotropic and hormonal drugs primarily affect foraging, hunting, escape behaviors, and the reproduction of organisms.[11] The reproductive cycle can be altered by the presence of steroids. In some cases, the sex of the organisms may even be reversed, leading to a reduction in the effective population.[12]

Exposure of microorganisms to antibiotics can lead to the development of resistance in these populations. This poses a significant threat to human health, as the emergence of antibiotic-resistant bacterial strains, even without direct human contact, is one of the major consequences of the indiscriminate use of antibiotics.[13] So-called superbugs are resistant to most antibiotics known to humans, and they are predicted to become one of the major public health challenges in the coming years.[14]

Finally, pesticides are chemical substances designed to optimize agricultural or livestock activities. There are various types of pesticides (*e.g.*, herbicides, insecticides, and fungicides) each with different chemical structures, modes of action, and applications.

These compounds tend to persist longer in soil than in plants or animals since chemical residues are more rapidly metabolized or diluted in actively growing living systems than in relatively static soil systems.[15] Although persistent pesticides tend to have long-term effectiveness in controlling pests, they can also have toxic and harmful effects on soil flora and fauna, and simultaneously contaminate the environment.

In this context, soil microorganisms respond differently to various types of pesticides depending on several factors, including the nature of the pesticide, the properties of the soil, and the groups of microorganisms present in the soil. The total numbers of bacteria, fungi, protozoa, and algae can increase or decrease depending primarily on the pesticide's characteristics, specifically its toxicity and potential as a nutrient or energy source. However, the overall structural and functional diversity of soil microbial populations is altered due to pesticide applications.[16]

As soil fertility depends not only on soil texture, as previously mentioned, but also on its biological capacity, microbial diversity can be altered after pesticide use, and such changes can affect soil fertility.[17] Therefore, it is evident that the relationships between pesticide structure and soil microorganisms are not easily predictable.

In addition to the previously mentioned organic compounds, another major concern involves contamination by heavy metals. From a biological perspective, "heavy" refers to a group of metals, and in some cases metalloids, that can be toxic to plants and animals even at low concentrations.[18]

Unlike organic pollutants, heavy metal pollution is insidious, persistent, and irreversible, and not only degrades the quality of water bodies, the atmosphere, and food crops but also poses a significant threat to the health and well-being of organisms and humans through bioaccumulation in the food chain.[19]

Soil pollution by heavy metals has become a global environmental issue. Recent advances in agriculture, industrialization, and urbanization have significantly contributed to the high levels of heavy metal contamination in soil. Anthropogenic activities such as mining and smelting,[20] fossil fuel combustion,[21] municipal waste disposal,[22] pesticide application,[23] sewage irrigation,[24] and fertilizer application[25] contribute to increasing heavy metal concentrations in the agricultural soil environment.

Thus, soil contamination with heavy metals negatively impacts the size of the soil microbial biomass and various measures of soil microbial activity, such as mineralization and nitrogen fixation. Additionally, heavy metals affect microbial populations by hindering their growth, altering their morphology, and disrupting their biochemical activities, and ultimately leading to a reduction in biomass and diversity. They can damage cell membranes, alter enzyme specificity, disrupt cellular functions, and damage DNA structure.[26–29]

Given the complexity of not only the matrix and pollutants but also the interactions between them, treatment techniques need to be extensively studied. Furthermore, considering that soils are non-renewable resources, it is necessary to reconcile the treatment of pollutants with the maintenance of vital soil functions. The generation of by-products must also be considered, as these can be more toxic than the original compounds.

15.3 Classic Treatment Systems

Typically, soil remediation processes can be carried out by *ex situ* or *in situ* technologies, depending on the characteristics of the site, as well as the type and degree of contamination. Techniques can be classified according to the method of action and are generally divided into four major groups: physical techniques, chemical techniques, physico–chemical techniques, and biological techniques. Some examples are provided below.

Ex situ remediation always involves removing the contaminated compartment by excavating the soil. It can be carried out "on-site" (the treatment involves moving soil, but is carried out within the contaminated site) or "off-site" (contaminated soil is transported for subsequent treatment in specialized facilities).

Among the *ex situ* processes, those of a biological nature, such as biopiles and landfarming, are particularly noteworthy. The biopile system can be considered a variation of the composting technique for organic materials, with the key difference being that it is characterized as a static process, without the usual turning involved in composting. Biopiles are typically arranged in waterproof locations, with oxygen supplied through a network of pipes connected to blowers or pumps.[30]

Landfarming, in turn, consists of disposing of waste on the surface layer of the soil, with periodic stirring to incorporate oxygen. pH, moisture, and nutrients can be adjusted to promote microbial activity. In this case, the method's efficiency depends on the ability of the local microbiota to immobilize or degrade the compounds of interest.[31,32]

Ex situ physical processes usually involve thermal desorption techniques: it is based on heating the soil, to separate contaminants with low boiling points, with subsequent treatment of the vapors generated. Although these methods are highly effective due to the greater control and homogeneity of the process, they have the significant disadvantage of being very expensive due to the need for excavation and transportation to the treatment site.

In situ soil treatment, on the other hand, represents a major technological challenge, due to the need to ensure the effective degradation of pollutants, without causing substantial changes to the remediated matrix. Classic examples of *in situ* remediation include air sparging, reactive barriers, bioremediation, and even natural attenuation.

The air-sparging method of remediation is used to extract volatile and semi-volatile compounds found in the saturated zone of the soil. It involves the injection of air to remove contaminants while also promoting the aerobic biodegradation of certain compounds, by increasing the amount of dissolved oxygen. Contaminants can be captured in the region immediately above, in the unsaturated zone of the soil, through combination with a vapor extraction system.[33]

Reactive barriers are permeable systems implemented underground to intercept the contamination plume in the flow of groundwater and treat dissolved contaminants when they come into contact with these barriers. They are formed by trenches excavated and filled with compounds that react with the contaminant dissolved in groundwater, degrading, adsorbing, or precipitating them.[33]

Bioremediation is the process by which living organisms such as microorganisms, fungi, plants, green algae, or their enzymes are used to reduce or remove/remediate contaminations in the environment.

In this context, monitored natural attenuation is a passive form of remediation, which involves microbial activity without any human intervention. It is a simple, typically slow, technique that depends on natural processes to reduce the mass, toxicity, and/or mobility of a contaminant in the soil.[34]

However, environmental conditions must be suitable to employ this technique. Furthermore, the presence of native microorganisms capable of biodegrading the contaminant present in the soil is necessary.[35] In general, microorganisms are present in the impacted area, and the longer their exposure to xenobiont compounds, the greater the probability that some will acquire the genetic capacity to produce enzymes that act in the degradation of these compounds.[36]

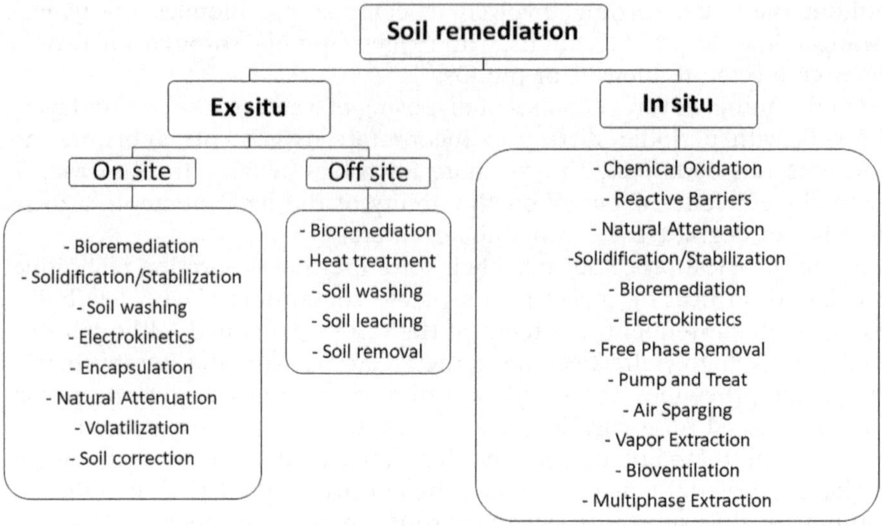

Figure 15.1 Main classical soil treatment techniques.

In general, *in situ* processes of a biological nature are quite widespread, with natural bio-attenuation processes standing out. These processes have a lower cost than *ex situ* processes, however, remediation efficiency is limited due to the difficulty of controlling the processes.

In addition to the methodologies mentioned, there are several other soil treatment techniques not explored here. For informational purposes, a schematic representation of the main soil treatment techniques can be seen in Figure 15.1.

Given the above, few technologies demonstrate good performance in the treatment of contaminated soil, particularly when the contamination involves chemical species resistant to available bio-attenuation techniques.

Therefore, the remediation of environmental matrices must be vigorous enough to degrade the pollutant load, but sufficiently mild to ensure their integrity. In this context, the study of remediation techniques based on colloid chemistry plays an important role.

15.4 Remediation Based on Colloid Chemistry

As previously reported, there are several remediation strategies for contaminated soils. In this context, washing processes using surfactants and biosurfactants are commonly employed in decontamination. The use of surfactants and biosurfactants in remediation processes is frequently cited in the literature;[37–43] however, a novel approach to using surfactants for the remediation of impacted soils involves the use of microfoams, replacing aqueous surfactant solutions, which will be described below.

15.4.1 Colloidal Gas Aphrons (CGAs)

Within the realm of colloidal-level techniques, the application of Colloidal Gas Afrons (CGAs) can be highlighted. First described by Sebba (1987),[44] it is a microfoam system, possessing some colloidal properties and consisting of spherical microbubbles, typically greater than 25 µm in diameter, and classified as kugelschaums or ball foam.

They consist of gas bubbles covered by a monolayer of surfactant molecules immersed in the solution containing micelles. The repulsion between similarly charged surfaces of the microbubbles prevents their coalescence. The CGA system traps approximately 65% of the air by volume and has a specific gravity of only about 0.35. CGA bubbles retain the same charge as the surfactant solution from which they are generated.[44] A schematic representation of the CGAs can be seen in Figure 15.2.

CGAs were first generated by stirring an aqueous surfactant solution contained in a baffled vessel using a spinning disk impeller at high speeds (5000–10 000 rpm).[45] In addition to the classical method of producing microbubbles and CGAs developed by Sebba (1985),[45] many other methods have been investigated to obtain bubbles with even smaller sizes, including dispersed air flotation[46], electroflotation,[47,48] dissolved air flotation,[48–50] gas sparger,[51] electrostatic spraying of air,[48,52,53] spherical microbubble generator[54], shirasu-porous-glass (SPG) membrane generator,[55] sonication,[56] homogenization,[56] and lotus ceramics generators.[57–60]

The main characteristics of CGAs that enable their diverse applications, are their large interfacial area, the adsorption of particles at microbubble interfaces, and their stability, which enhances mass transfer.[61] Furthermore, they can be pumped through pipes, columns, and connections, making them easier to apply in various processes.

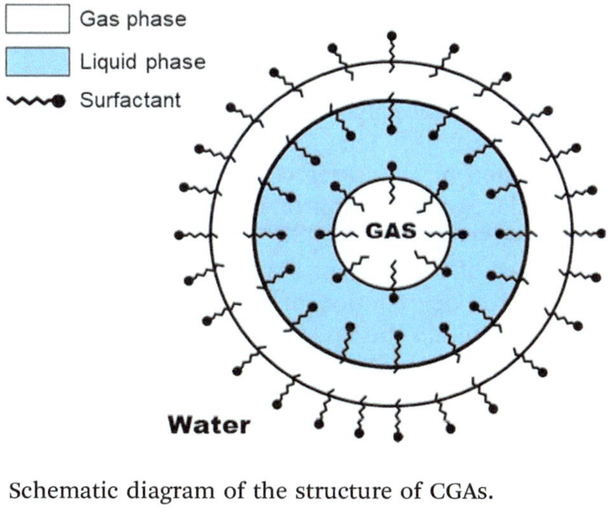

Figure 15.2 Schematic diagram of the structure of CGAs.

Among these characteristics, stability is considered the most important, as it depends on the thickness and viscosity of the encapsulating shell. Aphrons are unstable if the shell is thinner than 4 μm or thicker than 10 μm.[62] In the case of ionic surfactants, the stability of CGA dispersions is mainly due to electrostatic interactions and surface forces.

CGAs have a wide size range (10–100 μm). Large aphrons survive longer than smaller ones.[63] They exhibit the same physical phenomena that occur in foams, including liquid drainage, gas diffusion between bubbles, and Gibbs-Marangoni effects.[64] Due to the spherical shape of the microbubbles, effects like capillary drainage or plateau edge suction, which occur in polyhedral foams, are not observed.

Furthermore, due to the small size of the microbubbles of CGAs that act as colloidal particles, the buoyancy effect caused by the density difference between the gas and liquid phases is relatively limited. Another important parameter to be considered is gas retention – a measure of the air volume incorporated into the dispersion and the hold-up just before collapse[65]. Gas hold-up can be described by eqn (15.1):

$$\varepsilon = \frac{V_g}{V_{a0}} = \frac{(V_{a0} - V_{10})}{V_{a0}} \frac{(h_0 - h_{10})}{h_0} \tag{15.1}$$

where V_g is the gas volume, V_{a0} is the dispersion final volume, V_{10} is the initial volume of the liquid, h_0 is the height of the dispersion, and h_{10} is the height of the initial volume of liquid.

CGAs have been widely studied for their potential in soil remediation; for example, they can physically remove organic pollutants and heavy metals from contaminated soil, as well as carry and disperse oxygen, nutrients, bacteria, and dissolved materials for soil remediation. A schematic illustration of the use of CGAs to flush contaminated soil can be seen in Figure 15.3.

According to Tao *et al.* (2020),[66] the following specific mechanisms are involved in pollutant removal by CGAs. First, there is a phase separation of CGAs. During the transport of CGAs in the soil matrix, CGAs are separated into gaseous and liquid phases, the volatile contaminates can be removed with the gas phase, while the base/neutral pollutants are more likely to be removed by the liquid phase. Then, the displacement and dispersion of pollutants are facilitated by the lowered surface tension in the presence of CGAs. Finally, micellar solubilization or increased interfacial area leads to closer contact between remediation fluid and contaminant promoting mobilization and removal of the contaminant.[66,67]

Several studies on the degradation of organic and metallic pollutants by CGAs in soils are available in the literature. Table 15.1 provides a summary of some of these studies.

According to the cited studies, the efficiency of pollutant removal by CGAs is influenced by the following parameters: surfactant type and concentration, flushing mode, contaminant structure and concentration, and soil properties.

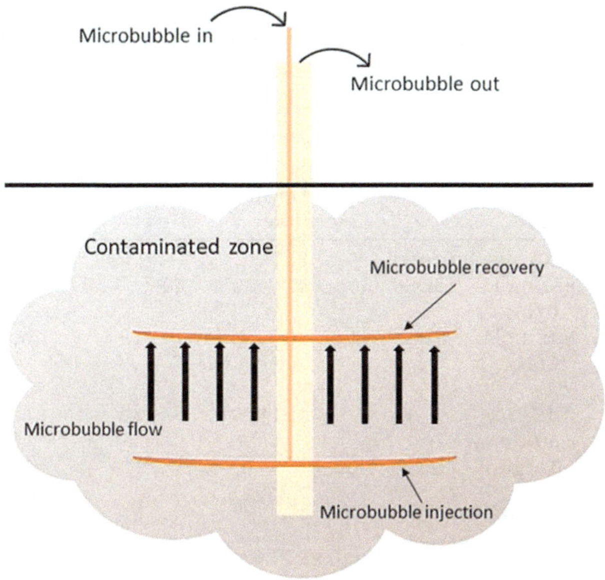

Figure 15.3 Schematic illustration of the use of CGAs to flush contaminated soil.

15.4.2 Permeable Reactive Barrier (PRB) and Colloidal Activated Carbon (CAC)

One of the methods used to help contain the spread of contaminants in the environment is the permeable reactive barrier (PRB), which can be made of different materials. A PRB creates a physical barrier downstream (after, following the direction of flow) of the contamination plume, aiming to filter the contaminants that pass through it and promote treatment through chemical and/or biological reactions, promoting actions that mitigate the contaminant load to an environmentally acceptable form.[77–80]

Different materials can be used in PRBs, such as granular activated carbon (GAC) or powdered activated carbon (PAC). Within the scope of techniques at the colloidal level, one novel material is colloidal activated carbon (CAC), which comprises activated carbon particles with diameters of 1–2 μm suspended in a solution with polymers, which can therefore be injected into the subsurface.[81]

Several studies have shown the high efficiency of CAC in treating per- and polyfluoroalkyl substances (PFAS)[82–86] – a large group of emerging anthropogenic organic contaminants.

The use of this material as a barrier was evaluated in a case study by Niarchos *et al.* (2023).[87] A barrier of CAC was constructed using direct-push technology to prevent the spread of PFAS in the groundwater and further to nearby surface water bodies. The application of the barrier resulted in an overall 52% reduction of Σ_{11}PFAS in groundwater, while reduction rates

Table 15.1 Efficiency of CGAs for pollutant removal. SDS: sodium dodecyl sulfate; SN: soapnut; CTAB: cetyltrimethylammonium bromide; RML: rhamnolipid.

Pollutant	Soil flushing agent	Percentage removal of pollutants at different flow modes: Up-flow	Down-flow	Horizontal flow	Ref.
2,4-D	CGA suspension	—	80	—	Roy *et al.* (1992)[68]
Oily waste	8 mmol L^{-1} SDS CGAs	56	47	49	Roy *et al.* (1994)[69]
Automatic transmission fluid (ATF)	8 mmol L^{-1} SDS CGAs	41	50	—	Roy *et al.* (1995)[70]
Hexachloro-benzene (HCB)	0.15, 20, 50 mmol L^{-1} Tergitol CGAs	—	1, 11, 51	—	Kommalapati *et al.* (1997)[71]
	0.5%, 1% SN solution CGAs	—	7.55, 15.1	—	Huang and Chang 2000[72]
n-Pentadecane	0.5%, 1% SN suspension CGAs	—	0.5, 1.7	—	Huang and Chang (2000)[72]
Diesel oil	SDS CGAs	—	96	—	Couto *et al.* (2009)[37]
	2.3 g L^{-1} SDS CGAs	—	62.9	—	da Rosa *et al.* (2015)[73]
	0.1, 1.0 g L^{-1} RML CGAs	—	22.1, 44.7	—	
	0.34 g L^{-1} CTAB CGAs	—	31.2	—	
Pyrene	3%, 7% BioNonex CGAs	—	53, 78	—	Boonamnuayvitaya *et al.* (2009)[74]
Petroleum	2.3 g L^{-1} SDS CGAs	—	44.26	—	da Rosa *et al.* (2015)[73]
	0.1, 1.0 g L^{-1} RML CGAs	—	46.2, 42.1	—	
	0.34 g L^{-1} CTAB CGAs	—	14.41	—	
Arsenic	1% SN solution CGAs	65.5	38	—	Mukhopadhyay *et al.* (2015)[75]
	20 mM SDS CGAs	35.4	31	—	
	20 mM SDS CGAs pH	~40	~30	—	Mukhopadhyay *et al.* (2017)[76]
	1% SN solution CGAs	70	41.3	—	
	1% SN + 100 mM phosphate CGAs	93	78	—	

reached 69% and 58% for perfluorooctanoic acid (PFOA) and perfluorooctane sulfonic acid (PFOS), respectively. The reduction was higher for long-chain PFAS and perfluorosulfonates (PFSAs) compared to other PFAS, in agreement with the results of other studies on CAC. The initial results showed a high drop in groundwater concentrations (76% for Σ_{11}PFAS); however, a rebound in concentrations was observed 6 months after treatment. The main contributing factor to the rebound was considered to be the partial bypass of the PRB. Therefore, long-term monitoring is essential to understand the effectiveness of stabilization treatment with CAC. The longevity of a CAC barrier is directly proportional to the CAC fraction applied.[86,87]

Carey *et al.* (2024)[86] evaluated the viability of a CAC remedy for a scenario involving aqueous film-forming foam (AFFF) containing PFAS. The study evaluates the benefits and limitations of placing a CAC remedy in three potential locations: (1) a source area (herein referred to as a 'grid' application), (2) a mid-plume PRB, and (3) a downgradient plume PRB near the property boundary. For this hypothetical scenario used, according to the authors, source or midplume treatment did not provide a significant benefit; the downgradient PRB had the greatest benefit concerning preventing PFAS migration beyond the site boundary. This indicates that for sites with a long distance between the source area and a downgradient property boundary, or slow PFAS of concern travel times, source control may not be beneficial when a downgradient PRB is implemented; the best performance in protecting downgradient receptors appears to be a downgradient PRB which has the most immediate benefit in reducing mass flux off-site.[86]

In addition to these applications, studies that combine other techniques also show significant potential. Sun *et al.* (2017)[88] studied a surfactant-enhanced electrokinetics (EK) coupled with a Fe/C-PRB to treat phenanthrene (PHE) and 2,4,6-trichlorophenol (TCP) co-contaminated clay soil. Five bench-scale tests were conducted to investigate the performance of EK-PRB for PHE and TCP removal from soil as well as the impact of pH and rhamnolipid concentration. The highest removal efficiency of PHE in the soil column was five times that of the control group, where only EK was applied (49.89 *versus* 9.40%). The highest removal efficiency of TCP in the soil column was 4.5 times higher than that of the control group (64.60 *versus* 14.30%). The combination of EK and a Fe/C-PRB proved to be efficient and promising for the removal of persistent organic pollutants from contaminated soil with the enhancement of rhamnolipid.[88]

Xiao *et al.* (2020)[89] used citric acid (CA) mixed with ferric chloride ($FeCl_3$) as a composite electrolyte and investigated the feasibility of electrokinetic (EK) coupled with Fe/C-PRB on the remediation of red soil artificially contaminated with U(VI). Under optimized conditions, after a 120-hour remediation experiment, the maximum removal efficiency of U(VI) was $80.58 \pm 0.99\%$.[89]

Therefore, studies using CAC and combined techniques are very promising, but there are still relatively few targeted investigations. It is a field with a lot to be explored and improved.

15.4.3 Other Techniques

Several reported techniques using colloids for the removal and remediation of drugs in soil include (i) adsorption onto various types of materials, (ii) use of magnetism, (iii) column drainage, and (iv) use of reverse micelles.

The adsorption mechanism is highly dependent on pH and the removal efficiency varies significantly depending on the type of sample.[90] It is possible to identify the optimal material for each drug according to the target pH of the sample, its composition, and other characteristics of the medium. However, the high variability of environmental conditions makes it difficult to predict adsorption interactions during treatment, which reduces the accuracy of the method. Additionally, treating large quantities of soil in this manner can be logistically challenging, costly, and inefficient.

Magnetic separation involves the adsorption of drugs onto molecules that respond to a magnetic field, followed by exposure of this complex to a magnet. The efficiency of this method, however, is dependent on the high specificity between the magnetic ligand and the target drug.[91] Therefore, the method is inefficient in removing multiple types of drugs at the same time with high efficiency. The need for exposure to a magnet also reduces the amount of soil that can be treated at one time, reducing the method's applicability for large-scale site remediation.

Drainage involves applying the sample containing the drugs to a column, followed by a separation technique. Depending on the chemical properties of the drug, the medium used, and the column, drainage is mediated by the mobility of colloids in the soil.[92]

The use of reverse micelles consists of encapsulating drugs in anionic surfactant micelles. As there are strong electrostatic and hydrophobic interactions between the micelle and the drugs, reverse micelles can be effectively employed. Micelles with a mixture of ionic and non-ionic surfactants are then used, which aggregate spontaneously in a non-polar solvent.[93] They have a significant capacity to retain water internally and can hold drugs without denaturing them. The primary advantages of this method are its high efficiency in drug removal and high selectivity. However, its applicability is limited due to the need to solubilize the soil.

Some methods, however, show greater potential as soil contamination prevention techniques and should be applied to treat water or sludge before landfilling. This is the case of nanofibers and hydrogels.

The use of nanofibers involves a new material prepared based on carbon nanofibers and carbon nanoparticles. This material offers a greater surface area, higher pore volume, and an abundant mesoporous structure. The advantage of this method is the high efficiency in drug removal and the high regenerability of the material.[94]

The use of hydrogels involves an amphiphilic biopolymer gel as a sorbent phase for solid-phase extraction. The gel is prepared in a disc format and inserted into a syringe. An aqueous sample can then be vacuum-pumped through this syringe, and the drug content is adsorbed onto the gel matrix.[95] The advantages of the method are high drug recovery, low cost, material reusability, and pH independence for the recovery of tested drugs.

Furthermore, one of the natural routes of drug degradation is direct or indirect photolysis. Natural colloidal particles in treatment systems can either increase or decrease the rate of degradation. When drugs are associated with high surface area colloids, their degradation rate is enhanced. Photochemical reactions can be accelerated in an acidic environment and slowed in an alkaline environment, with the main mechanisms of drug transformation being hydroxylation, demethylation, dehydration, and deamination.[96]

When it comes to the remediation of pesticides in the soil, the use of nanotechnology can be cited. Although this technique is widely studied for the treatment of liquid effluents, recent studies suggest its potential for the remediation of pesticides in soil as well.

Yuan *et al.* (2012)[97] used Pb/Fe nanoparticles stabilized in carboxymethyl cellulose to treat pentachlorophenol in soil. The researchers achieved up to 70% removal of the pollutant.[97]

In another study, El-Temsah and Joner (2013)[98] evaluated the effect of zero-valent iron nanoparticles with the pesticide DDT (dichlorodiphenyltrichloroethane). The studies were conducted in two types of soil (sandy and aged), varying the iron concentration (1–10 $g\,kg^{-1}$) and incubation time (up to 30 days). Generally, the addition of any concentration of nanoparticles led to the degradation of DDT, with better results for sandy soil than aged soil.[98]

Iron was also used by Zhu *et al.* (2010)[99] in the form of Cu/Fe nanoparticles to perform the dechlorination of hexachlorobenzene (HCB), a known persistent organic pollutant. The dechlorination process resulted in a less toxic product compared to the starting compound. An almost complete reduction in the pollutant was observed after 48 hours of treatment.[99]

Overall, several techniques show potential for remediating contaminated soils, with many areas still to be explored. Combining different techniques also appears to be a promising approach.

15.5 Conclusion

It is evident that colloid chemistry has significant potential for the remediation of soils contaminated with various anthropogenic pollutants. These techniques, when combined with classical treatment methods, have proven effective.

However, it appears that this potential has not yet been fully realized, with several gaps remaining to be addressed. This field is highly promising, but there are aspects needing improvement and numerous parameters yet to be evaluated.

References

1. D. L. Sparks, *Environmental Soil Chemistry*, Academic Press, Oxford, 2nd edn, 2003.
2. F. V. Molina, *Soil Colloids Properties and Ion Binding*, CRC Press, Boca Raton, 2014.
3. D. L. Johnson, S. H. Ambrose, T. J. Bassett, M. L. Bowen, D. E. Crummey, J. S. Isaacson, D. N. Johnson, P. Lamb, M. Saul and A. E. Winter-Nelson, *J. Environ. Qual.*, 1997, **26**, 581.

4. S. E. Manahan, *Environmental Chemistry*, CRC Press, Boca Raton, 8th edn, 2005.
5. R. Gothwal and T. Shashidhar, *Clean: Soil, Air, Water*, 2015, **43**, 479.
6. Y. Xing, X. Chen, J. Zhuang and X. Chen, *Ecotoxicology*, 2015, **24**, 2100.
7. H. E. Gall, S. A. Sassman, L. S. Lee and C. T. Jafvert, *Environ. Sci. Technol.*, 2011, **45**, 8597.
8. Q. Qin, X. Chen and J. Zhuang, *Crit. Rev. Environ. Sci. Technol.*, 2015, **45**, 1379.
9. O. Borgman and B. Chefetz, *Water Res.*, 2013, **47**, 3431.
10. A. B. A. Boxall, D. W. Kolpin, B. Halling-Sørensen and J. Tolls, *Environ. Sci. Technol.*, 2003, **37**, 286A.
11. E. Fabbri and S. Franzellitti, *Environ. Toxicol. Chem.*, 2016, **35**, 799.
12. J. M. Bergeron, D. Crewsa and J. A. McLachlan, *Environ. Health Perspect.*, 1994, **102**, 780.
13. S. Shao, Y. Hu, J. Cheng and Y. Chen, *Crit. Rev. Biotechnol.*, 2018, **38**, 1195.
14. R. N. Wang, Y. Zhang and Z. H. Cao, *et al.*, *Sci. Total Environ.*, 2019, **651**, 1946.
15. C. A. Edwards, *Pure Appl. Chem.*, 1975, **42**, 39.
16. P. Prashar and S. Shah, in *Sustainable Agriculture Reviews*. ed. E. Lichtfouse, Springer, Cham, 2016, vol. 8, pp. 331–361.
17. L. Chi-Chu, *J. Environ. Sci. Health, Part B*, 2010, **45**, 348.
18. N. Rascio and F. Navari-Izzo, *Plant Sci.*, 2011, **180**, 169.
19. H. I. Kankia and Y. Abdulhamid, *Arch. Appl. Sci. Res.*, 2014, **6**, 80.
20. H. Chen, Y. Teng, S. Lu, Y. Wang and J. Wang, *Sci. Total Environ.*, 2015, **512**, 143.
21. F. Muradoglu, M. Gundogdu, S. Ercisli, T. Encu, F. Balta, H. Z. E. Jaafar and M. Zia–Ul–Haq, *Biol. Res.*, 2015, **48**, 1.
22. I. Khan, A. Ghani, A. U. Rehman, S. A. Awan, A. Noreen and I. Khalid, *J. Clin. Toxicol.*, 2016, **6**, 1.
23. M. O. Ogunlade and S. O. Agbeniyi, *Afr. J. Agri. Res.*, 2011, **6**, 3725.
24. W. H. Sun, Y. X. Jiang and X. Li, *Appl. Mech. Mater.*, 2013, **295–298**, 1594.
25. Z. Atafar, A. Mesdaghinia, J. Nouri, M. Homaee, M. Yunesian, M. Ahmadimoghaddam and A. H. Mahvi, *Environ. Monit. Assess.*, 2010, **160**, 83.
26. T. M. Roane and I. L. Pepper, *Microb. Ecol.*, 1989, **38**, 358.
27. P. C. Brookes and S. P. McGrath, *Eur. J. Soil Sci.*, 1984, **35**, 341.
28. V. N. Rathnayake, M. Megharaj, N. Bolan and R. Naidu, *Int. J. Civ. Environ. Eng.*, 2010, **2**, 191.
29. N. A. Abdu, A. Abdullahi and A. Abdulkadir, *Environ. Chem. Lett.*, 2017, **15**, 65.
30. P. N. Seabra, in *Microbiologia ambiental*, Embrapa Meio Ambiente Jaguariúna, 2nd edn, 2008, vol. 24, p. 647.
31. C. M. R. Coneglian, *et al.*, *Holos Environ.*, 2006, **6**, 106.
32. A. C. L. Rizzo *et al.*, *Biorremediação de solos contaminados por petróleo: ênfase no uso de biorreatores*, CETEM/MCT, Rio de Janeiro, 2006.

33. S. R. L. Tavares, *Remediação de solos e águas contaminadas por metais pesados: conceitos básicos e fundamentos*, Clube de Autores, Joinville, 2013.
34. K. M. Scow and K. A. Hicks, *Curr. Opin. Biotechnol.*, 2005, **16**, 246.
35. S. J. Baptista, PhD thesis, UFRJ, 2007.
36. A. E. X. O. Dias, in *Resíduos Sólidos, Ambiente e Saúde: uma visão multidisciplinar*, ed. C. L. S. Sisinno and R. M. Oliveira, Editora Fiocruz, Rio de Janeiro, 2000, p. 142.
37. H. J. B. Couto, G. Massarani, E. C. Biscaia and G. L. Sant'Anna, *J. Hazard. Mater.*, 2009, **164**, 1325.
38. M. Han, G. Ji and J. Ni, *Chemosphere*, 2009, **76**, 579.
39. L. M. Whang, P. W. G. Liu, C. C. Ma and S. S. Cheng, *J. Hazard. Mater.*, 2008, **151**, 155.
40. S. S. Cameotra and P. Singh, *Int. Biodeterior. Biodegrad.*, 2008, **62**, 274.
41. H. Fu, G. Zeng, H. Zhong, X. Yuan, W. Wang, G. Huang and J. Li, *Colloids Surf., B*, 2007, **58**, 91.
42. L. M. M. Santa Anna, A. U. Soriano, A. C. Gomes, E. P. Menezes, M. L. E. Gutarra, D. M. G. Freire and N. Pereira Jr, *J. Chem. Technol. Biotechnol.*, 2007, **82**, 687.
43. K. Urum, T. Pekdemir and M. Gopur, *Process Saf. Environ. Prot.*, 2003, **81**, 203.
44. F. Sebba, *Foams and Biliquid Foams – Aphrons*, John Wiley & Sons, New York, 1987.
45. F. Sebba, *Sep. Purif. Methods*, 1985, **14**, 127.
46. N. Ahmed and G. J. Jameson, *Int. J. Miner. Process.*, 1985, **14**, 195.
47. D. R. Ketkar, R. Mallikarjunan and S. Venkatachalam, *Int. J. Miner. Process.*, 1991, **31**, 127.
48. S. E. Burns, S. Yiacoumi and C. Tsouris, *Sep. Purif. Technol.*, 1997, **11**, 221.
49. S. E. Rijk, J. H. J. Graaf and J. G. Blanken, *Water Res.*, 1994, **28**, 465.
50. J. P. Malley and J. K. Edzwald, *J. - Am. Water Works Assoc.*, 1991, **83**, 56.
51. S. Huang, H. Ho, Y. Li and C. Lin, *Environ. Sci. Technol.*, 1995, **29**, 1802.
52. K. J. Ptasinski, F. L. S. Geurts, A. J. P. M. Staring, E. J. M. V. Heesch and P. J. A. M. Kerkhof, *Sep. Sci. Technol.*, 1995, **30**, 2127.
53. W. T. Shin, S. Yiacoumi and C. Tsouris, *Ind. Eng. Chem. Res.*, 1997, **36**, 3647.
54. M. Sadatomi, A. Kawahara, K. Kano and A. Ohtomo, *Exp. Therm. Fluid Sci.*, 2005, **29**, 615.
55. M. Kukizaki and M. Goto, *J. Membr. Sci.*, 2006, **281**, 386.
56. Q. Xu, M. Nakajima, S. Ichikawa, N. Nakamura and T. Shiina, *Innovative Food Sci. Emerging Technol.*, 2008, **9**, 489.
57. J.-Y. Park, Y.-J. Choi, S. Moon, D.-Y. Shin and K. Nam, *J. Hazard. Mater.*, 2009, **163**, 761.
58. H. Nakamura, A. Hamasaki, M. Tsuji and E. Goto, *Technical Report Mitsubishi Heavy Industries*, 2005, 166.
59. K. Okada, M. Shimizu, T. Isobe, Y. Kameshima, M. Sakai and A. Nakajima, *et al.*, *J. Eur. Ceram. Soc.*, 2010, **30**, 1245.

60. K. Okada, S. Uchiyama, T. Isobe, Y. Kameshima, A. Nakajima and T. Kurata, *J. Eur. Ceram. Soc.*, 2009, **29**, 2491.
61. M. D. Bredwell and R. M. Worden, *Biotechnol. Prog.*, 1998, **14**, 31.
62. C. D. Ivan, J. L. Quintana and L. D. Blake, *Presented in Part at Annual Technical Conference and Exhibition SPE*, New Orleans, Louisiana, 2001.
63. A. Belkin, M. Irving, R. O'Connor, M. Fosdick, T. Hoff and F. B. Growcock, *Presented in part at SPE Annual Technical Conference and Exhibition*, Dallas, Texas, 2005.
64. A. Bhakta and E. Ruckenstein, *Adv. Colloid Interface Sci.*, 1997, **70**, 124.
65. P. Jauregi and J. Varley, *Trends Biotechnol.*, 1999, **17**, 389.
66. W. Tao, C. Mei and N. Hamzah, *J. Contam. Hydrol.*, 2020, **231**, 103620.
67. J. Kilbane, B. Liu, J. Conrad and V. Srivastava, *Topical Report, April 1992–September 1993*, Institute of gas Technology, Des Plaines, IL, United States, 1997.
68. D. Roy, A. Tamyo and K. T. Valsaraj, *Sep. Sci. Technol.*, 1992, **27**, 1555.
69. D. Roy, K. Valsaraj, W. Constant and M. Darji, *J. Hazard. Mater.*, 1994, **38**, 127.
70. D. Roy, R. R. Kommalapati, K. T. Valsaraj and W. D. Constant, *Water Res.*, 1995, **29**, 589.
71. R. R. Kommalapati, K. T. Valsaraj, W. D. Constant and D. Roy, *Water Res.*, 1997, **31**, 2161.
72. C. W. Huang and C. H. Chang, *Colloids Surf., A*, 2000, **173**, 171.
73. C. F. C. da Rosa, D. M. G. Freire and H. C. Ferraz, *J. Environ. Chem. Eng.*, 2015, **3**, 89.
74. V. Boonamnuayvitaya, P. Jutaporn, S. Sae-ung and S. Jarudilokkul, *Sep. Purif. Technol.*, 2009, **68**, 411.
75. S. Mukhopadhyay, S. Mukherjee, M. A. Hashim and B. Sen Gupta, *Chemosphere*, 2015, **119**, 355.
76. S. Mukhopadhyay, S. Mukherjee and M. A. Hashim, *Bull. Environ. Contam. Toxicol.*, 2017, **98**, 366.
77. C. H. R. Costa, MSc dissertation, Universidade do Vale do Rio dos Sinos - UNISINOS, 2006.
78. J. R. Craig, A. J. Rabideau and R. Suribhatlaet, *Adv. Water Resour.*, 2006, **29**, 99.
79. K. U. Mayer, S. G. Benner and D. W. Bloweset, *J. Contam. Hydrol.*, 2006, **85**, 195.
80. F. Ahmad, P. S. Schnitker and C. J. Newellet, *J. Contam. Hydrol.*, 2007, **90**, 1.
81. K. Mackenzie, A. Schierz, A. Georgi and F. D. Kopinke, *Global NEST J.*, 2008, **10**, 54.
82. G. Niarchos, A. Lutz, D. B. Kleja and F. Fagerlund, *Environ. Pollut.*, 2022, **308**, 119667.
83. M. Sorengard, D. B. Kleja and L. Ahrens, *J. Environ. Manage.*, 2019, **249**, 109345.
84. R. McGregor, *Biorem. J.*, 2020, **30**, 39.
85. R. McGregor and L. Benevenuto, *Front. Environ. Chem.*, 2021, **2**, 729779.

86. G. R. Carey, *et al.*, *Remediation*, 2024, **34**, 21772.
87. G. Niarchos, *et al.*, *Remediation*, 2023, **33**, 101.
88. Y. Sun, K. Gao and Y. Zhang, *et al.*, *Environ. Sci. Pollut. Res.*, 2017, **24**, 28142.
89. J. Xiao, Z. Pang and S. Zhou, *et al.*, *Sep. Purif. Technol.*, 2020, **244**, DOI: 10.1016/j.seppur.2020.116667.
90. S. S. Priya and K. V. A. Radha, *Chem. Eng. Commun.*, 2017, **204**, 821.
91. I. Ihara, K. Toyoda, N. Beneragama and K. Umetsu, *J. Phys.: Conf. Ser.*, 2009, **156**, 12034.
92. S.-C. Kim, J. E. Yang, Y.-S. Ok and K. Carlson, *Water Qual. Res. J.*, 2010, **45**, 275.
93. S. H. Mohd-Setapar, S. N. Mohamad-Aziz and C. S. Chuong, *et al.*, *Chem. Eng. Commun.*, 2014, **201**, 1664.
94. S. Li, Y. Zhang and Q. You, *et al.*, *Colloids Surf., A*, 2018, **558**, 392.
95. N. M. F. M. Sampaio, N. D. B. Castilhos, B. C. da Silva, I. C. Riegel-Vidotti and B. J. G. Silva, *Molecules*, 2018, **24**, 40.
96. F. Liu, X. Liu and S. Zhao, *et al.*, *Chemosphere*, 2019, **235**, 867.
97. S. Yuan, H. Long, W. Xie, P. Liao and M. Tong, *Geoderma*, 2012, **185–186**, 18.
98. Y. S. El-Temsah and E. J. Joner, *Chemosphere*, 2013, **92**, 131.
99. N. Zhu, H. Luan, S. Yuan, J. Chen, X. Wu and L. Wang, *J. Hazard. Mater.*, 2010, **176**, 1101.

CHAPTER 16

Aspects of Colloidal Technologies in Soil Nutrition

S. BIANCHI,[a,b] R. R. DOS SANTOS,[a,b] C. F. SOUZA[c] AND R. FAEZ*[a]

[a] Laboratory of Polymeric Materials and Biosorbents, Federal University of São Carlos, UFSCar-CCA, Araras, SP, Brazil; [b] Graduate Program in Materials Science and Engineering, University of Sao Paulo, USP-FZEA, Pirassununga, SP, Brazil; [c] Research Group on Water, Soil and Environment Engineering, Federal University of São Carlos, UFSCar-CCA, Araras, SP, Brazil
*Email: faez@ufscar.br

16.1 Fertilizers: Significance and Challenges

Agrochemicals in agriculture represent a revolution by enabling more efficient land use, expanding food production in response to population growth, and shrinking agricultural space. Agrochemicals encompass pesticides, regulators, chemical fertilizers, soil conditioners, and substances used in animal husbandry.[1] Fertilizers, for instance, are inorganic materials with a defined composition used to supply macro and micronutrients essential for plant growth.[1] Macronutrients like nitrogen (N), phosphorus (P), potassium (K), calcium (Ca), magnesium (Mg), and sulfur (S) are required in larger quantities and micronutrients such as boron (B), chlorine (Cl), copper (Cu), iron (Fe), manganese (Mn), molybdenum (Mo), and zinc (Zn) are crucial in smaller amounts but equally vital. Fertilizer formulation combines these nutrients in specific ratios tailored to each crop's nutritional needs.[2] However, unrestrained use of chemical fertilizers has become an environmental threat, prioritizing short-term gains over long-term consequences. According to the Intergovernmental Panel on Climate Change, agriculture accounts for

Contribution of Colloidal Materials to Air, Water and Soil Environmental Sustainability: From Basic Concepts to Advanced Applications
Edited by Izabel C. Riegel-Vidotti, Lizandra Maria Zimmermann and Heloise R. de Barros

Published by the Royal Society of Chemistry, www.rsc.org

approximately 24% of greenhouse gas emissions, with nitrogen fertilizers being significant contributors, notably through nitrous oxide (N_2O) emissions.[3] Environmental issues arising from fertilizer use, particularly nitrate (NO_3^-) leaching, are significant concerns due to high soil mobility and easy leaching, which reduce soil fertility and nutrient efficiency while threatening the environment and human health.[4] Aquifer contamination due to this leaching can render them unusable for decades, with contaminants migrating to rivers and lakes *via* the hydrological cycle.[5] It is not just a matter of choice but a necessity to optimize fertilizer management practices since 30 to 50% of applied fertilizer is lost.[6] Therefore, there is a pressing need to rethink fertilizer management in agriculture, aiming to enhance efficiency while minimizing environmental impacts. Nevertheless, understanding these chemicals' lifecycle and environmental impacts is essential to mitigate adverse effects such as bioaccumulation and ecosystem contamination.[3] Faced with these challenges, the search for alternative methods of managing nutrients in the soil has become essential. In recent years, fertilizer delivery systems have become a promising solution to improve various aspects of fertilization. These inputs are applied to increase agricultural productivity and reduce production costs while contributing to environmental conservation.

Figure 16.1 represents the impact of fertilization in an environment using conventional and environmentally friendly fertilizers (EFF), highlighting the risks and advantages of each approach. The figure illustrates several critical issues associated with the use of conventional fertilizers. These include problems such as non-renewable resource depletion, water contamination, nutrient leaching, and the emission of toxic nitrogen gases. Conventional fertilizers often rely on synthetic chemicals that, while effective in providing immediate nutrient availability, contribute to long-term environmental harm. For example, nutrient leaching can contaminate nearby streams and rivers, causing eutrophication and harming aquatic ecosystems. Additionally, the volatilization of nitrogen from these fertilizers can release greenhouse gases, contributing to air pollution and climate change. In contrast, using EFF offers several advantages, as shown in Figure 16.1. These formulations are designed to improve soil organic matter and promote a healthier soil biota, creating a more sustainable and productive agricultural environment. By utilizing natural sources of nutrients, these fertilizers often require fewer applications, reducing the overall input costs for farmers. Furthermore, environmentally friendly fertilizers are formulated to enhance plants' nutrient absorption, leading to higher efficiency and reduced waste. This supports plant growth and minimizes the environmental impact of farming practices. Using such fertilizers can significantly contribute to the soil's long-term health, ensuring its fertility for future generations.

16.2 Fertilization Process

Fertilization is the process of adding nutrients to the soil or directly to plants to promote their growth and increase productivity. This process can involve

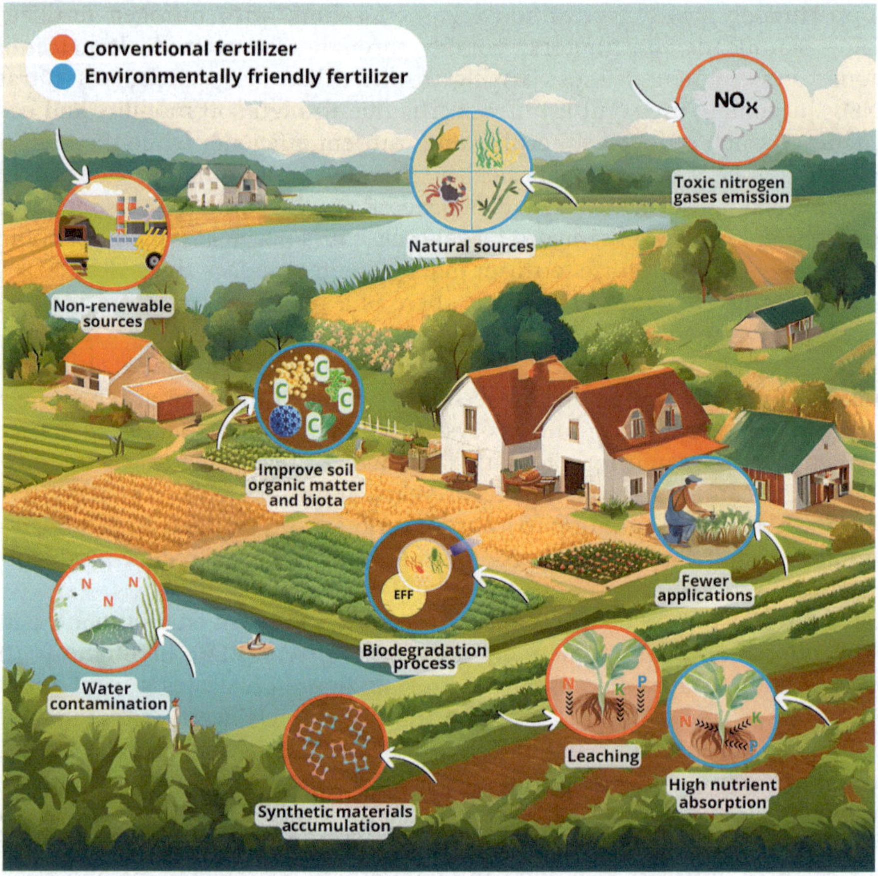

Figure 16.1 Impacts of conventional and environmentally friendly fertilizer.

applying organic fertilizers, such as compost and manure, or mineral fertilizers, which are chemical products formulated to provide specific nutrients. In plant fertilization, nutrients can be applied directly through soil and foliar fertilization. Soil fertilization involves applying fertilizers to the plant's root zone, while foliar fertilization applies nutrients directly to the leaves. Both methods aim to ensure that plants receive the essential nutrients for healthy and productive development.

Although "fertilization" is often used synonymously with "manuring" in agriculture, it is broader and refers to different processes of providing plant nutrients. Manuring, on the other hand, refers to adding manures or fertilizers to the soil or plants to ensure the supply of necessary nutrients for plant growth. This includes applying organic fertilizers (such as compost and manure) and mineral fertilizers (such as chemical fertilizers). The primary goal of manuring is to improve soil fertility and ensure that plants receive essential nutrients, including macronutrients (N, P, K, Ca, Mg and S) and micronutrients (Fe, Mn, Zn, Cu, B, Mo and Cl). Manuring should be conducted based on the plants' specific needs and the soil characteristics to ensure optimal and sustainable growth.

There are different types of manuring: organic manuring uses organic materials, such as compost and manure, to gradually improve soil structure and provide nutrients. Green manuring involves growing plants that, when decomposed, enrich the soil with nutrients. These plants, known as "green manures," can improve soil fertility and increase water-holding capacity. Additionally, foliar fertilization applies nutrients directly to the leaves of the plants and is an effective way to supply nutrients when plants have difficulty absorbing them from the soil. On the other hand, mineral manuring involves using chemical fertilizers that contain specific nutrients in concentrated forms, which can be either fast-release, such as water-soluble fertilizers, or slow-release.

Fertilizer application systems vary widely between crops and agricultural practices, reflecting diverse global traditions and technologies. Two main examples of fertilizer application systems are conventional and modern advanced systems. Conventional manuring refers to the traditional method of applying fertilizers to the soil to provide essential nutrients to plants. This method is characterized by its simplicity and widespread use in agriculture and gardening. In conventional manuring, fertilizers are spread evenly over the soil surface and then incorporated into the soil using methods such as plowing or harrowing. This improves nutrient solubilization and facilitates absorption by plant roots.

The advantages of conventional manuring include ease of implementation and relatively low cost. It is a well-established method suitable for a variety of crops and soil conditions. However, if not properly managed, it may have disadvantages such as nutrient losses due to volatilization, surface runoff, and leaching. Excessive application of fertilizers can also cause environmental impacts, such as water pollution from eutrophication. Despite these issues, conventional manuring remains a common practice due to its simplicity and effectiveness, often complemented by modern methods to enhance efficiency and minimize environmental impacts.

Examples of modern advanced systems include fertigation and enhanced efficiency fertilizers (EEFs). The production, application, advantages, and terminology of EEFs will be covered in this chapter. Fertigation is a technique that integrates irrigation with fertilization by dissolving fertilizers in irrigation water and delivering them directly to plants *via* an irrigation system. This method enhances nutrient delivery efficiency by targeting the root zone, reducing waste, and improving fertilizer use efficiency. Key advantages include precision in nutrient application, minimizing nutrient losses through volatilization, runoff, and leaching, and contributing to water conservation by using localized irrigation methods like drip irrigation. Fertigation is particularly beneficial in controlled environments such as greenhouses and high-tech agriculture, as well as in water-scarce regions where maximizing productivity is essential. By combining water and nutrients into a single process, it simplifies management and optimizes plant growth throughout different developmental stages.

In summary, fertilization plays a crucial role in agriculture by ensuring that plants receive the necessary nutrients for healthy growth and high productivity. From applying organic and mineral fertilizers to advanced methods such as fertigation and EEFs, each technique offers specific benefits that can be tailored to the needs of the plants and soil conditions. While manuring is fundamental for improving soil fertility and ensuring a continuous supply of nutrients, the choice of application method can significantly impact efficiency, sustainability, and cost. Integrating traditional and modern methods allows for optimizing fertilization practices, ensuring crop health, and long-term productivity. Therefore, the continuous evolution of techniques and adaptation to the specific needs of plants and the environment are essential for achieving more efficient and sustainable cultivation.

16.3 EEFs: A Modern Advanced Fertilizer System

16.3.1 Key Terminologies

An EEF is a system that reduces the use of agrochemicals, providing a level of control that minimizes environmental and economic impacts. Figure 16.2 illustrates the performance of different types of EEFs: controlled-release, slow-release, and environmentally friendly fertilizers. In a slow-release fertilizer (SRF) system, nutrients are gradually released more slowly than in common fertilizers. However, the release rate, pattern, and duration could be better controlled and are influenced by usage conditions and soil type. Conversely, controlled-release fertilizers (CRF) have settings that determine

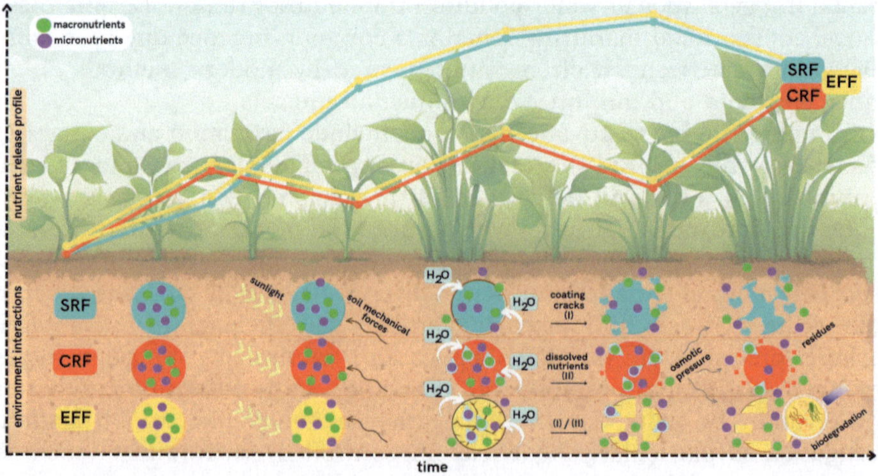

Figure 16.2 Performance of different delivery fertilizer systems.

the rate and release pattern, which can be managed during use.[7–9] To qualify as an EFF, the materials used for coating, encapsulating, or dispersing the fertilizer must degrade without harming the environment after nutrient release. Although slow and controlled-release fertilizers do not necessarily use biodegradable polymers, the most effective EEF systems combine slow or controlled-release with sustainability considerations, providing a reassuring level of predictability in sustainable agriculture.

16.3.2 Advantages and Disadvantages

EEFs offer a range of advantages in agriculture, economics, social considerations, and environmental sustainability. These materials provide a steady and continuous supply of nutrients to plants, reducing the frequency of soil applications. This cuts labor costs and minimizes nutrient loss through leaching, volatilization, and immobilization. By mitigating root damage caused by high salt concentrations and offering easier handling, EEFs improve ecological health. Their potential to significantly reduce environmental pollution is a beacon of hope for the future of agriculture.

However, despite these benefits, EEFs are tightly regulated due to their higher production costs compared to conventional agrochemicals, besides some notable disadvantages that need to be addressed. For instance, EEFs are prone to releasing nutrients prematurely in what is known as the burst effect, which is a critical issue that needs immediate attention to prevent potential damage to plants or failure to release nutrients during periods of high crop demand. Moreover, residues of synthetic materials left in the soil can alter soil acidity and contribute to microplastic accumulation, presenting additional challenges.

Another significant challenge with EEFs is the need for standardized methods to validate nutrient release profiles from laboratory scale to field conditions, a crucial aspect that cannot be overlooked. Equally important are the difficulties in measuring their impacts on soil health. Addressing this issue is paramount to ensuring the effective and safe use of EEFs in agriculture. Various simulations and experimental tests have been employed to model release mechanisms, with factors like diffusion coefficients and coat thickness (governed by Fick's first law of diffusion) influencing nutrient release rates.

In these challenges, EEFs show promise in advancing agricultural management. Brands like Nutricote®, Osmocote®, Polyon®, and Basacote® are prominent in the agricultural market, playing a significant role in the adoption and use of EEFs. These materials are formulated using non-biodegradable synthetic polymers such as polyolefins, poly(vinylidene chloride), thermoset dicyclopentadiene copolymer, polyurethane, and acrylic emulsions.[10] Then, for fertilizer delivery systems to significantly improve nutrient efficiency and reduce environmental impacts, it is essential to address various regulatory, technical, and operational challenges. Overcoming these challenges is crucial for maximizing the benefits of sustainable agriculture.

16.3.3 Types of Materials Used to Prepare EEFs

The latest literature categorizes materials for producing EEFs into three main groups.[11–13] The first group utilizes organic compounds such as urea-formaldehyde, isobutylene-diurea, and urea acetaldehyde/urea cycle,[11–13] while the second group incorporates inorganic compounds known for their low solubility, such as metallic ammonium phosphates,[14–16] to enhance fertilizer efficiency. The third group consists of encapsulated fertilizers shielded by polymeric matrices, sulfur, and minerals, serving as carriers for active ingredients.[17–19] Despite this classification, since the term SRF was introduced in the 1960s, there has been ongoing debate regarding categorizing this type of material. Classifying a fertilizer as SRF is straightforward: according to the European Standardization Committee (CEN), a fertilizer qualifies as slow-release if it meets specific criteria.[20] For instance, it should not release more than 15% of its content within 24 h and no more than 75% within 28 days. However, categorizing the various types of SRF remains challenging for researchers. Some classifications consider the matrix-nutrient interaction,[21] while others focus on the mode of dissolution release[11] or the production process,[22] among other factors.

Regardless of the classification used, the development of SRF production has primarily centered on the type of material employed as a matrix or coating and the methods used to obtain it, as illustrated in Figure 16.3. This figure presents the evolution of materials used as matrices or coating to produce EEFs over time, showing materials for each decade from 1960 until 2020. It highlights the transition from simple organic and inorganic compounds to more complex polymeric and mineral-based encapsulations. Notably, natural polymers began to be used in the 1990s, and in recent years, there has been a preference for biodegradable materials, reflecting the growing emphasis on sustainability. EEFs offer advantages in terms of cost-effectiveness

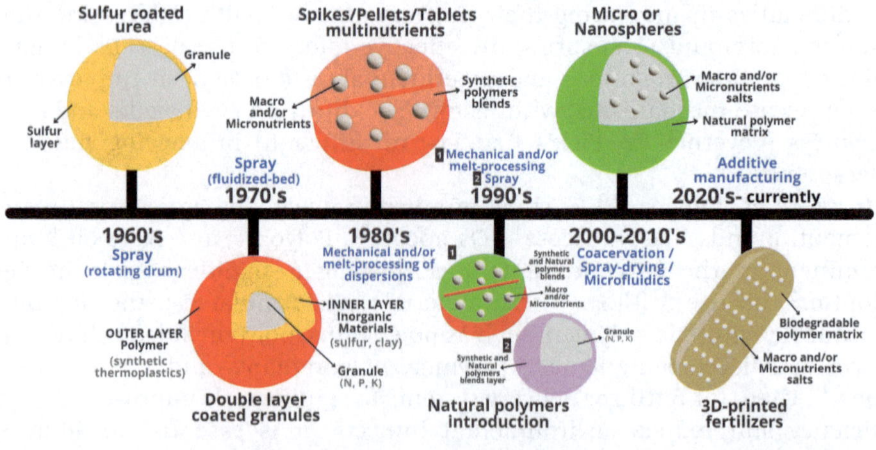

Figure 16.3 Evolution of materials used as matrices or coatings and methods for obtaining EEFs over time.

and practicality during field application and reduce environmental damage.[12,23] Despite their market prevalence, EEFs that rely on sulfur or synthetic polymers have inherent disadvantages. The extensive application of sulfur-based materials can increase soil acidity,[24] whereas the synthetic polymers used as barriers are largely non-biodegradable.[10,25] This has raised concerns within the scientific community about the potential accumulation of these substances in the soil over multiple crop cycles, leading to risks such as microplastic contamination.[26] Therefore, the need for continuous research and development to refine materials and formulations that balance effectiveness and environmental sustainability is urgent and important. The shift towards biodegradable and biomass polymer-based EEFs has great promise for fostering sustainable agricultural practices in the future. This transition is justified by the need to mitigate the environmental impact of traditional fertilizers, ensuring that agricultural advancements are in harmony with ecological preservation. Additionally, using biodegradable polymers can improve soil biota, enhancing soil health and fertility by supporting a more diverse and active microbial community.[27]

16.4 Polymers: Unlocking the Potential for Enhanced Nutrient Efficiency

Polymers can be natural, synthetic, biodegradable, or not, and these characteristics favor different applications in the vast field of applications they have. When covering fertilizers, some factors are essential, such as cost, being hydrophobic to better protect the fertilizer, and biodegradable so as not to generate waste after use. However, these characteristics cannot be found in a single material, as generally, more hydrophobic polymers have low biodegradation potential. Thus, a balance must be found between the desired properties and the physicochemical characteristics of the polymers.

Polysaccharides, such as starch, alginates, and chitosan, are natural polymers that have advantages over others since they have a lower cost, high availability, low toxicity, and, mainly, biodegradability.[28] Another class that possesses these properties and is already utilized for nutrient encapsulation is lignocellulosic materials, the most abundant natural polymers on Earth.[29] In addition to those of natural origin, the use of synthetic polymers that have biodegradability potential – such as poly(lactic acid) (PLA) and poly(vinyl alcohol) (PVA) – has also been included in EEF formulations. PVA is an example of a semi-crystalline synthetic polymer capable of biodegradation and suitable for melt processing, with potential for use as a matrix in EEFs.[30]

Based on the properties mentioned, EEFs derived from these polymers demonstrate promising features, including widespread availability, diverse chemical compositions (such as the complex structure of lignocellulosic materials or the simple structure of chitosan), potential biodegradability, and the ability to adapt to various production techniques.[31] Subsequently, selecting the polymer hinges on accessible features and desired material properties, such as sensitivity to pH and temperature and the presence of functional groups for nutrient interaction.[32] The choice of polymer is linked to the method used to obtain the final product, that is, during the processing stage. Combining the matrix and the nutrient through a feasible method compatible with the materials used and the desired objectives is necessary to obtain the EEF. The combination type will directly affect the nutrient release profile, so choosing a technique in which the matrix involves, disperses, or encapsulates the highly soluble nutrient is extremely important.[33] Coating,[34,35] coacervation,[36] melt-processing,[37,38] and spray-drying [39,40] are some of the techniques used, alone or together, to obtain EEFs in the form of gel, film, tablet, microparticles (formed by microspheres or microcapsules), among others.

The motivation for choosing the ideal polymer and obtaining method is to produce an EEF that enhances agrochemical efficiency and sustains a waste-free life cycle. One of the main steps to verify whether this objective has been achieved is evaluating the materials within the water–soil–plant system. Potential biodegradable polymers play a crucial role in this system, as they regulate the release of nutrients, ensuring optimal plant uptake and minimizing environmental impact.

16.5 Interactions Between Nutrients and Soil

Developing EEFs necessitates a comprehensive understanding of their nutrient release performance within soil or soil-plant systems, which remains one of the primary challenges in this field. This challenge is compounded by the multiple variables inherent in these systems, such as soil type, climate, geographical zone, experiment size (statistically representative), presence or absence of crops, and the choice between destructive or non-destructive testing techniques. Besides, soil-specific interactions between particles and nutrient ions, influenced by various physical and chemical factors, play a significant role in nutrient mobility. Physical characteristics include pore size and saturation level,[41] while chemical factors encompass soil solution concentration, pH, cation exchange capacity, and dissolution/precipitation reactions.[42,43] Soil release tests are crucial for determining the efficiency of EEF materials, as water tests cannot replicate the conditions that influence fertilizer release in real-world applications. Additionally, the lack of information on the dynamics of the water–soil–solute relationship has hindered the widespread adoption of EEFs, preventing the full realization of their potential.

Time-domain reflectometry (TDR) is a technique for quickly and accurately determining nutrient release profiles in soil or soil-plant systems.[44] Electromagnetic determination of soil water content: measurements in a coaxial transmission line.[45] TDR operates by emitting high-frequency electromagnetic waves through probes inserted into the soil, enabling indirect estimation of soil water content and electrical conductivity.[46,47] This allows for real-time, non-destructive monitoring of nutrient release, offering high repeatability and the ability to track the movement of water and ions in the soil.

Therefore, understanding the nutrient dynamics of EEFs in soil and plants is essential for their appropriate utilization. By addressing these challenges through advanced techniques and comprehensive soil release testing, we can optimize EEF formulations to enhance agricultural productivity while minimizing environmental impact.

16.5.1 Case Study 1: Interpretation of Nutrient Release from Ionic Conductivity and Soil-saturated Solution

Chiaregato and Faez (2021) evaluated the release kinetics and mobility in the soil of EEFs based on PVA/macro and PVA/starch/micro using the TDR techniques.[48] The authors used probes arranged in the soil container's center, side, and bottom to measure conductivity, Figure 16.4. In addition to the results obtained by the probes, solutions were periodically extracted from the containers to measure the concentration of specific ions over time. Figure 16.4 shows the conductivity and concentration of macro (A) and micronutrient (B) ions as a function of time. Figure 16.4(A) shows a significantly faster initial rise in conductivity measured by the central probe compared to the others, which is explained by the material positioned in the center of the container. As conductivity increases proportionally with ion concentration, this central rise suggests a higher initial concentration of

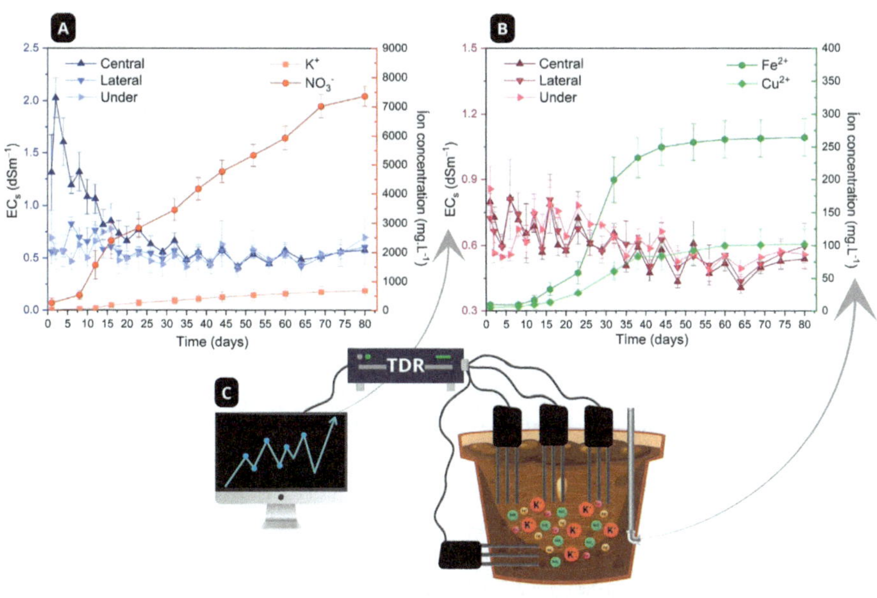

Figure 16.4 Electrical conductivity and ion concentration of (A) macro and (B) micronutrients from EEFs based on PVA/macro and PVA/starch/micro in the soil for 80 days, and (C) TDR probes and signal illustration.

ions around the central probe. Around day 12, an increase in NO_3^- concentration was observed, which likely caused the rise in conductivity measured by the central probe. This difference in behavior between macronutrients, potassium ion (K^+) and NO_3^-, highlights the impact of soil mobility. Since soil particles are negatively charged, NO_3^- interacts less electrostatically, leading to easier leaching. This explains the low K^+ concentration throughout the test, indicating stronger soil interaction and retention.

Micronutrients displayed consistently lower electrical conductivity values throughout the experiment, suggesting lower availability than macronutrients (Figure 16.4(B)). Their lower measured concentrations further support this. The starch coating on the PVA matrix may have hindered ion release. Figure 16.4(B) also reveals a lower Cu concentration than Fe throughout the test. Santos and Silva (2010) reported a decreasing order of micronutrient interaction with soil: Cu > Zn > Ni > Fe > Mn.[49] This is based on each ion's properties, particularly hydration radius and valence. Given the stronger interaction of Cu^{2+} with the soil, its lower concentration than Fe^{2+} is consistent. This finding aligns with the lower Cu^{2+} concentration in the extracted soil solution. In conclusion, this analysis demonstrates the varying release and movement of different nutrients in soil. It highlights the influence of coatings and electrostatic interactions on ion release and mobility dynamics.

16.5.2 Case Study 2: Interpretation of Nutrient Release from Biodegradation Aspects

According to França *et al.* (2021), the relationship between ionic conductivity and biodegradation can help comprehend nutrient release and ion mobility.[50] Figure 16.5 shows the nutrient release data in soil, the CO_2 values released during biodegradation, and the SEM images of the material before and after the biodegradation evaluation of EEFs based on poly(3-hydroxybutyrate) (PHB), thermoplastic starch (TPS), cellulose and potassium nitrate (KNO_3). EEFs were placed near the central probes to evaluate nutrient release (mobility of K^+ and NO_3^- ions). Initially, there was little ionic mobility when the ions were close to the central probes, but a peak near the 3rd day indicates ion release. Then, a slowdown in the central probe and a peak on the 15th day indicate ionic mobility as the ions moved toward the lateral probes. The first CO_2 evolution peak on the 7th day suggests biodegradation. This behavior indicates that the process begins with nutrient release and is subsequently followed by polymer degradation.

Additionally, when the conductivity of the central probe stabilized, biodegradation remained at higher levels compared to the earlier period. Furthermore, the conductivity of the lateral probes stabilized at higher values than the central probe after the 50th day, indicating increased ion mobility. The mechanism appeared cyclical, as biodegradation rates fluctuated, and conductivity levels stabilized on the 15th, 45th, and 72nd days, approximately every 30 days.

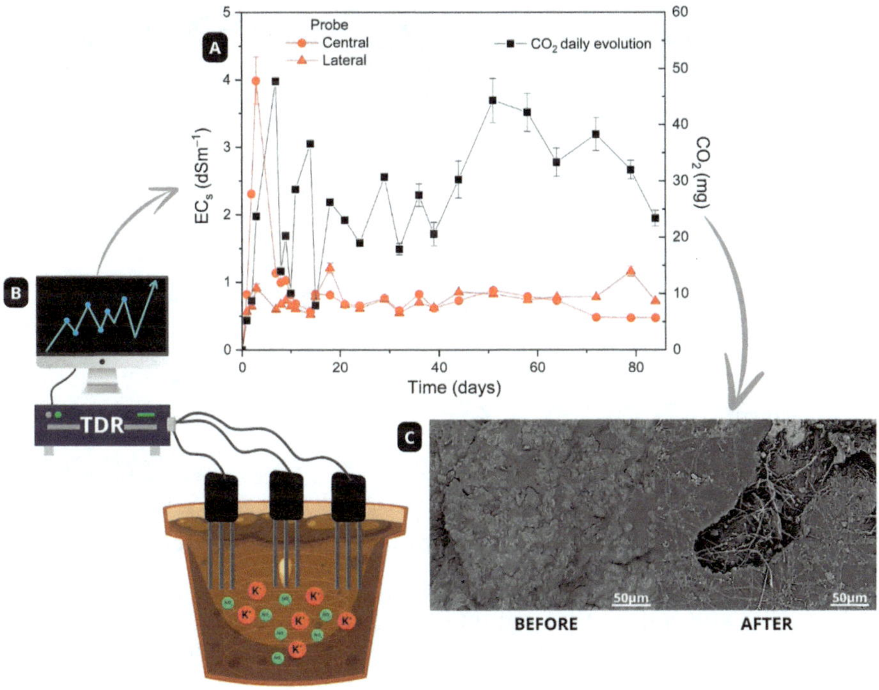

Figure 16.5 (A) Ion nutrient mobility into the soil during the release test and CO_2 evolution, (B) TDR probes and signal illustration, and (C) SEM image of PHB.TPS/CNF.KNO_3 before and after the biodegradation set.

16.5.3 Case Study 3: Dynamics of Nutrients from EEF in Soil

Faez *et al.* (2024) evaluated the water–soil dynamics of EEF fertilizer by analyzing the performance of polymer-coated KNO_3 under drip irrigation in sandy soil conditions.[51] Using the sensors in different layers and the TDR technique, a method for monitoring nutrient release, Figure 16.6, the authors observed that nutrient release initiates only after the fourth week. This delay is due to the permeation of water into the polymer, which changes the osmotic pressure of the system. The resulting pressure difference between the interior of the polymer and the environment causes the release of nutrients. In sandy soils characterized by a low cation exchange capacity (CEC),[52] there is less interaction between nutrients and soil colloids, facilitating nutrient loss through processes like leaching.[53] The properties of the ions, particularly hydration radius and valence, also influence their mobility, as they can enhance interaction with soil particles.[54,55] In this context, K^+ is more attracted to soil colloids than NO_3^- due to its smaller ionic radius. K^+ is the cation most vulnerable to leaching in soils with low CEC and good drainage, such as sandy soils.[56] Regarding valence, opposite charges repel each other, explaining the mobility of NO_3^- in the soil solution as the CEC repels its negative charge.

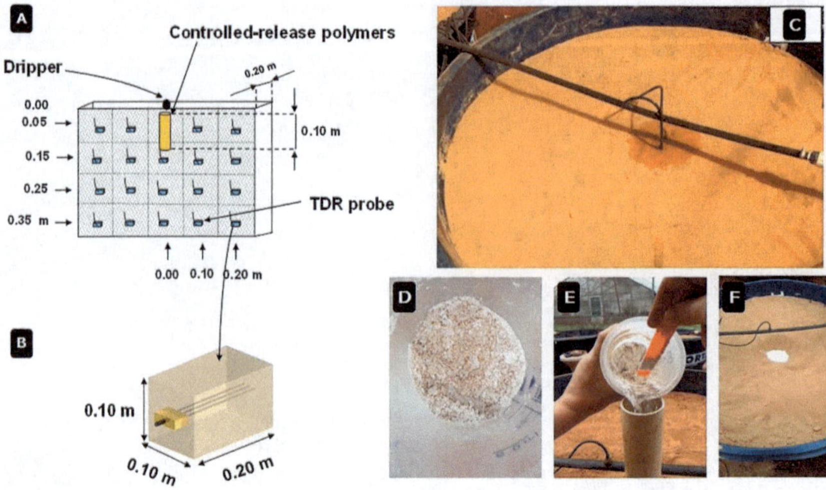

Figure 16.6 (A) Layout of the TDR probes inside the soil boxes, (B) soil volume monitored by a single probe, (C) installation of the irrigation system with detail for the line positioning, (D) detail of the polymers mixed with soil, and (E) and (F) method of inserting the polymers into the containers. Adapted from ref. 52 with permission from the Author, Copyright 2024.

16.6 Mechanism Models of Nutrient Release

Mathematical models are essential in formulating, assessing nutrient release processes, and enhancing new system designs. However, experimental validation greatly enhances their reliability, which provides a detailed account of nutrient release over time.[57] Developing a mathematical model is complex because they are designed to represent specific aspects of reality, particularly the factors that govern release kinetics.[58] Processes like dissolution, diffusion, segmentation, osmosis, swelling, and erosion can control nutrient release from modified systems.[59] The structure of the diffusion path significantly affects the diffusion of the active agent, with many models predicting diffusion equations. Hydrophilic agents generally diffuse more easily through a matrix than hydrophobic ones. In polymer matrices, hydrophobic agents often require swelling or matrix erosion to release, and their behavior varies with solubility factors.[58] Some models have been used to evaluate the experimental nutrient release data: zero-order, first-order, Higuchi, Korsmeyer–Peppas, Peppas–Sahlin, Hixson–Crowell, and Hopfenberg (Table 16.1).

Based on the descriptions of each model, it is clear that several variables can influence the nutrient release profile in water or soil. One of the most significant factors is the nutrient and matrix interaction. For instance, in SRFs, nutrients are typically incorporated into a matrix with low solubility, such as a polymer coating, which acts as a physical barrier. In contrast, CRFs feature nutrients encapsulated by the matrix, such as a porous polymer, creating chemical and physical bonds.[59] Another crucial factor is the environment in which the fertilizers are evaluated, including whether it is in water or soil, as well as the pH, temperature, and soil type (sandy or clayey). Since the water present in the soil is the primary factor influencing nutrient delivery, the release profiles of tests conducted in both water and soil (using the same material) tend to align and fit within the same mathematical model. However, the volume of water present in the medium can still impact this profile.

Additionally, the data collection methods significantly affect comparing a specific material's release profile with existing mathematical models. Detailed information on the amount and form in which the nutrient was released over time is essential to compare or predict results accurately. This information is crucial for understanding the nutrient release profile and its comparison with mathematical models.

When the methodologies used for data collection are consistent regarding the aspects mentioned above, it becomes possible to compare specific characteristics of the materials, such as the polarity and hydrophobicity. For example, in the case of a poorly soluble nutrient coated by a matrix or pure nutrients that do not degrade, the most suitable release profile model is zero-order.[60] The first-order model typically best fits more soluble nutrients covered within porous matrices. The Higuchi model assumes one-way diffusion of nutrients with minimal swelling and dissolution of the matrix and is most suitable for describing fertilizers where nutrients are incorporated into solid or semi-solid matrices like montmorillonite and biochar.[61] Similarly, the nutrient release from other types of materials can follow specific models, making correlations more easily.

16.6.1 Case Study 4: Potassium Release Mechanism in Different Polymer Matrices

As detailed in Table 16.1, the semi-empirical Korsmeyer–Peppas and Peppas–Sahlin models describe the release of nutrients from polymeric materials as matrices. The value of the release exponent (n) in the Korsmeyer–Peppas equation indicates whether the nutrient release is Fickian (diffusion through the matrix) or non-Fickian (relaxation/erosion of the matrix). The Peppas–Sahlin model allows for calculating the contribution from each mechanism: diffusional (k_1t^m) or relaxational (k_2t^{2m}). Since the coefficients m and n are linked to the Fickian diffusion of nutrients in polymeric systems of any shape (such as films, spheres, and capsules), the Korsmeyer–Peppas and Peppas–Sahlin models have been widely utilized to characterize the release of various types of EEFs.[59]

Four studies involving EEFs based on natural polymer composites and KNO_3 have utilized the Peppas–Sahlin model to help explain the release mechanism of the K macronutrient. Given that this model is suitable for

Table 16.1 Mathematical models used to represent nutrient release kinetics.

Model	Description	Equation	Terms meaning
Zero-order	The nutrient release depends only on time and occurs at a constant rate, independent of concentration.[38]	$F = F_0 + k_0 t$	F: nutrient amount released F_0: initial nutrient amount (usually zero) k_0: release constant t: time
First-order	Concentration-dependent release. Active agents released from a porous matrix decrease over time as their concentration within the matrix diminishes.[38]	$\mathrm{d}F/\mathrm{d}t = -kt$	F: nutrient amount released k: constant rate
Higuchi	It describes the release mechanism of homogeneous particles in a matrix undergoing diffusion, assuming high initial active concentration, unidirectional diffusion, and minimal matrix swelling or dissolution.[62,63]	$F = k_H t^{0.5}$	F: nutrient amount released k_H: kinetic constant
Hixson–Crowell	It relates nutrient release to the surface and volume changes of the polymer matrix. This pertains to tablet dissolution, where surface area decreases proportionally over time while maintaining its geometric form.[64]	$3\sqrt{F_0} = 3\sqrt{F_i} + k_{HC} t$	F_0: initial nutrient amount F_i: remaining amount at time t K_{HC}: surface–volume relationship constant
Hopfenberg	It explains nutrient release from surface-erodible polymers with different shapes, focusing on matrix erosion as the primary limiting factor independent of time-dependent diffusion resistances.[64]	$F = 100[1 - (1 - K_{HB} t)^n]$	F: nutrient amount released K_{HB}: the erosion rate constant n: shape dependence (1 – slabs, 2 – cylinders, 3 – spheres)
Korsmeyer–Peppas and Peppas–Sahlin	Semi-empirical models that describe nutrient release from polymeric systems, especially when multiple release phenomena are involved.[65–67]	Korsmeyer–Peppas: $F = k_{KP} t^n$ Peppas–Sahlin: $F = k_1 t^m + k_2 t^{2m}$	F: nutrient amount released k_{KP}: kinetic constant n: diffusion exponent k_1: Fickian contribution k_2: case-II relaxational contribution $m = n$: when relaxational effects are negligible

release in polymeric systems, a more detailed analysis can be conducted to understand, based on the coefficient of determination (r^2) values, which type of matrix and format contributes the most to diffusional and/or relaxational K release. The works of Pereira *et al.* (2021), Chiaregato *et al.* (2023), and França *et al.* (2018, 2021) carried out nutrient release tests in water following the same methodology and, subsequently, the first 60% of the collected data were adjusted in the Peppas–Sahlin model equation.[32,50,68,69]

As shown in Table 16.2, the materials from all studies yielded r^2 values greater than 0.9000, indicating that K release occurs through multiple mechanisms. The composite material from Pereira *et al.* (2021), consisting of a multilayer film of chitosan (Cs), carboxymethyl cellulose (CMC), and zeolites (Ze), exhibited a lower r^2 value (0.9575).[68] This lower value may be related to a lower swelling degree of the Cs film (around the CMC/Ze/K layer) to the other biopolymers, leading to the release of K mainly by diffusion. Additionally, unlike the other EEFs, the material from Pereira *et al.* (2021) also includes a ceramic component (Ze), which could cause the K release profile to deviate slightly from the Peppas–Sahlin model.[68] The material from França *et al.* (2018), which also contains Cs in its composition, exhibited the second-lowest r^2 value (0.9668).[69] In addition to Cs, the microcapsules contain sodium tripolyphosphate (TPP) as a crosslinker, which may contribute to matrix swelling and, consequently, to a less diffusional release of K.

The other two materials, from França *et al.* (2021) and Chiaregato *et al.* (2023), achieved very high r^2 values and are both in tablet form, produced by melt-mechanical mixing followed by hot pressing.[32,50] In addition to the nutrient release mechanism being influenced by the shape, the type of polymer used in the coating matrix significantly affects release through diffusional and relaxational mechanisms. The polymer used as a matrix by Chiaregato *et al.* (2023)—poly(vinyl alcohol) —was also crosslinked, this time with glycerol (Gly).[32] As mentioned, crosslinking facilitates the relaxation of the polymer chains, enhancing the release of K^+ through this mechanism. In França *et al.* (2021), the blend of starch (St) and PHB used as a matrix also facilitated the release of K through the mechanism described by the Peppas–Sahlin model.[50]

This comparative analysis of K^+ release mechanisms from different polymeric materials was possible because the nutrient release was

Table 16.2 Coefficient of determination (r^2) of the Peppas–Sahlin model equation obtained by the K^+ release from different polymeric materials.

Ref.	Matrix	Shape	r^2 (Peppas–Sahlin)
Pereira *et al.* (2021)[68]	Cs/CMC/Ze	Multilayer films	0.9575
Chiaregato *et al.* (2023)[32]	PVA/Gly	Tablets	0.9900
França *et al.* (2021)[50]	PHB/St	Tablets	0.9901
França *et al.* (2018)[69]	Cs/TPP	Microcapsules	0.9668

evaluated under consistent conditions. In water, the medium changed at predetermined intervals and by quantifying the ions in the solution. Without these controlled conditions, nutrient release values could vary, potentially affecting the r^2 values of the mathematical model equations. Analyses like this are crucial for selecting appropriate materials for specific applications, such as fertilizers that do not release their nutrients solely through diffusion.

16.7 Future Perspectives on Fertilizer Technology and Soil Interactions

The potential of materials that encapsulate, coat, or disperse fertilizers in interactions with the soil is a fascinating area yet to be fully explored. Current research primarily focuses on different types of materials, leading to various formulation techniques. There is a strong emphasis on ensuring that carrier materials are environmentally friendly. However, there is a lack of information on how these materials interact with the soil and what characteristics are necessary for specific soil and plant conditions. Future research adopting multidisciplinary approaches will provide a holistic understanding of material–soil interactions and their effects on nutrient release and soil health. This approach will allow the development of more efficient, sustainable fertilizer formulations adapted to different types of soil and crops. The potential of these perspectives to advance sustainable agriculture by maximizing the efficiency of agricultural inputs while minimizing environmental impacts is inspiring. Studies should focus on the properties of encapsulating materials, such as biodegradability, cation exchange capacity, water retention, and interaction with soil microorganisms. Developing new materials or composites that respond to specific soil and plant conditions is also crucial. Innovative materials that react to pH, temperature, and humidity changes can optimize nutrient release.

Research into how encapsulating materials affect microbial activity, nutrient mineralization, and overall soil health is crucial. However, it is equally important to understand the complex process of nutrient delivery. This involves understanding how nutrients move from the encapsulating material into the soil and then within the soil to the plants. In addition to biodegradation, it is essential to study how materials affect the physical and chemical properties of the soil, such as pH, soil structure, water retention, and cation exchange capacity. This ensures that EEFs efficiently deliver nutrients and improve or maintain soil characteristics. Furthermore, understanding the biodegradation processes and the long-term impact of residues left by EEFs on soil ecosystems is essential. In the face of climate change and environmental depletion, the EU has introduced the European Green Deal, a plan to promote more sustainable agriculture to reach food security. This plan is anchored in Regulation 2019/1009, which stipulates that from 2026, polymers applied to agriculture must be wholly biodegradable, and their by-products of degradation must be non-toxic. This requirement underscores the need to balance using biodegradable and non-toxic materials with environmental compatibility and nutrient release efficiency. Comparative studies can be crucial in identifying sustainable options that meet these criteria without compromising agronomic efficiency.

Abbreviations

PVA	Poly(vinyl alcohol)
PHB	Poly(3-hydroxybutyrate)
CNF	Cellulose nanofibril
EEF	Enhanced efficiency fertilizer
EFF	Environmentally friendly fertilizer
SRF	Slow release fertilizer
CRF	Controlled release fertilizer
TDR	Time domain reflectometry
CEC	Cation exchange capacity
Cs	Chitosan
CMC	Carboxymethyl cellulose
Ze	Zeolites
TPP	Tripolyphosphate
Gly	Glycerol
St	Starch
TPS	Thermoplastic starch

References

1. P. Koli, N. R. Bhardwaj and S. K. Mahawer, in *Climate Change and Agricultural Ecosystems*, ed. K. K. Choudhary, A. Kumar and A. K. Singh, Elsevier, United Kingdom, 2019, p. 209.
2. N. K. Fageria, V. C. Baligar and C. A. Jones, *Growth and Mineral Nutrition of Field Crops*, CRC Press, New York, 2010.
3. Food and Agriculture Organization (FAO), *Presented in part at FAO's work on climate change*, United Nations, 2019.
4. F. M. Padilla, M. Gallardo and F. Manzano-Agugliaro, *Sci. Total Environ.*, 2018, **643**, 400.
5. N. M. Burri, R. Weatherl, C. Moeck and M. Schirmer, *Sci. Total Environ.*, 2019, **684**, 136.
6. A. Jarosiewicz and M. Tomaszewska, *J. Agric. Food Chem.*, 2003, **51**, 413.
7. M. E. Trenkel, *Controlled Release and Stabilized Fertilizers in Agriculture*, International Fertilizer Industry Association, Paris, 1997.
8. A. Shaviv, *Advances in Agronomy*, 2001, vol. 71, p. 1.
9. M. E. Trenkel, *Slow-and Controlled-Release and Stabilized Fertilizers: An Option for Enhancing Nutrient Use Efficiency in Agriculture*, International Fertilizer Industry Association, Paris, 2010.

10. Z. Majeed, N. K. Ramli, N. Mansor and Z. Man, *Rev. Chem. Eng.*, 2015, **31**, 69.
11. B. Azeem, K. KuShaari, Z. B. Man, A. Basit and T. H. Thanh, *J. Controlled Release*, 2014, **181**, 11.
12. E. Sugawara and H. Nikaido, *Antimicrob. Agents Chemother.*, 2014, **58**, 7250.
13. L. Wu and M. Liu, *Carbohydr. Polym.*, 2008, **72**, 240.
14. H. Arslanoglu, *Chemosphere*, 2019, **217**, 393.
15. G. L. Bridger, M. L. Salutsky and R. W. Starostka, *J. Agric. Food Chem.*, 1962, **10**, 181.
16. L. E. de-Bashan and Y. Bashan, *Water Res.*, 2004, **38**, 4222.
17. G. M. Blouin, D. W. Rindt and O. E. Moore, *J. Agric. Food Chem.*, 1971, **19**, 801.
18. R. Bortoletto-Santos, F. Ploteghe r, V. F. Majaron, M. G. da Silva, W. L. Polito and C. Ribeiro, *Appl. Clay Sci.*, 2020, **199**, 105874.
19. M. Mann, J. E. Kruger, F. Andari, J. McErlean, J. R. Gascooke, J. A. Smith, M. J. H. Worthington, C. C. C. McKinley, J. A. Campbell, D. A. Lewis, T. Hasell, M. V. Perkins and J. M. Chalker, *Org. Biomol. Chem.*, 2019, **17**, 1929.
20. J. Fu, C. Wang, X. Chen, Z. Huang and D. Chen, *Commun. Soil Sci. Plant Anal.*, 2018, **49**, 2219.
21. H. Z. Yang, G. L. Zheng, H. L. Liu, Q. H. Lin and W. Luo, *Chin. J. Trop. Agric.*, 2016, **36**, 21.
22. Z. Chen, *et al.*, *Chem. Rev.*, 2012, **112**(7), 3833.
23. A. C. de Campos Bernardi, J. C. Polidoro, M. B. de Melo Monte, E. I. Pereira, C. R. de Oliveira and K. Ramesh, *Adv. Chem. Eng. Sci.*, 2016, **06**, 295.
24. B. Wiedenfeld, *J. Plant Nutr.*, 2011, **34**, 1003.
25. M. Calabi-Floody, J. Medina, C. Rumpel, L. M. Condron, M. Hernandez, M. Dumont and M. de la L. Mora, *Adv. Agron.*, 2018, **147**, 1.
26. J. Stubenrauch and F. Ekardt, *Environments*, 2020, **7**, 38.
27. J. Chen, S. Lü, Z. Zhang, X. Zhao, X. Li, P. Ning and M. Liu, *Sci. Total Environ.*, 2018, **613**, 829.
28. E. V. R. Campos, J. L. de Oliveira, L. F. Fraceto and B. Singh, *Agron. Sustainable Dev.*, 2015, **35**, 47.
29. G. Izydorczyk, D. Skrzypczak, M. Mironiuk, K. Mikula, M. Samoraj, F. Gil, R. Taf, K. Moustakas and K. Chojnacka, *Sci. Total Environ.*, 2024, **923**, 171343.
30. M. Aslam, M. A. Kalyar and Z. A. Raza, *Polym. Eng. Sci.*, 2018, **58**, 2119.
31. A. Sikder, A. K. Pearce, S. J. Parkinson, R. Napier and R. K. O'Reilly, *ACS Appl. Polym. Mater.*, 2021, **3**, 1203.
32. C. G. Chiaregato, C. F. Souza and R. Faez, *ACS Agric. Sci. Technol.*, 2023, **3**, 322.
33. P. Milani, D. França, A. G. Balieiro and R. Faez, *Polimeros*, 2017, **27**, 256.
34. D. Qiao, H. Liu, L. Yu, X. Bao, G. P. Simon, E. Petinakis and L. Chen, *Carbohydr. Polym.*, 2016, **147**, 146.

35. J. Treinyte, V. Grazuleviciene, R. Paleckiene, J. Ostrauskaite and L. Cesoniene, *J. Polym. Environ.*, 2018, **26**, 543.
36. L. Messa, J. D. Froes, C. Souza and R. Faez, *Quim. Nova*, 2016, **39**, 1215.
37. C. G. Chiaregato and R. Faez, *Carbohydr. Polym.*, 2021, **271**, 118419.
38. Y. Xiang, X. Ru, J. Shi, J. Song, H. Zhao, Y. Liu and G. Zhao, *J. Agric. Food Chem.*, 2018, **66**, 7606.
39. L. M. Angelo, D. França and R. Faez, *Carbohydr. Polym.*, 2021, **257**, 117635.
40. D. França, L. L. Messa, C. F. Souza and R. Faez, in *Polymers for Agri-Food Applications*, ed. T. J. Gutiérrez, Springer International Publishing, Springer, Buenos Aires, 2019, p. 29.
41. T. S. Dierolf, L. M. Arya and R. S. Yost, *Agron. J.*, 1997, **89**, 572.
42. M. Ishiguro, K. C. Song and K. Yuita, *Soil Sci. Soc. Am. J.*, 1992, **56**, 1789.
43. O. O. Akinremi and C. M. Cho, *Soil Sci. Soc. Am. J.*, 1991, **55**, 1282.
44. G. C. Topp, J. L. Davis and A. P. Annan, *Water Resour. Res.*, 1980, **16**, 574.
45. G. C. Topp, J. L. Davis and A. P. Annan, *Soil Sci. Soc. Am. J.*, 1982, **46**, 678.
46. C. F. Souza, D. Or and E. E. Matsura, *Soil Sci. Soc. Am. J.*, 2004, **68**, 25.
47. C. F. Souza and E. E. Matsura, *Eng. Agríc.*, 2004, **26**, 282.
48. C. G. Chiaregato and R. Faez, *Carbohydr. Polym.*, 2021, **271**, 118419.
49. L. S. Dos Santos and D. R. Da Silva, *Rev. Bras. Cienc. Solo*, 2010, **46**, e0210091.
50. D. França, L. M. Angelo, C. F. Souza and R. Faez, *ACS Appl. Polym. Mater.*, 2021, **3**, 3227.
51. R. Faez, L. L. Messa and C. F. Souza, *Rev. Cienc. Agron.*, 2024, **55**, e20228509.
52. H. M. de P. Santana, E. E. Sano, M. P. de Oliveira Júnior, M. P. C. Lacerda and J. V. Malaquias, *Rev. Bras. Cienc. Solo*, 2015, **39**, 1172.
53. C. O. Da Ros, T. B. Silvestrin, L. Somavilla, E. R. Perrando and R. F. da Silva, *Floresta e Ambiente*, 2017, **24**, e20160081.
54. G. K. Donagemma, *Rev. Bras. Cienc. Solo*, 2008, **32**, 2493.
55. J. B. Basso, M. A. A. Soto and H. K. Chang, *Águas Subterrâneas*, 2017, **31**, 52.
56. W. D. C. Mendes, *Irriga*, 2015, **1**, 47.
57. N. A. Peppas and B. Narasimhan, *J. Controlled Release*, 2014, **190**, 75.
58. M. L. Bruschi, in *Strategies to Modify the Drug Release from Pharmaceutical Systems*, ed. M. L. Bruschi, Elsevier, United Kingdom, 2015, p. 63.
59. N. Lakshani, H. S. Wijerathne, C. Sandaruwan, N. Kottegoda and V. Karunarathne, *ACS Agric. Sci. Technol.*, 2023, **3**, 939.
60. Y. P. Timilsena, R. Adhikari, P. Casey, T. Muster, H. Gill and B. Adhikari, *J. Sci. Food Agric.*, 2015, **95**, 1131.
61. M. P. Paarakh, P. A. Jose, C. M. Setty and G. V. Peterchristoper, *Int. J. Pharm. Res. Technol.*, 2018, **8**(1), 12.
62. H. Omidian and K. Park, in *Biomedical Applications of Hydrogels Handbook*, ed. R. M. Ottenbrite, K. Park and T. Okano, Springer, New York, 2010, p. 1.

63. J. Siepmann, *Adv. Drug Delivery Rev.*, 2001, **48**, 139.
64. P. Costa and J. M. Sousa Lobo, *Eur. J. Pharm. Sci.*, 2001, **13**, 123.
65. R. W. Korsmeyer, R. Gurny, E. Doelker, P. Buri and N. A. Peppas, *Int. J. Pharm.*, 1983, **15**, 25.
66. N. A. Peppas and J. J. Sahlin, *Int. J. Pharm.*, 1989, **57**, 169.
67. P. L. Ritger and N. A. Peppas, *J. Controlled Release*, 1987, **5**, 37.
68. T. dos Santos Pereira, L. S. Fernandes, C. F. Souza and R. Faez, *ACS Agric. Sci. Technol.*, 2021, **1**, 131.
69. D. França, Â. F. Medina, L. L. Messa, C. F. Souza and R. Faez, *Carbohydr. Polym.*, 2018, **196**, 47.

Section IV: Conclusion

CHAPTER 17

Concluding Remarks

I. C. RIEGEL-VIDOTTI,[a] L. M. ZIMMERMANN[b] AND H. B. BARROS*[c]

[a] Department of Chemistry, Federal University of Paraná, Brazil;
[b] Department of Chemistry, Regional University of Blumenau (FURB), Brazil;
[c] Department of Applied Science and Technology (DISAT), Politecnico di Torino, Italy
*Email: heloise.barros@polito.it

The current global environmental challenges underscore the criticality of this topic for addressing real-world solutions to a more sustainable world. This book has explored both fundamental concepts and practical applications involving colloidal materials, demonstrating that these technologies have a significant impact not only on air, soil, and water environments but also on ecosystems. Current knowledge and promising perspectives using colloidal materials were highlighted to leverage potential advancements in integrated sustainability management. Throughout this book, multifaceted contributions of colloidal systems in combating pollution across air, water, and soil were explored from multiple perspectives. Synthesis and manipulation of colloids have proven to be powerful tools in addressing the complex and interconnected challenges intensified by human activity and industrialization.

Importantly, this book underscored the pivotal role of innovative agrochemical technologies in addressing the intertwined challenges of food security and environmental sustainability. Through exploring the significance of advanced delivery systems composed of colloidal technologies, the necessity for tailored nutrient release mechanisms that enhance soil health and optimize agricultural productivity was emphasized. In other words, this

Contribution of Colloidal Materials to Air, Water and Soil Environmental Sustainability: From Basic Concepts to Advanced Applications
Edited by Izabel C. Riegel-Vidotti, Lizandra Maria Zimmermann and Heloise R. de Barros

Published by the Royal Society of Chemistry, www.rsc.org

book highlights the critical intersection of agriculture, technology, and sustainability in an integrative approach.

The atmospheric environment represents a complex system, comprising a diverse array of gases, airborne particles, viruses, and bacteria. In urban settings, the concentration of anthropogenic particulate matter (PM) poses significant challenges, as these pollutants can traverse vast distances and affect even remote ecosystems such as the Amazon and Antarctica, which is an alarming global concern. To protect the whole ecosystem, the fundamental step is obtaining effective methods for atmospheric sensing and monitoring. The development of robust detection mechanisms for PM, gases, and volatile compounds remains a critical area of ongoing research, given the inherent complexity of these systems. In this context, colloids serve as useful tools in monitoring and sensing technologies due to their high surface-to-volume ratio and the feasibility of surface functionalization. However, ensuring the sensitivity and specificity of these applications continues to be a focal point for scientific findings. In terms of effectively addressing climate change challenges, mitigation strategies using colloid materials have been facilitating further advancements in CO_2 removal technologies that are essential for this goal. Advancing further, novel photovoltaic devices are drawing attention to this field as they efficiently convert solar energy into electricity through methods that are cost-effective, scalable, and have an abundance of raw materials. Particularly, photovoltaic technology based on organic photovoltaics (OPVs) has emerged as a promising eco-friendly alternative, appealing for its use of green solvents in processing. Miniemulsion, a widely used technique in colloid chemistry, has been explored to perform the synthesis of organic semiconductor nanoparticles affecting the performance of OPVs. Continued research and innovation in this field will be crucial for developing effective strategies to protect our environment and public health from the widespread impacts of pollution.

In the water ecosystem, the removal and treatment of contaminants present significant challenges. These challenges stem from the diverse origins and impacts of the pollutants and involve both regulatory frameworks and innovative remediation strategies. Throughout this book, various chapters have highlighted key advancements, such as self-assembled structures, carbon quantum dots (CQDs), and magnetic nanosorbents, which collectively demonstrate the potential for the development of smart devices for pollutant sensing, removal, and remediation efforts. Researchers have made substantial progress in developing advanced water purification systems that are not only efficient but also scalable, more durable, and reusable.

Certainly, colloids are essential for developing advanced delivery systems in soil that enable more precise targeting of molecules like those in fertilizers and pesticides. These molecules are critical for enhancing agricultural productivity and food security. Colloid-based systems can improve nutrient release efficiency, reduce waste, and address the inefficiencies commonly seen in traditional agrochemical applications. For this reason, understanding how nutrients are released from fertilizers, how they interact with the soil, and the physical and chemical factors that affect nutrient mobility is very important. Moreover, smart colloidal materials that respond to environmental changes, such as pH, temperature, and soil composition, based on specific plant needs, are crucial for optimizing nutrient release, ensuring a good balance of the ecosystem. Additionally, we emphasized the importance of understanding the long-term effect of agrochemicals on soil ecosystems to minimize any environmental impact.

Currently, the integration of cutting-edge technologies such as artificial intelligence (AI) and machine learning presents promising opportunities for real-time monitoring and adaptive management of agrochemicals and environmental predictions. Future research could further leverage digital twins—virtual models that simulate nutrient release and other dynamics—to optimize formulations based on real-time data, thereby enhancing the effectiveness of agrochemical use. As we progress towards a greener future, these advancements are essential for ensuring that agricultural practices remain both efficient and ecologically responsible.

As we move forward, embracing these collaborative efforts will be crucial for achieving sustainable agricultural practices that protect our environment while meeting the legislation demand in the context of a growing population and increasing urbanization. From the perspectives highlighted here, it is evident that bridging the gap between interdisciplinary research, theoretical studies, and advancements in nanoengineering will be essential in developing effective and sustainable solutions for cleaner water, air, and soil. Furthermore, the solutions presented in this book are aligned with the urgency to combat climate change since it has presented a myriad of solutions to preserve natural resources and, at the same time, protect biodiversity. New businesses should emerge and benefit from what is discussed in this book, leading to the development of smart and responsible solutions. By integrating these insights, we hope to contribute to the continued innovation of colloidal materials toward environmental sustainability in air, water, soil, and most importantly, the broader ecosystem.

Subject Index